Springer Series on Atomic, Optical, and Plasma Physics

Volume 86

The Springer Series on Atomic, Optical, and Plasma Physics covers in a comprehensive manner theory and experiment in the entire field of atoms and molecules and their interaction with electromagnetic radiation. Books in the series provide a rich source of new ideas and techniques with wide applications in fields such as chemistry, materials science, astrophysics, surface science, plasma technology, advanced optics, aeronomy, and engineering. Laser physics is a particular connecting theme that has provided much of the continuing impetus for new developments in the field, such as quantum computation and Bose-Einstein condensation. The purpose of the series is to cover the gap between standard undergraduate textbooks and the research literature with emphasis on the fundamental ideas, methods, techniques, and results in the field.

More information about this series at http://www.springer.com/series/411

Markus Kitzler · Stefanie Gräfe
Editors

Ultrafast Dynamics Driven by Intense Light Pulses

From Atoms to Solids, from Lasers to Intense X-rays

Springer

Editors
Markus Kitzler
Photonics Institute
Vienna University of Technology
Vienna
Austria

Stefanie Gräfe
Institute for Physical Chemistry
University of Jena
Jena
Germany

ISSN 1615-5653 ISSN 2197-6791 (electronic)
Springer Series on Atomic, Optical, and Plasma Physics
ISBN 978-3-319-20172-6 ISBN 978-3-319-20173-3 (eBook)
DOI 10.1007/978-3-319-20173-3

Library of Congress Control Number: 2015943437

Springer Cham Heidelberg New York Dordrecht London

Printed on acid-free paper

Springer International Publishing AG Switzerland is part of Springer Science+Business Media (www.springer.com)

Preface

The properties of matter are ultimately determined by its electronic structure. A suitable perturbation of the electrons' distribution from its equilibrium configuration in a system such as an atom, a molecule, or a solid, can therefore initiate certain dynamical processes. Intense and ultrashort light pulses are ideal tools for that purpose as their electric fields couple directly to the electrons. Therefore, they are not only able to distort the equilibrium electron distribution but, even allow driving a system on sub-femtosecond timescales with the light's Petahertz field-oscillations. This has been realized first in experiments that studied atoms in strong laser fields some 25 years ago. These experiments revealed a wealth of fundamentally important phenomena that are strictly timed to the laser-field oscillations such as the release of electrons by tunneling through the field-distorted Coulomb binding potential, or field-driven (re-)collisions of the released electrons with the atomic ion. The latter, in turn, led to the discovery of a range of essential secondary processes such as the generation of very high orders of harmonics of the driving light with photon energies that can extend into the X-ray range.

Since then, thanks to a true revolution in laser technology, tremendous progress has been made in the field. Laser science has now reached a level of perfection where it is possible to produce intense light pulses with durations down to a single oscillation cycle and with virtually arbitrary evolution of the electric field in a wide range of frequencies. The availability of such field transients enabled a number of exciting possibilities, such as control over the breakage of selected chemical bonds in molecules by directly driving the molecular valence electrons that actually form the bond, or the production of coherent attosecond pulses in the soft X-ray wavelength range that can be used for probing or initiating dynamics on time-intervals during which the electronic distribution in the system under study stays essentially frozen.

The recent years have seen a particularly vivid progress in the research of using ultrashort intense light pulses for controlling and probing ultrafast dynamics. On the one hand, a number of groups have extended the research field to systems with a much increased complexity and have studied and controlled field-induced dynamics in large polyatomic molecules, cluster complexes, bio-matter, nanoparticles and crystals, and also in condensed phase systems such as solid surfaces, nanostructures, and bulk solids. On the other hand, the availability of new, coherent light sources in both the very short (X-rays) and very long (mid-infrared) wavelength ranges have allowed for the production and application of short and intense pulses in previously unexplored regimes.

Owing to this large diversity of pulsed sources and dynamical systems studied with them it becomes inherently difficult to provide a unique definition and sharp boundaries for this field of research. This is also reflected by the considerable variety of titles used for conferences in this field. Obviously the same difficulty arises in providing a clear and well-defined but at the same time comprehensive title for a book on this research field. Although *ultrafast dynamics* is a relative term and covers a large range of dynamical processes and timescales, including rotational and vibrational dynamics, we would like to define this term here as electronic dynamics and processes that result from an essentially instantaneous distortion of the equilibrium electronic structure in a given system.

With this book we have tried, by carefully picking 14 examples of cutting-edge scientific research, grouped in four areas, to provide a comprehensive overview not only over the current state of the research field that uses ultrashort intense light pulses and light sources based on these pulses for initiating, driving, controlling, and probing ultrafast dynamics, but also over its recent tremendous and exciting developments. With the selection of the four areas we have attempted to provide the broadest possible overview over the such defined research field by covering essentially all currently studied physical systems from individual atoms and molecules to nanostructures and bulk macroscopic media, and all available ultrafast pulse sources from the mid-infrared to the X-ray range. Of particular importance for us was to highlight the possibilities that are opened up by the availability of new light sources, and the new research questions that arise by pushing research toward new systems with increased complexity such as nanostructures and bulk macroscopic media. Also, we have tried to provide both an experimental and theoretical perspective on the research field. The book is structured as follows.

The first part of research that is discussed during the first four chapters shall provide an overview of the possibilities that a strong laser field opens up for controlling electronic processes in atoms, molecules, nanostructures, and solids. *The second part* of the book is dedicated to the application of intense laser pulses in combination with attosecond pulses, obtained by the laser-driven process of high-harmonic generation, for triggering and probing ultrafast dynamics. *The third part* of the book discusses in four chapters the only very recently opened research route of using ultrashort intense laser pulses for driving electronic dynamics on surfaces, in nanostructures and in solids. While on the one hand this type of research is interesting from a fundamental point of view as it investigates the interaction of light and matter in completely new regimes of parameters where collective effects, material parameters, and system geometries start to play a role, this research also comprises considerable potential for applications in that it could be used for, e.g., fabrication of devices for information transmission or fast switching. A particularly important process in this context is the excitation of collective surface electron oscillations called a surface plasmon. *The fourth and final part* of the book is dedicated to the exciting possibilities that are opened up by the availability of intense light pulses in the X-ray wavelength regime that can be produced by free-electron lasers, the first of which have started their operation just a few years ago. The two chapters in this part discuss applications of such pulses depicted by examples of research performed at the free-electron lasers in Hamburg (FLASH) and Stanford (Linac Coherent Light Source, LCLS), respectively.

We hope that this book will be equally inspiring and helpful for young researchers, who would like to step into this field, and for experienced researchers who may enjoy the exhaustive discussion that covers the research on essentially all currently studied objects and with all available ultrafast pulse sources in the field that uses ultrashort light pulses for controlling and probing ultrafast dynamics.

Vienna Markus Kitzler
Jena Stefanie Gräfe

Contents

Contributors

Jakob Andreasson Molecular Biophysics Department of Cell and Molecular Biology (ICM), Uppsala University, Uppsala, Sweden; ELI-Beamlines, Institute of Physics, Academy of Sciences of the Czech Republic, Prague, Czech Republic

Thomas Barillot Institut Lumière Matière ILM, CNRS, Villeurbanne CEDEX, France

Thomas Baumert Universität Kassel, Institut für Physik und CINSaT, Kassel, Germany

Tim Bayer Carl von Ossietzky Universität Oldenburg, Institut für Physik, Oldenburg, Germany

Jérémie Caillat UPMC, UMR 7614, Laboratoire de Chimie Physique—Matière et Rayonnement 11, Paris Cedex 05, France; CNRS, UMR 7614, Laboratoire de Chimie Physique—Matière et Rayonnement 11, Paris Cedex 05, France

Francesca Calegari Institute of Photonics and Nanotechnologies, IFN-CNR, Milano, Italy

J. Marcus Dahlström Department of Physics, Stockholm University, AlbaNova University Center, Stockholm, Sweden; Max-Planck Institute for the Physics of Complex Systems, Dresden, Germany

Victor Despré Institut Lumière Matière ILM, CNRS, Villeurbanne CEDEX, France

Péter Dombi MTA "Lendület" Ultrafast Nanooptics Group, Wigner Research Centre for Physics, Budapest, Hungary

Reinhard Dörner Institut Für Kernphysik, Goethe Universität, Frankfurt, Germany

K.E. Echternkamp IV. Physical Institute—Solids and Nanostructures, University of Göttingen, Göttingen, Germany

U. Eichmann Max-Born-Institute, Berlin, Germany

Jason Greenwood Centre for Plasma Physics, School of Maths and Physics, Queen's University Belfast, BT7 1NN, UK

Markus Gühr PULSE, SLAC National Accelerator Laboratory and Stanford University, Menlo Park, CA, USA

Stefan Haessler Photonics Institute, Vienna University of Technology, Vienna, Austria; LOA, ENSTA ParisTech, CNRS, Ecole Polytechnique, Université Paris-Saclay, Palaiseau Cedex, France

G. Herink IV. Physical Institute—Solids and Nanostructures, University of Göttingen, Göttingen, Germany

Ferenc Krausz Max-Planck-Institut für Quantenoptik, Hans-Kopfermann-Straße 1, Garching, Germany; Ludwig-Maximilians-Universität, Am Coulombwall 1, Garching, Germany

Maria Krikunova Institut Für Optik und Atomare Physik, Technische Universität Berlin, Berlin, Germany

Stanislav Yu. Kruchinin Max-Planck-Institut für Quantenoptik, Hans-Kopfermann-Straße 1, Garching, Germany

Franck Lépine Institut Lumière Matière ILM, CNRS, Villeurbanne CEDEX, France

Candong Liu State Key Laboratory of High Field Laser Physics, Shanghai Institute of Optics and Fine Mechanics, Chinese Academy of Sciences, Shanghai, China

Vincent Loriot Institut Lumière Matière ILM, CNRS, Villeurbanne CEDEX, France

Matteo Lucchini Department of Physics, ETH Zurich, Zürich, Switzerland

Mattia Lupetti Ludwig Maximilians Universität, Munich, Germany

Alfred Maquet UPMC, UMR 7614, Laboratoire de Chimie Physique—Matière et Rayonnement 11, Paris Cedex 05, France; CNRS, UMR 7614, Laboratoire de Chimie Physique—Matière et Rayonnement 11, Paris Cedex 05, France

Alexandre Marciniak Institut Lumière Matière ILM, CNRS, Villeurbanne CEDEX, France

Dejan B. Milošević Faculty of Science, University of Sarajevo, Sarajevo, Bosnia and Hercegovina; Academy of Sciences and Arts of Bosnia and Hercegovina, Sarajevo, Bosnia and Hercegovina; Max-Born-Institut, Berlin, Germany

Mauro Nisoli Institute of Photonics and Nanotechnologies, IFN-CNR, Milano, Italy; Department of Physics, Politecnico di Milano, Milano, Italy

Tim Paasch-Colberg Max-Planck-Institut für Quantenoptik, Hans-Kopfermann-Straße 1, Garching, Germany

Maurizio Reduzzi Department of Physics, Politecnico di Milano, Milano, Italy

C. Ropers IV. Physical Institute—Solids and Nanostructures, University of Göttingen, Göttingen, Germany

Arnaud Rouzée Max-Born Institute (MBI), Berlin, Germany

Giuseppe Sansone Institute of Photonics and Nanotechnologies, IFN-CNR, Milano, Italy; Department of Physics, Politecnico di Milano, Milano, Italy

Armin Scrinzi Ludwig Maximilians Universität, Munich, Germany

M. Sivis IV. Physical Institute—Solids and Nanostructures, University of Göttingen, Göttingen, Germany

Mark I. Stockman Center for Nano-Optics (CeNO) and Department of Physics and Astronomy, Georgia State University (GSU), Atlanta, GA, USA

Nicusor Timneanu Molecular Biophysics Department of Cell and Molecular Biology (ICM), Uppsala University, Uppsala, Sweden; Department of Physics and Astronomy, Uppsala University, Uppsala, Sweden

Xiao-Min Tong Division of Materials Science, Faculty of Pure and Applied Sciences, University of Tsukuba, Tsukuba, Ibaraki, Japan; Center for Computational Sciences, University of Tsukuba, Tsukuba, Ibaraki, Japan

Nobuyuki Toshima Division of Materials Science, Faculty of Pure and Applied Sciences, University of Tsukuba, Tsukuba, Ibaraki, Japan

Andrea Trabattoni Department of Physics, Politecnico di Milano, Milano, Italy

Morgane Vacher Department of Chemistry, Imperial College London, London, United Kingdom

Marc. J.J. Vrakking Max-Born Institute (MBI), Berlin, Germany

L. Wimmer IV. Physical Institute—Solids and Nanostructures, University of Göttingen, Göttingen, Germany

Matthias Wollenhaupt Carl von Ossietzky Universität Oldenburg, Institut für Physik, Oldenburg, Germany

Jian Wu State Key Laboratory of Precision Spectroscopy, East China Normal University, Shanghai, China

Vladislav S. Yakovlev Max-Planck-Institut für Quantenoptik, Hans-Kopfermann-Straße 1, Garching, Germany; Ludwig-Maximilians-Universität, Am Coulombwall 1, Garching, Germany

S.V. Yalunin IV. Physical Institute—Solids and Nanostructures, University of Göttingen, Göttingen, Germany

Part I
Control of Electronic Processes with Strong Laser Fields

Chapter 1
Strong-Field Induced Atomic Excitation and Kinematics

U. Eichmann

Abstract Frustrated tunneling ionization (FTI) has recently been found to be an important exit channel of atomic strong-field ionization models such as the simple man's or rescattering model if one considers the Coulomb field explicitly. It leads to the population of bound excited states rather than to ionization after the electron has tunneled and quivered in the laser field. In this chapter we introduce the FTI model and describe experiments whose outcome supports its importance. In particular, we focus on strong-field excitation of atoms and the observation of neutral (ionic) excited fragments with high kinetic energy in strong-field fragmentation and Coulomb explosion of small molecules. Furthermore, we present experiments in which a direct position sensitive detection of excited neutral atoms reveals the exceptionally high acceleration of atoms in short pulsed strongly focused laser fields and discuss possible applications.

1.1 Introduction

The understanding of strong-field ionization dynamics of atoms and molecules rests a great deal on the seminal tunneling picture introduced by Keldysh [1]. A linearly polarized pulsed strong laser field is considered as a classical electric field $\mathbf{F}(t) = \mathbf{F}_L(t)\cos(\omega t)$ with $\mathbf{F}_L(t) = f(t)F_0\hat{\mathbf{e}}_\mathbf{x}$, where $f(t)$ is a slowly time-varying pulse envelope, F_0 is the field amplitude and ω is the angular frequency of the laser. All equations throughout the paper are given in atomic units unless otherwise stated. If an atom in its ground state with an ionization potential I_P is exposed to such a laser field, the atomic Coulomb potential is periodically bent up and down and allows for tunneling of an electron at certain phases of the laser field. The pure tunneling picture describes ionization extremely well, particularly in those situations, where a fairly large number of photons is necessary to overcome the binding energy. The Keldysh parameter $\gamma = \sqrt{I_P/2U_p}$, where $U_p = F^2/4\omega^2$ is the ponderomotive potential, is

U. Eichmann (✉)
Max-Born-Institute, Max-Born-Strasse 2a,12489 Berlin, Germany
e-mail: eichmann@mbi-berlin.de

M. Kitzler and S. Gräfe (eds.), *Ultrafast Dynamics Driven by Intense Light Pulses*, Springer Series on Atomic, Optical, and Plasma Physics 86, DOI 10.1007/978-3-319-20173-3_1

used to distinguish the tunneling regime $\gamma < 1$ from the multiphoton regime $\gamma > 1$. It has been found, however, that the subsequent dynamics of the electron in the laser field is of decisive importance. It results in secondary processes, which are embraced in the simple man's model [2–4] and in the famous three-step or rescattering model [5–7]. The simple man's model concentrates on the cycle-averaged energy a liberated electron can extract from the classical laser field, neglecting any interaction with the parent ion in the first place. The rescattering model, on the other hand, focuses on the processes initiated after the first return(s) of the electron to the core, where particularly electrons liberated in a certain phase range after a field cycle maximum provide high kinetic energies at the return. These processes include high-order above threshold ionization (HATI) [8], non-sequential double or non-sequential multiple-ionization (NSDI) by collision [9–12] and radiative recombination generating high harmonics (HHG) [13, 14].

We will concentrate on the dynamics of electrons, which tunnel around the field cycle maximum thus avoiding substantial energy transfer during rescattering with the parent ion. Astonishingly, the physical consequences that arise in the context of these electrons have not been considered coherently before, neither in the simple man's model nor in the rescattering model. By taking into account the Coulomb field explicitly, one finds that ionization of the atom, expected to follow the tunneling process of the electron, is frustrated under certain circumstances. This exit channel leads to the population of excited states. The process, which has been dubbed frustrated tunneling ionization (FTI), describes quantitatively (quasi multi-photon) excitation within the tunneling picture [15].

To put the FTI model into perspective one has to mention that since the early days of optical strong-field physics, experiments have shown beside multiphoton ionization also multiphoton excitation [16–19]. In the multiphoton picture, it was argued qualitatively that a Rydberg state is excited at the beginning of the laser pulse. Similar to a free electron the quasi free Rydberg electron does not absorb energy from the electromagnetic field and remains bound. The picture of Rydberg state excitation was strongly supported by the observation of strong enhancements in the above threshold ionization (ATI) electron spectra, which were explained in terms of transient Freeman resonances [20]. These are Rydberg states that are shifted ponderomotively into resonance with the laser field at particularly intensities during the rise of the laser pulse and subsequently ionized. To explain why an atom in a Rydberg state is finally stable against ionization, different stabilization mechanisms [21] such as interference stabilization at lower intensities [22–27] or strongly reduced ionization rates due to high angular momentum [17, 28] have been suggested. An easy quantitative explanation of excitation in the multiphoton picture, however, has not been achieved.

We remark that population trapping in excited states should not be misinterpreted with atomic stabilization in strong laser fields. Simply speaking the strong-field actually stabilizes the atom by reducing the overlap of the laser driven electronic wave function with the ionic core. This phenomenon, usually associated with the situation that a single-photon absorption is in principle sufficient to ionize, was first predicted about a quarter century ago [29–32]. Since then, the subject has been extensively discussed theoretically for the last two decades [33, 34] with newly increased interest [35, 36]. Stabilization of a single prepared low lying Rydberg state in a moderately strong laser field without any remaining loop holes such as state redistribution has been observed in impressive experiments [37, 38].

The qualitative arguments expressed within the multiphoton picture were also applied to explain excitation of atoms in the strong-field tunneling regime. In [18, 39] high-lying Rydberg states after strong-field interaction were observed by subsequent field ionization and in [40], e.g., excitation was deduced from structures in the measured ionization yields. First trials to explain excitation in the strong-field tunneling regime have been given within the simple man's model [2–4]. Early investigations on stabilization of atoms using classical Monte Carlo analysis [41] found as an alternate way to stabilization that the quivering electrons land on Rydberg states after the laser pulse has terminated. Yudin and Ivanov reported transient Rydberg trajectories [42] in the tunneling regime and finally Muller concluded from extensive quantum mechanical calculation excitation in the tunneling regime [43]. Only recently excited states of He have been observed [15], in which the intensities were well in the tunneling regime of strong-field physics. The obvious question that arose in this context was whether there is a way to comprehend strong-field excitation purely within the tunneling picture without invoking the multiphoton picture. The solution lies in the frustrated tunneling ionization model, which describes astonishingly well observed features.

In the following sections we will elucidate in detail how frustrated tunneling ionization leads to excited states and we will discuss consequences for strong-field physics. Besides important ramifications in atomic strong-field physics we also find FTI at work in strong-field dissociation of molecules. Finally, FTI establishes the basis to explain observed kinematic effects of strong inhomogeneous fields on neutral atoms.

1.2 Strong Field Excitation of Atoms by Frustrated Tunneling Ionization (FTI)

1.2.1 Linearly Polarized Laser Fields

In tunneling models, the tunneling process is mostly regarded to be tantamount to tunneling ionization. This is correct, if no attractive potential, whatsoever, is explicitly considered in the first place. However, by taking into account the Coulomb potential of the parent ion, frustrated tunneling ionization might happen. Assuming that the tunneling process is instantaneous at some time t_t, the electron is then located at the tunnel exit, which is only a few Bohr radii away from the ionic core. At this point, the electron has a high negative potential energy of a few eV. The electron quivers in the strong laser field and, whenever the tunneled electron does not gain enough energy during the interaction with both the laser field and the Coulomb field to finally overcome the attractive Coulomb force [15], it has not been set free at all. Therefore, one expects FTI to occur mainly for electrons that tunnel in the vicinity of a field cycle maximum of a linearly polarized laser field at a phase $\phi_t = \omega t_t \approx 0$ (or a multiple integer of π). In this case, the laser induced drift energy, given by $E_{dr} = 2U_p \sin^2 \phi_t$ [19], is close to zero and also the energy upon the first rescattering on the parent ion is rather moderate and typically less than the potential energy in the Coulomb field.

Decisive, whether ionization occurs after tunneling, or not, is the total energy T of the electron at a time, when the laser pulse is over. If it is positive, the electron motion is eventually unbound, if it is negative, the electron will firmly relax into a bound state. Again, the notion in the rescattering model that the "electron tunnel ionizes in the first step" is correct as long as no attractive potential is considered. In the presence of an attractive potential one can assert that ionization has happened only after the laser pulse is over . Consequently, thinking in terms of the rescattering model, where it is common agreement that an electron is considered to be "ionized" after the tunneling process, the term frustrated tunneling ionization is meaningful and justified. However, it is most important to emphasize that an electron that has undergone frustrated tunneling ionization, has in fact never been unbound in the sense of a strict definition of ionization.

In the following we will explore the parameter range for producing bound excited states through FTI. For given laser pulse parameters and a specified atom, which will be Helium in the present examples, the position and momentum of the electron, and the phase ϕ_t of the oscillating laser field at the instant of tunneling are crucial [15, 44]. To calculate trajectories leading to frustrated tunneling ionization we solve the classical Newton equations for an electron in a combined pure Coulomb potential $V_c(r) = -1/r$, where $r = \sqrt{x^2 + y^2 + z^2}$, and the electric field $\mathbf{F}(t)$.

$$\ddot{\mathbf{x}}(\mathbf{t}) = -\mathbf{F}(t) - \nabla V_c(r(t)) \tag{1.1}$$

The initial conditions at t_t are obtained from the tunneling ionization model, which locate the tunneling exit in a linearly polarized laser field at $x(t_t) = -\left(I_P + [I_P^2 - 4|F(t_t)|]^{1/2}\right)/2F(t_t)$, and $y(t_t) = z(t_t) = 0$. Furthermore, at t_t, the longitudinal momentum of the electron along the polarization axis is $p_x(t_t) = 0$. The initial momentum perpendicular to the field axis $p_\perp(t_t) = \sqrt{p_y^2 + p_z^2}$ is a parameter. To get an overview over bound and unbound trajectories we exploit the fact that trajectories are planar and symmetric with respect to the field axis. We thus vary p_y and take $p_z = 0$.

In Fig. 1.1 we show the occurrence of bound trajectories as a function of the parameters $\phi_t = \omega t_t$ and p_y. For clarity we restrict the electron to tunnel only in the vicinity of the field cycle maximum at the laser pulse envelope maximum. The calculations are performed with a linearly polarized laser pulse with 8 fs (FWHM) pulse duration and field strengths of $F_0 = 0.0755$ a.u. and $F_0 = 0.169$ a.u., Fig. 1.1a, b, respectively, and with 29 fs (FWHM) pulse duration and the same field strengths as before, Fig. 1.1c, d, respectively. The Keldysh parameters associated with the two field strengths are $\gamma = 1$ and $\gamma = 0.44$, respectively. Obviously, there are only

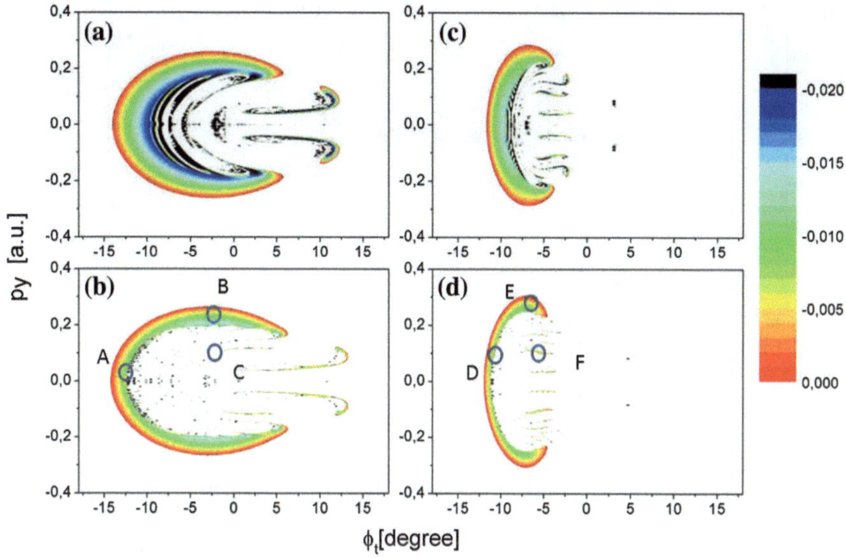

Fig. 1.1 Occurence of bound states after tunneling. Laser parameters: $F_0 = 0.0755$ a.u. (2×10^{14} W cm^{-2}) and **a** 8 fs (FWHM) and **b** 29 fs (FWHM) pulse duration. Laser parameters: $F_0 = 0.169$ a.u. (10^{15} W cm^{-2}) and **c** 8 fs and **d** 29 fs pulse duration. The *circles* indicate initial parameters for calculated trajectories shown in Fig. 1.2. The phase $\phi_t = \omega t_t$ is indicated with respect to the field cycle maximum at the maximum of the laser pulse envelope. Final negative total energy T of the electron is color coded. *White areas* stand for trajectories with positive total energy corresponding to strong field ionization

certain well defined regions of the parameters, where frustrated tunneling ionization prevails. Inspecting Fig. 1.1a, b one finds that the parameter space for bound trajectories is much larger for electrons starting before the maximum than for electrons starting after it. Particularly, for the short laser pulses, Fig. 1.1a, one finds a relatively large region of parameters allowing for bound states [44]. In this case the laser drift momentum the electron acquires is opposite to the Coulomb force. If the electron starts after the field cycle maximum, the recollision with the ionic core is likely, which obviously counteracts formation of bound states. At longer pulse duration, Fig. 1.1b, on the other hand, the allowed parameter range is reduced and is characterized by distinct isolated areas. Apparently, bound states are no longer populated due to the higher probability of a fatal encounter of the electron with the ionic core at longer pulse durations. If we increase the field amplitude and use otherwise identical laser parameters, the parameter space for bound trajectories shrinks substantially, Fig. 1.1c, d. Most striking is that bound states are no longer found for electrons that tunnel after the field cycle maximum.

The interesting question that arises is how important is the influence of the Coulomb potential on the electron dynamics during the laser pulse? It is well known that most of the strong-field physics associated with the rescattering model

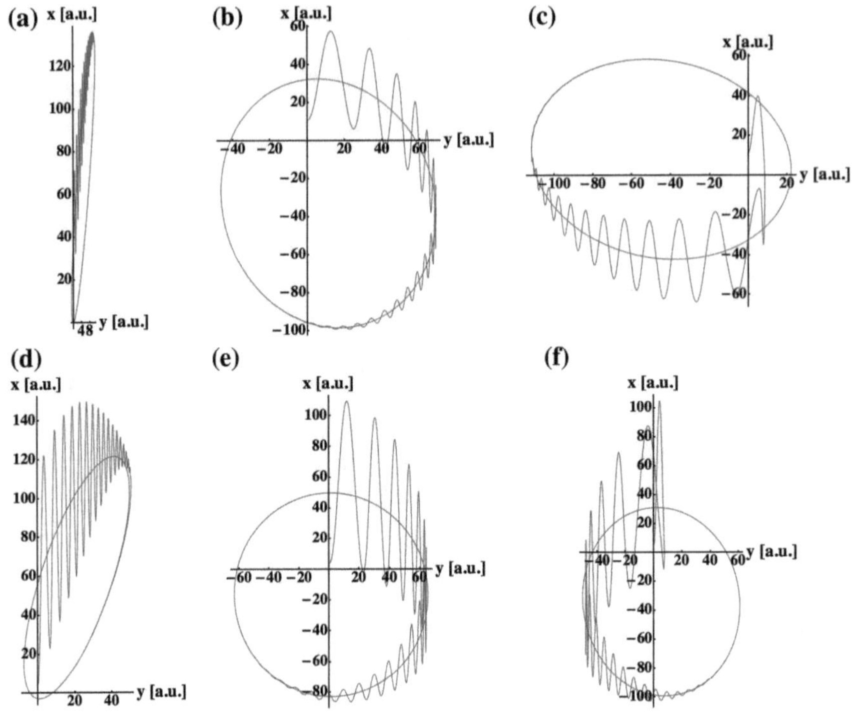

Fig. 1.2 Selected trajectories for initial conditions indicated in Fig. 1.1

is well described without invoking the Coulomb potential. To shed more light on the dynamics, it is very instructive to visualize and study electron trajectories at specific (ϕ_t, p_y) parameters. We show different trajectories along an energy contour line with $E_b \approx -7.8 \times 10^{-3}$ a.u. corresponding to roughly a principal quantum number $n = 8$. The initial conditions for trajectories shown in Fig. 1.2a–c are indicated in Fig. 1.1b by A–C, resp., and the initial conditions for trajectories shown in Fig. 1.2d–f are indicated in Fig. 1.1d by D–F, resp. The trajectories are calculated for several revolutions, so that the wiggly motion at early times during the laser pulse superimposes the final bound orbit at later times.

Trajectories, shown in (a) and (d), and in (b) and (e), were calculated using similar initial conditions, respectively. However, the quiver amplitude has more than doubled in (b) and (d), as expected for the higher field strength. In (a) and (c) the final bound orbits are oriented along the laser polarization with low angular momentum owing to the low initial perpendicular momentum. The role of the Coulomb field is important and by no means negligible, since the maximum excursion of the electron after the laser pulse is smaller than expected on the basis of a Coulomb field free motion. The same is true for the orbits shown in Fig. 1.2b, e. Here, the drift motion solely due to the laser field is weak, so that the electron is dominantly under the influence of the strong Coulomb field. Furthermore, the quiver motion averaged over one laser period follows astonishingly well the final Coulomb orbit almost right from the start indicating that the Coulomb potential plays a significant role at each stage of the orbit. In Fig. 1.2c we show a trajectory, where the tunneling process started after the field cycle maximum so that the laser drift momentum pushes the electron initially towards the core. It is obvious that in this case the average over the wiggly trajectory at short times does not follow the orbital motion, but due to the close encounter with the ionic core the electron uses a complicated shortcut at some point, before it behaves regularly again. Overall, the bound orbits on an isoenergy shell behave regularly. They show a decrease of the eccentricity e of the orbit equivalent to an increase in angular momentum l, with increasing initial lateral momentum. Orbits calculated with parameters of isolated areas in the parameter space around the field cycle maximum show initially strongly irregular behavior. The occurrence of bound trajectories with initial parameters taken from the inner parameter space region strongly depends on the number of laser cycles following the tunneling event, as the electron dynamics is dominated by strong rescattering at the core. This is reflected in irregular behavior during the laser pulse, before the orbit finally merges into a stable orbit, as can be seen in Fig. 1.2f.

To summarize the semiclassical analysis, we conclude that the population of bound excited states stems from an interplay of the Coulomb field and the laser field on equal footing. Often, the averaged motion of the electron during the laser pulse follows already nicely the trajectory of the final orbit it will merge into. Consequently, any approximation of the Coulomb interaction during the laser field is not a priori justified. Although the Coulomb field influences the overall trajectory decisively, the amplitude of the electron's quiver motion seems to be still in accord with the assumption of a quasi free electron quivering in the laser field.

To calculate the yield of excited states one has to consider the probability for tunneling of an electron in the ground state with the magnetic quantum number $m = 0$, which is given within the simplest approximation of the tunneling model by [42, 45, 46]

$$w_0 \propto \left(\frac{2(2I_P)^{3/2}}{|F(t_t)|} \right)^{\left(\frac{2}{\sqrt{2I_P}} - 1\right)} \exp\left[-2(2I_P)^{3/2}/3\,|F(t_t)| \right]. \tag{1.2}$$

The probability to find a certain lateral momentum is given by [45]

$$w_\perp \propto \exp\left[-p_\perp^2 \sqrt{2I_P}/\,|F(t_t)| \right]. \tag{1.3}$$

In semiclassical Monte-Carlo simulations one uses (1.2) and (1.3) to randomly pick initial conditions and start typically about 10^5–10^6 trajectories. The total energy $T = p(t_{fin})^2/2 + V(r(t_{fin}))$ of each trajectory is evaluated at time t_{fin} shortly after the laser pulse is over. As an example, we evaluate trajectories for He exposed to a laser field with intensities between (10^{14} and 10^{15} W cm^{-2}) and a laser pulse duration (τ = 27 fs). We find $\approx$10 % of all launched trajectories to be bound at the lower

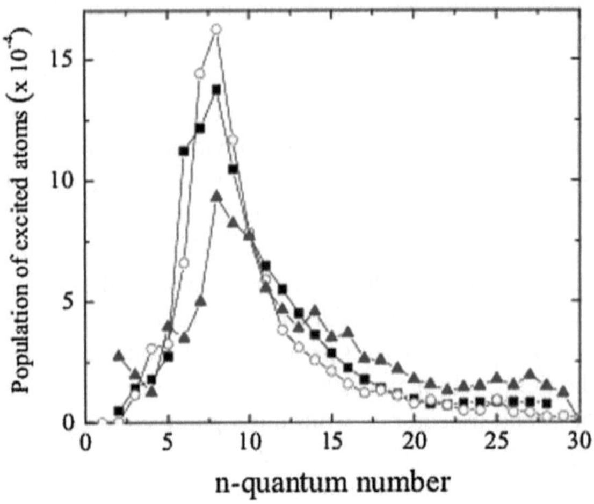

Fig. 1.3 n distribution of the population of excited states from a MC simulation (*red circle*), a quasi-one-electron (*black square*) and a full two-electron quantum mechanical calculation (*blue triangle*) at a laser intensity 10^{15} W cm^{-2}. The MC simulation has been normalized to the quasi-one-electron calculation at $n = 10$. From [15]

intensity 10^{14} W cm^{-2} and $\approx$2.5 % at 10^{15} W cm^{-2}, see Fig. 1 of [15], confirming the FTI process to be an important exit channel.

One can then assign an effective quantum number ν to each trajectory with negative total energy, determined by the Rydberg formula $T = -1/2\nu^2$. By sorting the effective quantum numbers into integer bins one obtains an n distribution, as shown in Fig. 1.3 and which is taken from [15]. The maximum of the distribution is around $n = 8$ and the probability drops steeply towards both sides, but allowing for Rydberg states with higher n quantum numbers. An analysis of the classical angular momentum reveals that for states with $n < 9$, the probability to find a specific angular momentum within a fixed n shell increases strongly towards the maximum allowed value of $l = n - 1$. For larger n states, we find the same distribution with no angular momentum states higher than $l = 9$.

It is also instructive to look at the wavelength dependence of the yield of bound states. The drift energy and the quiver amplitude both scale $\propto \omega^{-2}$ at constant laser intensity. Consequently, for shorter wavelengths the range of the parameter space for bound states increases and, with it, also the number of surviving trajectories. Particularly, population of lower n states with low radial extension benefits from the reduced quiver amplitude. On the other hand, increasing the wavelength results in the opposite behavior. Due to higher drift energy and a larger quiver motion, the parameter space for allowed trajectories shrinks and only bound trajectories with sufficient radial extension survive. The number of atoms surviving in lower n states is strongly diminished, while the number of atoms in higher n states remains the same. Consequently, the percentage of surviving atoms decreases. Finally, we mention that the FTI mechanism should also be in effect for strong-field excitation of ionic systems, see also Sect. 1.3.3.

The distribution shown in Fig. 1.3 has been confirmed by Monte-Carlo simulations of other groups [44, 47]. Most astounding, however, is the outcome of a direct comparison of the FTI results with TDSE calculations. In [15] the classical predictions were compared with results using the single-active electron (SAE) model and a full two-electron calculation, performed in the group of A. Saenz. The composition of the resulting Rydberg wave packet is in striking agreement, Fig. 1.3. The results demonstrate the predictive power of the frustrated tunneling ionization model. Similar results for the n distribution for different atoms and conditions have been obtained from other TDSE calculations [48–51] showing a maximum in the vicinity of $n = 8$. Only recently, we succeeded in measuring the n distribution by using a state selective field ionization method on the surviving neutral excited atoms. The results nicely confirm the theoretical predictions [52].

1.2.2 Elliptically Polarized Laser Fields

So far, we have described the FTI process in linearly polarized laser light. Hereby, most electrons tunnel in the vicinity of the field cycle maximum, so that the associated laser induced drift energy $E_{dr} \approx 0$. In an elliptically polarized laser field given by

$$\mathbf{F}(t) = F_0 f(t) \left(\cos(\omega t)\hat{\mathbf{e}}_x + \epsilon \sin(\omega t)\hat{\mathbf{e}}_y\right) / \sqrt{1+\epsilon^2}, \tag{1.4}$$

however, an additional laser induced drift momentum in y direction (lateral direction) arises, which reads for $\phi_t \approx 0$ $p_{dr,y} = -\epsilon F_0/\omega\sqrt{\epsilon^2+1}$. $p_{dr,y}$, and thus the total drift energy, are always nonzero for electrons that tunnel at $\phi_t \approx 0$ and increase strongly with ellipticity ϵ. It is this strong drift momentum that reduces substantially the population of bound states in elliptically polarized light. To include the polarization dependence, one might require in a simple approach that $p_{dr,y}$ must be compensated by an appropriate initial lateral momentum, $p_\perp = -p_{dr,y}$ to obtain bound states. Inserting this into (1.3), $w(\epsilon)$ is then given by [53]

$$w(\epsilon) \approx \exp -\frac{F_0(2I_p)^{1/2}}{\omega^2}\epsilon^2. \tag{1.5}$$

The results of this equation are in very good agreement with the experiment [15], which measured the total excited neutral yield as a function of ellipticity as shown in Fig. 1.4. We note that more extensive results have been obtained [47, 53] by solving (1.1) with the elliptically polarized laser field (1.4).

1.2.3 Intermediate Conclusion

The systematic investigation of electron trajectories emerging in a time window around a field cycle maximum revealed that the explicit inclusion of the Coulomb field in the rescattering model or simple man's model leads to bound states. Although bound or transient bound electron trajectories have been displayed occasionally in

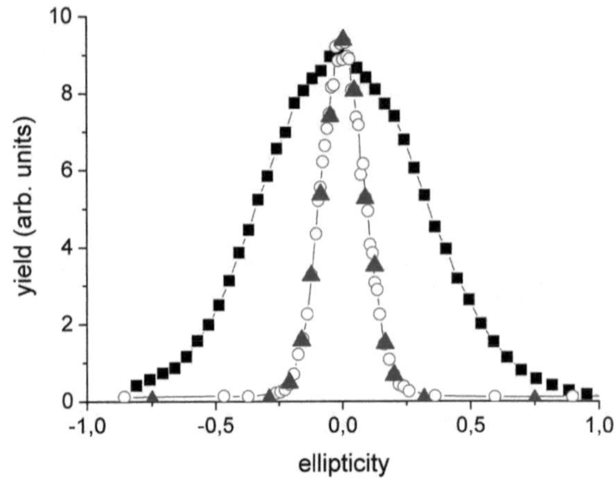

Fig. 1.4 Dependence of the He^+ (*black square*) and He* (*red circle*) yield on the ellipticity at fixed laser intensity of 10^{15} W cm^{-2}. For better comparison the height of both measurements have been set equal at $\epsilon = 0$, from [15]. Theoretical data points (*blue triangle*) are taken from [53]

the literature, the systematic study has shaped a clear picture of excitation in the tunneling regime. The FTI model also facilitates the interpretation of more complex experiments involving bound states and has led to a simple way to make quantitative predictions, in particular on the total yield of bound states and their n distribution. The FTI process shows qualitatively a similar behavior on ellipticity as the rescattering related processes HHG, NSDI, and HATI. In the latter cases, however, elliptical polarization reduces the rescattering probability with the ionic core, while in the FTI process formation of bound states is suppressed due to the high drift momentum. Nevertheless, one can consider the FTI process as a relevant additional exit channel in the rescattering model. It is important to note that if one uses the correct definition of the term ionization, which has no strict meaning unless the laser pulse is off, an atom finally in a bound excited state has never become ionized in the first place. This, in turn, implies that the process is not a classical electron recombination process involving emission of radiation as has been misleadingly argued occasionally in the literature [49, 50].

1.2.4 Detection of Excited Atoms

The most prominent detection schemes in strong-field physics involve electron and ion detection by means of time of flight (TOF) techniques, velocity map imaging (VMI) [54], or a reaction microscope [55], as well as detection of high harmonic radiation. Usually, excited states of an atom are detected by subsequent photoionization or state selective field ionization, occasionally also by fluorescence measurements. In the scope of the present investigations experiments are based on a direct measurement. In Fig. 1.5 the sketch of an experimental setup shows how it is realized.

A well collimated thermal effusive beam of atoms, which is directed towards a position-sensitive multichannel plate (MCP) detector, is crossed by an intense focused laser beam. Charged particles created by the laser-matter interaction can be refrained from reaching the detector by means of small electric fields. Excited and ground state atoms reach the detector after a mean time of flight that is determined by the drift distance d to the detector, the temperature and the mass of the atoms (for He atoms at room temperature with $d = 0.38\,\mathrm{m}$ it is $\approx 200\,\mu\mathrm{s}$). The detector, however, is only able to detect excited atoms with sufficient internal energy. Considering the comparably long time of flight, radiative decay and black-body induced redistribution of the initial excited state distribution is unavoidable. Although the decay to the ground state is dominant, especially for low n and low l states, a substantial fraction of atoms decays to long-lived metastable states present in all rare gas atoms. Thus, an efficient detection is assured [52].

Without additional forces, as they will be discussed in Sect. 1.4.1, one expects to measure a pattern reflecting the excited atom probability distribution in the laser beam. While the radial dimension of the Gaussian laser beam is only on the order of the beam waist of a few tens of μm, which is small compared to the extension in laser beam direction on the order of the Rayleigh length of a few mm, excitation occurs within an elongated cylindrical volume along the laser beam axis. This is projected onto the detector, as can be seen for a measurement of excited Ar* atoms hitting the detector, see Fig. 1.5a. Furthermore, neutral excited atom detection furthermore allows for a partially position dependent analysis of frustrated ionization dynamics within the focused volume. As will be discussed in Sect. 1.3, the detection technique allows also for the study of excited neutral atomic fragments from a strong-field dissociation process. Excited neutral fragment detection is well known in molecular physics, but has only been introduced in strong-field physics recently. The kinetic energy of the fragments is typically much higher than in an effusive beam, thus reducing the time of flight substantially. Consequently, excited states can be detected more likely in their initially populated states.

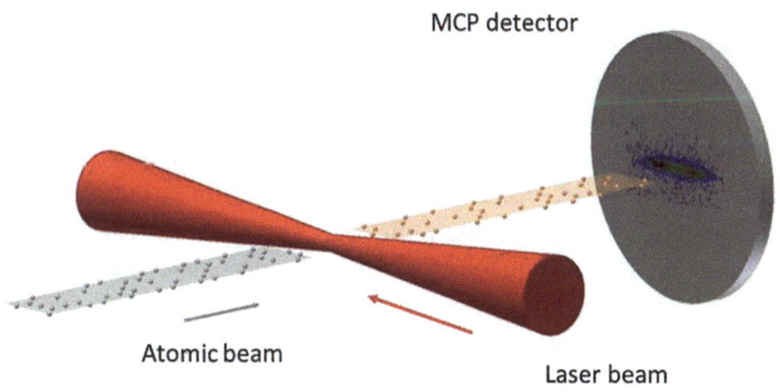

Fig. 1.5 Sketch of the experimental set up, as described in the text. The detector shows a typical distribution of strong-field excited atoms

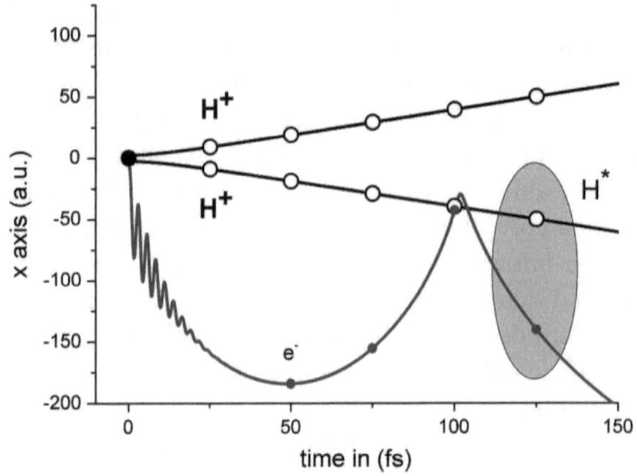

Fig. 1.6 Sketch of the molecular FTI in H_2^+. The *black dot* designate the two protons initially located too close together to be resolved. See text for more explanation

1.3 Frustrated Tunneling Ionization in Strong-Field Fragmentation of Molecules

1.3.1 Hydrogen Molecule

The process of frustrated tunneling ionization is not restricted to atoms but can also be observed in the strong-field dissociation of molecules [56], as first explored in H_2. The fragmentation of the simplest molecules H_2 and H_2^+ in strong laser fields has been well studied over the last decades. Fundamental fragmentation processes such as bond softening within the laser field and Coulomb explosion have been revealed [57]. Particularly, the process of Coulomb explosion, which is the repulsion of two protons, possible after the removal of the electrons in the strong field, can be exploited to study time and position dependent ionization [58]. The kinetic energy of the fragments gives detailed information on the nuclear distance at the instant of "ionization". Since a tunneling process might precede both the fragmentation and also the Coulomb explosion, it is very tempting to investigate the fragmentation process in view of the FTI process.

To elucidate the physics, one can start by considering strong-field tunneling in the H_2^+ molecular ion instead of the molecule. The scenario can be calculated by solving the coupled Newton equations for the three particles. One set of trajectories calculated for appropriate initial conditions is depicted in Fig. 1.6.

After tunneling the liberated electron quivers in the laser field sufficiently far away from the protons, so that it does not influence their acceleration process. After the laser pulse is over, the electron follows a trajectory that stays near one of the accelerated protons forming eventually an excited state of H. Consequently, the neutral fragment has virtually the same kinetic energy as the proton. In fact, starting from H_2, one can consider this process as Coulomb explosion without double ionization. Based on the detection method presented in Sect. 1.2.4, experiments become possible, where one

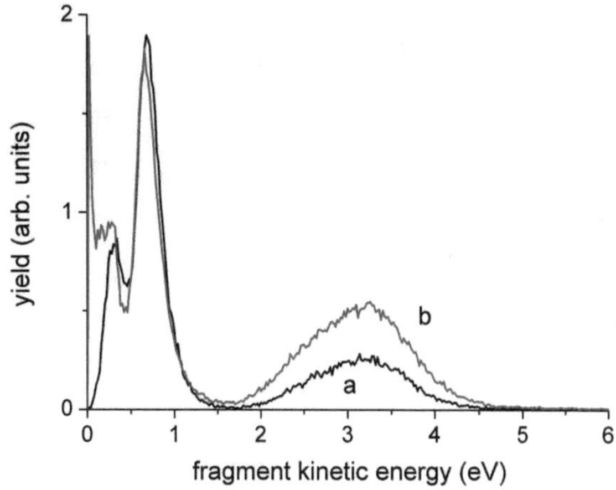

Fig. 1.7 Kinetic energy distribution of ionic and excited neutral fragments after strong field interaction at an intensity of $I = 3 \times 10^{14}$ W/cm^2. Kinetic energy distribution of **a** H^+ (*black curve*) and **b** H^* (*red curve*). From [56]

specifically detects excited neutral fragments from the strong-field fragmentation of H_2 together with an ionic fragment spectrum. The detection of excited neutral fragments is facilitated by the fact that, owing to the high kinetic energy, the travel times towards the detector are substantially shorter than for excited atoms in an effusive beam. Consequently, the loss of excited atoms due to radiative decay to the ground state as in the atomic case is much less. In Fig. 1.7 the yield of excited neutral and charged fragments as a function of their kinetic energy is shown. As can be seen, both spectra are perfectly linked indicating that excited neutral fragments occur with the same kinetic energy as the charged ones. This obviously confirms that the Coulomb explosion involves finally a neutral excited state as sketched in Fig. 1.6. Moreover, also fragmentation processes such as the bond-softening, which are the origin of the peak at lower kinetic energy release (KER), are partially accompanied by FTI leading to low energy excited neutral fragments. To finally prove that the neutral fragments with high energy originate in a Coulomb explosion process, a coincidence measurement of the fast excited neutral and of the ionic fragment at 3 eV has been performed using a reaction microscope. The resulting correlated signal proves that both fragments stem from a single fragmentation process as suggested by the extended FTI model [56]. Comprehensive semiclassical calculations essentially confirm the validity of the picture [59, 60].

1.3.2 Small Molecules

The molecular FTI process is not restricted to the hydrogen molecule. The FTI process has also been confirmed to be in effect in the strong-field fragmentation of D_2^+ [61]. Extensive experimental studies, in which the yields of excited fragments have been measured as a function of the laser intensity, the laser pulse duration and the laser polarization, reveal an overall satisfying modeling of the data within the

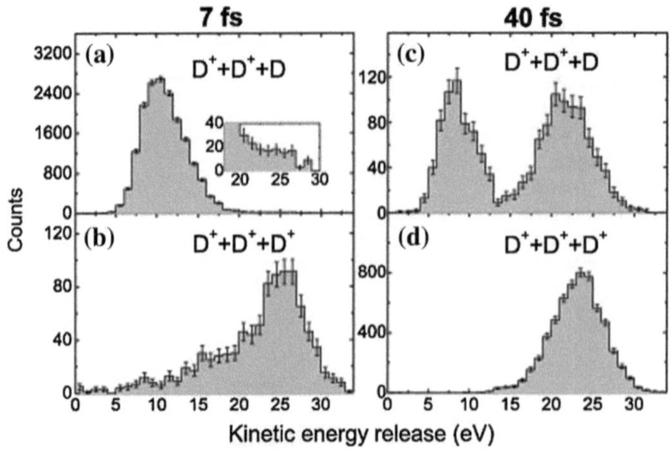

Fig. 1.8 KER distributions for the three-body breakup of D_3^+ at 10^{16} W cm^{-2}, 790 nm. Fragmentation with 7 fs pulses for **a** $D^+ + D^+ + D$ and **b** $D^+ + D^+ + D^+$ channels (inset of (**a**) shows an expanded vertical scale of the high-KER range). Fragmentation with 40 fs pulses for **c** $D^+ + D^+ + D$ and **d** $D^+ + D^+ + D^+$ channels. Note the high-KER feature between 15 and 30 eV in panel (**c**), the main subject of discussion. From [63]

FTI process. An alternate process, namely re-collision excitation, has been discussed to possibly also contribute to excited fragments. However, it was concluded that it is by far not sufficient to explain the overall set of data.

Moreover, neutral excited fragments after strong-field dissociation have also been observed for N_2 [62]. The N_2 molecule provides a rich kinetic energy spectrum of fragments originating from many possible fragmentation channels involving single and also higher charged N^{n+} fragments. Analysis of the kinetic energy spectrum of excited neutrals shows that all fragmentation channels containing singly charged fragments allow for the FTI process. Based on this example one may conclude that the FTI process is a quite general phenomenon in strong-field dissociation of molecules.

Evidence of FTI in a polyatomic system has been found by the Kansas group, which studied the fragmentation process of $D_3^+ \rightarrow D^+ + D^+ + D^*$ [63]. Surprisingly, at a laser intensity of 10^{16} W cm^{-2}, they observe high kinetic energy of the fragments that mimics the behavior of the $D^+ + D^+ + D^+$ channel and which can be finally attributed to frustrated tunneling ionization, see Fig. 1.8. Furthermore, they also find evidence of FTI in both two-body and three-body break up. More elaborate calculations on the fragmentation process are performed by Lötstedt [64].

1.3.3 Dimers

Frustrated tunneling ionization works also in strong-field multiple ionization of loosely bound noble gas dimers, and has been specifically studied in Ar_2 [65–69],

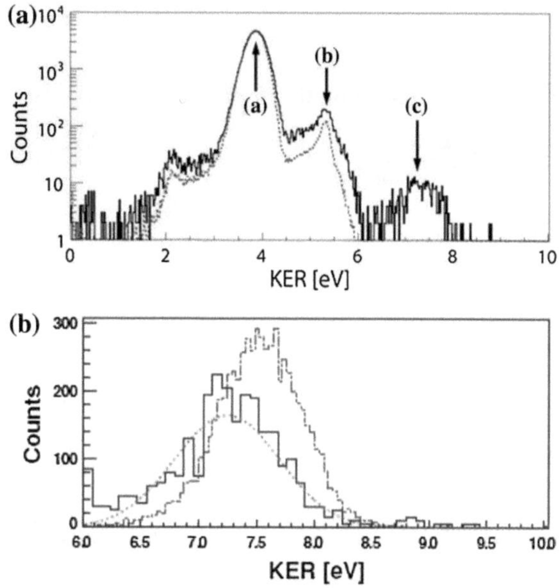

Fig. 1.9 **a** The ion pair KER after strong-field double ionization of argon dimers at a laser pulse peak electric-field strength of 0.1 a.u. *Solid* (*black*) *line*, linear polarization; *dashed* (*red*) *line*, circular polarization of the laser beam, where the signal vanishes as expected for the FTI process (**a** similar spectrum is shown in [66]). **b** KER distribution for the Coulomb explosion channel $Ar^{2+} + Ar^{+}$ after triple ionization of the Ar_2 dimer [*dashed* (*blue*) *line*]. The spectrum was taken simultaneously with the corresponding KER spectrum for the Coulomb explosion channel $Ar^{+} + Ar^{+}$, see (**a**). From [65]

where the FTI mechanism has been observed to be in effect in ionic fragments. Triple ionization of the Ar_2 dimer, where altogether three electrons of the dimer (tunnel) ionize, leads to the Coulomb explosion of Ar^{2+} and Ar^{+} fragments with high kinetic energy. Rather than detecting a fast Ar^{2+} ion fragment, FTI of one of the electrons reduces the charge by unity by trapping the electron in an excited ionic Rydberg state. This is shown in Fig. 1.9a. The process is thus similar to FTI in H_2, but involves higher charge states, and rather than finding a neutral excited fragment the process results in an excited singly charged fragment. In Fig. 1.9b it is shown that the kinetic energy of the fragments in the FTI process is slightly reduced compared to the pure $Ar^{+}-Ar^{+}$ fragmentation channel providing information about the final n and l states populated in the excited fragment. Furthermore, frustrated tunneling ionization of two electrons have been postulated by [60]. This process has been observed and studied in Ar dimers [67]. Finally, in mixed and pure rare gas dimers the FTI process might initiate ion core charge oscillations, which eventually result in ionization of the populated Rydberg states giving rise to a specific low energy feature in the photoelectron spectrum [68, 69].

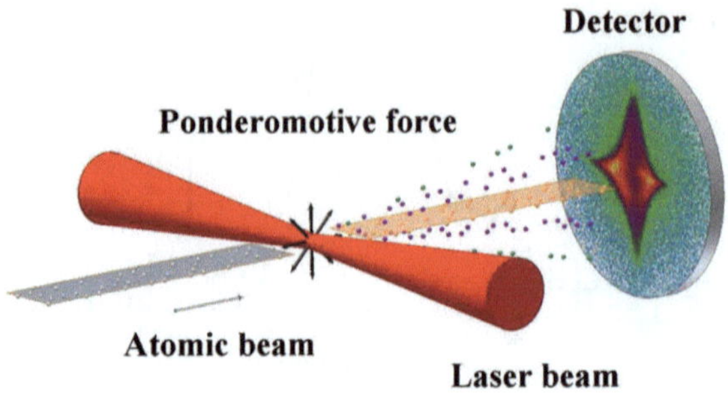

Fig. 1.10 Influence of the ponderomotive force on the measured spatial distribution of strong-field excited atoms

1.4 Kinematic Effects on Atoms

1.4.1 Acceleration of Neutral Atoms in Strong Laser Fields

Gradient (dipole) forces acting on matter in the focus of a continuous wave laser field are well known and extensively exploited to trap macroscopic neutral particles (optical tweezer) or to confine cold quantum gases in light lattices [70]. In contrast to recoil forces, which are in effect, e.g., in laser cooling, and are also present in plane-wave fields, gradient forces are directly proportional to the light intensity gradient. Kinematic manipulation of neutral atoms in inhomogeneous laser fields is essentially based on the Lorentz force acting on a polarizable atomic system. The laser field causes only a minor perturbation and the static polarizability of the atom can be employed in the case that the laser photon energy is sufficiently below the excitation energy of the first excited state. In the course of the work on atomic excitation in strong laser fields, it turns out that a similar force acts on the surviving neutral atoms. In this case, however, the light intensities exceed the threshold intensity for ionization by far and constitute by no means a small perturbation.

In fact, using He atoms, a strong radial ponderomotive force (schematically indicated by the black arrows in Fig. 1.10b was identified that accelerates the excited atoms with an unprecedented rate and causes a measurable radial deflection [71] despite the fact that the force acts only on extremely short times scales. We note that the intensity gradient in a focused laser beam is strongest in radial direction and negligible along the z direction, which is the propagation direction of the laser beam. The radial intensity gradient is strongest at the focal plane at half the beam spot size and drops symmetrically along the laser beam axis. It gives rise to a characteristic detection pattern. In Fig. 1.11 we show an experimental result, where the distribution of laser excited He atoms in a beam with a position sensitive detector located 0.3 m downstream from the interaction region was measured. The momentum transfer to the neutral atoms during the short interaction time is by far stronger than what one would expect from the dipole force acting on ground state atoms.

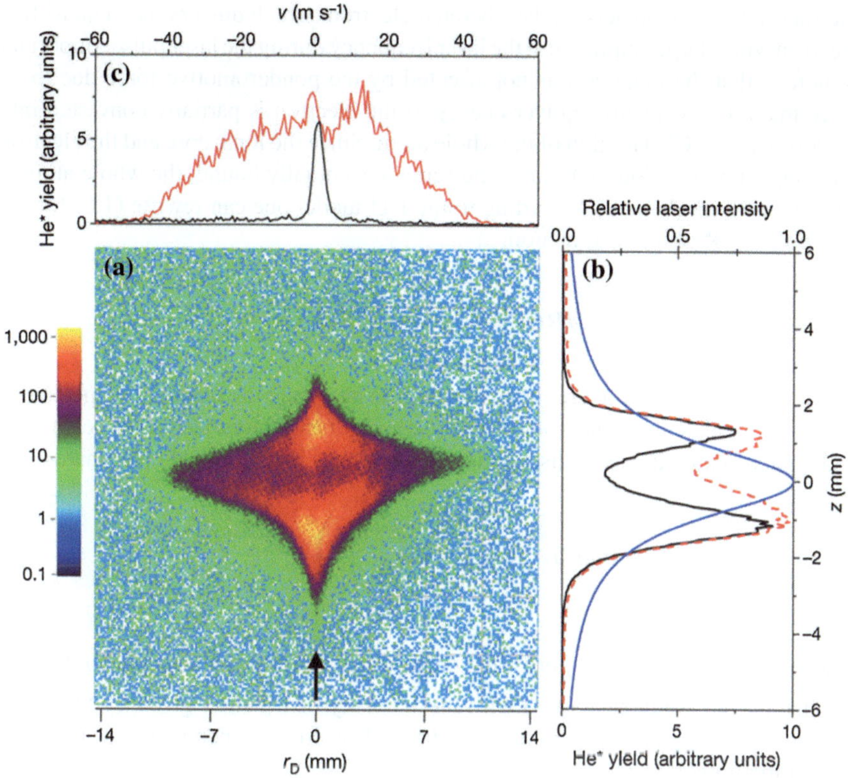

Fig. 1.11 Deflection of neutral He atoms after interaction with a focused laser beam. **a** Distribution of excited He* atoms on the detector. The laser beam direction is indicated by the *arrow*. **b** Cut through the atom distribution along the laser beam axis (z axis) at $r_D = 0$ mm (*black curve*) and full projection on z axis (*dashed red curve*). Intensity along the z axis in units of the laser peak intensity $I_0 = 6.9 \times 10^{15}$ W cm^{-2} (*blue curve*). **c** Cuts through the distribution at $z = 0$ mm (*red curve*) and $z = -2.7$ mm (*black curve*). The *black curve* shows the velocity distribution of excited neutral atoms at a position unaffected by the ponderomotive force, showing quasi the "natural" velocity spread, while the *red curve* shows the velocity gain through the ponderomotive force. From [71]

The FTI picture helps to grasp the idea of how atomic acceleration proceeds in strong laser fields. We first recall that the ponderomotive force $\mathbf{F_p}$ on a free electron (charged particle) exposed to an inhomogeneous electromagnetic field is given by

$$\mathbf{F_p} = -\frac{q^2}{2m\omega^2}\nabla\overline{|\mathbf{F}(\mathbf{r},\mathbf{t})|^2}, \tag{1.6}$$

Here, m and q are the mass and the charge of the particle, respectively, and the linearly polarized electric field now additionally depends on spatial coordinates $\mathbf{F}(\mathbf{r}, t) = \mathbf{F_0}(\mathbf{r}) f(t) \cos\omega t$, with $\mathbf{F_0}(\mathbf{r}) = F_0(\mathbf{r})\hat{\mathbf{e}}_\mathbf{x}$. Hence, transferred to the FTI model, we assume that the ponderomotive force due to the strong radial intensity gradient in the focused laser field acts on the liberated electron, which quivers like a quasi free electron with a large amplitude at the laser frequency during the laser pulse. Important to note is that the ionic core is not affected by the ponderomotive force due to its large mass. Consequently, quiver energy of the electron is partially converted into center-of-mass (CM) motion of the whole atom. Since the ionic core and the electron are coupled by the Coulomb force and remain eventually bound, the whole atom is substantially accelerated. According to the FTI model one can rewrite (1.6) for the CM position **R** of the neutral atom

$$M\ddot{\mathbf{R}}(t) = -\frac{e^2}{2m_e\omega^2}\nabla\overline{|\mathbf{F}(\mathbf{r},\mathbf{t})|^2} \tag{1.7}$$

Here, M and $m_e = 1$ a.u. are the masses of the atom and the electron, respectively. One can calculate the ponderomotive force for a linearly polarized laser beam with a Gaussian spatial intensity distribution, which reads in cylindrical coordinates

$$I(\mathbf{r}) = |\mathbf{F_0}(\mathbf{r})|^2 = I_0\left(1+\left(\frac{z}{z_0}\right)^2\right)^{-1}\exp\frac{-2r^2}{r_0^2}, \tag{1.8}$$

where $r_0 = w_0\sqrt{1+\left(\frac{z}{z_0}\right)^2}$, w_0 is the beam waist and I_0 is the laser peak intensity. Evaluating the gradient in (1.7) with the intensity distribution given by (1.8) one obtains for the radial component r_c of the CM position perpendicular to the laser beam direction

$$\ddot{r_c}(t) = \frac{I(\mathbf{R})}{M\omega^2}\frac{r_c(t)}{r_0^2}f(t) \tag{1.9}$$

One may take $f(t) = \exp(-t^2/\tau^2)$, where τ is the pulse width.

From (1.9) one finds that the maximum force along the radial direction scales as r_0^{-1}. Similarly, one can show that it scales as z_0^{-1} along the laser beam direction. Since the Rayleigh length z_0 is typically a factor 100 larger than the beam waist r_0, the gradient and thus the ponderomotive force in laser beam direction is much smaller than in the radial direction and can be neglected. Assuming that the neutral atom does not move during the laser pulse, one can set $r_c(t) = r_c$ on the right hand side of the equation. This allows one to solve (1.9) analytically for any initial position of an atom in the laser beam by just time integrating over laser pulse envelope. Considering the instant of tunneling with respect to the pulse envelope maximum, t_s, one obtains $S(t_s) = \frac{\sqrt{\pi}\tau}{2}\operatorname{erfc}(t_s/\tau)$, where erfc denotes the complementary error function. Atoms located at half beam waist $r_0/2$ experience the maximum force. The appropriate velocity $v_{\max}(z)$ is given by

$$v_{\max}(z) = \frac{I_0}{2M\omega^2 w_0} \frac{\exp(-0.5)}{\sqrt{1+\left(\frac{z}{z_0}\right)^2}^3} S(t_s) \tag{1.10}$$

Under the assumption that the initial tunneling process occurs early in the pulse, the experimental results agree very nicely with this formula. For He atoms exposed to a focused laser beam at maximum intensity (I = 7 × 10^{15} W cm^{-2}) one obtains from the data a maximum velocity of about 55 m/s. This, in turn, results from an acceleration of about 2 × 10^{14} g, where g is Earth's acceleration. This is one of the highest direct acceleration rate of neutral atoms ever achieved.

In order to derive acceleration of the CM of the neutral atom within the FTI model without field cycle averaging the full coupled full Lorentz equations, including the magnetic field, for the ionic core and the electron need to be solved. Most important, one finds bound excited states with the CM velocity confirming basically the results of (1.10). Observed deflections for He and Ne atoms for different laser parameters are in very good agreement with simple theoretical predictions [71] and more extended calculations [72].

Finally, we mention, that the relatively weak intensity gradient, which is on the laser beam waist length scale, is not strong enough to change the electron dynamics on an atomic scale and the predicted n state distribution remains unchanged. This situation might change significantly, however, if one studies neutral atom acceleration in a short-pulse intense standing wave. The strong periodic intensity gradients on the scale of the laser wavelength might introduce non dipole effects in the electron dynamics. First results of this Kapitza-Dirac scattering of neutral He atoms in an intense standing wave indicates this [73].

1.4.2 Rydberg Atoms in Strong Laser Fields

From the experiments described before, one can infer that an atom surviving the interaction with a strong laser field carries information about the laser intensity it has interacted with. This twist has been exploited in an experiment studying the survival and ionization of Rydberg atoms in a strong laser field [74]. We note that the ionization of a Rydberg atom is expected to neither follow the picture of tunneling nor the picture of a multi-photon process. The very successful semi-classical Keldysh theory for strong-field ionization [1] and with it, the Keldysh parameter γ, are not applicable. Although Rydberg atoms are expected to be rather stable against ionization, experimental studies are scarce. The experimental setup is as follows: A Mach-Zehnder interferometer provides two time delayed laser pulses with different polarization and very good spatial overlap in the focus [75]. With the first linearly polarized strong laser pulse with an intensity of 2.7 × 10^{15} W cm^{-2} a Rydberg wavepacket is excited in He and accelerated in the laser focus, as described before. After a time delay of 500 fs, a second elliptically polarized laser pulse (ellipticity $\epsilon = 0.66$) with an intensity of 3.8 × 10^{15} W cm^{-2} is applied. As outlined in

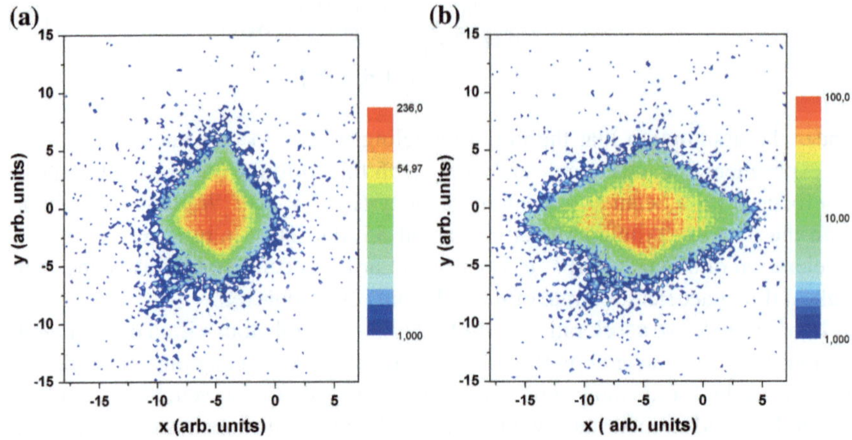

Fig. 1.12 **a** Excitation of He Rydberg atoms and deflection after interaction with a linearly polarized laser field, $I = 2.7 \times 10^{15}$ W cm^{-2}. **b** Interaction of the produced Rydberg population with an elliptically polarized laser, $I = 3.8 \times 10^{15}$ W cm^{-2}, $\epsilon = 0.66$

Sect. 1.2.2, sufficient elliptical polarization ensures that no further excitation from the ground state takes place [15]. Consequently, any Rydberg atom, that interacts with the second laser pulse and survives it, is additionally accelerated, provided, it is located in a nonzero intensity gradient. Hence, the deflection of Rydberg atoms, which can be measured as described before, verifies the interaction with the strong laser field.

We show the measured deflection of the Rydberg atoms excited by the first laser pulse in Fig. 1.12a. Figure 1.12b displays Rydberg atoms surviving the second laser pulse, which also accelerates them additionally. It has been found that the detected population after the second pulse drops at most by 15 % indicating for the first time an experimental proof of the exceptional stability of Rydberg atoms at laser intensities above 10^{15} W cm^{-2}. This means that the Rydberg atoms withstand corresponding field amplitudes of more than 1 GV/cm, which exceed the thresholds for static-field ionization by more than six orders of magnitude. Further analysis reveals that Rydberg atoms are predominantly excited within a radius half of the focus size $w_0/2$ of the second laser pulse. For this regime there is a unique relation between deflection of the atom and the intensity it has interacted with. Thus, surviving atoms have seen 60–95 % of the maximum intensity of the second pulse. As a result of quantum mechanical calculations, mainly low order processes are in effect, even at highest intensities. However, the calculations show important deviations from simple low-order perturbation theory at intensities, where the quiver amplitude is comparable or larger than the classical inner turning point of the Rydberg orbit. Finally, we remark that the high survival rate can be fruitfully exploited to steer neutral atoms by varying the spatial overlap of the two laser pulses, which breaks the radial symmetry of the field gradient [76].

1.5 Summary and Outlook

In conclusion we describe the process of frustrated tunneling ionization (FTI), a recently found new exit channel in the strong-field tunneling-plus-rescattering model that leads to the population of excited states, if the Coulomb field is taken into account explicitly. The FTI process allows for a qualitative and, even more important, also for a quantitative understanding of how excitation proceeds in the tunneling regime of strong-field physics. We demonstrate the ubiquity of FTI in strong-field physics by presenting various examples in atomic and molecular strong-field dynamics and dissociation. In all cases the concept of FTI facilitates the comprehension of observed phenomena. Furthermore, FTI also grants access to the understanding of acceleration of surviving excited atoms in spatially non-uniform strong laser fields such as a single focused laser field or two overlapping co- and counter-propagating laser fields. Observation of the resulting atomic deflection becomes possible by exploiting a position sensitive detection scheme for neutral excited atoms. This enables one to test the stability of Rydberg atoms in strong laser fields, where the experiment not only proves the high survival rate of Rydberg states, but also show unambiguously that the information, which intensity a Rydberg atom has survived, is encoded in its kinematics.

Future experiments on atomic and molecular excitation and acceleration will rely on and benefit from newly available laser sources in the mid infrared range. Particularly interesting is to study whether the interplay of stronger ponderomotive forces and the modified bound state formation through the strong acceleration and large excursion of the electron (non-dipole effects) leads to stronger neutral atomic acceleration. Furthermore, investigation of bound state formation in H-like and He-like ions is suited to explore the limits of the frustrated tunneling ionization model at higher laser intensities.

Acknowledgments The author thanks W. Becker, S. Eilzer, T. Nubbemeyer, H. Rottke, A. Saenz, W.Sandner, H. Reiss, and H. Zimmermann for fruitful discussions.

References

1. L.V. Keldysh, Sov. Phys. JETP **20**, 1307 (1965)
2. H.B. van Linden van den Heuvell, H.G. Muller, in *Multiphoton Processes*, ed. by S.J. Smith, P.L. Knight (Cambridge University Press, Cambridge, 1988)
3. T.F. Gallagher, Phys. Rev. Lett. **61**, 2304 (1988)
4. P.B. Corkum, N.H. Burnett, F. Brunel, Phys. Rev. Lett. **62**, 1259 (1989)
5. P.B. Corkum, Phys. Rev. Lett. **71**, 1994 (1993)
6. K.C. Kulander, K.J. Schafer, J.L. Krause, *Super-Intense Laser-Atom Physics* (Plenum, New York, 1993)
7. K.J. Schafer, B. Yang, L.F. DiMauro, K.C. Kulander, Phys. Rev. Lett. **70**, 1599 (1993)
8. G.G. Paulus, W. Nicklich, H. Xu, P. Lambropoulos, H. Walther, Phys. Rev. Lett. **72**, 2851 (1994)

9. B. Walker, B. Sheehy, L.F. DiMauro, P. Agostini, K.J. Schafer, K.C. Kulander, Phys. Rev. Lett. **73**, 1227 (1994)
10. R. Moshammer, B. Feuerstein, W. Schmitt, A. Dorn, C.D. Schröter, J. Ullrich, H. Rottke, C. Trump, M. Wittmann, G. Korn, K. Hoffmann, W. Sandner, Phys. Rev. Lett. **84**, 447 (2000)
11. T. Weber, M. Weckenbrock, A. Staudte, L. Spielberger, O. Jagutzki, V. Mergel, F. Afaneh, G. Urbasch, M. Vollmer, H. Giessen, R. Dörner, Phys. Rev. Lett. **84**, 443 (2000)
12. W. Becker, X. Liu, P.J. Ho, J.H. Eberly, Rev. Mod. Phys. **84**, 1011 (2012)
13. M. Lewenstein, P. Balcou, M.Y. Ivanov, A. L'Huillier, P.B. Corkum, Phys. Rev. A **49**, 2117 (1994)
14. F. Krausz, M. Ivanov, Rev. Mod. Phys. **81**, 163 (2009)
15. T. Nubbemeyer, K. Gorling, A. Saenz, U. Eichmann, W. Sandner, Phys. Rev. Lett. **101**, 233001 (2008)
16. M.P. de Boer, H.G. Muller, Phys. Rev. Lett. **68**, 2747 (1992)
17. M.P. de Boer, L.D. Noordam, H.G. Muller, Phys. Rev. A **47**, R45 (1993)
18. R.R. Jones, D.W. Schumacher, P.H. Bucksbaum, Phys. Rev. A **47**, R49 (1993)
19. L. Dimauro, P. Agostini, Adv. At. Mol. Opt. Phys. **35**, 79 (1995)
20. R.R. Freeman, P.H. Bucksbaum, H. Milchberg, S. Darack, D. Schumacher, M.E. Geusic, Phys. Rev. Lett. **59**, 1092 (1987)
21. A. Scrinzi, N. Elander, B. Piraux, Phys. Rev. A **48**, R2527 (1993)
22. M.V. Fedorov, A. Movsesian, J. Phys. B: At. Mol. Opt. Phys. **21**, L155 (1988)
23. R.R. Jones, P.H. Bucksbaum, Phys. Rev. Lett. **67**, 3215 (1991)
24. L.D. Noordam, H. Stapelfeldt, D.I. Duncan, T.F. Gallagher, Phys. Rev. Lett. **68**, 1496 (1992)
25. J.G. Story, D.I. Duncan, T.F. Gallagher, Phys. Rev. Lett. **70**, 3012 (1993)
26. J.H. Hoogenraad, R.B. Vrijen, L.D. Noordam, Phys. Rev. A **50**, 4133 (1994)
27. O.V. Tikhonova, E.A. Volkova, A.M. Popov, M.V. Fedorov, Phys. Rev. A **60**, R749 (1999)
28. F. Benvenuto, G. Casati, D.L. Shepelyansky, Phys. Rev. A **45**, R7670 (1992)
29. W.C. Henneberger, Phys. Rev. Lett. **21**, 838 (1968)
30. M. Pont, N.R. Walet, M. Gavrila, C.W. McCurdy, Phys. Rev. Lett. **61**, 939 (1988)
31. K.C. Kulander, K.J. Schafer, J.L. Krause, Phys. Rev. Lett. **66**, 2601 (1991)
32. J.H. Eberly, K.C. Kulander, Science **262**, 1229 (1993)
33. M. Gavrila, J. Phys. B: At. Mol. Opt. Phys. **35**, R147 (2002)
34. A.M. Popov, O.V. Tikhonova, E.A. Volkova, J. Phys. B: At. Mol. Opt. Phys. **36**, R125 (2003)
35. F. Morales, M. Richter, S. Patchkovskii, O. Smirnova, Proc. Nat. Acad. Sci. **108**, 16906 (2011)
36. M. Richter, S. Patchkovskii, F. Morales, O. Smirnova, M. Ivanov, New J. Phys. **15**, 083012 (2013)
37. M.P. de Boer, J.H. Hoogenraad, R.B. Vrijen, L.D. Noordam, H.G. Muller, Phys. Rev. Lett. **71**, 3263 (1993)
38. N.J. van Druten, R.C. Constantinescu, J.M. Schins, H. Nieuwenhuize, H.G. Muller, Phys. Rev. A **55**, 622 (1997)
39. E. Wells, I. Ben-Itzhak, R.R. Jones, Phys. Rev. Lett. **93**, 023001 (2004)
40. A. Talebpour, C.Y. Chien, S.L. Chin, J. Phys. B: At. Mol. Opt. Phys. **29**, 5725 (1996)
41. M. Gajda, J. Grochmalicki, M. Lewenstein, K. Rzążewski, Phys. Rev. A **46**, 1638 (1992)
42. G.L. Yudin, M.Y. Ivanov, Phys. Rev. A **63**, 033404 (2001)
43. H.G. Muller, Phys. Rev. Lett. **83**, 3158 (1999)
44. N. Shvetsov-Shilovski, S. Goreslavski, S. Popruzhenko, W. Becker, Laser Phys. **19**, 1550 (2009)
45. N.B. Delone, V.P. Krainov, J. Opt. Soc. Am. B **8**, 1207 (1991)
46. D. Comtois, D. Zeidler, H. Pepin, J.C. Kieffer, D.M. Villeneuve, P.B. Corkum, J. Phys. B: At. Mol. Opt. Phys. **38**, 1923 (2005)
47. K. Huang, Q. Xia, L. Fu, Phys. Rev. A **87**, 033415 (2013)
48. S. Chen, X. Gao, J. Li, A. Becker, A. Jaroń Becker. Phys. Rev. A **86**, 013410 (2012)
49. E. Volkova, A. Popov, O. Tikhonova, J. Exp. Theor. Phys. **113**, 394 (2011)
50. T. Morishita, C.D. Lin, Phys. Rev. A **87**, 063405 (2013)
51. Q. Li, X.M. Tong, T. Morishita, H. Wei, C.D. Lin, Phys. Rev. A **89**, 023421 (2014)
52. H. Zimmermann, J. Buller, S. Eilzer, U. Eichmann, Phys. Rev. Lett. **114**, 123003 (2015)

53. A.S. Landsman, A.N. Pfeiffer, C. Hofmann, M. Smolarski, C. Cirelli, U. Keller, New J. Phys. **15**, 013001 (2013)
54. A.T.J.B. Eppink, D.H. Parker, Rev. Sci. Instrum. **68**, 3477 (1997)
55. R. Dörner, V. Mergel, O. Jagutzki, L. Spielberger, J. Ullrich, R. Moshammer, H. Schmidt-Böcking, Phys. Rep. **330**, 95 (2000)
56. B. Manschwetus, T. Nubbemeyer, K. Gorling, G. Steinmeyer, U. Eichmann, H. Rottke, W. Sandner, Phys. Rev. Lett. **102**, 113002 (2009)
57. J.H. Posthumus, Rep. Progr. Phys. **67**, 623 (2004)
58. S. Chelkowski, P.B. Corkum, A.D. Bandrauk, Phys. Rev. Lett. **82**, 3416 (1999)
59. A. Emmanouilidou, C. Lazarou, A. Staudte, U. Eichmann, Phys. Rev. A **85**, 011402 (2012)
60. A. Emmanouilidou, C. Lazarou, New J. Phys. **14**, 115010 (2012)
61. J. McKenna, S. Zeng, J.J. Hua, A.M. Sayler, M. Zohrabi, N.G. Johnson, B. Gaire, K.D. Carnes, B.D. Esry, I. Ben-Itzhak, Phys. Rev. A **84**, 043425 (2011)
62. T. Nubbemeyer, U. Eichmann, W. Sandner, J. Phys. B: At. Mol. Opt. Phys. **42**, 134010 (2009)
63. J. McKenna, A.M. Sayler, B. Gaire, N.G. Kling, B.D. Esry, K.D. Carnes, I. Ben-Itzhak, New J. Phys. **14**, 103029 (2012)
64. E. Lötstedt, T. Kato, K. Yamanouchi, Phys. Rev. Lett. **106**, 203001 (2011)
65. B. Manschwetus, H. Rottke, G. Steinmeyer, L. Foucar, A. Czasch, H. Schmidt-Böcking, W. Sandner, Phys. Rev. A **82**, 013413 (2010)
66. B. Ulrich, A. Vredenborg, A. Malakzadeh, M. Meckel, K. Cole, M. Smolarski, Z. Chang, T. Jahnke, R. Dörner, Phys. Rev. A **82**, 013412 (2010)
67. J. Wu, A. Vredenborg, B. Ulrich, L.P.H. Schmidt, M. Meckel, S. Voss, H. Sann, H. Kim, T. Jahnke, R. Dörner, Phys. Rev. Lett. **107**, 043003 (2011)
68. A. von Veltheim, B. Manschwetus, W. Quan, B. Borchers, G. Steinmeyer, H. Rottke, W. Sandner, Phys. Rev. Lett. **110**, 023001 (2013)
69. M. Li, H. Liu, C. Wu, Y. Deng, C. Wu, Q. Gong, Y. Liu, Phys. Rev. A **89**, 025402 (2014)
70. R. Grimm, M. Weidemüller, Y.B. Ovchinnikov, Adv. At. Mol. Opt. Phys. **42**, 95 (2000)
71. U. Eichmann, T. Nubbemeyer, H. Rottke, W. Sandner, Nature **461**, 1261 (2009)
72. Q.Z. Xia, L.B. Fu, J. Liu, Phys. Rev. A **87**, 033404 (2013)
73. S. Eilzer, H. Zimmermann, U. Eichmann, Phys. Rev. Lett. **112**, 113001 (2014)
74. U. Eichmann, A. Saenz, S. Eilzer, T. Nubbemeyer, W. Sandner, Phys. Rev. Lett. **110**, 203002 (2013)
75. T. Nubbemeyer, U. Eichmann, Eur. Phys. J. Spec. Top. **222**, 2267 (2013)
76. S. Eilzer, U. Eichmann, J. Phys. B: At. Mol. Opt. Phys. **47**, 204014 (2014)

Chapter 2
Few-Cycle-Laser-Pulse Induced and Assisted Processes in Atoms, Molecules, and Nanostructures

Dejan B. Milošević

Abstract Dynamics of atomic and molecular processes in a strong few-cycle laser field is analyzed with emphasis on the dependence on the carrier-envelope phase. From the laser-induced processes we considered above-threshold ionization and high-order harmonic generation, while from the laser-assisted processes we considered electron–atom potential scattering and electron–ion recombination. All these processes can be treated theoretically using the introduced phase space path-integral formalism.

2.1 Introduction

Photons and electrons have played a crucial role in the scientific and technological revolution of the twentieth century. They are still a workhorse of modern science – in particular the attoscience is able to describe electronic motion in atoms, molecules, and nanostructures. The electrons and coherent ultrashort soft X-ray pulses can be emitted in the laser-matter interactions. The control of these processes on the attosecond time scale can be achieved using the few-cycle laser pulses. The basic parameter of few-cycle laser pulses which is used for such control is the carrier-envelope (CE) phase. A change in the CE phase changes significantly temporal evolution of the

D.B. Milošević (✉)
Faculty of Science, University of Sarajevo,
Zmaja od Bosne 35, 71000 Sarajevo, Bosnia and Hercegovina
e-mail: milo@bih.net.ba

D.B. Milošević
Academy of Sciences and Arts of Bosnia and Hercegovina,
Bistrik 7, 71000 Sarajevo, Bosnia and Hercegovina

D.B. Milošević
Max-Born-Institut, Max-Born-Strasse 2a, 12489 Berlin, Germany

M. Kitzler and S. Gräfe (eds.), *Ultrafast Dynamics Driven by Intense Light Pulses*, Springer Series on Atomic, Optical, and Plasma Physics 86, DOI 10.1007/978-3-319-20173-3_2

few-cycle laser field. In turn, this change affects the laser-matter interaction on the atomic scale. This can be utilized for measuring and controlling the CE phase and thus opening a route to light waveform control [1].

In this chapter we will consider the role of CE phase in various laser-induced and laser-assisted processes. Namely, the processes which take place in a powerful laser field can be divided into two main groups [2]: laser-assisted processes and laser-induced processes. For the first group a basic process exists in the absence of the laser field and it is modified by the simultaneous interaction with laser radiation. On the other hand, the laser-induced processes very often require a threshold value for the number of absorbed photons from the laser field to take place at all. Examples of various such processes for long pulses which are not affected by the CE phase can be found in references [2, 3].

We will firstly give definition of few-cycle pulses, CE phase, and related parameters. Next, in the theoretical part of this chapter, we derive an expression for the time-evolution operator obtained using phase space path-integral method [4]. This result is general and can be used to obtain transition matrix elements for all considered laser-induced and laser-assisted processes.

In Sect. 2.4 we give a historical introduction about the first theoretical and experimental discoveries of the CE effect in laser-induced processes such as high-order harmonic generation and above-threshold ionization (ATI). The emphasis will be on the so-called stereo-ATI with the CE phase stabilized few-cycle laser pulses [5]. In particular, we will use the results of references [6–8] to present more recent achievements in the CE-phase control of high-order ATI in atomic and molecular systems.

In the next section we will consider high-order harmonic generation (HHG) process. We will focus on the CE-phase control of plasmonic field enhanced HHG [9]. We will show that, contrary to the homogeneous field case, for inhomogeneous field the difference between the maximum and minimum HHG cutoff remains substantial even for pulses longer than ten optical cycles.

Section 2.5 is devoted to few-cycle-laser-pulse-assisted scattering [10] and recombination [11] processes. We will show common features of these processes and explore their dependence on the CE phase.

In concluding section we will briefly mention some other processes for which the CE phase is an important controlling parameter. We use atomic units so that $\hbar = 1, m = 1, |e| = 1$, and $4\pi\epsilon_0 = 1$ throughout this chapter.

2.2 Definition of Few-Cycle Laser Pulse Parameters

The pulse envelope is defined via the electric-field vector $\boldsymbol{E}(t)$ [5, 12–14].[1] We assume that the total pulse duration T_{p} is equal to an integer number n_{p} of optical cycles $T = 2\pi/\omega$, $T_{\mathrm{p}} = n_{\mathrm{p}}T$, and we denote the CE phase by ϕ. The laser field is

[1] A different definition of a few-cycle pulse, based on the vector-potential pulse envelope, was considered in [15, 16].

elliptically-polarized in the xy plane, having the ellipticity ε. We adopt the notation $\varepsilon_1 = 1/\sqrt{1+\varepsilon^2}, \varepsilon_2 = \varepsilon\varepsilon_1, \phi_1 = \phi+\pi/2, \phi_2 = \phi, \hat{\boldsymbol{e}}_1 = \hat{\boldsymbol{e}}_x, \hat{\boldsymbol{e}}_2 = \hat{\boldsymbol{e}}_y, \varphi_j = \omega t+\phi_j$, $\varphi_{ij} = \omega_i t + \phi_j$ $(i, j = 1, 2)$, and $\omega_0 = \omega, \omega_{1,2} = \omega \pm \omega_\mathrm{p}$, where $\omega_\mathrm{p} = \omega/n_\mathrm{p}$. Then, the electric-field vector having a sine square pulse envelope can be written in the form

$$\boldsymbol{E}(t) = E_0 \sin^2(\omega_\mathrm{p} t/2) \sum_{j=1,2} \varepsilon_j \hat{\boldsymbol{e}}_j \sin\varphi_j = \sum_{j=1,2} \varepsilon_j \hat{\boldsymbol{e}}_j \sum_{i=0,1,2} \mathcal{E}_i \sin\varphi_{ij} \tag{2.1}$$

for $0 \leq t \leq T_\mathrm{p}$ and $\boldsymbol{E}(t) = \boldsymbol{0}$ outside this interval. We also defined $\mathcal{E}_0 = E_0/2$ and $\mathcal{E}_i = -\mathcal{E}_0/2$ $(i = 1, 2)$. The above relation shows that the field (2.1) with a sine-square envelope is equivalent to a three-color field with constant envelope for each color within $0 \leq t \leq T_\mathrm{p}$. The electron ponderomotive energy in few-cycle pulse (2.1) is $U_\mathrm{p} = \sum_{i=0,1,2} \mathcal{E}_i^2/(4\omega_i^2)$ [5].

Examples of the above-defined pulses are presented in Fig. 2.1. The common nomenclature of cosine-like and sine-like pulses is explained in the upper panels where the electric field vector of a 6-cycle linearly polarized pulse is shown. For circular polarization we have $\varepsilon = 1$ and $\varepsilon_1 = \varepsilon_2 = 1/\sqrt{2}$. Examples of 2-cycle circularly polarized pulse are shown in lower panels of Fig. 2.1. One can see that for an arbitrary value of the CE phase ϕ, the curves have to be rotated about the origin $(E_x(t), E_y(t)) = (0, 0)$ counterclockwise by the angle ϕ.

Note that $\boldsymbol{E}(0) = \boldsymbol{E}(T_\mathrm{p}) = \boldsymbol{0}$. The vector potential $\boldsymbol{A}(t) = -\int \mathrm{d}t \boldsymbol{E}(t)$ is, in general, different from zero for $t \leq 0$ and $t \geq T_\mathrm{p}$. However, if the total pulse duration equals an integer number of laser cycles, we have $\boldsymbol{A}(0) = \boldsymbol{A}(T_\mathrm{p})$, so that the integral over the electric field is zero, $\int_0^{T_\mathrm{p}} \mathrm{d}t \boldsymbol{E}(t) = -[\boldsymbol{A}(T_\mathrm{p}) - \boldsymbol{A}(0)] = \boldsymbol{0}$, that is, the electric field has no dc component [5]. The fact that the vector potential is different from zero at the beginning and at the end of the pulse has a consequence that $\boldsymbol{A}(0)$ $[\boldsymbol{A}(T_\mathrm{p})]$ should be subtracted from the electron momenta which define initial (final) continuum states of the considered process.

We define the pulse duration as the total pulse length $T_\mathrm{p} = n_\mathrm{p} T$. Experimentalists usually define the pulse duration τ by the full width at half maximum of the intensity. In [5] it was shown that these two pulse duration times are connected by the relation $\tau = 0.36406\ T_\mathrm{p}$. For a laser with the wavelength 800 nm we get $\tau = 0.97149\ n_\mathrm{p}$ fs, while for the wavelength 1800 nm we have $\tau = 2.186\ n_\mathrm{p}$ fs.

2.3 Phase Space Path Integral and Transition Matrix Element

We will use the phase space path-integral approach [17, 18] and start by considering the momentum-space matrix element of the total time-evolution operator $U(t_\mathrm{f}, t_\mathrm{i})$. For our problem the corresponding Hamiltonian is $H(t) = H_0 + V_\mathrm{A}(\boldsymbol{r}) + V_\mathrm{L}(t)$, where $H_0 = -\boldsymbol{\nabla}^2/2$, $\boldsymbol{\nabla} \equiv \partial/\partial\boldsymbol{r}$, and V_A is the potential of the effective interaction

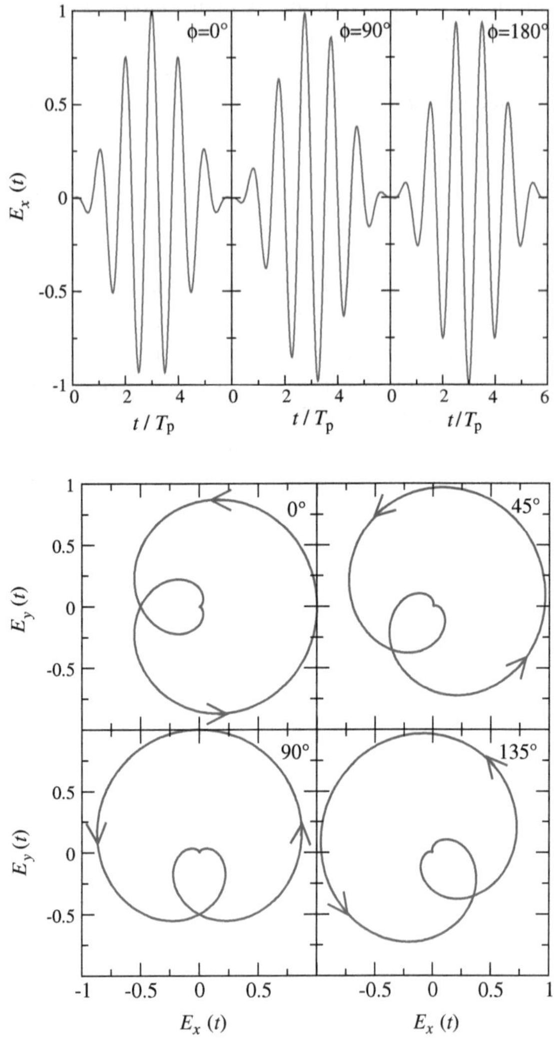

Fig. 2.1 *Upper panels* An example of a linearly polarized 6-cycle laser field: $\boldsymbol{E}(t) = \hat{\boldsymbol{e}}_x E_0 \sin^2(\omega_\mathrm{p} t/2)\cos(\omega t + \phi)$, for three values of the CE phase (the relative phase between the carrier wave and the pulse envelope): $\phi = 0°$, cosine-like pulse (*left-hand panel*), $\phi = 90°$, −sine-like pulse (*middle panel*), and $\phi = 180°$, −cosine-like pulse (*right-hand panel*). *Lower panels* The electric-field vector of a circularly polarized few-cycle laser pulse for $E_0 = \sqrt{2}$ a.u. and $n_\mathrm{p} = 2$. The four panels correspond to the CE phase $\phi = 0°$, 45°, 90°, and 135°, as denoted in the upper right corner of each panel. The *arrows* indicate the direction of increasing time from $t = 0$ to $t = n_\mathrm{p} T$

of one electron and the rest of the atom in the absence of the laser field (single-active-electron approximation). In length gauge and dipole approximation the laser field—electron interaction is $V_\mathrm{L}(t) = \boldsymbol{r}\cdot\boldsymbol{E}(t)$. By slicing $U(t_\mathrm{f}, t_\mathrm{i})$ into $N+1$ time-evolution operators, each acting across an infinitesimal time slice of width $\varepsilon = t_n - t_{n-1} = (t_\mathrm{f} - t_\mathrm{i})/(N+1)$, denoting $\boldsymbol{p}' = \boldsymbol{p}_{N+1}$, $\boldsymbol{p}'' = \boldsymbol{p}_0$, $t_{N+1} = t_\mathrm{f}$, $t_0 = t_\mathrm{i}$, we obtain

$$\langle \boldsymbol{p}'|U(t_\mathrm{f}, t_\mathrm{i})|\boldsymbol{p}''\rangle = \prod_{n=1}^{N}\left[\int \mathrm{d}\boldsymbol{p}_n\right]\prod_{n=1}^{N+1}\left[\int \frac{\mathrm{d}\boldsymbol{r}_n}{(2\pi)^3}\right]\mathrm{e}^{\mathrm{i}\mathcal{A}_N}, \tag{2.2}$$

where $\mathcal{A}_N$ is the sum

$$\mathcal{A}_N = \sum_{n=1}^{N+1} \Big\{ -(\boldsymbol{p}_n - \boldsymbol{p}_{n-1}) \cdot \boldsymbol{r}_n - \varepsilon \Big[\boldsymbol{p}_n^2/2 + \boldsymbol{r}_n \cdot \boldsymbol{E}(t_n) + V_\mathrm{A}(\boldsymbol{r}_n) \Big] \Big\}. \tag{2.3}$$

In the path-integral notation [18] these relations can be rewritten as

$$\langle \boldsymbol{p}' | U(t_\mathrm{f}, t_\mathrm{i}) | \boldsymbol{p}'' \rangle = \int_{(\boldsymbol{p}'', t_\mathrm{i}) \to (\boldsymbol{p}', t_\mathrm{f})} \mathcal{D}' \boldsymbol{p} \int \frac{\mathcal{D}\boldsymbol{r}}{(2\pi)^3} \mathrm{e}^{\mathrm{i}\mathcal{A}[\boldsymbol{p},\boldsymbol{r}]}, \tag{2.4}$$

$$\mathcal{A}[\boldsymbol{p}, \boldsymbol{r}] = \int_{t_\mathrm{i}}^{t_\mathrm{f}} \mathrm{d}t \left[-\dot{\boldsymbol{p}}(t) \cdot \boldsymbol{r}(t) - H(\boldsymbol{p}, \boldsymbol{r}, t) \right], \tag{2.5}$$

$$H(\boldsymbol{p}, \boldsymbol{r}, t) = \boldsymbol{p}^2(t)/2 + \boldsymbol{r}(t) \cdot \boldsymbol{E}(t) + V_\mathrm{A}(\boldsymbol{r}(t)). \tag{2.6}$$

The interpretation of this formula is the following: integration over all paths corresponds to summation over all histories along which a physical system can possible evolve when going from $(\boldsymbol{p}'', t_\mathrm{i})$ to $(\boldsymbol{p}', t_\mathrm{f})$.

Using the relation

$$\boldsymbol{A}(t_{n-1}) - \boldsymbol{A}(t_n) = -(t_n - t_{n-1}) \frac{\boldsymbol{A}(t_n) - \boldsymbol{A}(t_{n-1})}{t_n - t_{n-1}} = (t_n - t_{n-1}) \boldsymbol{E}(t_n) = \varepsilon \boldsymbol{E}(t_n), \tag{2.7}$$

the sum $\mathcal{A}_N$, (2.3), can be rewritten as

$$\mathcal{A}_N = -\sum_{n-1}^{N+1} \Big\{ \boldsymbol{Q}_n \cdot \boldsymbol{r}_n + \varepsilon \Big[\boldsymbol{p}_n^2/2 + V_\mathrm{A}(\boldsymbol{r}_n) \Big] \Big\}, \tag{2.8}$$

$$\boldsymbol{Q}_n = \boldsymbol{p}_n - \boldsymbol{A}(t_n) - \boldsymbol{p}_{n-1} + \boldsymbol{A}(t_{n-1}). \tag{2.9}$$

Introducing the Fourier transform of the potential $V_\mathrm{A}(\boldsymbol{r})$,

$$V_\mathrm{A}(\boldsymbol{k}) = \int \frac{\mathrm{d}\boldsymbol{r}}{(2\pi)^3} V_\mathrm{A}(\boldsymbol{r}) \mathrm{e}^{\mathrm{i}\boldsymbol{k}\cdot\boldsymbol{r}}, \quad V_\mathrm{A}(\boldsymbol{r}) = \int \mathrm{d}\boldsymbol{k} V_\mathrm{A}(\boldsymbol{k}) \mathrm{e}^{-\mathrm{i}\boldsymbol{k}\cdot\boldsymbol{r}}, \tag{2.10}$$

we obtain the following expansion in powers of εV_A:

$$\exp\left[-\mathrm{i}\varepsilon \sum_{n=1}^{N+1} V_\mathrm{A}(\boldsymbol{r}_n) \right] = 1 + \sum_{m=1}^{\infty} \frac{(-\mathrm{i}\varepsilon)^m}{m!} \prod_{j=1}^{m} \left[\int \mathrm{d}\boldsymbol{k}_j V_\mathrm{A}(\boldsymbol{k}_j) \sum_{n_j=1}^{N+1} \mathrm{e}^{-\mathrm{i}\boldsymbol{k}_j \cdot \boldsymbol{r}_{n_j}} \right]. \tag{2.11}$$

The sliced momentum-space matrix element of the total time-evolution operator can be written in the form

$$\langle \boldsymbol{p}'|U(t_\mathrm{f},t_\mathrm{i})|\boldsymbol{p}''\rangle = \sum_{m=0}^{\infty} U^{(m)}_{\boldsymbol{p}'\boldsymbol{p}''}(t_\mathrm{f},t_\mathrm{i}), \tag{2.12}$$

$$U^{(m)}_{\boldsymbol{p}'\boldsymbol{p}''}(t_\mathrm{f},t_\mathrm{i}) = \frac{(-\mathrm{i}\varepsilon)^m}{m!}\prod_{j=1}^{m}\left[\int \mathrm{d}\boldsymbol{k}_j V_\mathrm{A}(\boldsymbol{k}_j)\sum_{n_j=1}^{N+1}\right]\prod_{n=1}^{N}\left[\int \mathrm{d}\boldsymbol{p}_n\right]\exp\left(-\mathrm{i}\varepsilon\sum_{n=1}^{N+1}\boldsymbol{p}_n^2/2\right)$$
$$\times\prod_{n=1}^{N+1}\left[\int\frac{\mathrm{d}\boldsymbol{r}_n}{(2\pi)^3}\right]\exp\left(-\mathrm{i}\sum_{n=1}^{N+1}\boldsymbol{Q}_n\cdot\boldsymbol{r}_n - \mathrm{i}\sum_{j=1}^{m}\boldsymbol{k}_j\cdot\boldsymbol{r}_{n_j}\right). \tag{2.13}$$

The integral over $d\boldsymbol{r}_n$ in the last row gives the delta function

$$\delta\left(\boldsymbol{Q}_n + \sum_{j=1}^{m}\boldsymbol{k}_j\delta_{n_j,n}\right), \tag{2.14}$$

which cancels the integral over $d\boldsymbol{p}_n$, leaving only one delta function with the argument $\boldsymbol{p}_{N+1} - \boldsymbol{A}(t_{N+1}) - (\boldsymbol{p}_0 - \boldsymbol{A}(t_0)) + \sum_{j=1}^{m}\boldsymbol{k}_j$, where we have used the relation $\sum_{l=1}^{N+1}\sum_{j=1}^{m}\boldsymbol{k}_j\delta_{n_j,l} = \sum_{j=1}^{m}\boldsymbol{k}_j$. Having also in mind that $\boldsymbol{p}_{N+1} - \boldsymbol{A}(t_{N+1}) = \boldsymbol{p}' - \boldsymbol{A}(t_\mathrm{f}) \equiv \tilde{\boldsymbol{p}}'$ and $\boldsymbol{p}_0 - \boldsymbol{A}(t_0) = \boldsymbol{p}'' - \boldsymbol{A}(t_\mathrm{i})$, we can write

$$U^{(m)}_{\boldsymbol{p}'\boldsymbol{p}''}(t_\mathrm{f},t_\mathrm{i}) = \frac{(-\mathrm{i}\varepsilon)^m}{m!}\prod_{j=1}^{m}\left[\int \mathrm{d}\boldsymbol{k}_j V_\mathrm{A}(\boldsymbol{k}_j)\sum_{n_j=1}^{N+1}\right]\delta\left(\tilde{\boldsymbol{p}}' - \boldsymbol{p}'' + \boldsymbol{A}(t_\mathrm{i}) + \sum_{j=1}^{m}\boldsymbol{k}_j\right)\mathrm{e}^{-\mathrm{i}\varepsilon K_N}, \tag{2.15}$$

where

$$K_N = \sum_{n=1}^{N+1}\boldsymbol{p}_n^2/2, \quad \boldsymbol{p}_n = \tilde{\boldsymbol{p}}' + \boldsymbol{A}(t_n) + \sum_{j=1}^{m}\boldsymbol{k}_j\left(1 - \sum_{l=1}^{n}\delta_{n_j,l}\right), \tag{2.16}$$

and we used relation (2.14) with (2.9). In the continuum limit ($N \to \infty$, $\varepsilon \to 0$), the sums in (2.15) become integrals, i.e.,

$$\varepsilon\sum_{n_j=1}^{N+1} f(t_{n_j}) \to \int_{t_\mathrm{i}}^{t_\mathrm{f}} \mathrm{d}\tau_j f(\tau_j), \quad j = 0, 1, \ldots, m. \tag{2.17}$$

Taking into account relation (2.16) it is convenient to write the sum K_N in the exponent in (2.15) in the form of a sum of partial sums and then use the continuum limit

$$\varepsilon\sum_{n=n_j}^{n_{j+1}-1} g(t_n) \to \int_{\tau_j}^{\tau_{j+1}} \mathrm{d}t\, g(t). \tag{2.18}$$

The result (2.12) with (2.15) and (2.16) is an expansion in powers of V_{A}. We have eliminated all integrals over $\mathcal{D}'\boldsymbol{p}\mathcal{D}\boldsymbol{r}$ and we can now analyze this perturbation series term by term. For the zeroth-order term, $m = 0$, we obtain

$$U^{(0)}_{\boldsymbol{p}'\boldsymbol{p}''}(t_{\mathrm{f}}, t_{\mathrm{i}}) = \delta(\tilde{\boldsymbol{p}}' - \boldsymbol{p}'' + \boldsymbol{A}(t_{\mathrm{i}}))\exp(\mathrm{i}\mathcal{A}^{(0)}_N),$$
$$\mathcal{A}^{(0)}(\boldsymbol{p}'; t_{\mathrm{i}}) = -\int_{t_{\mathrm{i}}}^{t_{\mathrm{f}}} \mathrm{d}t \left[\tilde{\boldsymbol{p}}' + \boldsymbol{A}(t)\right]^2/2. \tag{2.19}$$

For the first-order term, $m = 1$, the δ function cancels the integral over $\int \mathrm{d}\boldsymbol{k}_1$ and, denoting $\tau' \equiv \tau_1$, we obtain

$$U^{(1)}_{\boldsymbol{p}'\boldsymbol{p}''}(t_{\mathrm{f}}, t_{\mathrm{i}}) = -\mathrm{i}\langle\tilde{\boldsymbol{p}}'|V_{\mathrm{A}}|\boldsymbol{p}'' - \boldsymbol{A}(t_{\mathrm{i}})\rangle \int_{t_{\mathrm{i}}}^{t_{\mathrm{f}}} \mathrm{d}\tau' e^{\mathrm{i}\mathcal{A}^{(1)}}, \tag{2.20}$$

where $\mathcal{A}^{(1)}$ is the continuum limit of

$$-\varepsilon K_N = -\frac{\varepsilon}{2}\left\{\sum_{n=1}^{n_1-1}\left[\boldsymbol{p}'' - \boldsymbol{A}(t_{\mathrm{i}}) + \boldsymbol{A}(t_n)\right]^2 + \sum_{n=n_1}^{N+1}\left[\tilde{\boldsymbol{p}}' + \boldsymbol{A}(t_n)\right]^2\right\}, \tag{2.21}$$

so that, in accordance with (2.18), we have

$$\mathcal{A}^{(1)} \equiv \mathcal{A}^{(1)}(\boldsymbol{p}'; \tau', \boldsymbol{k}; t_{\mathrm{i}}) = -\int_{\tau'}^{t_{\mathrm{f}}} \mathrm{d}t \left[\tilde{\boldsymbol{p}}' + \boldsymbol{A}(t)\right]^2/2 \quad \int_{t_{\mathrm{i}}}^{\tau'} \mathrm{d}t \left[\boldsymbol{k} + \boldsymbol{A}(t)\right]^2/2, \tag{2.22}$$

where $\boldsymbol{k} = \boldsymbol{p}'' - \boldsymbol{A}(t_{\mathrm{i}})$. Applying (2.15)–(2.18) and denoting $\boldsymbol{k}_1 = \boldsymbol{p}'' - \boldsymbol{A}(t_{\mathrm{i}})$, for the second-order term, $m = 2$, we obtain

$$U^{(2)}_{\boldsymbol{p}'\boldsymbol{p}''}(t_{\mathrm{f}}, t_{\mathrm{i}}) = (-\mathrm{i})^2 \int \mathrm{d}\boldsymbol{k}_2 \langle\tilde{\boldsymbol{p}}'|V_{\mathrm{A}}|\boldsymbol{k}_2\rangle\langle\boldsymbol{k}_2|V_{\mathrm{A}}|\boldsymbol{k}_1\rangle \int_{t_{\mathrm{i}}}^{t_{\mathrm{f}}} \mathrm{d}\tau_1 \int_{\tau_1}^{t_{\mathrm{f}}} \mathrm{d}\tau_2 e^{\mathrm{i}\mathcal{A}^{(2)}}, \tag{2.23}$$

$$2\mathcal{A}^{(2)} \equiv -\int_{\tau_2}^{t_{\mathrm{f}}} \mathrm{d}\tau' \left[\tilde{\boldsymbol{p}}' + \boldsymbol{A}(\tau')\right]^2 \int_{\tau_1}^{\tau_2} \mathrm{d}\tau' \left[\boldsymbol{k}_2 + \boldsymbol{A}(\tau')\right]^2 - \int_{t_{\mathrm{i}}}^{\tau_1} \mathrm{d}\tau' \left[\boldsymbol{k}_1 + \boldsymbol{A}(\tau')\right]^2, \tag{2.24}$$

with $\mathcal{A}^{(2)} \equiv \mathcal{A}^{(2)}(\boldsymbol{p}'; \tau_2, \boldsymbol{k}_2; \tau_1, \boldsymbol{k}_1; t_{\mathrm{i}})$. Higher-order terms can be obtained in the same way.

2.4 Above-Threshold Ionization by Few-Cycle Pulses

The few-cycle laser pulses powerful enough to drive strong-field processes were generated in 1997 [19]. This discovery has stimulated scientists to consider the effect of such fields on laser-induced processes. It was shown that high-order harmonic generation is very sensitive to the CE phase [15, 16] (we will consider this process in Sect. 2.5). The experimental verification of the CE-phase effect in strong-field processes was prevented by the fact that the CE phase was not stabilized, i.e., its value changed randomly from one shot to the other during the experiments in which data were accumulated for many thousands of laser pulses.

The first experimental evidence of the CE-phase effects has been achieved in an ATI experiment [20]. The ATI, discovered in 1979 [21], is a process in which more photons are absorbed from the laser field than is necessary for ionization. For a monochromatic infinitely long laser pulse the electric field is identical in each optical cycle so that the interference of cycles' contributions to the electron energy spectrum results in a series of peaks separated by the photon energy ω. Mathematically, this is expressed through the delta function whose argument accounts for the energy-conserving condition: the ionized electron energy $E_{\boldsymbol{p}} = \boldsymbol{p}^2/2$ is equal to the sum of the ionization potential I_{p}, ponderomotive energy U_{p}, and the energy of n absorbed photons, $E_{\boldsymbol{p}} = I_{\text{p}} + U_{\text{p}} + n\omega$. For a few-cycle laser pulse the contribution of different cycles to the spectrum is different and the spectrum is continuous. With increasing the number of cycles the peaks separated by ω become more and more pronounced (see Fig. 2.3).

The probability amplitude for detecting an ATI electron with momentum $\boldsymbol{p}$ after the laser pulse is gone is

$$M_{\boldsymbol{p}i}(T_{\text{p}}) = \langle \psi_{\boldsymbol{p}}^{(-)}(T_{\text{p}})|U(T_{\text{p}}, 0)|\psi_i(0)\rangle. \tag{2.25}$$

Here $\psi_i(t) = \psi_i \exp(\mathrm{i}I_{\text{p}}t)$ and $\psi_{\boldsymbol{p}}^{(-)}(t) = \psi_{\boldsymbol{p}}^{(-)} \exp(-\mathrm{i}E_{\boldsymbol{p}}t)$, where ψ_i is the bound and $\psi_{\boldsymbol{p}}^{(-)}$ is the continuum eigenstate of the Hamiltonian $H_{\text{A}} = H_0 + V_{\text{A}}$ in the absence of the laser field. Using the following integral equation

$$U(t, t') = U_{\text{A}}(t, t') - \mathrm{i} \int_{t'}^{t} \mathrm{d}\tau U(t, \tau) V_{\text{L}}(\tau) U_{\text{A}}(\tau, t'), \tag{2.26}$$

where U_{A} is the time-evolution operator of the Hamiltonian H_{A}, the identity $U(T_{\text{p}}, \tau) = \int \mathrm{d}\boldsymbol{p}'|\boldsymbol{p}'\rangle\langle\boldsymbol{p}'|U(T_{\text{p}}, \tau) \int \mathrm{d}\boldsymbol{p}''|\boldsymbol{p}''\rangle\langle\boldsymbol{p}''|$, and replacing $\langle\psi_{\boldsymbol{p}}^{(-)}(T_{\text{p}})|$ by the plane wave $\langle\boldsymbol{p}|$, from (2.25) to (2.19) with $t_{\text{i}} = 0$ and $t_{\text{f}} = T_{\text{p}}$, we obtain (up to a phase factor)

$$M_{\boldsymbol{p}i}^{(0)}(T_{\text{p}}) = \int_0^{T_{\text{p}}} \mathrm{d}\tau \mathrm{e}^{\mathrm{i}\mathcal{A}^{(0)}(\tau, \boldsymbol{p})} \langle \tilde{\boldsymbol{p}} + \boldsymbol{A}(\tau)|V_{\text{L}}(\tau)|\psi_i(\tau)\rangle. \tag{2.27}$$

This amplitude describes the direct ionization. Notice that in this expression the momentum $\tilde{\boldsymbol{p}} = \boldsymbol{p} - \boldsymbol{A}(T_{\text{p}})$ assures that we calculate the transition amplitude to a final state with momentum $\boldsymbol{p}$ at the detector outside the laser field. This result explicitly follows from our phase space path-integral formalism, while in previous S-matrix consideration of the strong-field approximation (SFA) the substitution $\boldsymbol{p} \to \tilde{\boldsymbol{p}}$ has been done "by hand" [5]. Higher-order corrections for the transition amplitude can be obtained using relations (2.20)–(2.24). For example, first-order correction gives the improved SFA (ISFA)

$$M_{pi}^{(1)}(T_{\mathrm{p}}) = \int \mathrm{d}\boldsymbol{k}\langle\tilde{\boldsymbol{p}}|V_{\mathrm{A}}|\boldsymbol{k}\rangle \int_0^{T_{\mathrm{p}}} \mathrm{d}\tau \int_\tau^{T_{\mathrm{p}}} \mathrm{d}\tau' \mathrm{e}^{\mathrm{i}\mathcal{A}^{(1)}(\boldsymbol{p};\tau',\boldsymbol{k};0)} \langle\boldsymbol{k}+\boldsymbol{A}(\tau)|V_{\mathrm{L}}(\tau)|\psi_i(\tau)\rangle, \tag{2.28}$$

which corresponds to the ATI with a rescattering. Next-order correction corresponds to the ATI with double scattering [4] etc.

Let us elaborate the correlation technique used in [20]. It was known that even for infinitely long multicolor pulses the electron spectra generated by atoms may exhibit a lack of backward-forward (left-right) symmetry [22]. This means that detectors placed opposite each other in the direction of the laser polarization record different spectra. The same is valid for few-cycle pulses [see (2.1) which essentially represents a three-color field]. The so-called stereo-ATI experiment [20] has been performed with circularly polarized non-phase-stable laser pulses. Each laser pulse has resulted in a pair of integers representing the number of electrons detected for this pulse by the left and the right photoelectron detector. Presenting the number of electrons detected by the right detector as a function of the number of electrons detected by the left detector, for 10^5–10^6 pulses, the so-called contingency map is generated. An example of this map for ATI of krypton atoms is shown in Fig. 2.2. The results are obtained by theoretical modeling introduced in [13, 14]. Both for circular and linear polarization anti-correlations show up as structures perpendicular to the diagonal. Namely, if a certain pulse (with fixed but unknown CE phase) produces many electrons emitted to the left, then there is a high probability that only few electrons will be emitted to the right, and vice versa. These contingency maps were investigated for various pulse durations and laser intensities [5, 13, 14].

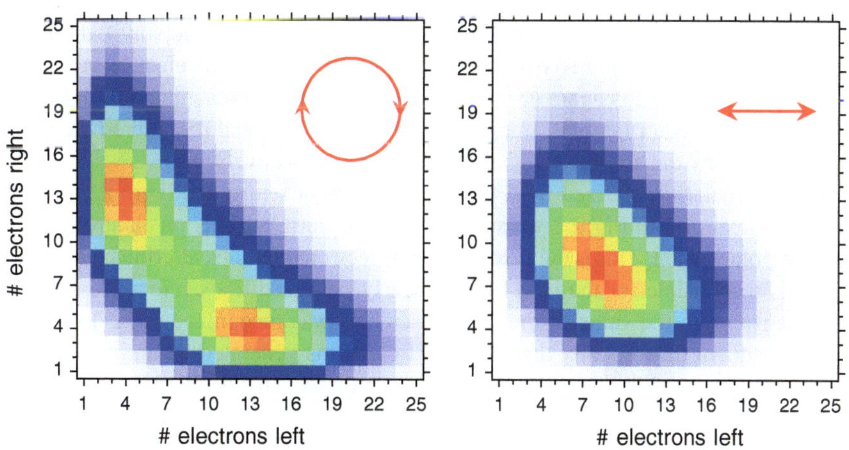

Fig. 2.2 Comparison of the *left-right* asymmetry in a stereo-ATI-experiment simulation for circularly (*left-hand panel*) and linearly (*right-hand panel*) polarized laser pulses having duration of four cycles, the wavelength 800 nm, and the intensity 4×10^{13} W/cm^2

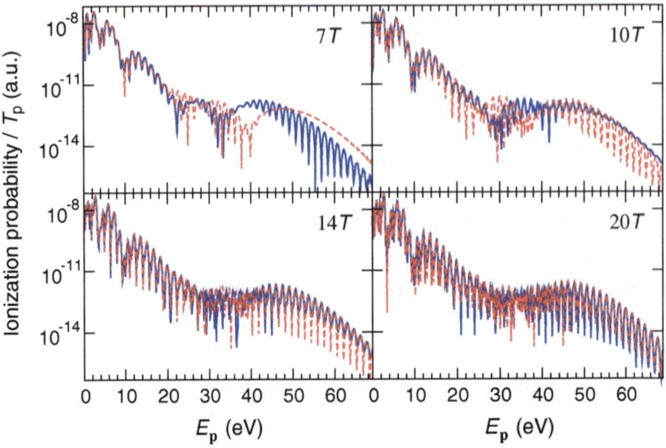

Fig. 2.3 Differential ionization probabilities of argon atoms as functions of the electrons kinetic energy E_p, for emission in the direction $\theta = 0°$ (*solid blue curves*) and $\theta = 180°$ (*dashed red curves*), for a linearly-polarized few-cycle sine-square pulse with the CE phase $\phi = 0°$. The laser-field intensity is 8×10^{13} W/cm^2 and the wavelength 800 nm. The number of optical cycles is denoted in the right upper corner of each panel. The results are obtained using the method described in [5]

In 1990 s high-order ATI (HATI) process was discovered [23, 24]. It was shown that the low-energy spectrum, caused by the so-called direct electrons which after ionization are not affected by the binding atomic potential, is followed by a plateau where the ionization probability, lower by several orders of magnitude than that of the direct electrons, is approximately independent of the electron energy. This plateau finishes by a cutoff after which the ionization probability decreases exponentially. The appearance of the plateau and the cutoff in the high-energy spectrum is explained by the following three-step mechanism: the electron is ionized (step one), the laser-field accelerated electron revisits its parent ion (step 2), and this electron rescatters off the parent ion and is again accelerated by the laser field reaching a high kinetic energy at the detector (step 3). In our formalism the direct and rescattered electrons are described by the amplitudes (2.27) and (2.28), respectively.

For a circularly polarized field the electron cannot return and there is no high-energy plateau. From Fig. 2.2 we see that the anti-correlation is much stronger for circular polarization. This is the case for direct electrons. The rescattering mechanism which is responsible for the HATI by a linearly polarized few-cycle pulse is much more sensitive to the value of the CE phase. This was recognized in [25] where it was first shown theoretically that the HATI electrons generated by an intense few-cycle laser pulse exhibit a strong left-right asymmetry. This is clearly visible in the upper left panel of Fig. 2.3, where the spectrum for the electrons registered by the right (left) detector is shown by a solid blue (dashed red) curve. The difference between these two spectra is significant for the plateau and cutoff electrons. The first experimental analysis of such left-right asymmetry is given in [26]. For realization of this experiment crucial was the possibility of CE-phase control, first verified by its effect on the HHG spectrum [27]. The stabilization of the CE-phase is related to the stabilization of the femtosecond laser's optical frequency comb, a technique which is important for laser-based precision spectroscopy. For their contribution to this field J.L. Hall and T.W. Hänsch have obtained one half of the 2005 Nobel Prize in Physics. There are many further publications in which the role of the CE phase in stereo-HATI experiment was analyzed [28]. We mention here an appealing explanation of the CE-phase effect in terms of quantum orbits [29]. From the fundamental point of view a very important paper is [30]. The experiment with interference of one electron on two spatial slits is connected with paradoxes and mysteries of quantum mechanics. According to journals Physics World and New York Times this experiment has been chosen as the most beautiful scientific experiment in history. The experiment [30] is an analog of this experiment, but in time instead of space. The above-mentioned results were summarized in the review article [5]. Let us further analyze the example shown in Fig. 2.3. The ATI peaks in the upper left panel are well developed for $\theta = 0°$, while for $\theta = 180°$ the peaks are not visible in the cutoff region. An explanation of this effect in terms of interference of quantum-orbit contributions is given in [5, 29]. For the pulse duration of ten optical cycles (upper right panel) the situation is changed: the peaks are well developed for $\theta = 180°$, while for $\theta = 0°$ they are much less visible. The reason is that the pulse shape is different for the pulse duration equal to the even and to the odd number of optical cycles. With an increase of the pulse duration the spectra registered by the left-hand and the right-hand detector become almost identical as it can be seen in the lower panels of Fig. 2.3.

Let us now show some more recent results for the stereo-HATI experiment [6–8]. For this purpose we define the normalized spatial asymmetry coefficient

$$A(E_{\mathbf{p}}, \phi) = \frac{W_{\uparrow}(E_{\mathbf{p}}, \phi) - W_{\downarrow}(E_{\mathbf{p}}, \phi)}{W_{\uparrow}(E_{\mathbf{p}}, \phi) + W_{\downarrow}(E_{\mathbf{p}}, \phi)} \cong a(E_{\mathbf{p}}) \sin(\phi + \phi_0(E_{\mathbf{p}})), \tag{2.29}$$

where a is the asymmetry amplitude and ϕ_0 is an offset. $W_{\uparrow}$ and $W_{\downarrow}$ are the probability of detection of the electron, having the energy $E_{\mathbf{p}}$, in the laser polarization direction ($W_{\uparrow}$) and opposite to it ($W_{\downarrow}$). In Fig. 2.4 we presented the coefficient A in the $(E_{\mathbf{p}}, \phi)$ plane in the form of an asymmetry map in false colors. The results are obtained using the ISFA for atoms [6] and molecules [8] which includes averaging over the laser intensity distribution in the focus. For atoms the slopes in the asymmetry oscillations with energy agree reasonable well with the experimental data for middle and higher energies [6]. The phase jump around 30–35 eV is explained by the interference of quantum-orbit contributions [29, 31, 32] to the ionization probability which changes as function of $E_{\mathbf{p}}$ and ϕ. Similar results are obtained in the molecular case (right-hand panel of Fig. 2.4). It is interesting that the mentioned interference has survived focal and molecular orientation averaging. It should also be mentioned that in [8] the electron–molecular positive ion differential scattering cross sections were extracted from the theoretical and the experimental photoelectron spectra.

The stereo-HATI measurement can serve as a CE-phase meter, a powerful, real-time diagnostic tool with attosecond sensitivity. The design of a CE-phase meter is

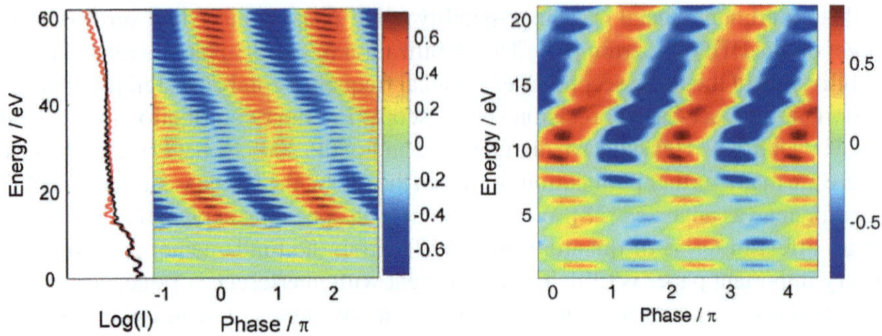

Fig. 2.4 Asymmetry maps for ATI by a linearly polarized 6-cycle laser pulse for argon atoms (*left-hand panel* the laser intensity is $I = 10^{14}\,\mathrm{W/cm^2}$ and the wavelength $\lambda = 760\,\mathrm{nm}$) and for randomly oriented nitrogen molecules (*right-hand panel* $I = 4.3 \times 10^{13}\,\mathrm{W/cm^2}$ and $\lambda = 750\,\mathrm{nm}$). The results are obtained using the ISFA with focal averaging as described in [6, 8]. In the *left-hand panel* the spectra for $\phi = 0°$ and $\theta = 0°$ (*black curve*) and $\theta = 180°$ (*red curve*) are also shown

Fig. 2.5 An example of two-dimensional parametric asymmetry plot [7]

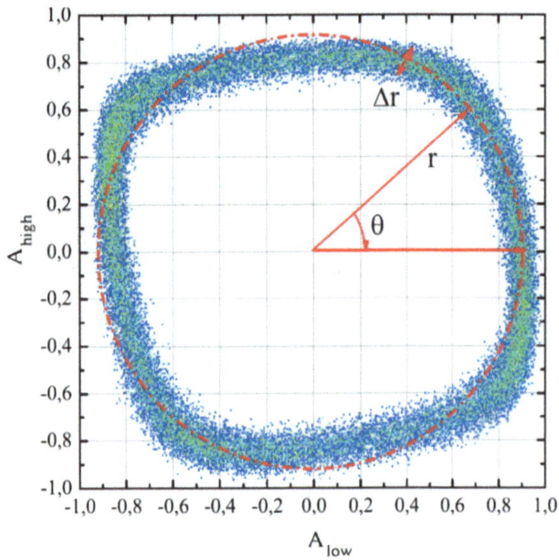

described in more recent review article [7]. We have seen that the yield of plateau and cutoff electrons is much more sensitive to the value of the CE phase, but that this yield is few orders of magnitude lower than that of the low-energy electrons. In order to improve the quality of the CE-phase meter, the low-energy direct electrons are suppressed by applying a static repelling voltage. As we can see from Fig. 2.4, the CE dependence of the asymmetry is sinusoidal with an energy dependent offset ϕ_0. One can choose two energy regions for rescattered electrons, a low-energy region for which the integrated asymmetry coefficient is $A_{\mathrm{low}} \cong \sin(\phi + \phi_0)$ and a high-energy region for which the offset is shifted by 90° so that $A_{\mathrm{high}} \cong \cos(\phi + \phi_0)$. The two integrated asymmetry coefficients are plotted against one another as it is shown in

Fig. 2.5. This is the so-called two-dimensional parametric asymmetry plot in which the polar angle $\theta \cong \phi + \phi_0$ can be used to determine the CE phase, while the radius r of this plot is related to the few-cycle laser pulse length [7]. The radius is bigger for shorter pulses. The reason is that for shorter pulses the CE-phase effect is stronger (see Fig. 2.3) and the asymmetry is larger.

2.5 High-order Harmonic Generation by Few-Cycle Pulses

In previous section we have shown how the CE phase affects the spectra of electrons ionized by a strong few-cycle laser pulse. We have seen that the rescattered electrons are very sensitive to the value of the CE phase. It is known that the third step of the mentioned three-step model, instead of rescattering, can be the recombination of the ionized electron with the parent ion. The energy released in this process can be used for emission of lights in the form of a high-energy photon. This is the so-called high-order harmonic generation, a process in which photons of energy $\Omega = n\omega$, with n approaching few hundreds, or even few thousands [33], are emitted. The emitted coherent light is in the form of ultrashort soft X-ray pulses on the attosecond scale (see the review articles [1–3, 34, 35]). Similarly as the spectra of HATI electrons,

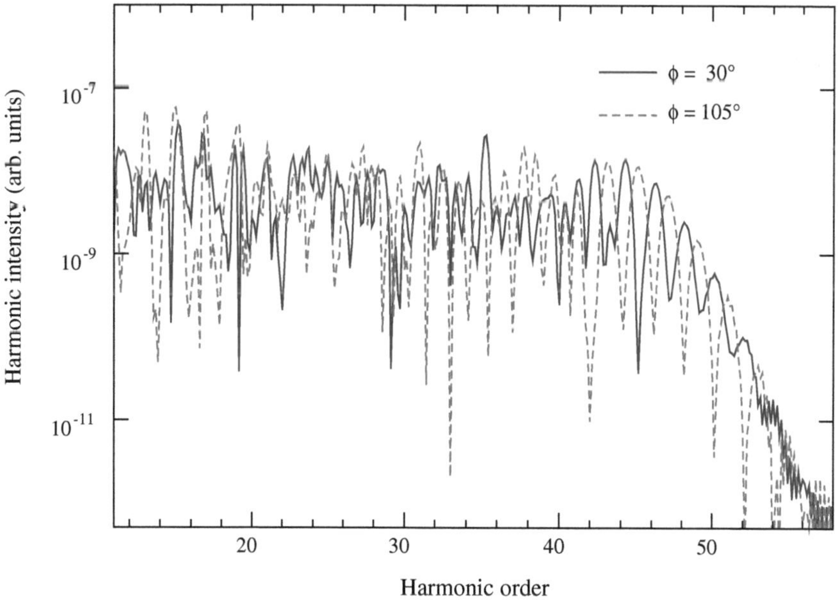

Fig. 2.6 Harmonic intensities as functions of the harmonic order for high harmonics generated by hydrogen atoms, exposed to a 7-cycle linearly polarized laser pulse with cosine squared envelope having the intensity 3×10^{14} W/cm^2 and wavelength 800 nm. The CE phase is $\phi = 30°$ (*solid blue curve*) and $\phi = 105°$ (*dashed red curve*). The results obtained using the SFA [2]

the spectra of high-order harmonics are also characterized by a plateau and a cutoff. Since similar quantum orbits are responsible both for HATI and HHG [31, 32], the CE-phase effect is very pronounced for the plateau and cutoff harmonics. This was first shown in [15]. An example of the HHG spectra for high harmonics generated by hydrogen atoms, exposed to a seven-cycle laser pulse, is shown in Fig. 2.6. We see that, for two chosen values of the CE phase, the positions of minima and maxima in the high-energy region are at different harmonic orders. Theoretical method for calculation of the HHG spectra, based on the SFA, is reviewed in [2]. The same results can be obtained using the formalism described in Sect. 2.3, having in mind that for HHG both the initial and final states are bound states. The CE-phase dependence of the HHG spectra was first experimentally observed in [27].

High-harmonic EUV and XUV coherent light has found many applications in attoscience. It is beyond the scope of this article to analyze all of them (see the above-mentioned review articles). Instead of this, the remaining part of this section is devoted to the CE-phase control of plasmonic-field enhanced HHG [9, 36, 37]. Namely, a method of HHG that used the enhancement of the local field within a metallic nanostructure induced by resonant plasmons has been presented in [36] (for a critical study of this process see the chapter by Sivis et al. in this book). Nanoplasmonics is an interesting area of research per se [38]. The ultrafast strong-laser-pulse driven nonlinear phenomena, analyzed in this chapter, connect nanoplasmonics and attoscience.

Method of calculation and numerical results for the plasmonic-field enhanced HHG spectra are presented in [9, 37]. A qualitative picture of this process can be obtained using the three-step model, generalized to the case of the inhomogeneous field (see the Appendix B in [37]). The electron starts its motion from the tunnel exit point x_e at the time t_0 with the velocity $v_{x,e}$ and follows the classical trajectory, which

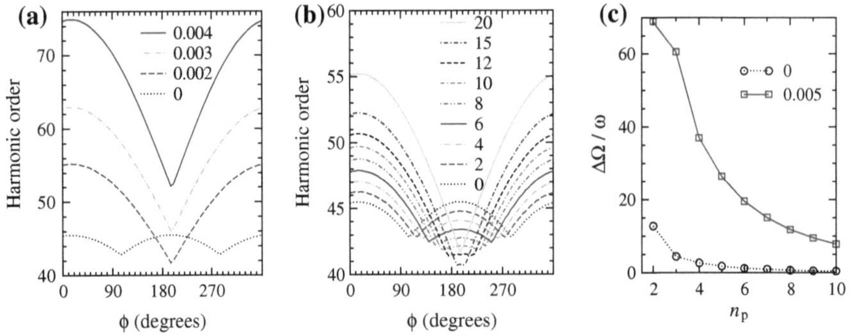

Fig. 2.7 Harmonic order $\Omega_{\max}/\omega$ as a function of the CE phase ϕ for different values of the inhomogeneity parameter β. Panel (**a**) $\beta = 0$ (*black dotted curves*), 0.002 (*red dashed curves*), 0.003 (*green dot-dashed curves*), and 0.004 a.u. (*blue solid curves*). Panel (**b**) the value of β (in 0.0001 a.u.) is denoted in the legend. Panel (**c**) shows the modulation depth of the cutoff, $\Delta\Omega/\omega$, as a function of the number of optical cycles in the pulse, for $\beta = 0$ (*black dotted curve* with *circles*) and 0.005 a.u. (*red solid curve* with *squares*). The field intensity is 3×10^{14} W/cm^2 and its wavelength is 800 nm

is obtained as numerical solution of the Newton equation $\ddot{x}(t) = -\partial V_{\rm int}(\boldsymbol{r}, t)/\partial x = -E(t)[1 + 2\beta x(t)]$, with the initial condition $x(t_0) = x_{\rm e}$ and $\dot{x}(t_0) = v_{x,\rm e}$. The electron propagates and may return to the nucleus, recombine at the time $t_{\rm r}$, and emit a high harmonic of frequency $\Omega = I_{\rm p} + v_x^2(t_{\rm r})/2$. The kinetic energy of the returning electron is determined as the solution of the above differential equation with the condition $x(t_{\rm r}) = 0$. The corresponding cutoff of the HHG spectrum is $\Omega_{\rm max} = \max_{t_0}\{\Omega\}$. Using this method we have calculated and presented in Fig. 2.7 the cutoff position as a function of the CE phase for different values of the inhomogeneity parameter β. For homogeneous field ($\beta = 0$) $\Omega_{\rm max}$ slowly decreases from 45ω for $\phi = 10°$ to 43ω for $\phi = 105°$. For $\phi > 105°$ the cutoff position increases up to a maximum at 45ω for $\phi = 190°$ and then again decreases. The HHG spectra for homogeneous field are determined by the CE phase modulo π. However, for inhomogeneous field the plasmonic-field enhanced HHG spectra depend on the CE phase modulo 2π (this was also the case for HATI process). Only one minimum in the harmonic order as a function of the CE phase, $0° \le \phi \le 360°$, appears for $\phi = 195°$. The position of this minimum is the same for all values of β and at the same place both the ionization and recombination times exhibit a jump. With the increase of β from zero, the minima move towards each other and merge for $\beta = 0.0017$ a.u. and $\phi = 196°$ (see Fig. 2.7b). For larger values of β we have only one jump and one minimum.

Next, we define the modulation depth of the cutoff by $\Delta\Omega = \max_\phi \Omega_{\rm max} - \min_\phi \Omega_{\rm max}$. In Fig. 2.7c, for $\beta = 0$ and 0.005 a.u., we show $\Delta\Omega/\omega$ as a function of the number of optical cycles in the pulse. For homogeneous field $\Delta\Omega$ quickly becomes negligible with the increase of $n_{\rm p}$. However, for inhomogeneous field, $\Delta\Omega$ is very large (above 60ω for $n_{\rm p} \le 3$) and remains substantial with the increase of $n_{\rm p}$ ($\Delta\Omega = 10\omega$ for $n_{\rm p} = 9$). This implies that the determination of the cutoff position of HHG spectra generated by inhomogeneous field is particularly suitable for measurement of the CE phase of longer pulses.

2.6 Few-Cycle-Laser-Pulse Assisted Processes

We will consider two examples of laser-assisted processes: electron–atom potential scattering and electron–ion recombination. Both processes can happen in the absence of the laser field. However, in the presence of the laser field these processes are modified. First experimental evidence of nonlinear multiphoton phenomena in which a strong laser field plays a crucial role is given in 1977 [39]. In this laser-assisted scattering experiment, in the so-called free-free transitions, n photons were exchanged with the laser field so that the final electron kinetic energy $E_{\rm f} = \boldsymbol{p}_{\rm f}^2/2$ is connected with the initial electron kinetic energy $E_{\rm i} = \boldsymbol{p}_{\rm i}^2/2$ by the formula $E_{\rm f} = E_{\rm i} + n\omega$. In the following decades this process was further investigated (see [10] for a list of more recent results).

The few-cycle-laser-pulse-assisted electron–atom scattering was first analyzed in [10], motivated by the discovery of few-cycle strong laser pulses in mid-infrared region [33]. It was shown that the transition matrix element from an initial plane wave state $|\boldsymbol{p}_\mathrm{i}\rangle$ to the final state $|\boldsymbol{p}_\mathrm{f}\rangle$ is given by the momentum space matrix element of the total time-evolution operator, $\langle \boldsymbol{p}_\mathrm{f}|U(T_\mathrm{p},0)|\boldsymbol{p}_\mathrm{i}\rangle$. Since we have already presented the result for this matrix element in Sect. 2.3, we immediately obtain the required result. According to (2.19) the zeroth-order term $U^{(0)}_{\boldsymbol{p}_\mathrm{f}\boldsymbol{p}_\mathrm{i}}(T_\mathrm{p},0)$ can be neglected since in this case the momentum remains unchanged. The sum of the first- and second-order terms, $U^{(1)}_{\boldsymbol{p}_\mathrm{f}\boldsymbol{p}_\mathrm{i}}(T_\mathrm{p},0)+U^{(2)}_{\boldsymbol{p}_\mathrm{f}\boldsymbol{p}_\mathrm{i}}(T_\mathrm{p},0)$, according to (2.20) and (2.22)–(2.24) is proportional to the integral from $t=0$ to $t=T_\mathrm{p}$ of the following expression

$$\langle\tilde{\boldsymbol{p}}_\mathrm{f}|V_\mathrm{A}|\tilde{\boldsymbol{p}}_\mathrm{i}\rangle e^{\mathrm{i}\mathcal{A}^{(1)}(\boldsymbol{p}_\mathrm{f};t,\tilde{\boldsymbol{p}}_\mathrm{i};0)} - \mathrm{i}\int d\boldsymbol{k}\langle\tilde{\boldsymbol{p}}_\mathrm{f}|V_\mathrm{A}|\boldsymbol{k}\rangle\langle\boldsymbol{k}|V_\mathrm{A}|\tilde{\boldsymbol{p}}_\mathrm{i}\rangle\int_t^{T_\mathrm{p}} d\tau e^{\mathrm{i}\mathcal{A}^{(2)}(\boldsymbol{p}_\mathrm{f};\tau,\boldsymbol{k};t,\tilde{\boldsymbol{p}}_\mathrm{i};0)}, \tag{2.30}$$

where $\tilde{\boldsymbol{p}}_\mathrm{f}=\boldsymbol{p}_\mathrm{f}-\boldsymbol{A}(T_\mathrm{p})$ and $\tilde{\boldsymbol{p}}_\mathrm{i}=\boldsymbol{p}_\mathrm{i}-\boldsymbol{A}(0)$. Similarly as for the HATI process, the shift of the momenta by $-\boldsymbol{A}(0)=-\boldsymbol{A}(T_\mathrm{p})$ assures that we calculate the probability of transition from a state with the initial momentum $\boldsymbol{p}_\mathrm{i}$ to a state with the momentum $\boldsymbol{p}_\mathrm{f}$ at the detector outside the laser field. The first term in (2.30) describes the process in which the few-cycle-laser-field-dressed electron scatters once at the potential V_A at the time t (this is the so-called direct scattering or the first Born approximation). This term is integrated over the whole pulse duration. The second term describes the process in which the electron scatters on the potential V_A at the time t, moves in the laser field and finally rescatters off the potential V_A at the time τ (this is the second Born approximation for the few-cycle-laser-pulse-assisted electron–atom potential scattering). The integration over time t is over whole pulse duration, while the integration over time τ is from $\tau=t$ to $\tau=T_\mathrm{p}$. The method of calculation of this transition matrix element and a detailed analysis of the dependence of the differential scattering cross section on the incident and final electron energies, scattering angle θ of the final state electron, laser pulse duration, CE phase, chosen atomic system, etc., is given in [10]. We decided to show here the result pre-

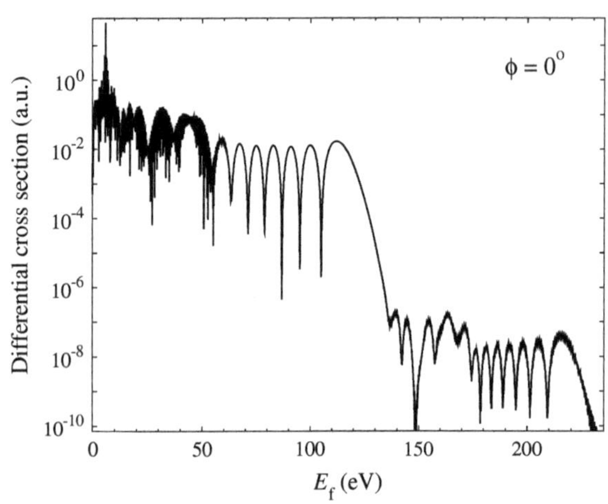

Fig. 2.8 The differential cross section for potential scattering of electrons on neon atoms in the presence of a linearly polarized four-cycle laser pulse, as a function of the final electron energy E_f. The laser wavelength and intensity are 3100 nm and 2.5×10^{13} W/cm^2, respectively, while the CE phase is $\phi=0°$. The initial electron energy is $E_\mathrm{i}=6$ eV and the scattering angle of the final-state electrons is $\theta=0°$

sented in Fig. 2.8. The initial electron momentum $\boldsymbol{p}_{\mathrm{i}}$ is in the direction of the linearly polarized laser-field polarization vector, while the initial electron energy is $E_{\mathrm{i}} = 6\,\mathrm{eV}$ and the scattering angle is $\theta = 0°$. The laser wavelength is 3100 nm, its intensity is $2.5 \times 10^{13}\,\mathrm{W/cm^2}$, and the pulse duration is four optical cycles. The differential cross section as a function of the final electron energy for the CE phase $\phi = 0°$ is presented. One can notice two plateaus. The low-energy plateau corresponds to direct scattering, while the high-energy plateau, which is five orders of magnitude lower, corresponds to the scattering with a rescattering. Similarly as in the cases of HHG and HATI processes, both plateaus finish with an abrupt cutoff. The oscillatory structure of the spectra was explained in [10]. It should be understood in terms of Feynman's path integral interpretation of quantum physics [32]: The probability amplitude of quantum-mechanical scattering process can be represented as a coherent superposition of all possible spatio-temporal paths that connect the initial and the final state of the system [40]. The interference of these contributions is responsible for the mentioned oscillatory structure. In the energy region below 60 eV more direct scattering solutions contribute, which results in a much reacher oscillatory structure. However, in the region between 60 and 120 eV only two stationary-phase solutions contribute and oscillations are much less pronounced. Finally, for higher energies (beyond the direct scattering cutoff) only the double scattering solutions contribute. The above-mentioned solutions are obtained solving the integrals in (2.30) using the stationary phase method [41]. This leads to the energy conserving conditions at the first and second scattering times, which allows a classical analysis of the obtained results. In this way one can calculate the maximum value of the final electron energy for direct scattering and rescattering as function of the CE phase. The obtained energies of double scattering electrons can be much higher than that of the single scattering electrons. The length of both plateaus can change by few tens of electronvolts with the change of the CE phase [10].

Electron–ion radiative recombination, which is important in plasma physics and astrophysics [2], also belongs to the class of laser-assisted processes. In this process the energy is transferred from the free electron to the photon, which acts as a third body in the process. In the laser-assisted radiative electron–ion recombination the incident electron may absorb the energy from the laser field, before it recombines with the target ion, and, as a result, a high-energy photon is emitted. This is the process of direct laser-assisted recombination (LAR). It can also happen that the incident electron scatters on the ion and that, driven by the laser field, this electron returns to the same ion and recombines with it (this process we denote as SLAR, where "S" stands for "scattering"). In the case of few-cycle laser pulses the LAR process was investigated in [42–44], while the SLAR process was first analyzed in [11]. In both processes the shift of the initial electron momentum by $-\boldsymbol{A}(0)$ appears, while the final state is bound. The recombination process is the third step of the HHG process and the derivation of the expression for the transition amplitude and the differential power spectrum is similar (see, for example, the review article [2] and the paper [11]). From the results of [11] we decided to present an example of the differential power spectrum as a function of the emitted X-ray photon energy $\omega_{\boldsymbol{K}}$ for the CE phase $\phi = 0°$. The result shown in Fig. 2.9 is analogous to the laser-assisted scattering result

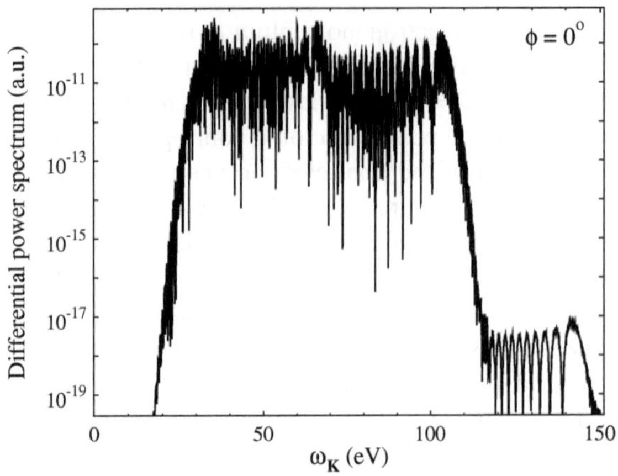

Fig. 2.9 The differential power spectrum for the laser-assisted radiative recombination of electrons with He^+ ions in the presence of a linearly polarized four-cycle laser pulse, as a function of the emitted X-ray energy ω_K. The laser wavelength and intensity are 3700 nm and 3×10^{13} W/cm^2, respectively, while the CE phase is $\phi = 0°$. The incident electron energy and angle are $E_p = 5$ eV and $\theta = 90°$, respectively

of Fig. 2.8. Two plateaus, for direct recombination and for recombination preceded by a scattering, are clearly visible and the oscillatory spectrum is determined by the number of stationary-phase contributions.

2.7 Concluding Remarks

In this concluding section we will mention some other processes which can be controlled by the few-cycle laser pulses. Above-threshold detachment of electrons from negative ions is an analog of atomic ATI and can be studied using the same approximate theoretical method (see [45] and references therein). For a comparison with the exact solution of the time-dependent Schrödinger equation see [46, 47]. Nonsequential double or multiple ionization is another laser-induced process which was studied in the presence of few-cycle laser pulses. It was found that this process is also sensible to the variation of the CE phase (see the review articles [48, 49] and references therein; for nonsequential double ionization of molecules see [50]). The CE phase affects the ultra-fast strong-field ionization dynamics of multielectron systems (see examples of Xe and Cs_2 in [51]) and the ionization dynamics of clusters (see [52] for the case of Xe_{400} clusters). More recent results about CE-phase effect in photodissociation can be found in [54, 55], while the attosecond-recollision-controlled selective fragmentation of polyatomic molecules was considered in [56].

Above-threshold photoemission, which is the solid-state analog of the ATI, is another process in which the CE-phase effects are important (for a review see [53]).

Strong-field photoemission from sharp metal tips using near- and mid-infrared pulses has been studied in the chapter by Sivis et al. in this book. The CE-phase control was used to study ultra-fast dynamics of processes in condensed matter in the chapter by Heinzmann. The attosecond streak camera and a method to get the CE phase of attosecond pulses is mentioned in the chapter by Calegari et al. (see also [57] and references therein).

From the laser-assisted processes we mention X-ray–atom scattering [2]. This process has not yet been studied for the few-cycle pulse case, but it was shown that it is sensitive to the relative phase of a bichromatic field [58]. It was also shown that the plateau length of the high-energy scattered X-ray photons can be extended by an order of magnitude by adding a week static field to the laser field [59] (see also the review article [60]).

Our conclusion is that the CE phase is an important parameter for control of laser-induced and laser-assisted processes in various media. This opens perspectives and new challenges for attoscience.

References

1. F. Krausz, M. Ivanov, Rev. Mod. Phys. **81**, 163–234 (2009)
2. D.B. Milošević, F. Ehlotzky, Adv. At. Mol. Opt. Phys. **49**, 373–532 (2003)
3. W. Becker, F. Grasbon, R. Kopold, D.B. Milošević, G.G. Paulus, H. Walther, Adv. At. Mol. Opt. Phys. **48**, 35–98 (2002)
4. D.B. Milošević, J. Math. Phys. **54**, 042101 (2013)
5. D.B. Milošević, G.G. Paulus, D. Bauer, W. Becker, J. Phys. B **39**, R203–R262 (2006)
6. M.F. Kling, J. Rauschenberger, A.J. Verhoef, E. Hasović, T. Uphues, D.B. Milošević, H.G. Muller, M.J.J. Vrakking, New J. Phys. **10**, 025024 (2008)
7. T. Rathje, N.G. Johnson, M. Möller, F. Süßmann, D. Adolph, M. Kübel, R. Kienberger, M.F. Kling, G.G. Paulus, A.M. Sayler, J. Phys. B **45**, 074003 (2012)
8. A. Gazibegović-Busuladžić, E. Hasović, M. Busuladžić, D.B. Milošević, F. Kelkensberg, W.K. Siu, M.J.J. Vrakking, F. Lépine, G. Sansone, M. Nisoli, I. Znakovskaya, M.F. Kling, Phys. Rev. A **84**, 043426 (2011)
9. B. Fetić, D.B. Milošević, J. Mod. Opt. **60**, 1466 (2013)
10. A. Čerkić, D.B. Milošević, Phys. Rev. A **87**, 033417 (2013)
11. A. Čerkić, D.B. Milošević, Phys. Rev. A **88**, 023414 (2013)
12. J.-C. Diels, W. Rudolph, *Ultrashort Laser Pulse Phenomena* (Academic Press, San Diego, 1996)
13. D.B. Milošević, G.G. Paulus, W. Becker, Phys. Rev. Lett. **89**, 153001 (2002)
14. D.B. Milošević, G.G. Paulus, W. Becker, Laser Phys. **13**, 948 (2003)
15. A. de Bohan, Ph. Antoine, D.B. Milošević, B. Piraux, Phys. Rev. Lett. **81**, 1837 (1998)
16. A. de Bohan, Ph. Antoine, D.B. Milošević, G.L. Kamta, B.Piraux, Laser Phys. **9**, 175 (1999)
17. L.S. Schulman, *Techniques and Applications of Path Integration* (Wiley, New York, 1981)
18. H. Kleinert, *Path Integrals in Quantum Mechanics, Statistics, Polimer Physics, and Financial Markets*, 5th edn. (World Scientific, Singapore, 2009)
19. M. Nisoli, S. De Silvestri, O. Svelto, R. Szipöcs, K. Ferencz, Ch. Spielmann, S. Sartania, F. Krausz, Opt. Lett. **22**, 522 (1997)
20. G.G. Paulus, F. Grasbon, H. Walther, P. Villoresi, M. Nisoli, S. Stagira, E. Priori, S. De Silvestri, Nature (London) **414**, 182 (2001)
21. P. Agostini, F. Fabre, G. Mainfray, G. Petite, N.K. Rahman, Phys. Rev. Lett. **42**, 1127 (1979)

22. F. Ehlotzky, Phys. Rep. **345**, 175 (2001)
23. B. Yang, K.J. Schafer, B. Walker, K.C. Kulander, P. Agostini, L.F. DiMauro, Phys. Rev. Lett. **71**, 3770 (1993)
24. G.G. Paulus, W. Nicklich, H. Xu, P. Lambropoulos, H. Walther, Phys. Rev. Lett. **72**, 2851 (1994)
25. D.B. Milošević, G.G. Paulus, W. Becker, Opt. Express **11**, 1418 (2003)
26. G.G. Paulus, F. Lindner, H. Walther, A. Baltuška, E. Goulielmakis, M. Lezius, F. Krausz, Phys. Rev. Lett. **91**, 253004 (2003)
27. A. Baltuška, Th Udem, M. Uiberacker, M. Hentschel, E. Goulielmakis, Ch. Gohle, R. Holzwarth, V.S. Yakovlev, A. Scrinzi, T .W. Hänsch, F. Krausz, Nature (London) **421**, 611–615 (2003)
28. D.B. Milošević, G.G. Paulus, W. Becker, Laser Phys. Lett. **1**, 93 (2004)
29. D.B. Milošević, G.G. Paulus, W. Becker, Phys. Rev. A **71**, 061404(R) (2005)
30. F. Lindner, M.G. Schätzel, H. Walther, A. Baltuška, E. Goulielmakis, F. Krausz, D.B. Milošević, D. Bauer, W. Becker, G.G. Paulus, Phys. Rev. Lett. **95**, 040401 (2005)
31. R. Kopold, D.B. Milošević, W. Becker, Phys. Rev. Lett. **84**, 3831 (2000)
32. P. Salières, B. Carré, L. Le Déroff, F. Grasbon, G.G. Paulus, H. Walther, R. Kopold, W. Becker, D.B. Milošević, A. Sanpera, M. Lewenstein, Science **292**, 902 (2001)
33. T. Popmintchev, M.-C. Chen, D. Popmintchev, P. Arpin, S. Brown, S. Ališauskas, G. Andriukaitis, T. Balčiunas, O.D. Mücke, A. Pugzlys, A. Baltuška, B. Shim, S.E. Schrauth, A. Gaeta, C. Hernández-García, L. Plaja, A. Becker, A. Jaron-Becker, M.M. Murnane, H.C. Kapteyn, Science **336**, 1287 (2012)
34. M. Nisoli, G. Sansone, Prog. Quantum Electron. **33**, 17–59 (2009)
35. M.C. Kohler, T. Pfeifer, K.Z. Hatsagortsyan, C.H. Keitel, Adv. At. Mol. Opt. Phys. **61**, 159–208 (2012)
36. S. Kim, J. Jin, Y.-J. Kim, In.-Y. Park, Y. Kim, S.-W. Kim, Nature (London) **453**, 757 (2008)
37. B. Fetić, K. Kalajdžić, D.B. Milošević, Ann. Phys. (Berlin) **525**, 107 (2013)
38. M.I. Stockman, Phys. Today **64**, 39 (2011)
39. A. Weingartshofer, J.K. Holmes, G. Caudle, E.M. Clarke, H. Kruger, Phys. Rev. Lett. **39**, 269 (1977)
40. D.B. Milošević, D. Bauer, W. Becker, J. Mod. Opt. **53**, 125 (2006)
41. A. Čerkić, D.B. Milošević, Phys. Rev. A **73**, 033413 (2006)
42. J.Z. Kamiński, F. Ehlotzky, Phys. Rev. A **71**, 043402 (2005)
43. S. Bivona, R. Burlon, G. Ferrante, C. Leone, Opt. Express **14**, 3715 (2006)
44. S. Bivona, R. Burlon, C. Leone, Laser Phys. Lett. **4**, 44 (2007)
45. S.F.C. Shearer, M.R. Monteith, Phys. Rev. A **88**, 033415 (2013)
46. B. Fetić, D.B. Milošević, W. Becker, J. Mod. Opt. **58**, 1149 (2011)
47. Ph.A. Korneev, S.V. Popruzhenko, S.P. Goreslavski, W. Becker, G.G. Paulus, B. Fetić, D.B. Milošević, New J. Phys. **14**, 055019 (2012)
48. C. Figueira de Morisson, X. Liu, J. Mod. Opt. **58**, 1076–1131 (2011)
49. W. Becker, X. Liu, P.J. Ho, J.H. Eberly, Rev. Mod. Phys. **84**, 1011–1043 (2012)
50. M. Kübel, N.G. Kling, K.J. Betsch, N. Camus, A. Kaldun, U. Kleineberg, I. Ben-Itzhak, R.R. Jones, G.G. Paulus, T. Pfeifer, J. Ullrich, R. Moshammer, M.F. Kling, B. Bergues, Phys. Rev. A **88**, 023418 (2013)
51. D. Mathur, K. Dota, A.K. Dharmadhikari, J.A. Dharmadhikari, Phys. Rev. Lett. **110**, 083602 (2013)
52. G. Mishra, N.K. Gupta, J. Phys. B **46**, 125602 (2013)
53. M. Krüger, M. Schenk, M. Förster, P. Hommelhoff, J. Phys. B **45**, 074006 (2012)
54. T. Rathje, A.M. Sayler, S. Zeng, P. Wustelt, H. Figger, B.D. Esry, G.G. Paulus, Phys. Rev. Lett. **111**, 093002 (2013)
55. N.G. Kling, K.J. Betsch, M. Zohrabi, S. Zeng, F. Anis, U. Ablikim, B. Jochim, Z. Wang, M. Kübel, M.F. Kling, K.D. Carnes, B.D. Esry, I. Ben-Itzhak, Phys. Rev. Lett. **111**, 163004 (2013)
56. X. Xie, K. Doblhoff-Dier, S. Roither, M.S. Schöffler, D. Kartashov, H. Xu, T. Rathje, G.G. Paulus, A. Baltuška, S. Gräfe, M. Kitzler, Phys. Rev. Lett. **109**, 243001 (2012)

57. C. Liu, M. Reduzzi, A. Trabattoni, A. Sunilkumar, A. Dubrouil, F. Calegari, M. Nisoli, G. Sansone, Phys. Rev. Lett. **111**, 123901 (2013)
58. D.B. Milošević, A.F. Starace, J. Phys. B **32**, 1831 (1999)
59. D.B. Milošević, A.F. Starace, Phys. Rev. Lett. **81**, 5097 (1998)
60. D.B. Milošević, A.F. Starace, Laser Phys. **10**, 278–293 (2000)

Chapter 3
Angular Streaking for Strong Field Ionization of Molecules—Attosecond Physics Without Attosecond Pulses

Jian Wu and Reinhard Dörner

Abstract The angular streaking of a close-to-circular polarized ultrahsort laser pulse provides a powerful tool and opens new possibilities for attosecond physics without the using of attosecond pulses. For strong field ionization of molecules we discuss in this review how this technique allows to uncover the intrinsic phase-dependent directional molecular bond breaking in a symmetric multicycle laser pulse (Sect. 3.2), how it allows probing the tunnelling site of the freed electron in the electron localization-assisted enhanced ionization of molecules near the critical internuclear distance (Sect. 3.3), how it reveals the orientation-dependent tunnelling ionization rate of polar molecules without pre-alignment of the molecules (Sect. 3.4), and how to use it to sequence the release order of the freed electrons in multielectron ionization of a multicenter molecular dimer (Sect. 3.5).

3.1 Coincidence Angular Streaking

To probe attosecond (10^{-18} s) dynamics of electrons in atoms and molecules the most common approach is to combine attosecond pulses [1, 2] in extreme ultraviolet (XUV) spectrum region produced by high-harmonic generation (HHG) with streaking the energy of a released electron by an intense phase-locked near-infrared (IR) laser pulses [3–6]. Alternatively, the angular streaking [7] by a circularly polarized near-IR ultrashort laser pulse provides a powerful tool and opens new possibilities for attosecond physics in strong laser field even without the using of attosecond pulses. While the first works along this line were performed for atoms [7–9] we here demon-

J. Wu · R. Dörner (✉)
State Key Laboratory of Precision Spectroscopy, East China Normal University, Shanghai 200062, China
e-mail: doerner@atom.uni-frankfurt.de

J. Wu
e-mail: jwu@phy.ecnu.edu.cn

R. Dörner
Institut Für Kernphysik, Goethe Universität, 60438 Frankfurt, Germany

M. Kitzler and S. Gräfe (eds.), *Ultrafast Dynamics Driven by Intense Light Pulses*, Springer Series on Atomic, Optical, and Plasma Physics 86, DOI 10.1007/978-3-319-20173-3_3

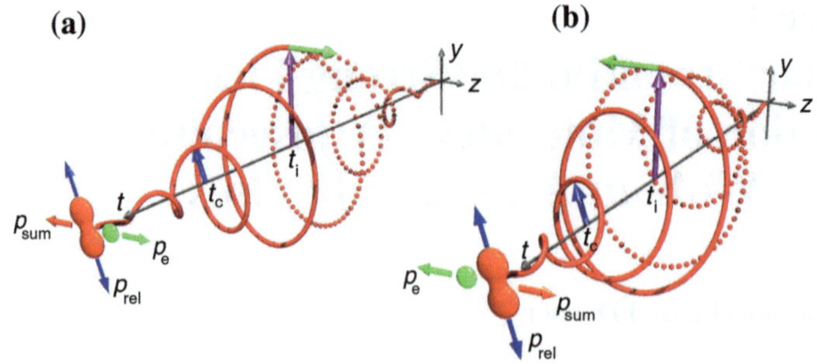

Fig. 3.1 Coincidence angular streaking for strong field ionization of molecules by circularly polarized ultrashort laser pulse (*red helix*). The freed electron (*green ball*) acquires a final momentum p_e (*green arrow*) perpendicular to the laser field vector (*purple arrow*) at the ionization instant t_i owing to the driving of the **a** counter-clockwise **b** clockwise rotating laser fields (*solid red helix*). The measurement of the relative momentum (*blue arrows*) of the ionic fragments (*yellow balls*) provides the molecular orientation, which reveals the laser phase at the dipole-allowed transition t_c for laser-coupled dissociation. Due to the momentum conservation, the electron momentum can be deduced from the ion sum-momentum (*yellow arrow*) $p_{sum} = -p_e$

strate the power of this technique for studies of molecular processes. The enabling technology for this route of attosecond science is COLTRIMS [11, 12] (COLd Target Recoil Ion Momentum Spectroscopy) which allows for the coincidence imaging of momentum vectors of electrons and fragment ions.

As shown in Fig. 3.1, the circularly polarized laser field (solid red helix) drives the freed electron. It acquires a final momentum approximately perpendicular to the instantaneous laser field vector at the moment of ionization, which is termed "angular streaking". This maps the instant of tunnelling of the electron within one laser cycle ($T_p \sim 2.6$ fs at 790-nm) to the 2π interval of electron emission directions ϕ_{el} ($\sim$7.3 as for 1°) in the polarization plane. Meanwhile, the laser field intensity at the instant of ionization is encoded in the magnitude of the freed electron momentum. Therefore, by measuring the momentum of the freed electron, including the emission direction and magnitude, the phase and intensity of the laser field at the ionization can been retrieved, which marks the time in the laser pulse with attosecond resolution. The angular streaking has been used to probe the tunnelling interval of electrons from atoms [8, 9] by measuring the momenta of the freed electrons (or the correlated atomic ion) and adopting either the major axis of elliptically polarized pulse or the carrier-envelope phase (CEP) of a CEP-stabilized pulse to provide the reference time.

As compared to strong field ionization of atoms, the molecular axis offers the angular streaking a new coordinate to probe the ultrafast dynamics of electron in the molecular frame. For circularly polarized laser pulse, following the ultrafast ionization, the launched nuclear wave packet on the bound state of a molecular ion may be photon-coupled to a dissociative state. The direction of dissociation gives the laser field vector (phase) at the instant of the photon transition. The laser phases at the ionization and transition thus act as two time markers of an ultrafast stopwatch, which are mapped to the emission directions of the freed electron ϕ_{el} and ionic fragment

ϕ_{ion}, respectively. We therefore can probe the ultrafast dynamics of electrons in the self-referenced molecular frame by coincidently detecting the ejected electron and ionic fragments in circularly polarized pulses, an approach we name "coincidence angular streaking" [10]. Experimentally, this coincidence measurement is performed in a COLTRIMS reaction microscope [11, 12], where the intense femtosecond laser pulses are focused onto a supersonic molecular beam. The photon ionization created ions and electrons are accelerated by a weak homogeneous electric field (several V/cm) and detected by two time- and position-sensitive microchannel plate detectors [13] at the opposite ends of the spectrometer. A weak magnetic field (several Gauss) produced by a pair of Helmholtz coils is used collect electron with 4π solid angle. The three-dimensional momentum vectors of the correlated electrons and ions are retrieved from the measured time-of-flights and positions of impact on the detectors during the off-line analysis.

3.2 Phase-Dependent Directional Molecular Bond Breaking in a Symmetric Laser Pulse

By steering the attosecond dynamics of a bound electron during dissociative ionization, controllable directional chemical bond breaking has been achieved in phase controlled asymmetric CEP-stabilized few-cycle [14–16] or two-color multicycle laser pulses [17, 18], which is a significant step toward the control of chemical reaction. In all these experiments the phase difference between the ionization event and the dissociation was determined by breaking the symmetry of the light by optical means. However for symmetric laser field, no directional molecular bond breaking was expected until our recent experiments [10]. By resolving the laser phases at ionization θ_i and dipole-allowed transitions between the bound and dissociative states θ_c using the coincidence angular streaking, we demonstrated in this work that asymmetric molecular bond breaking indeed naturally exists even in symmetric multicycle laser pulses.

We study the dissociative single ionization of the simplest molecule of H_2, i.e. $H_2 + n\hbar\omega \rightarrow H + p + e$ [denoted $H_2(1, 0)$], which has been intensively explored to understand directional molecular bond breaking observed in phase controlled asymmetric laser pulses [14–18]. As schematically illustrated in Fig. 3.2, a circularly polarized (Fig. 3.2 only shows the 1D projection) IR pulse is employed to singly ionize H_2 at time t_i. The succeeding part of the laser pulse drives the motion of the remaining bound electron in the H_2^+ until it finally localizes during the dissociation at one of the protons leading to $H_2(1, 0)$. By controlling (via measurement of the streaking angle of the freed electron) the laser phase at t_i, which is similar to the CEP of an asymmetric laser pulse [14–18], the dynamics of electron localization can be controlled with attosecond precision, leading to directional ejection of the proton p. In addition to this intuitive picture, the directional bond breaking can be understood as the coherent superposition of the dissociating nuclear wave packet moving on

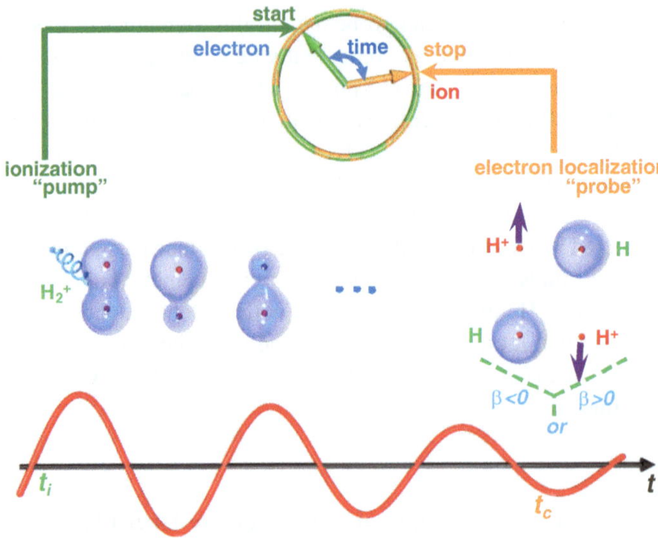

Fig. 3.2 An ultrafast stopwatch by coincidence measurement of two particles. The start (t_i) and stop (t_c) times of the stopwatch are retrieved by coincident detection of the emitted electron and ion from a breaking molecular bond of H_2^+. The laser field (*red curve*), based on its instantaneous direction, drives an electron (*transparent sky-blue blob*) up and down during the breaking of the molecular bond. The electron eventually localizes at one nucleus when the laser field is too weak to drive the electron over the barrier between the well-separated H^+ nuclei (*red dots*)

two potential energy curves of different inversion symmetries [14–22]. By releasing one electron from H_2, a nuclear wave packet Ψ_0 is launched on the $1s\sigma_g^+$ bound state of H_2^+ at time t_i. During the outward propagation of Ψ_0, there are at least three pathways: the wave packet can undergo (a) a dipole-allowed transition by single photon absorption (denoted 1ω pathway) or (b) three photon transition to the $2p\sigma_u^+$ dissociative state and (c) the three photon transition to the $2p\sigma_u^+$ can be followed by a one photon transition back to the $2p\sigma_g^+$ ground state (denoted net-2ω pathway). The superposition of the dissociated nuclear wave packets along the 1ω and net-2ω pathways results in directional ejection of proton, which is strongly linked to the laser phases at the ionization and dipole-allowed transitions between the gerade and ungerade states.

Regardless of the optical symmetry of the employed laser pulse, the laser phase-dependent driven electron localization or coherent superposition of the dissociating nuclear wave packet should naturally exit even in a symmetric single-color multicycle laser pulse [10]. In all exist experiments where only a sole proton is detected this localization is however hidden. Our coincidence angular streaking [10] by coincidently measuring the ejected electron and proton provides us both "start" and "stop" time markers to pump-probe this ultrafast electron localization dynamics within a single multicycle laser pulse.

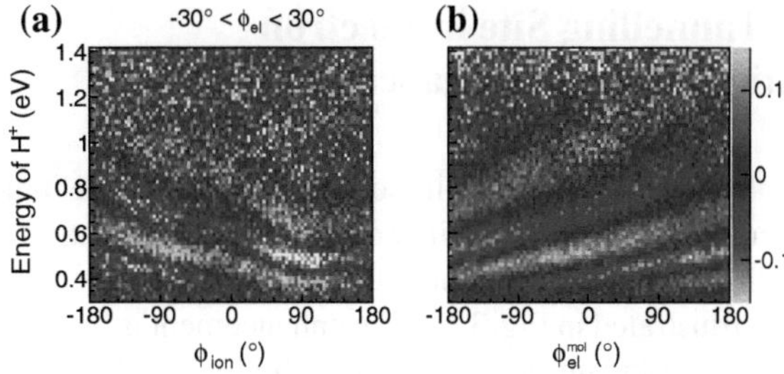

Fig. 3.3 **a** 2D map of the asymmetry $\beta_{lab} = [N_l(\phi_{el} = \phi, \phi_{ion}, E_k) - N_l(\phi_{el} = \phi + 180°, \phi_{ion}, E_k)]/[N_l(\phi_{el} = \phi, \phi_{ion}, E_k) + N_l(\phi_{el} = \phi + 180°, \phi_{ion}, E_k)]$ in the laboratory frame, where $N_l(\phi_{el}, \phi_{ion}, E_k)$ is the H^+ yield at the proton kinetic energy E_k and the emission angle of the ion ϕ_{ion} for the emission angles of the coincident electron in the laboratory frame gated to lie in the intervals $-30° < \phi_{el} < 30°$. **b** 2D map of the asymmetry $\beta_{mol} = [N_m(\phi_{el}^{mol} = \phi, E_k) - N_m(\phi_{el}^{mol} = \phi + 180°, E_k)]/[N_m(\phi_{el}^{mol} = \phi, E_k) + N_m(\phi_{el}^{mol} = \phi + 180°, E_k)]$ in the molecular frame, where $N_m(\phi_{el}^{mol}, E_k)$ is the H^+ yield at the proton kinetic energy E_k and the emission angle of the correlated electron in the molecular frame ϕ_{el}^{mol}. A 35-fs, 800-nm, circularly polarized laser pulse with a peak intensity of $I_0 = 2.3 \times 10^{14}$ W/cm^2 was employed in the experiments. (adapted from [10])

We can either scan the "probe" time for a given "pump" time by testing the proton ejection direction ϕ_{ion} for a given electron emission direction ϕ_{el} in the laboratory frame, or scan the "pump" time for a given "probe" time by testing the electron angular distribution in molecular frame $\phi_{el}^{mol} = \phi_{el} - \phi_{ion}$. It corresponds to the fixing of the laser phase at the ionization θ_i and dipole-allowed transitions θ_c, respectively. The former case is actually similar to the two-color XUV-attosecond-pump and IR-femtosecond-probe experiment [19–22] where the electron was controlled to localize at either the left or right protons by adjusting the time delay between the pump and probe pulses. Here we pump and probe this attosecond electron localization dynamics in a single multicycle femtosecond laser pulse. Figure 3.3 displays the measured asymmetry β_{lab} in the laboratory frame (and β_{mol} in molecular frame) as a function of the proton kinetic energy release (KER) and the ϕ_{ion} (or ϕ_{el}^{mol}) in our counter-clockwise rotating circularly polarized field, which stands for the laser phase-dependent directional breaking of the molecular bond. We note that the here revealed asymmetry is about five time larger than the recent two-color pump-probe experiment [22].

We experimentally confirmed our observations by switching the circular polarized pulse from counter-clockwise to clockwise, which mirrored the KER-ϕ_{ion} (or -ϕ_{el}^{mol})-dependent asymmetries of β_{lab} (or β_{mol}) [10]. It was also well reproduced by the quantum and semi-classical simulations [10].

The two-particle coincidence angular streaking not only clearly reveals the naturally exist laser-phase-dependent directional molecular bond breaking in a symmetric multicycle laser pulse due to the electron localization, but also most interestingly provides us with a robust and powerful tool to study attosecond dynamics of molecules in strong laser field without using of attosecond pulse.

3.3 Electron Tunnelling Site in Electron Localization-Assisted Enhanced Ionization

The electron localization also plays an important role in multiple ionization of molecules. The molecular ion shows a significantly enhanced ionization rate compare to an atomic ion with the same binding energy. The underlying so called enhanced ionization scenario is illustrated in Fig. 3.4a. The enhancement is believed to result from the electron localized at the up-field potential well [23–26]. It is generally observed in multielectron ionization of molecules orientating along the laser field direction [27–32]. After the removal of one or few electrons in the first ionization step, the molecular ion stretches and near a critical internuclear distance the remaining electrons localizes at the nuclei. When in the oscillating laser field this electron is on the uphill site it only needs to tunnel through the narrow inner-barrier between the nuclei to escape resulting in a significantly enhanced total ionization rate.

However, the underlying physical picture that it is the up-field electron which boosts the total ionization probability had not been experimentally verified until our recent work [33]. To probe it, the molecular orientation, the laser field vector at the ionization instant, and the atomic core from which the electron is freed must be simultaneously determined in one measurement. We test this scenario by using the angular streaking of an elliptically polarized laser pulse and studying the double ionization of a van der Waals (vdW) dimer of ArXe.

We start with randomly orientated molecules without pre-alignment. The molecular orientation of the individual molecule at the instant of ionization is measured from the relative momentum vector of the ionic fragments following Coulomb explosion (CE) [34, 49] of a doubly ionized molecule. The angular streaking by the close-to-circular polarized pulse maps the laser field vector at the instant of ionization to the emission direction of the freed electron. Since the electron tunnelling probability strongly depends on the field strength, as shown in Fig. 3.4a, the ionization mostly occurs when the laser field points to $\pm y$ (the major axis of our elliptically polarized pulse), leading to a final momentum of the released electron along $\mp z$ ($\mp pz_e$). Due to momentum conservation, the correlated ionic fragments receive a sum-momentum recoil along $\pm z$ ($\pm pz_{sum}$). Without direct measurement of the freed electrons, we therefore can recover the laser field vector along $+y$ or $-y$ at the instant of ionization from the sum-momentum of the ionic fragments [35]. Finally one knows that, for two-site double ionization of ArXe dimer by an elliptically polarized pulse, the first and second electrons are sequentially released, the first one from Xe ($I_{p,Xe} = 11.9$ eV) and the second one from the Ar ($I_{p,Ar} = 15.9$ eV) sites because of their largely different binding energies. For the ground state of ArXe, the influence of the molecular orbital on the ionization dynamics is avoided as the orbitals are close to the atomic orbitals. The ArXe therefore provides us a "clean" target to probe the scenario of the electron localization-assisted enhanced ionization.

Figure 3.4b depicts the measured ion sum-momentum distributions ($pz_{sum} = pz_{Ar^+} + pz_{Xe^+}$) of the Ar^+ + Xe^+ breakup [denoted ArXe(1,1)], which can be understood as the convolution of the momentum distributions of two sequentially

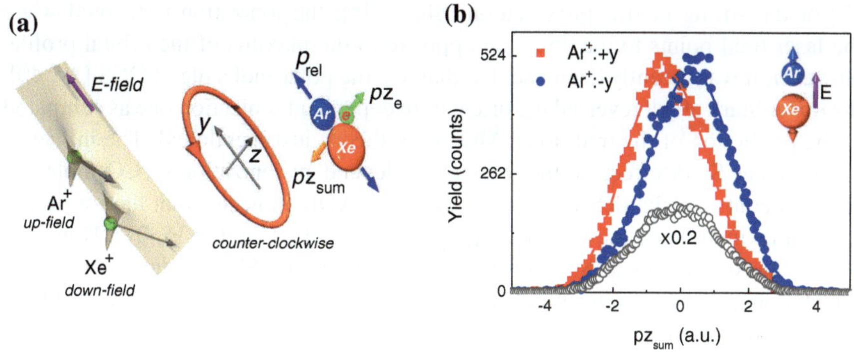

Fig. 3.4 **a** Schematic illustration of the electron localization-assisted enhanced ionization of $ArXe^+$ by elliptically polarized laser field (*red ellipse*). The electron localized at the up-field potential well is much easier to be freed by tunnelling through the narrow inner-barrier between the nuclei. Due to the angular streaking of the rotating laser field, the freed electron (*green ball*) receives a final momentum (*green arrow*) perpendicular to the laser field vector (*purple arrow*) at the ionization moment. **b** Ion sum-momentum distribution of ArXe(1,1). The *red squares* and *blue circles* are events where Ar^+ fly in the $+y$ and $-y$ directions, and the *grey squares* are the sum of them. The peak intensity and ellipticity of the employed 35-fs, 790-nm laser pulse were measured to be $I_0 = 4.3 \times 10^{14}$ W/cm^2 and $\varepsilon \sim 0.68$, respectively. (adapted from [33])

released electrons with four possibilities of $pz_{sum,ij} = -(ipz_{e1} + jpz_{e2})$ $(i, j = +$ or$-$ account for the electrons flying to $\pm z$, pz_{e1} and pz_{e2} are the momenta of the freed first and second electron along z-axis). The overall ionization probability of randomly orientated ArXe in symmetric multicycle laser pulse is the same for the laser field pointing $\pm y$, resulting mirror symmetric distribution of pz_{sum} [gray open circles in Fig. 3.4b]. However, an enhanced rate at pz_{sum} (or $+pz_{sum}$) emerges by gating on Ar^+ flying to $+y$ (or $-y$). Since there is no field direction-dependent ionization of the first electron from Xe site, the observed asymmetric distribution of pz_{sum} shown in Fig. 3.4b for our counter-clockwise elliptical light indicates that the release of the second electron from the Ar site is favored when the laser field points from Xe^+ to Ar. These data for the first time experimentally verifies [33] that the electron localized at the up-field potential well is preferred to be released as compare to that at the down-field, which is the cornerstone of the scenario of the electron localization-assisted enhanced ionization [23–32] in multielectron ionization of molecules by strong laser field.

3.4 Orientation-Dependent Single Ionization of CO Molecule

For molecules close to their ground state configuration the profile of the molecular orbital plays an important role in strong-field single ionization of molecules. As described by the molecular Ammosov-Delone-Krainov (MO-ADK) theory [36,

37] or the strong field approximation (SFA) [38], the ionization is favored when the laser field points to the direction opposite to the maxima of the orbital profile. However, it was recently demonstrated that, for the polar molecule of OCS [39, 40], the linear Stark effect reversed the orientation dependent ionization rate as compared to the prediction of the traditional MO-ADK theory just mentioned. The influence of the linear Stark effect on the orientation-dependent ionization of CO molecule is still a controversial debate [41–46]. The MO-ADK indicates that the ionization rate is maximal when the laser field points from nuclei C to O [36, 37]. However, by considering the linear Stark effect, the MO-ADK and SFA predict the opposite orientation-dependent ionization rates [45], i.e. the ionization is favored by laser field pointing from nuclei O to C for Stark-corrected MO-ADK and from nuclei C to O for Stark-corrected SFA, respectively.

To resolve this controversy we experimentally probe [47] the orientation-dependent ionization rate of CO by measuring the molecular-frame photoelectron angular distribution (MFPAD) [48], i.e. ϕ_{el}^{mol} =ϕ_{el}-ϕ_{ion}, of the dissociative single ionization channel of CO + $n\hbar\omega \rightarrow C^+ + O + e$ [denoted CO(1, 0)] using the angular streaking of a circularly polarized ultrashort laser pulse. It allows us to reveal the orientation-dependent ionization rate of molecules without pre-orientation of the molecules. As displayed in Fig. 3.5a (gray curve), the count rate of the MFPAD of CO(1, 0) is higher at $\phi_{el}^{mol} = 90°$ as compared to $\phi_{el}^{mol} = 270°$, where the C^+ orients along 0°. For the counter-clockwise circular polarized light used in this experiment, it indicates that the CO is favored to be ionized by laser field pointing from nuclei C to O. This implies that the linear Stark effect plays a minor role, if any, in strong field ionization of CO in the present experiments, and in contrast the ionization is ruled by the geometric profile of the ionizing orbital.

As shown in Fig. 3.5b, there is a broad KER distribution of CO(1, 0). Interestingly, as plotted in Fig. 3.5a, a larger asymmetry in MFPAD of CO(1, 0) is observed at high KERs (KER$>$ 0.75 eV, red curve) than that at low KERs (KER$<$ 0.75

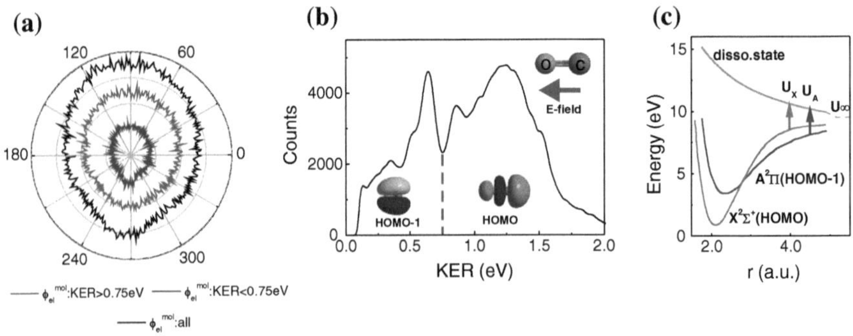

Fig. 3.5 Measured **a** MFPADs and **b** KER of CO(1,0). The *inserts* of **b** show the profiles of the HOMO and HOMO-1 of CO, where the nuclei O and C lie on the *left* and *right* sides, respectively. The peak intensity of the 790-nm, 35-fs circularly polarized pulse is $I_0 = 4 \times 10^{14}$ W/cm^2. (adapted from [47]) **c** The potential energy surfaces of CO^+ adapted from [51]

eV, blue curve). Since the single ionization of CO in the present experiments is dominated by the geometric profile of the ionizing orbital, the different asymmetries of the MFPADs indicate the contribution of various orbitals. As illustrated in Fig. 3.5c, the ionization of the highest occupied molecular orbital (HOMO) and the next lower-lying orbital of HOMO-1 populate the $X^2\Sigma^+$ and $A^2\Pi$ states of CO^+, followed by photon-coupled dissociation into CO(1, 0) fragments [50, 51] with high and low KERs (KER $\approx U_{X(A)} - U_\infty$), respectively. As expected from the geometric profile of the HOMO and HOMO-1 [insets of Fig. 3.5b], a smaller asymmetry of MFPAD at low KERs is observed by ionizing the HOMO-1 than that at high KERs originating from the ionization of the HOMO. Although the binding energy of the HOMO-1 is about 2.5 eV higher than that of the HOMO, about 30 % contribution of the HOMO-1 to the CO(1, 0) is observed. The multiorbital contribution beyond the HOMO is actually quite general in strong field ionization of molecules [50, 52–54].

3.5 Sequencing Multiple Ionization of a Multicenter Molecular Cluster

The vast majority of molecules are nonlinear and polyatomic. For all these molecules the molecular geometry plays an important role in the complex ionization dynamics. It might overrule the effects of the binding energy and the profile of the molecular orbital. Here, by taking the vdW dimer of N_2Ar as a simple yet rich example, we demonstrate [55] that the angular streaking of an elliptically polarized ultrashort laser pulse can be used to sequence the releasing order of electrons from a complex molecule.

The T-shaped N_2Ar dimer [56, 57] contains a weak vdW bond ($R_{Ar-N_2} \approx 3.7$ Å) between Ar and N_2 and a orthogonally orientated strong covalent bond in N_2 ($R_{N-N} \approx 1.2$ Å). The electrons mostly localize around the centers of Ar and N_2 [55]. In our experiment we study in which order and from which center the electrons are sequentially released during a multielectron ionization by strong laser fields.

To learn about the ionization sequence of the first two electrons, we start with the coulomb exploding two-site double ionization of $N_2Ar + n\hbar\omega \rightarrow N_2^+ + Ar^+ + 2e$ [denoted as $N_2Ar(1,1)$] with the breakage of the vdW bond, which is favored when the vdW bond is along the major axis (y-axis) of the elliptic light [55]. As discussed above, we study the ionization dynamics by analyzing the ion sum-momentum pz_{sum} of the ionic fragments, instead of direct measurement of the freed electron. Figure 3.6a shows the measured pz_{sum} of the $N_2Ar(1,1)$ channel. The symmetric three-peak structure of pz_{sum} (gray circles) reflects the overall equal probability of the electrons freed by laser field pointing to $\pm y$ in the multicycle symmetric laser pulse. Interestingly, by gating the departing direction of Ar^+ to $\pm y$, noticeable enhanced rate at $\mp pz_{sum}$ is observed, as shown in Fig. 3.6a. It indicates a laser field direction-dependent ionization rate, which is most likely a result of the second ionization step since there is no real molecular orbital between N_2 and Ar. The

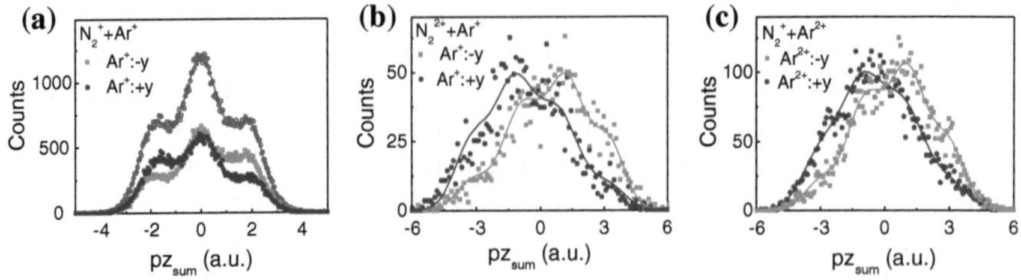

Fig. 3.6 Ion sum-momentum distribution pz_{sum} of **a** $N_2^+ + Ar^+$, b $N_2^{2+} + Ar^+$, c $N_2^+ + Ar^{2+}$ following two-sites double and triple ionization of N_2Ar. The peak intensity and ellipticity of the employed 790-nm, 35-fs elliptically polarized pulse is $I_0 = 1.2 \times 10^{15}$ W/cm^2 and $\varepsilon \sim 0.9$, respectively. (adapted from [55])

vdW bond orientation-dependent asymmetric distributions of pz_{sum} indicate that the second electron is favored to be freed when the laser field points from Ar to N_2.

One possible underlying mechanism could be the electron localization-assisted enhanced ionization [33] when the vdW bond is along the laser field. In this case, the first electron should be freed from Ar site and then the second electron from the N_2 site when it is on the up-field site. Although the binding energy of Ar is similar to N_2, the Ar shows a higher single ionization rate at the laser intensity we used here [58]. The single ionization probability of N_2 is about 3 times lower for a molecule oriented perpendicular to the laser field than that parallel to it [59]. The T-shaped N_2Ar therefore increases the difference of the ionization probability of N_2 and Ar, where the axis of N_2 is orthogonal to the laser field when the vdW bond is along the laser field. Due to the different ionization potentials, the first and second ionization steps occur at different intensities in the laser pulse, which is encoded in the magnitudes of the final momenta of the freed electrons. We obtain the momenta of the first and second electrons to be $pz_{e1} = 0.77$ a.u. and $pz_{e2} = 1.09$ a.u., respectively using the following fitting procedure. The ion sum-momentum distribution is the convolution of the the momentum distribution of two sequentially released electrons [33], i.e. it is given by

$$pz_{sum,12}(pz) = \frac{1}{\sqrt{2\pi}} \sum_{i,j=+,-} \frac{A_{12ij}}{\sigma_{12ij}} \exp[-0.5(pz - pz_{sumij})^2/\sigma_{12ij}^2]. \quad (3.1)$$

Here, the momentum distribution of each electron is assumed to have a Gaussian shape, i.e. $A_k/(\sigma_k\sqrt{2\pi}) \times \exp[-0.5(pz - pz_{ek})^2/\sigma_k^2]$, $A_{12ij} = A_{1i} \times A_{2j}$, $\sigma_{12ij}^2 = \sigma_{1i}^2 + \sigma_{2j}^2$, and $pz_{sumij} = -(ipz_{e1} + jpz_{e2})$ accounts for four possible ion sum-momenta. Therefore, by sequentially releasing the first electron from Ar and the second electron from N_2, the two-site double ionization in producing $N_2Ar(1,1)$ is dominated by the orientation of the vdW bond owing to the mechanism of electron localization-assisted enhanced ionization [33].

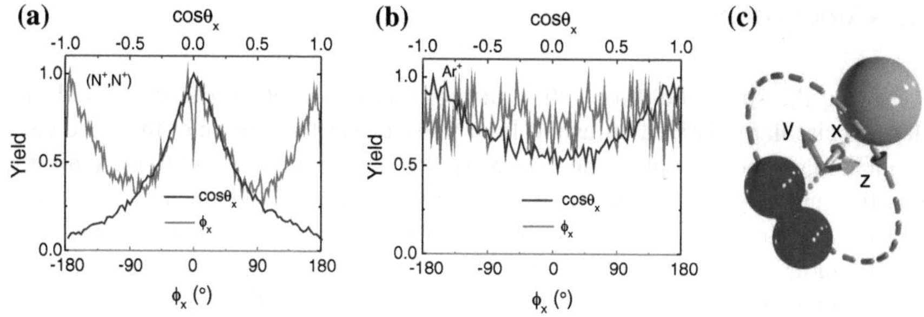

Fig. 3.7 Measured angular distribution of **a** the (N^+,N^+) pair and **b** the Ar^+, and **c** the favored orientation of N_2Ar for the three-body breakup channel of N^++N^++Ar^+ in our elliptically polarized pulse (*dashed red ellipse*). (adapted from [55])

Following two-site double ionization, the third electron could be freed either from the Ar^+ or the N_2^+ sites, leading to the N_2^+ + Ar^{2+} [denoted as $N_2Ar(1,2)$] or the N_2^{2+} + Ar^+ [denoted as $N_2Ar(2,1)$]. As shown in Fig. 3.6b, c, enhanced count rate at $\pm pz_{sum}$ are observed for both $N_2Ar(2,1)$ and $N_2Ar(1,2)$ channels when the Ar^+ or Ar^{2+} depart to $\mp y$. This indicates that the third electron is favored to be freed by laser field pointing from Ar^+ to N_2^+. The electron localization-assisted enhanced ionization therefore also plays an important role in producing $N_2Ar(2,1)$ by releasing the third electron from N_2^+ when it is up-field. On the other hand, for the $N_2Ar(1,2)$, the multielectron effect with intramolecular electron transfer [60] from Ar^+ to N_2^{2+} may play a role after the third ionization from the N_2^+ site.

Another interesting question to ask for the T-shaped N_2Ar is how the electrons are released when both the vdW and covalent bonds are broken leading to the three-body breakup channel of N^++N^++Ar^+ [denoted as $N_2Ar(1,1,1)$]. As shown in Fig. 3.7a, b, the measured angular distributions of the (N^+,N^+) pair and Ar^+ indicate that the $N_2Ar(1,1,1)$ is most likely to be produced when the covalent bond in N_2 is along the major axis of our elliptical light and the vdW bond is perpendicular to the polarization plane (see Fig. 3.7c). Here, the breakage of the covalent bond in N_2 occurs around the equilibrium internuclear distance of N_2, where the molecular orbital plays an important role. It is much shorter than that generally observed in the strong field exploding double ionization of N_2 monomer [23, 24, 26] where the electron localization-assisted enhanced ionization plays an important role at a large internuclear distance. The production of $N_2Ar(1,1,1)$ channel is therefore dominated by the orientation of the covalent bond by firstly releasing two electron from the N_2, followed by the third ionization from the Ar site.

The multielectron ionization dynamics of a nonlinear polyatomic molecule with increased geometry is governed by the collective effects of the binding energy, the geometric profile of the ionizing orbital, and the orientation-dependent electron localization-assisted enhanced ionization. The angular streaking of an elliptically polarized pulse allows us to sequence the complex ionization dynamics and distinguish the various mechanisms.

3.6 Conclusions

Following the strong field ionization of molecules, the rotating electric field of a close-to-circular polarized ultrashort laser pulse drives the motion of the freed electron and maps the laser phase and intensity at the ionization instant to the emission direction and magnitude of the final momentum of the released electron. By coincidently measuring the ejected electrons and ionic fragments, the ionization dynamics can be studied in the molecular frame. This "coincidence" measurement can further be achieved by solely detecting the ionic fragments in the experiments, where the momenta of the sequentially released electrons are encoded in the ion sum-momentum and can be retrieved by fitting the ion sum-momentum distribution. The coincidence angular streaking therefore provides us with a powerful and robust tool to probe the complex dynamics of molecules in strong laser fields. Beyond what we have demonstrated here, we believe that there are much more surprises await us in the near future.

Acknowledgments We would like to thank the many coworkers and colleagues who helped in the present work. We would like to mention M. Kunitski, A. Vredenborg, M. Meckel, M. Magrakvelidze, B. Ulrich, M. Schöffler, M. Pitzer, M. Richter, S. Voss, H. Sann, H. Kim, F. K. Amankona-Diawuo J. Lower, T. Jahnke, L.P.H. Schmidt, A. Czasch, U. Thumm, T. Pfeifer, T. Seideman, and H. Schmidt-Böcking. This work was partially supported by the Deutsche Forschungsgemeinschaft. J.W. acknowledges support by the Alexander von Humboldt Foundation, the "Eastern Scholar" Program, the NCET in University (NCET-12–0177), the Project from SSTC (13QH1401400), the "Shu Guang" project (12SG25), and the National Natural Science Fund (11425416 and 11374103).

References

1. M. Hentschel et al., Nature **414**, 509 (2001)
2. G. Sansone et al., Science **314**, 443 (2006)
3. E. Goulielmakis et al., Science **305**, 1267 (2004)
4. M. Drescher et al., Nature **419**, 803 (2002)
5. A.L. Cavalieri et al., Nature **449**, 1029 (2007)
6. S. Gräfe, V. Engel, MYu. Ivanov, Phys. Rev. Lett. **101**, 103001 (2008)
7. P. Eckle et al., Nat. Phys. **4**, 565 (2008)
8. P. Eckle et al., Science **322**, 1525 (2008)
9. A.N. Pfeiffer, C. Cirelli, M. Smolarski, R. Dörner, U. Keller, Nat. Phys. **7**, 418 (2011)
10. J. Wu et al., Nat. Commun. **4**, 2177 (2013)
11. R. Dörner et al., Phys. Rep. **330**, 95 (2000)
12. J. Ulrich et al., Rep. Rrog. Phys. **66**, 1463 (2003)
13. O. Jagutzki et al., Nucl. Instr. Meth. Phys. Res. A **477**, 244 (2002)
14. M.F. Kling et al., Science **312**, 246 (2006)
15. M. Kremer et al., Phys. Rev. Lett. **103**, 213003 (2009)
16. I. Znakovskaya et al., Phys. Rev. Lett. **108**, 063002 (2012)
17. D. Ray et al., Phys. Rev. Lett. **103**, 223201 (2009)
18. J. Wu et al., Phys. Rev. A **87**, 023406 (2013)
19. F. He, A. Becker, U. Thumm, Phys. Rev. Lett. **101**, 213002 (2008)
20. F. He, C. Ruiz, A. Becker, J. Phys. B **41**, 081003 (2008)

21. G. Sansone et al., Nature **465**, 763 (2010)
22. K.P. Singh et al., Phys. Rev. Lett. **104**, 023001 (2010)
23. K. Codling, L.J. Frasinski, P.A. Hatherly, J. Phys. B **22**, L321 (1989)
24. T. Seideman, MYu. Ivanov, P.B. Corkum, Phys. Rev. Lett. **75**, 2819 (1995)
25. T. Zuo, A.D. Bandrauk, Phys. Rev. A **52**, R2511 (1995)
26. S. Chelkowski, A.D. Bandrauk, J. Phys. B **28**, L723 (1995)
27. L.J. Frasinski et al., Phys. Rev. Lett. **58**, 2424 (1987)
28. W. Hill et al., Phys. Rev. Lett. **69**, 2646 (1992)
29. M. Schmidt, D. Normand, C. Cornaggia, Phys. Rev. A **50**, 5037 (1994)
30. E. Constant, H. Stapelfeldt, P.B. Corkum, Phys. Rev. Lett. **76**, 4140 (1996)
31. M.J. DeWitt, R.J. Levis, J. Chem. Phys. **108**, 7045 (1998)
32. A.N. Markevitch, D.A. Romanov, S.M. Smith, R.J. Levis, Phys. Rev. Lett. **92**, 063001 (2004)
33. J. Wu et al., Nat. Commun. **3**, 1113 (2012)
34. Z. Vager, R. Naaman, E.P. Kanter, Science **244**, 426 (1989)
35. J. Wu et al., Phys. Rev. Lett. **108**, 043002 (2012)
36. X.M. Tong, Z.X. Zhao, C.D. Lin, Phys. Rev. A **66**, 033402 (2002)
37. O.I. Tolstikhin, T. Morishita, L.B. Madsen, Phys. Rev. A **84**, 053423 (2011)
38. J. Muth-Böhm, A. Becker, F.H.M. Faisal, Phys. Rev. Lett. **85**, 2280 (2000)
39. L. Holmegaard et al., Nat. Phys. **6**, 428 (2010)
40. D. Dimitrovski et al., Phys. Rev. A **83**, 023405 (2011)
41. S. De et al., Phys. Rev. Lett. **103**, 153002 (2009)
42. M. Abu-samha, L.B. Madsen, Phys. Rev. A **82**, 043413 (2010)
43. K.J. Betsch, D.W. Pinkham, R.R. Jones, Phys. Rev. Lett. **105**, 223002 (2010)
44. H. Ohmura, N. Saito, T. Morishita, Phys. Rev. A **83**, 063407 (2011)
45. H. Li et al., Phys. Rev. A **84**, 043429 (2011)
46. X. Zhu, Q. Zhang, W. Hong, P. Lu, Z. Xu, Opt. Express 19, 24 198 (2011)
47. J. Wu et al., Phys. Rev. Lett. **108**, 183001 (2012)
48. A. Staudte et al., Phys. Rev. Lett. **102**, 033004 (2009)
49. X. Gong et al., Phys. Rev. A **88**, 013422 (2013)
50. I. Znakovskaya et al., Phys. Rev. Lett. **103**, 103002 (2009)
51. S. De et al., Phys. Rev. A **84**, 043410 (2011)
52. O. Smirnova et al., Nature **460**, 972 (2009)
53. B.K. McFarland, J.P. Farrell, P.H. Bucksbaum, M. Gühr, Science **322**, 1232 (2008)
54. H. Akagi et al., Science **325**, 1364 (2009)
55. J. Wu et al., Phys. Rev. Lett. **111**, 083003 (2013)
56. C.R. Munteanu, J.L. Cacheiro, B. Fernández, J. Chem. Phys. **121**, 10419 (2004)
57. J. Wu et al., J. Chem. Phys. **137**, 104308 (2012)
58. C. Guo, M. Li, J.P. Nibarger, G.N. Gibson, Phys. Rev. A **58**, R4271 (1998)
59. D. Pavičić et al., Phys. Rev. Lett. **98**, 243001 (2007)
60. V. Tagliamonti, H. Chen, G.N. Gibson, Phys. Rev. Lett. **110**, 073002 (2013)

Chapter 4
Control of Ultrafast Electron Dynamics with Shaped Femtosecond Laser Pulses: From Atoms to Solids

Matthias Wollenhaupt, Tim Bayer and Thomas Baumert

Abstract In this chapter, we present an introduction to the fundamentals of femtosecond pulse shaping and review recent demonstrations of coherent control by pulse tailoring. We portray control of three-dimensional free-electron wave packets, strong-field control by selective population of dressed states (SPODS) and control of ionization processes in dielectrics. Prototypical spectral phase masks such as polynomial- and sinusoidal functions are discussed and concepts of polarization shaping such as the instantaneous frequency and the instantaneous polarization state are introduced and illustrated on representative examples. In addition, experiments on coherent control are reviewed. Coherence transfer from light to matter is studied on the interference of free-electron wave packets. We analyze control and adaptive optimization of three-dimensional designer free-electron wave packets by polarization shaping. Strong-field control via SPODS is introduced and elucidated on specific realizations via rapid adiabatic passage and photon locking. This concept is extended to strong-field control of the concerted electron-nuclear dynamics in molecules. Finally, we present recent experiments on control of ionization processes in dielectrics.

M. Wollenhaupt (✉) · T. Bayer
Carl von Ossietzky Universität Oldenburg, Institut für Physik,
Carl-von-Ossietzky-Straße 9-11, 26129 Oldenburg, Germany
e-mail: matthias.wollenhaupt@uni-oldenburg.de

T. Bayer
e-mail: tim.bayer@uni-oldenburg.de

T. Baumert
Universität Kassel, Institut für Physik und CINSaT,
Heinrich-Plett-Str. 40, 34132 Kassel, Germany
e-mail: baumert@physik.uni-kassel.de

M. Kitzler and S. Gräfe (eds.), *Ultrafast Dynamics Driven by Intense Light Pulses*, Springer Series on Atomic, Optical, and Plasma Physics 86, DOI 10.1007/978-3-319-20173-3_4

4.1 Introduction

In coherent control shaped femtosecond laser pulses are used as a tool to steer the ultrafast light-induced dynamics of a quantum system from a given initial state to a predefined target state, preferably with high selectivity and high efficiency. Progress in this fast expanding research field is reported in recent text books [1–3], review articles [4–23] and special issues [24–29]. In this review, we focus on advanced control of ultrafast electron dynamics with shaped femtosecond laser pulses investigated by photo-ionization. In the studies presented, the role of photo-ionization is twofold: on the one hand, the details of the photo-ionization dynamics are explored, whereas, on the other hand, photoelectron spectra are employed as a powerful tool to monitor the controlled neutral dynamics.

A deeper understanding of the principles of pulse shaping and the ability to generate shaped femtosecond pulses with high precision in the laboratory are two important prerequisites to successfully carry out coherent control experiments. Therefore, we start this chapter with an introduction to the theoretical and experimental fundamentals of pulse shaping and elaborate on control of the time-dependent polarization profile of a femtosecond laser pulse, in addition to its temporal envelope and phase. Before we present some recent applications, we will clarify the essential concepts of polarization shaping including a characterization of the instantaneous frequency and the instantaneous polarization state and present several prototypical examples.

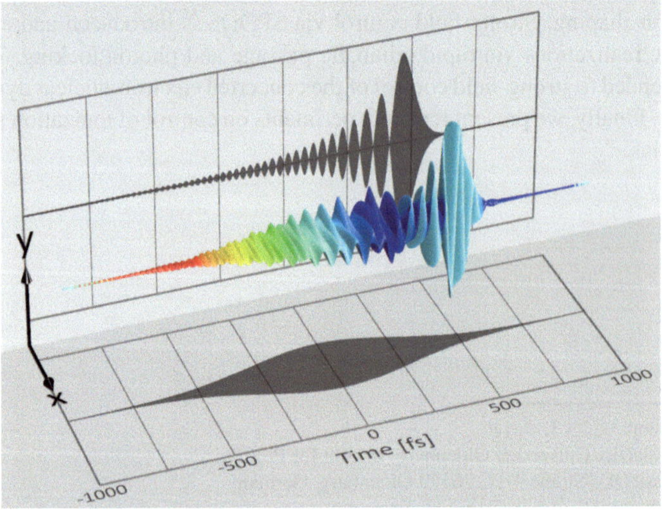

Because the emphasis of this review is placed on understanding the physical mechanisms of coherent control, we mainly focus our attention on isolated model systems in the gas phase. Since the physical basis of coherent control is the manipulation of the interferences of matter waves by laser radiation, we will start by investigating the transfer of the coherence properties from laser pulses to matter waves. To this end, we study the photoionization of potassium atoms with a phase-coherent double pulse sequence. Building on this simple example we discuss creation and measurement of three dimensional designer free electron wave packets as an example for complex but still comprehensible control by polarization-shaped pulses. In addition, two examples of adaptive optimization of photoionization processes by polarization-shaped femtosecond laser pulses are reported. New physical mechanisms to exert efficient control become available when intense femtosecond laser pulses are used. The non-perturbative interaction with shaped pulses entails novel dynamics such as complete population transfer, Rabi-oscillations or Photon Locking (see below) which are unavailable in weak fields and thus opens numerous novel control scenarios. We will portray some prototype examples highlighting specific control mechanisms such as non-resonant dynamic Stark shifts and bipolar energy shifts through resonant excitation. We introduce the concept of control by selective population of dressed states (SPODS) which is particularly suited to describe strong field interactions and provides a unified perspective on adiabatic techniques and ultrafast switching by phase jumps. Applying these techniques to selective excitation of molecules opens the fascinating subject of coherent control of the concerted nuclear and electronic dynamics which concludes the section on strong-field control. Finally, we present a short overview on recent work on control of ionization processes in dielectrics with applications in nano-materials processing.

4.2 Fundamentals of Femtosecond Pulse Shaping

In this introductory section we present the fundamentals of femtosecond pulse shaping in view of the applications discussed in this chapter, such as photo-ionization of atoms and molecules as well as material processing. In this theoretical section a mathematical description of femtosecond pulse shaping is presented along with basic definitions and relevant equations. Special emphasis is placed on polarization shaping which plays a central role in the creation of designer electron wave packets. In addition, the closed loop approach to find optimal laser pulses by adaptive feedback learning is summarized and a brief description of the experimental implementation of a liquid crystal based 4f pulse shaper used in these experiments is given.

4.2.1 Theoretical Description

In this section we introduce the basic physical quantities of the laser electric field which are required to understand the properties of shaped pulses [20, 30, 31]. We will concentrate on pulse shaping by spectral phase modulation with applications to control of ultrafast electron dynamics. Polynomial phase functions such as GDD and TOD and pulse trains created by sinusoidal phase modulation are investigated in some detail. In addition, we discuss spectral phase functions resulting from the combined action of two different modulation functions. The pulse duration is a central quantity to characterize femtosecond pulses and to classify the interaction in terms of the relevant physical time scales. Therefore, we generalize this concept towards shaped laser pulses and introduce statistical quantities based on the moments of the temporal pulse intensity distribution as a quantitative measure for the duration of a shaped pulse. A practical analytic expression for determining the pulse duration in frequency domain only using the spectral phase function is given. As a reference, analytic expressions for the temporal field envelopes and the corresponding pulse durations of the relevant shaped pulses are provided.

4.2.1.1 Definition of the Physical Quantities

We start by considering the linearly polarized real-valued time-dependent pulsed laser electric field (in units of Vm^{-1})

$$E(t) = A(t)\cos\left[\omega_0 t + \zeta(t)\right] \tag{4.1}$$

oscillating with the carrier frequency ω_0. Here $A(t)$, is the real valued envelope of the field and the temporal phase function $\zeta(t)$ denotes slowly varying temporal phase changes. Physically, the temporal phase function is responsible for variations in the instantaneous frequency, for instance frequency sweeps or phase discontinuities in pulse sequences. Defining the time-dependent detuning $\Delta(t)$ by the derivative of the temporal phase function

$$\Delta(t) = \frac{d\zeta(t)}{dt} \tag{4.2}$$

the instantaneous frequency $\omega(t)$ is given by the sum of the laser central frequency and the time-dependent detuning

$$\omega(t) = \omega_0 + \Delta(t) \tag{4.3}$$

Defining the slowly varying—and in general complex valued—function

$$\mathcal{E}(t) = A(t)\, e^{i\zeta(t)} \tag{4.4}$$

which is called the complex pulse envelope, (4.1) can be rewritten as

$$E(t) = Re\left[\mathcal{E}(t)\, e^{i\omega_0 t}\right] = \frac{1}{2}\left[\mathcal{E}(t)\, e^{i\omega_0 t} + \mathcal{E}^*(t)\, e^{-i\omega_0 t}\right], \tag{4.5}$$

where the symbol (*) denotes complex conjugation.

In general, the analytic signals $\mathcal{E}^+(t)$ and $\mathcal{E}^-(t)$ are introduced to characterize ultrashort shaped laser pulses [30]. However, since the SVEA (Slowly Varying Envelope Approximation) applies for the pulses considered here, the envelope function

$\mathcal{E}(t)$ defined by (4.4) is sufficient to describe the electric field. Combining (4.3) and (4.5) provides a useful expression to derive the instantaneous detuning directly from the complex envelope without resorting to inverse trigonometric functions and hence avoiding phase ambiguities [32]

$$\Delta(t) = \frac{i}{2}\frac{\varepsilon(t)\frac{d}{dt}\varepsilon^*(t) - \varepsilon^*(t)\frac{d}{dt}\varepsilon(t)}{\varepsilon(t)\varepsilon^*(t)} = -\frac{\mathrm{Im}\left[\varepsilon(t)\frac{d}{dt}\varepsilon^*(t)\right]}{|\varepsilon(t)|^2}, \tag{4.6}$$

where Im denotes the imaginary part and hence, $\Delta(t)$ is always a real quantity. The spectrum $\tilde{E}(\omega)$ (in units of Vsm^{-1}) of the laser electric field $E(t)$ is obtained by Fourier transform of $E(t)$

$$\tilde{E}(\omega) = \int_{-\infty}^{\infty} E(t)\, e^{-i\omega t} dt. \tag{4.7}$$

Because the field $E(t)$ is real-valued and oscillates with the carrier frequency ω_0, its spectrum is Hermitian $\tilde{E}(-\omega) = \tilde{E}^*(\omega)$ and has non-vanishing contributions only around $\omega = +\omega_0$ and $\omega = -\omega_0$. Analogously, the Fourier spectrum $\tilde{\mathcal{E}}(\omega)$ (also in units of Vsm^{-1}) of the complex valued envelope $\mathcal{E}(\omega)$ is given by

$$\tilde{\mathcal{E}}(\omega) = \int_{-\infty}^{\infty} \mathcal{E}(t)\, e^{-i\omega t} dt \tag{4.8}$$

and has only non-vanishing components around $\omega = 0$. For instance, the Gaussian

$$\mathcal{E}(t) = \frac{\varepsilon_0}{2} e^{-\ln 4\left(\frac{t}{\Delta t}\right)^2} \tag{4.9}$$

is a typical example of an unmodulated pulse envelope with an intensity FWHM (Full Width at Half Maximum of the intensity envelope $I(t) \propto \mathcal{E}^2(t)$) of Δt. The corresponding spectrum of the amplitude reads

$$\tilde{\mathcal{E}}(\omega) = \frac{\varepsilon_0 \Delta t}{2}\sqrt{\frac{\pi}{\ln 4}} e^{-\ln 4\left(\frac{\omega}{\Delta\omega}\right)^2}, \tag{4.10}$$

where the intensity FWHM of the spectrum is

$$\Delta\omega = \frac{4\ln 2}{\Delta t} \tag{4.11}$$

in accordance with the FWHM-bandwidth product for a Gaussian laser pulse

$$\Delta\nu\,\Delta t = \frac{\Delta\omega\,\Delta t}{2\pi} = \frac{4\ln 2}{2\pi} \approx 0.44. \tag{4.12}$$

4.2.1.2 Spectral Modulation

Because femtosecond laser pulses are too short to be directly shaped in the time domain they are modulated in the frequency domain by applying a linear spectral transfer function $\tilde{\mathcal{M}}(\omega)$ (see Sect. 4.2.2 for the experimental implementation of this approach). The modulated spectrum $\tilde{\mathcal{E}}_{mod}(\omega)$ of the shaped pulse is obtained by multiplication of the spectral modulation function $\tilde{\mathcal{M}}(\omega)$ with the initially unmodulated spectrum

$$\tilde{\mathcal{E}}_{mod}(\omega) = \tilde{\mathcal{M}}(\omega)\,\tilde{\mathcal{E}}(\omega)\,. \tag{4.13}$$

Inverse Fourier transform of (4.13) shows that the modulated complex-valued temporal field envelope $\mathcal{E}_{mod}(t)$ is obtained by convolution with the temporal modulation function

$$\mathcal{E}_{mod}(t) = \mathcal{M}(t) \otimes \mathcal{E}(t)\,, \tag{4.14}$$

where the modulation functions $\mathcal{M}(\omega)$ and $\tilde{\mathcal{M}}(\omega)$ are a Fourier pair and $\otimes$ denotes the convolution. In most cases, however, it is more practical to evaluate the expression for the modulated spectrum $\tilde{\mathcal{E}}_{mod}(\omega)$ in (4.13) and derive the shaped pulse in time domain directly from $\tilde{\mathcal{E}}_{mod}(\omega)$ by inverse Fourier transform

$$\mathcal{E}_{mod}(t) = \frac{1}{2\pi}\int_{-\infty}^{\infty} \tilde{\mathcal{E}}_{mod}(\omega)\, e^{i\omega t} d\omega. \tag{4.15}$$

In general, the spectral modulation function can be decomposed into a spectral amplitude modulation function $\tilde{\mathcal{R}}(\omega)$ and a spectral phase modulation function $\varphi(\omega)$ by

$$\tilde{\mathcal{M}}(\omega) = \tilde{\mathcal{R}}(\omega)\, e^{-i\varphi(\omega)}. \tag{4.16}$$

Some pulse shapes—such as double pulses [33–35]—require amplitude modulation leading to an attenuation of the laser beam. In all conventional implementations of pulse shaping the amplitude modulation function fulfils the condition $|\tilde{\mathcal{R}}(\omega)| \leq 1$ because linear spectral filters only allow for attenuation of spectral components. For pure phase modulation by optically transparent spectral filters, when none of the spectral amplitudes is attenuated ($\tilde{\mathcal{R}}(\omega) \equiv 1$), one obtains the basic relation of spectral phase modulation

$$\tilde{\mathcal{E}}_{mod}(\omega) = \tilde{\mathcal{E}}(\omega)\, e^{-i\,\varphi(\omega)}. \tag{4.17}$$

Physically it is clear, that in the case of pure phase modulation the pulse energy is conserved. Formally, this also follows by applying Parseval's theorem (see for example [36]) to the modulated field

$$\int_{-\infty}^{\infty} |\mathcal{E}_{mod}(t)|^2 \, dt = \frac{1}{2\pi}\int_{-\infty}^{\infty} \left|\tilde{\mathcal{E}}_{mod}(\omega)\right|^2 d\omega = \frac{1}{2\pi}\int_{-\infty}^{\infty} \left|\tilde{\mathcal{E}}(\omega)\right|^2 d\omega = \int_{-\infty}^{\infty} |\mathcal{E}(t)|^2 \, dt \tag{4.18}$$

where we made use of the relation $|\tilde{\mathcal{E}}_{mod}(\omega)| = |\tilde{\mathcal{E}}(\omega)|$ for pure phase modulation. It is remarkable that in the case of phase modulation, besides the conservation of the pulse energy shown in (4.18), the modulus of the pulse area $|\theta| = \left|\frac{\mu}{\hbar}\int_{-\infty}^{\infty} \mathcal{E}(t)\, dt\right|$ which is relevant to strong field excitation (see Sect. 4.3.3 and for example [37–39]), is also conserved. This is seen by considering

$$\theta_{mod} = \int_{-\infty}^{\infty} \mathcal{E}_{mod}(t)\, dt = \tilde{\mathcal{E}}_{mod}(0) = \tilde{\mathcal{E}}(0)\, e^{-i\varphi(0)} = e^{-i\varphi(0)} \int_{-\infty}^{\infty} \mathcal{E}(t)\, dt = e^{-i\varphi(0)}\theta, \tag{4.19}$$

where θ and θ_{mod} denote the pulse area of unmodulated and the modulated pulse respectively. Note that for simplicity both the dipole moment μ and Planck's constant $\hbar$ were set to unity here.

4.2.1.3 Combined Spectral Phase Modulation

If a sum of multiple spectral phase functions $\varphi_i(\omega)$ is applied, the combined action of the linear combination of those phase functions can be decomposed into subsequent executions of the corresponding individual phase functions:

$$\tilde{\mathcal{M}}(\omega) = e^{-i[\varphi_1(\omega)+\varphi_2(\omega)]} = e^{-i\varphi_1(\omega)} e^{-i\varphi_2(\omega)} = \tilde{\mathcal{M}}_1(\omega)\, \tilde{\mathcal{M}}_2(\omega)\,. \tag{4.20}$$

In the time domain, the multiplication of spectral modulation functions leads to a convolution of the temporal modulation functions. For example, combining sinusoidal- and quadratic spectral phase modulation yields a sequence of chirped pulses [40] as illustrated in Fig. 4.2c. In general, it is not always easy to predict the outcome of combinations of spectral phase functions on the temporal pulse shape due to intra-pulse interference effects of the subsequent modulations.

4.2.1.4 Pulse Durations

The pulse duration determines the relevant physical time scale for the light matter interaction and is therefore an important experimental parameter. For an unmodulated femtosecond pulse the intensity FWHM is often used as a quantitative measure for the pulse duration. However, the shape of phase modulated pulses is generally very complicated. For instance, shaped pulses can be temporally asymmetric, have several maxima or even consist of several distinct sub-pulses. In these circumstances, the FWHM—which is generally employed to characterize simple bell-shaped distributions—may no longer be a meaningful parameter to describe the shaped pulse. A general way to characterize the shape of a pulse by statistical methods is based on the moments of its intensity distribution. To this end, we define the shape function $f(t)$ of the pulse by

$$f(t) = \frac{\mathcal{E}(t)}{\sqrt{\int |\mathcal{E}(t)|^2 \, dt}} \tag{4.21}$$

which is independent of the pulse energy such that $f(t)$ is a normalized distribution function with

$$\int_{-\infty}^{\infty} |f(t)|^2 \, dt = 1. \tag{4.22}$$

The nth-order moments of the modulus squared of the shape function read

$$\langle t^n \rangle = \int_{-\infty}^{\infty} t^n \, |f(t)|^2 \, dt. \tag{4.23}$$

With the above definition of the moments, one might in principle evaluate all of the higher order moments of the distribution to fully characterize the pulse. For our purpose it is sufficient to define the variance σ^2 of the function (t)

$$\sigma^2 = \langle t^2 \rangle - \langle t \rangle^2 . \tag{4.24}$$

which is a convenient measure of the statistical pulse duration of a shaped pulse. For simple unmodulated pulse shapes the calculation of the variance is straightforward. For example the variance of the (normalized) Gaussian shaped laser pulse is

$$\sigma^2 = \frac{\Delta t^2}{8 \ln 2}. \tag{4.25}$$

This relation between the variance and the pulse duration will also be applied as a definition for the pulse duration from the variance of a modulated pulse

$$\Delta t_{\mathrm{mod}} = \sigma_{\mathrm{mod}} \sqrt{8 \ln 2}. \tag{4.26}$$

However, the analytic expression of the time profile of phase modulated pulses may be overly complicated or even unknown precluding the analytic calculation of the variance. In this case numerical techniques can be used to determine the pulse duration. It is more insightful to calculate the variance directly in the spectral domain only by referring to the spectral phase function $\varphi(\omega)$. To this end, we assume the shape function of the modulated pulse to be generated by spectral phase modulation in analogy to (4.17)

$$\tilde{F}_{mod}(\omega) = \tilde{F}(\omega) \, e^{-i\varphi(\omega)}. \tag{4.27}$$

If we further assume that the spectrum of the unmodulated function $f(t)$ is real-valued, i.e. it has no additional temporal phase, and the first moment of the unmodulated function $f(t)$ vanishes (which is always the case for an even unmodulated shape function and could be achieved by time-shifting of the unmodulated pulse otherwise), we obtain a convenient expression for the variance of the modulated pulse in the frequency domain

$$\sigma_{mod}^2 = \sigma^2 + \frac{1}{2\pi}\int_{-\infty}^{\infty}\left[\tilde{F}(\omega)\right]^2\left[\varphi'(\omega)\right]^2 d\omega - \left\{\frac{1}{2\pi}\int_{-\infty}^{\infty}\left[\tilde{F}(\omega)\right]^2 \varphi'(\omega)\, d\omega\right\}^2 \tag{4.28}$$

where σ^2 denotes the variance of the unmodulated pulse (see for example [41, 42] for a derivation). Equation (4.28) illustrates that the group delay function $\varphi'(\omega)$ determines the pulse duration and provides a powerful tool to calculate the statistical pulse duration even in cases where the pulse shape function is unknown.

4.2.1.5 Examples of Spectral Phase Functions

We now consider some examples of spectral phase modulation (GDD, TOD, sinusoidal modulation and combinations thereof) which are relevant to the experiments on control of electron dynamics with shaped femtosecond laser pulses presented in this chapter.

Polynomial Spectral Phase Modulation Functions

A simple approach to understand the physical significance of a spectral phase function $\varphi(\omega)$ with respect to the temporal pulse shape is based on its Taylor series resulting in a sum of polynomial phase functions

$$\varphi(\omega) = \phi_0 + \phi_1 \cdot \omega + \frac{\phi_2}{2!} \cdot \omega^2 + \frac{\phi_3}{3!} \cdot \omega^3 + \ldots . \tag{4.29}$$

The absolute phase, which relates the carrier oscillation to the envelope, is modulated if the first term ϕ_0 is non-zero. Although this type of modulation can be important for coherent control experiments [43] and plays an important role for experiments with Carrier-Envelope-Phase (CEP) stabilized femtosecond laser pulses [44–49], it does not influence the pulse envelope and is therefore not considered here. In accordance with the Fourier shift theorem, the linear term in the spectral modulation function $\phi_1 \cdot \omega$ is responsible for a time shift of the pulse envelope by $t = \phi_1$ with respect to the unmodulated pulse. However it should be noted that, unlike the case in interferometric setups, the envelope is shifted rather than a copy of the pulse implying a change of the CEP phase.

Second order spectral phase function

Quadratic phase modulation—also termed Group Delay Dispersion (GDD)—using the spectral phase function

$$\varphi_{GDD}(\omega) = \frac{\phi_2}{2!} \cdot \omega^2 \tag{4.30}$$

plays a major role in coherent control (see for example [50–56]). Often both the group delay dispersion function

$$GDD(\omega) = \frac{d^2}{d\omega^2}\varphi(\omega) \tag{4.31}$$

and the modulation parameter ϕ_2 are referred to as GDD. However, only in the special case of polynomial spectral phase modulation up to the second-order both quantities are identical. By applying $\varphi_{GDD}(\omega)$, the modulated spectrum reads

$$\tilde{\mathcal{E}}_{GDD}(\omega) = \tilde{\mathcal{E}}(\omega)\, e^{-i\varphi_{GDD}(\omega)}. \tag{4.32}$$

As a consequence, the pulse duration of the laser radiation is modified and a linear frequency sweep is introduced. Assuming a bandwidth limited Gaussian input pulse with a pulse duration Δt (see (4.9)) the chirped pulse remains Gaussian shaped and reads

$$\mathcal{E}_{GDD}(t) = \frac{\varepsilon_0}{2}\frac{1}{\gamma^{\frac{1}{4}}} e^{-\frac{t^2}{4\beta\gamma}} e^{i\left(\alpha t^2 - \epsilon\right)}, \tag{4.33}$$

where

$$\alpha = \frac{\phi_2}{8\beta^2\gamma},\ \beta = \frac{\Delta t^2}{8\ln 2},\ \gamma = 1 + \frac{\phi_2^2}{4\beta^2} \text{ and } \varepsilon = \frac{1}{2}\arctan\left(\frac{\phi_2}{2\beta}\right) \tag{4.34}$$

featuring an increased variance of

$$\sigma_{GDD}^2 = \sigma^2 + 2\ln(2)\left(\frac{\phi_2}{\Delta t}\right)^2 = \sigma^2 + \delta_{GDD}^2, \tag{4.35}$$

where δ_{GDD}^2 describes the additional lengthening due to GDD. The frequency sweep is characterized by a linear change of the instantaneous detuning

$$\Delta(t) = 2\alpha t, \tag{4.36}$$

where the parameter α

$$\alpha = \left\{ 2\phi_2 + \frac{\Delta t^4}{8\phi_2 \left[\ln 2\right]^2} \right\}^{-1} \tag{4.37}$$

describes the chirp rate. A positive value of ϕ_2 leads to an "up-chirp" characterized by an increase of the instantaneous frequency, whereas a negative ϕ_2 implies a decrease of $\Delta(t)$, known as "down-chirp". With increasing chirp parameter ϕ_2, the pulse duration increases according to (4.35) accompanied by a reduction in intensity (see Fig. 4.1a). Equation (4.37) shows that the chirp rate is a non-monotonic function of the chirp parameter ϕ_2.

Third order spectral phase function

Third Order Dispersion (TOD) given by the spectral phase function

$$\varphi_{TOD}(\omega) = \frac{\phi_3}{3!} \cdot \omega^3 \tag{4.38}$$

results in an asymmetric pulse shape and is analytically described by a damped Airy function [30, 57]. The analytic expression for the temporal profile reads

$$\mathcal{E}_{TOD}(t) = \frac{\varepsilon_0}{2} \sqrt{\frac{\pi}{2 \ln 2}} \frac{\Delta t}{|\vartheta|} e^{\frac{\ln 2}{2} \frac{\frac{2}{3}\tau - t}{\tau_{1/2}}} Ai\left(\frac{\tau - t}{\vartheta}\right) \tag{4.39}$$

$$\text{with} \ \ \vartheta = \sqrt[3]{\frac{|\phi_3|}{2}} \operatorname{sign}(\phi_3), \ \tau = \frac{\Delta t^4}{32\phi_3 (\ln 2)^2} \ \text{and} \ \tau_{1/2} = \frac{2\phi_3 (\ln 2)^2}{\Delta t^2}, \tag{4.40}$$

where $Ai(z)$ is the Airy function and the substitutions for ϑ, τ and $\tau_{1/2}$ are so chosen as to describe meaningful physically quantities in units of time. Equation (4.39) shows that the modulated temporal pulse shape $\mathcal{E}_{TOD}(t)$ is given by the product of an exponentially decaying function with a half-life time of $\tau_{1/2}$ and the Airy function shifted by a delay of τ and scaled by ϑ. Figure 4.1b shows an example of a pulse subjected to third order dispersion. TOD leads to a quadratic group delay $\varphi'(\omega) = \frac{\phi_3}{2}\omega^2$. Therefore, for a positive value of ϕ_3, the central frequency of the pulse arrives first, while frequency bands on either side of the central frequency arrive later. The two frequency bands which are symmetrically detuned cause beats in the temporal intensity profile explaining the oscillations after (or before if $\phi_3 < 0$) the main pulse. The beating is also responsible for the phase jumps of π which occur at the zeros of the Airy function. Most of the relevant properties of TOD modulation are determined by the parameter ϑ which is proportional to $\sqrt[3]{|\phi_3|}$. The ratio of $\vartheta/\Delta t$ determines whether the pulse is significantly modulated. As a rule of thumb, only if $|\vartheta/\Delta t| \gg 1$ a series of sub-pulses and phase jumps are observed. Also, the time shift of the most intense sub-pulse with respect to the unmodulated pulse and the FWHM of the sub-pulses are in the order of ϑ. The highly asymmetric TOD pulse is a good example to see that the FWHM is not a representative quantity. Instead, the statistical definition of the pulse duration (4.28) yields a formula similar to (4.35) for the chirp to describe the duration of the TOD modulated pulse

$$\sigma_{TOD}^2 = \sigma^2 + 2[\ln(2)]^2 \left(\frac{\phi_3}{\Delta t^2}\right)^2 = \sigma^2 + \delta_{TOD}^2. \tag{4.41}$$

For the applications in materials processing (see Sect. 4.4) the remarkable features of TOD are (i) temporal symmetry-breaking of the envelope implying control on the time-dependent energy flux onto the sample and (ii) the ability to produce a short intense pulse accompanied by a weak long pulse train. It has been demonstrated that this type of modulation is well suited to control basic ionization processes to manipulate the ablation dynamics [20, 58, 59].

Combined second and third order modulation

An example for the application of combined GDD and TOD modulation along with a derivation of the analytical expression for the field envelope $\mathcal{E}_{GDDTOD}(t)$ can be found in [32]. Here we report the result of the derivation

$$\mathcal{E}_{GDDTOD}(t) = \frac{\varepsilon_0}{2}\frac{\Delta t}{|\vartheta|}\sqrt{\frac{\pi}{2\ln 2}}\, e^{\frac{8}{3\Omega^6\phi_3^2} - \frac{2t}{\Omega^2\phi_3}}\, Ai\left(\frac{\frac{2}{\Omega^4\phi_3} - t}{\vartheta}\right) \tag{4.42}$$

$$\text{with } \frac{1}{\Omega^2} = \frac{\ln 4}{\Delta\omega^2} + i\frac{\phi_2}{2} \tag{4.43}$$

and ϑ defined in (4.40). Equation (4.42) shows that the GDD/TOD-modulated field is again a product of an exponential function and an Airy function. However, since Ω is complex valued both the exponential function and the Airy function have in general complex valued arguments. If $\phi_2 \to 0$, i.e. in the case of pure TOD modulation, the modulated field converges to the exponentially damped Airy function described above. Likewise, the chirped pulse is reproduced if $\phi_3 \to 0$. Combined modulation by GDD and TOD results in complicated pulse shapes with an asymmetric temporal intensity profile as well as a temporally asymmetric instantaneous detuning (see Fig. 4.1c). Coherent control by combined GDD and TOD spectral modulation was reported in a study on the enhancement of the three-photon absorption probability of iodine [60]. More recently, GDD/TOD modulation was used to exert efficient and robust strong-field control on the population transfer in a sensitizer dye [32]. In addition, coherent control of the fluorescence of colloidal semiconductor nanocrystals with combined GDD and TOD modulation has recently been reported [61]. In that publication, it was demonstrated experimentally that GDD/TOD modulated pulses are a versatile spectroscopic tool to probe and to control multi-photon processes. The analytical analysis of the higher-order spectrum reported in [61] revealed that combined GDD and TOD modulation allows to scan the second-order spectrum with a preselected narrow bandwidth by varying ϕ_2 at a given value of ϕ_3. Despite the

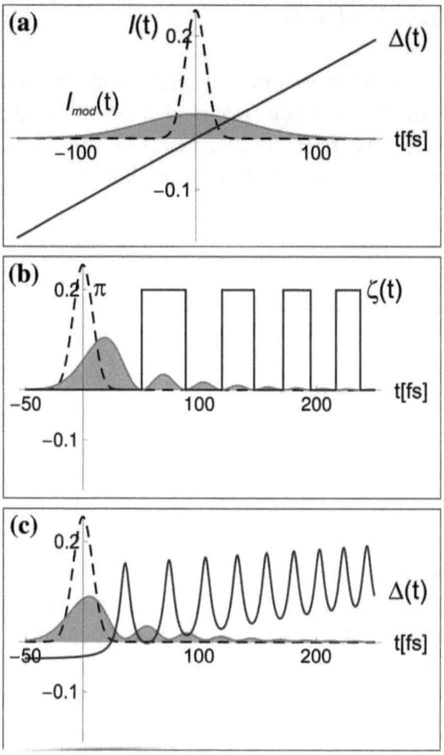

Fig. 4.1 Examples of a spectrally modulated 20 fs (intensity FWHM) pulse. For reference the intensity envelope $I(t)$ of the unmodulated pulse is depicted as *black dashed line*. The intensity profile I_{mod}(t) of the modulated pulse is shown in *red*. The *blue line* indicates the instantaneous detuning $\Delta(t) = \frac{d}{dt}\zeta(t) = \omega(t) - \omega_0$. **a** The second order spectral phase function $\varphi_{GDD}(\omega) = \frac{\phi_2}{2!}\omega^2$ with $\phi_2 = 750\,\text{fs}^2$ generates a chirped laser pulse with a temporally symmetric intensity distribution and linearly increasing instantaneous frequency. **b** Third Order Dispersion (TOD) $\varphi_{TOD}(\omega) = \frac{\phi_3}{3!}\omega^3$ with $\phi_3 = 20{,}000\,\text{fs}^3$ generates an asymmetric temporal intensity profile at the central frequency, i.e. $\Delta(t) \equiv 0$. Therefore, the temporal phase $\zeta(t)$ is plotted (*magenta*) rather than the instantaneous frequency. **c** Combination of both spectral modulation functions $\varphi(\omega) = \varphi_{GDD}(\omega) + \varphi_{TOD}(\omega)$ produces an asymmetric intensity profile with asymmetric instantaneous frequency evolution

complicated pulse structure, the pulse duration—calculated in frequency domain by (4.28)—is obtained by summation over the contributing additional terms of GDD and TOD suggesting that GDD and TOD modulation are somehow "independent"

$$\sigma^2_{GDDTOD} = \sigma^2 + 2\ln(2)\left(\frac{\phi_2}{\Delta t}\right)^2 + 2[\ln(2)]^2\left(\frac{\phi_3}{\Delta t^2}\right)^2 = \sigma^2 + \delta^2_{GDD} + \delta^2_{TOD}. \tag{4.44}$$

Periodic Spectral Phase Modulation

Periodic spectral phase modulation functions play an important role in coherent control. Numerous experiments and theoretical studies on coherent control by sinusoidal phase modulation have been carried out on atoms [38, 39, 43, 62, 63] and molecules [16, 17, 64–72]. Applying sinusoidal phase modulation

$$\varphi_{SIN}(\omega) = A \sin(\omega T + \phi) \tag{4.45}$$

to the spectrum generates the modulated field envelope

$$\mathcal{E}_{SIN}(t) = \sum_{n=-\infty}^{\infty} J_n(A)\, \mathcal{E}(t - n\,T)\, e^{-in\phi}, \tag{4.46}$$

where $\mathcal{E}(t)$ is the unmodulated field envelope, and J_n the Bessel function of the first kind and order n. Equation (4.46) shows that sinusoidal phase modulation in frequency domain produces a sequence of sub-pulses with a temporal separation determined by the parameter T and well defined relative temporal phases controlled by the absolute phase ϕ of the sine function. Provided the individual sub-pulses are temporally separated, i.e. $T \gg \Delta t$, the envelope of each sub-pulse is a scaled and shifted replica of the unmodulated pulse envelope. The amplitudes of the sub-pulses are controlled by the modulation parameter A via the Bessel function $J_n(A)$. If the temporal separation of the pulses is smaller than the pulse duration, i.e. $T \ll \Delta t$, the sub-pulses with different phases overlap giving rise to a complicated temporal profile and variations of the instantaneous frequency (see Fig. 4.2c). A more detailed description of the effect of sinusoidal phase modulation can be found in [39, 63]. Experiments on coherent control of ultrafast electron dynamics by intense sinusoidally modulated fields will be discussed in Sects. 4.3.3.4 and 4.3.3.5.

4.2.1.6 Polarization Shaping

A new degree of control became accessible with the ability to control the time-dependent polarization state of a femtosecond laser pulse in addition to its temporal structure [73]. The physical rational for the enhanced controllability is the manipulation of the vectorial properties of the light-matter interaction. Polarization shaping was for instance applied to coherent control of atomic and molecular dynamics [74–77], adaptive sub-wavelength control of nano-optical fields [78] and non-linear microscopy [79]. In addition, polarization-shaped laser pulses have been used to create polarization-shaped ultraviolet femtosecond pulses [80] and double pulse sequences with zeptosecond precision [35]. In general, any polarization state can be created by the superposition of two orthogonal linearly polarized fields with appropriate phases. Therefore, in order to describe polarization-shaped pulses, we

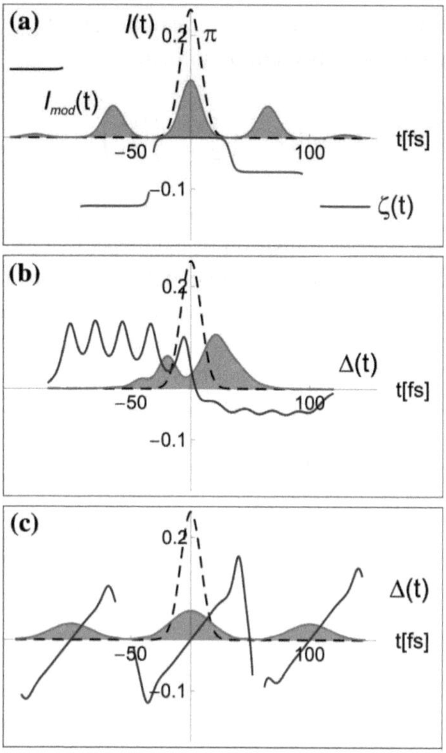

Fig. 4.2 Sinusoidal spectral phase modulation $\varphi_{SIN}(\omega) = A\sin(\omega T + \phi)$ produces pulse sequences with phase jumps between adjacent sub-pulses. **a** Sinusoidal phase modulation with the parameters $A = 1.2$, $T = 65$ fs, and $\phi = \pi/3$ leads to a sequence of separated pulses. Note that in (**a**) the temporal phase $\zeta(t)$ is plotted (*magenta*) rather than the instantaneous frequency. **b** Sinusoidal modulation with the parameters $A = 2.0$, $T = 20$ fs, and $\phi = \pi/3$ leads to overlapping sub-pulses. **c** Combination of sinusoidal and GDD modulation with the parameters $A = 1.2$, $T = 100$ fs, $\phi = \pi/4$ and $\phi_2 = 250\,\text{fs}^2$. This combination generates a sequence of chirped pulses. The frequency sweep of each sub-pulse is seen in the instantaneous detuning $\Delta(t)$

start by considering the time-dependent unmodulated field envelope perpendicular to the propagation direction

$$\vec{\mathcal{E}}(t) = \begin{pmatrix} \mathcal{E}_x(t) \\ \mathcal{E}_y(t) \end{pmatrix}. \tag{4.47}$$

Independent spectral phase modulation of two orthogonal components of the spectrum $\vec{\tilde{\mathcal{E}}}(\omega) = \left(\tilde{\mathcal{E}}_x(\omega), \tilde{\mathcal{E}}_y(\omega)\right)$ by the two phase functions $\varphi_x(\omega)$ and $\varphi_y(\omega)$ using

$$\begin{pmatrix} \tilde{\mathcal{E}}_{x,mod}(\omega) \\ \tilde{\mathcal{E}}_{y,mod}(\omega) \end{pmatrix} = \begin{pmatrix} \tilde{\mathcal{E}}_x(\omega)\, e^{-i\varphi_x(\omega)} \\ \tilde{\mathcal{E}}_y(\omega)\, e^{-i\varphi_y(\omega)} \end{pmatrix} \tag{4.48}$$

yields two linearly polarized modulated spectral light fields $\tilde{\mathcal{E}}_{x,mod}(\omega)$ and $\tilde{\mathcal{E}}_{y,mod}(\omega)$ resulting in the polarization-shaped temporal field

$$\vec{\mathcal{E}}_{mod}(t) = \begin{pmatrix} \mathcal{E}_{x,mod}(t) \\ \mathcal{E}_{y,mod}(t) \end{pmatrix}, \tag{4.49}$$

where the complex valued temporal envelope functions $\mathcal{E}_{x,mod}(t)$ and $\mathcal{E}_{y,mod}(t)$ are obtained by inverse Fourier transform of the respective modulated spectra [76]. Although both components of the time dependent envelope of the modulated field $\mathcal{E}_{x,mod}(t)$ and $\mathcal{E}_{y,mod}(t)$ fully describe the polarization-shaped femtosecond laser pulse, we introduce the instantaneous polarization state and the instantaneous frequency in order to better characterize its physical properties [73]. For a simplified notation we will momentarily suppress the index *mod* in (4.49) and use instead (4.47) to describe the modulated field as well. Similar to the linearly polarized case (4.5), the real valued fast oscillating field is obtained by

$$\vec{E}(t) = \begin{pmatrix} E_x(t) \\ E_y(t) \end{pmatrix} = Re[\vec{\mathcal{E}}(t)e^{i\omega_0 t}] = Re\left[\begin{pmatrix} \mathcal{E}_x(t) \\ \mathcal{E}_y(t) \end{pmatrix} e^{i\omega_0 t}\right]. \tag{4.50}$$

Instantaneous ellipticity

In order to characterize the instantaneous polarization state, we derive the parameters to define the instantaneous ellipticity. To this end, the slowly varying envelope function $\vec{\mathcal{E}}(t)$ is assumed to be constant over one optical period, thus the ellipticity within this time interval is well-defined. In order to determine the ellipticity parameters at a given time t_0 within the pulse, we start by considering the initial electric field amplitudes $A = |\mathcal{E}_x(t_0)|$ and $B = |\mathcal{E}_y(t_0)|$ with the respective temporal phases $\alpha = \mathrm{Arg}\,[\mathcal{E}_x(t_0)]$ and $\beta = \mathrm{Arg}\left[\mathcal{E}_y(t_0)\right]$. The ellipse is constructed by multiplication of the initial complex valued electric field vector

$$\vec{\mathcal{E}}(t_0) = \begin{pmatrix} \mathcal{E}_x(t_0) \\ \mathcal{E}_y(t_0) \end{pmatrix} = \begin{pmatrix} Ae^{i\,\alpha} \\ Be^{i\,\beta} \end{pmatrix} \tag{4.51}$$

with the phase factor $e^{i\chi}$, where χ sweeps through the interval of $(0, 2\pi)$. Accordingly, the real valued field reads

$$\vec{E}(\chi) = \begin{pmatrix} E_x(\chi) \\ E_y(\chi) \end{pmatrix} = Re\left[\begin{pmatrix} \mathcal{E}_x \\ \mathcal{E}_y \end{pmatrix} e^{i\,\chi}\right] = Re\left[\begin{pmatrix} Ae^{i\,\alpha} \\ Be^{i\,\beta} \end{pmatrix} e^{i\,\chi}\right] = \begin{pmatrix} A\cos(\chi+\alpha) \\ B\cos(\chi+\beta) \end{pmatrix} \tag{4.52}$$

and hence $\vec{E}(\chi)$ describes an ellipse around the origin (see Fig. 4.3). If $\alpha = \beta \pm n\pi$, the ellipse degenerates yielding linearly polarized light. Initially, the angle of the electric field vector $\vec{E}(0)$ is described by

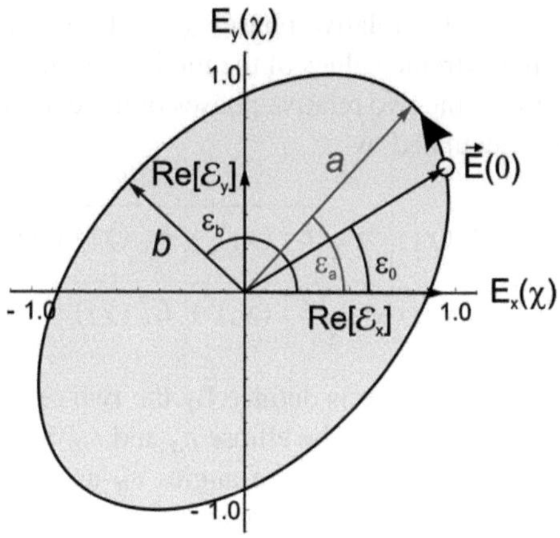

Fig. 4.3 Instantaneous polarization ellipse for one optical cycle. The ellipse is generated by the electric field vector $\vec{E}(\chi)$, where χ sweeps through the interval of $(0, 2\pi)$. Initially, the electric field vector starts at the position $(Re[\mathcal{E}_x], Re[\mathcal{E}_y])$ with an angle ε_0 and evolves in the direction of the *large arrow* on the ellipse. The instantaneous polarization state is characterized by the semi axes a and b and one of the corresponding angles ε_a and ε_b

$$\varepsilon_0 = \arctan\left(\frac{B\cos(\beta)}{A\cos(\alpha)}\right). \tag{4.53}$$

The phases χ_1 and χ_2 of the two semi axes relative to the initial phases α and β are given by

$$\chi_1 = \frac{1}{4i}\ln(z) \text{ and } \chi_2 = \chi_1 + \frac{\pi}{2}, \text{ with } z = \frac{A^2 e^{-2i\alpha} + B^2 e^{-2i\beta}}{A^2 e^{2i\alpha} + B^2 e^{2i\beta}} \tag{4.54}$$

being a complex number with the modulus of unity, i.e. $|z| = 1$. Therefore, the phases χ_1 and χ_2 are both real valued. Because $\chi_2 = \chi_1 + \frac{\pi}{2}$, it is sufficient to specify one of the two phases. The angle of both semi axes of the polarization ellipse is determined by adding the phases χ_1 and χ_2 to the initial phases α and β respectively, i.e. $\alpha + \chi_1$ and $\beta + \chi_1$ for one semi axis $\alpha + \chi_2$ and $\beta + \chi_2$ and for the other. The angles of both axes read explicitly

$$\varepsilon_a = \arctan\left(\frac{B\cos(\beta + \chi_1)}{A\cos(\alpha + \chi_1)}\right) \text{ and } \varepsilon_b = \arctan\left(\frac{B\cos(\beta + \chi_2)}{A\cos(\alpha + \chi_2)}\right). \tag{4.55}$$

The modulus of the vector $\vec{E}(\chi)$ describing the ellipse reads

$$\left|\vec{E}(\chi)\right| = \sqrt{E_x^2(\chi) + E_y^2(\chi)} = \sqrt{A^2\cos^2(\chi + \alpha) + B^2\cos^2(\chi + \beta)}. \tag{4.56}$$

A convenient way to derive the relative phases χ_1 and χ_2 of the two semi axes is based on determining the extreme values of the modulus of the field as a function of the phase χ. Equipped with the two relative phases of the extreme values, the length of the two semi axes is calculated by

$$a = \left|\vec{E}\,(\chi_1)\right| = \sqrt{E_x^2\,(\chi_1) + E_y^2\,(\chi_1)} \text{ and}$$
$$b = \left|\vec{E}\,(\chi_2)\right| = \sqrt{E_x^2\,(\chi_2) + E_y^2\,(\chi_2)}. \tag{4.57}$$

The instantaneous polarization state is defined by the two semi axes a and b along with the respective rotation angles of the ellipse ε_a and ε_b. Since the two semi axes of the ellipse are at right angle, one of the angles ε_a and ε_b is sufficient for the construction of the ellipse. Note that the mapping of a and b to the major and minor semi axis is not unique. However, the mapping is directly related to the initial field parameters by

$$c = a^2 - b^2 = A^2 cos\ [2(\chi_1 + \alpha)] + B^2 cos\ [2(\chi_1 + \beta)]\,. \tag{4.58}$$

If $c > 0$, then a is the major semi axes, otherwise, if $c < 0$, b is the major axis. The instantaneous polarization is circular if $c = 0$. In all figures showing polarization-shaped pulses, the ellipses with the two semi axes a and b rotated by the corresponding angle were used to represent the instantaneous ellipticity for the shaped pulses. The color coding for the ellipses is derived from the instantaneous frequency.

The above analysis yields explicit values for both semi-axes and the corresponding rotation angle. Alternatively, the vectors of both semi axes $\vec{a}$ and $\vec{b}$ can be derived employing *singular value decomposition* (SVD) of a matrix $\mathcal{M}$, which represents the ellipse defined in (4.52):

$$\vec{E}\,(\chi) = \begin{pmatrix} A\cos(\chi + \alpha) \\ B\cos(\chi + \beta) \end{pmatrix}.$$

Starting from the above representation we define the vectors

$$\vec{A} = \begin{pmatrix} A\cos\,(\alpha) \\ B\cos\,(\beta) \end{pmatrix} \tag{4.59a}$$

and

$$\vec{B} = \begin{pmatrix} A\sin\,(\alpha) \\ B\sin\,(\beta) \end{pmatrix} \tag{4.59b}$$

so as to obtain

$$\vec{E}(\chi) = \vec{A}\cos(\chi) - \vec{B}\sin(\chi) = \mathcal{M}\begin{pmatrix} \cos(\chi) \\ \sin(\chi) \end{pmatrix} \tag{4.60}$$

with the matrix

$$\mathcal{M} = \begin{pmatrix} A\cos(\alpha) & -A\sin(\alpha) \\ B\cos(\beta) & -B\sin(\beta) \end{pmatrix}. \tag{4.61}$$

Through SVD, the matrix $\mathcal{M}$ is decomposed into

$$\mathcal{M} = \mathcal{T}\Lambda\mathcal{S}^{\dagger} \tag{4.62}$$

where $\mathcal{T}$ and $\mathcal{S}^{\dagger}$ are unitary rotation matrices and Λ is a diagonal matrix containing the singular values of $\mathcal{M}$. The two semi axes $\vec{a}$ and $\vec{b}$ are then obtained by

$$\vec{a} = \mathcal{T}\Lambda\vec{e}_1 \tag{4.63a}$$

$$\vec{b} = \mathcal{T}\Lambda\vec{e}_2. \tag{4.63b}$$

Instantaneous frequency

In order to provide a unique color for each ellipse, we determine a single time dependent instantaneous frequency $\varpi(t)$ for the polarization-shaped pulse which is in general composed of two components with different time dependent instantaneous frequencies $\omega_x(t)$ and $\omega_y(t)$. We strive to obtain an expression for the instantaneous frequency which is derived without ambiguities from inverse trigonometric functions and which is physically plausible. To this end, we need to slightly generalize the ansatz made in (4.50) by writing

$$\vec{E}(t) = \begin{pmatrix} E_x(t) \\ E_y(t) \end{pmatrix} = Re\left[\begin{pmatrix} \mathcal{E}_x(t)\, e^{i\omega_x t} \\ \mathcal{E}_y(t)\, e^{i\omega_y t} \end{pmatrix}\right] \tag{4.64}$$

in order to allow for two different frequencies ω_x and ω_y. Here, we assume that the two frequencies ω_x and ω_y are time independent over one optical cycle and have already been determined from the complex fields, for example with the help of (4.6). By evaluating the real part in (4.64) and again assuming that A, B, α and β are constant over one optical period, we see that

$$\vec{E}(t) = \begin{pmatrix} A\cos(\omega_x t + \alpha) \\ B\cos\left(\omega_y t + \beta\right) \end{pmatrix} \tag{4.65}$$

and hence, we obtain for the modulus of $\vec{E}(t)$ similar to (4.56)

$$\left|\vec{E}(t)\right| = \sqrt{E_x^2(t) + E_y^2(t)} = \sqrt{A^2\cos^2(\omega_x t + \alpha) + B^2\cos^2\left(\omega_y t + \beta\right)}. \tag{4.66}$$

During one optical cycle the electric field vector sweeps twice through the apex giving rise to two maxima. Therefore, the time dependent modulus of the electric field oscillates with double the instantaneous frequency (see Fig. 4.4). The instantaneous frequency is determined by applying again (4.6) to the analytic signal corresponding to $|\vec{E}(t)|^2$

$$\frac{1}{4}A^2 e^{2i(\omega_x t+\alpha)} + \frac{1}{4}B^2 e^{2i(\omega_y t+\beta)} \tag{4.67}$$

and subsequent division by two. Ignoring the oscillatory terms, we obtain

$$\varpi(t) = \frac{A^4\omega_x(t) + B^4\omega_y(t)}{A^4 + B^4}, \tag{4.68}$$

i.e. the instantaneous frequency of the polarization-shaped laser pulse $\varpi(t)$ is the weighted average of the instantaneous frequencies $\omega_x(t)$ and $\omega_y(t)$ of the two electric field components where the weights are determined by the field amplitudes A^4 and B^4. Equation (4.68) displays the physically expected properties for the limiting values, for example $\varpi = \omega_x = \omega_y$ if both instantaneous frequencies are equal. If one component vanishes (say $B = 0$) the frequency is determined by the other component $\varpi = \omega_x$ and if both field components have the same amplitude $A = B$, the frequency is the arithmetic mean $\varpi = \frac{\omega_x+\omega_y}{2}$ of both frequencies.

Figure 4.4 shows the oscillations of $|\vec{E}(t)|$(thick black line) which are used to define the instantaneous frequency of a polarization-shaped femtosecond laser pulse. For clarity, ω_x and ω_y have been chosen time independent in this illustration. The comparison with the sinusoidal oscillation with the frequency $\varpi(t)$ shows good agreement. Because the field has two different instantaneous frequencies ω_x and ω_y, the field vector $\vec{E}(t)$ describes only approximately an ellipse during one optical

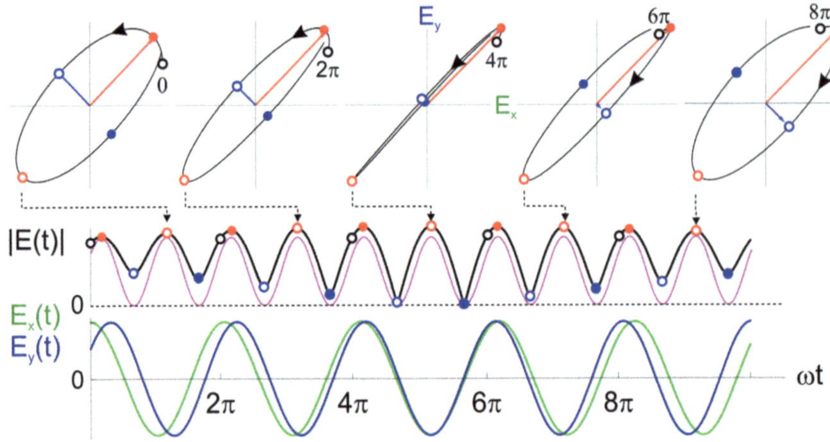

Fig. 4.4 The field $\vec{E}(t)$ (*upper panels*) with the two components $E_x(t)$ and $E_y(t)$ (*lower panel*) with different instantaneous frequencies ω_x and ω_y describes only approximate ellipses. The comparison between the oscillations of $|\vec{E}(t)|$ (*thick black line*) and the sinusoidal oscillation with a frequency of $\varpi(t)$ (*middle*) shows good agreement. *Red dots* indicate the positions of the major semi axis whereas the *blue dots* show the positions of the minor semi axis

cycle, i.e. the ellipses are "open". Due to the difference in frequency the semi axes change their length and orientation. The red dots in Fig. 4.4 indicate the positions of the major semi axis whereas the blue dots show the positions of the minor semi axis. Note that the field $|\vec{E}(t)|$ has two maxima and two minima for each oscillation as seen by comparison of $|\vec{E}(t)|$ with $E_x(t)$ and $E_y(t)$ in Fig. 4.4. Figure 4.5 shows a gallery of prototypical polarization-shaped laser pulses generated by the above procedure.

4.2.2 Experimental Implementation

Various experimental laser pulse shaping techniques are reviewed in the literature [9, 30, 31, 81–83]. Recently, precision pulse shaping down to the zeptosecond regime has been reported opening the route to control electron dynamics with unprecedented precision [35]. A spectral phase modulator generating quasi-arbitrary spectral phase functions $\varphi(\omega)$ can be experimentally implemented employing a $4f$-setup (see Fig. 4.6). In the setup shown in Fig. 4.6, the ultrashort laser pulse is spatially resolved in its spectral components by a grating and imaged onto a liquid crystal display (LCD) situated in the Fourier plane of the $4f$-setup. In the Fourier plane the spectral phase shaping is achieved by space-selective phase modulation via the LCD. Behind the LCD, the spectral components of the pulse are merged again by another grating. For highest possible throughput reflective optics are preferred. Especially the efficiency of gratings in terms of spectrally resolved reflectivity and/or transmittance of the laser radiation should be uniform over the laser spectrum. The LCD should feature a high damage threshold for optimal throughput and a large number of pixels (implying high spectral resolution) allowing for the production of complex shaped pulses ranging from the femtosecond to several tens of picoseconds. A design of a compact and robust spectral phase modulator was presented in [82] and extended towards a polarization shaper with reflective optics and transmission gratings [76].

4.2.3 Adaptive Optimization

Modern pulse shaping devices with typically 640 pixels for the modulation of both individual spectral polarization components make available a huge number of laser pulse shapes. Therefore, it is in practice very difficult to find an optimal laser pulse which guides the quantum system most efficiently from the initial state into a preselected final target state. If the relevant physical mechanisms are already identified, optimal light fields can be found by fine tuning the parameters of physically motivated pulse shapes [40, 64, 69, 84–86]. However, if the systems are too complex, adaptive feedback learning loops [87–93] provide a solution to this high dimensional search space problem. Genetic algorithms are often employed to carry out the optimization procedure [87, 88, 90, 91, 94]. In this iterative procedure the shape of the laser pulse is described by a set of parameters, similar to the genetic code of a biologic

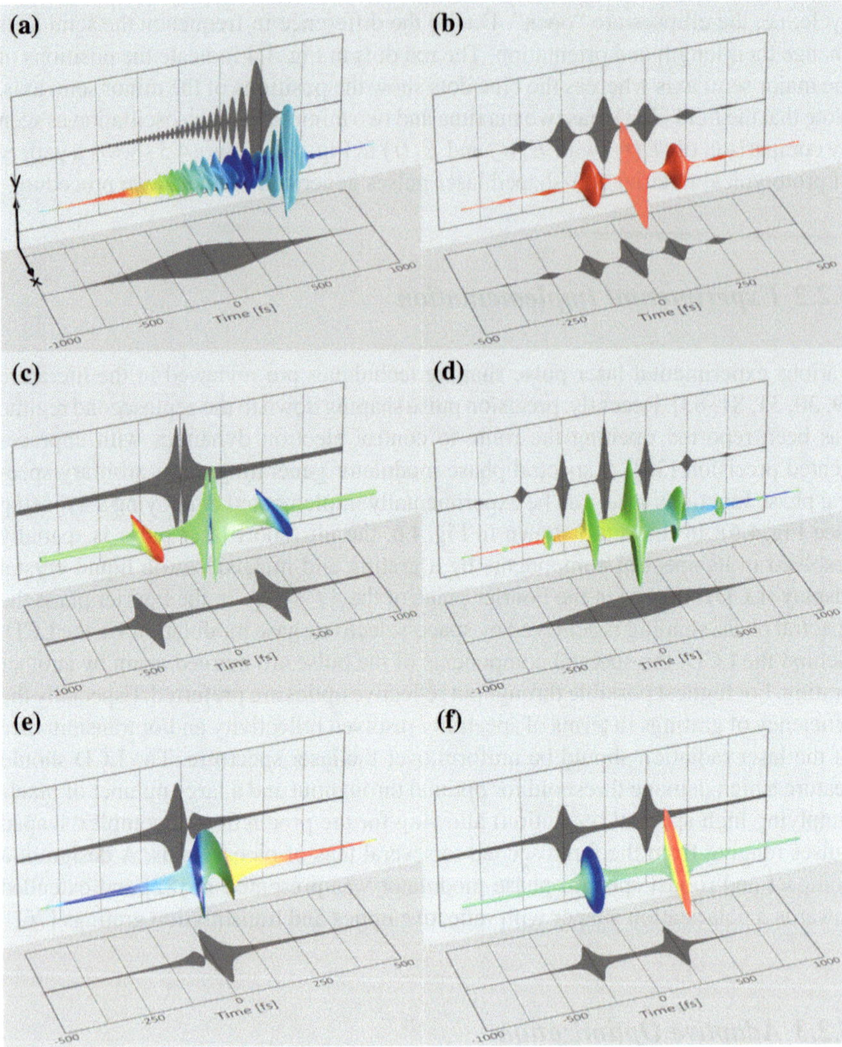

Fig. 4.5 Gallery of polarization-shaped laser pulses represented by ellipses with the semi axes a and b rotated by the respective angles ε_a and ε_b. The *color* represents the instantaneous frequency according to (4.68). The spectral phase functions are **a** $\varphi_x(\omega) = \frac{\phi_2}{2!}\omega^2$ and $\varphi_y(\omega) = \phi_1\omega + \frac{\phi_3}{3!}\omega^3$ with $\phi_1 = 500\,\text{fs}$, $\phi_2 = 5000\ \text{fs}^2$ and $\phi_3 = -32{,}000\,\text{fs}^3$, **b** $\varphi_x(\omega) = \sin(\omega T)$ and $\varphi_y(\omega) = \cos(\omega T)$ with $T = 125\,\text{fs}$, **c** $\varphi_x(\omega) = \tau|\omega|$ and $\varphi_y(\omega) = \frac{\pi}{2}\text{sign}(\omega)$ with $\tau = 375$ fs, **d** $\varphi_x(\omega) = \frac{\phi_2}{2!}\omega^2$ and $\varphi_y(\omega) = \sin(\omega T)$ with $\phi_2 = 5000\,\text{fs}^2$ and $T = 250\,\text{fs}$, **e** $\varphi_y(\omega) = -\frac{1}{4}\arctan\left(\frac{\omega+\delta\omega}{\Delta\omega}\right)$ and $\varphi_y(\omega) = -\frac{\pi}{2} + \frac{1}{4}\arctan\left(\frac{\omega-\delta\omega}{\Delta\omega}\right)$ with $\delta\omega = 1\ \text{mradfs}^{-1}$ and $\Delta\omega = 1.25\ \text{mradfs}^{-1}$, **f** $\varphi_x(\omega) = \tau\,|\omega - \delta\omega|$ and $\varphi_y(\omega) = \tau\,|\omega - \delta\omega| + \frac{\pi}{4}\text{sign}(\omega - \delta\omega)$ with $\tau = -250\,\text{fs}$ and $\delta\omega = 5\ \text{mradfs}^{-1}$

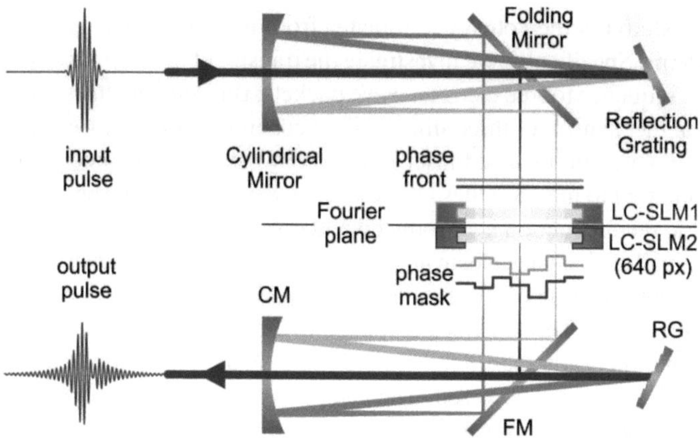

Fig. 4.6 Schematic setup of a compact and robust spectral phase modulator based on a Liquid Crystal Display (LCD) situated in the Fourier plane of a 4f zero dispersion compressor. The incoming laser beam is diffracted and spatially dispersed by the first reflection grating (RG). A Cylindrical Mirror (CM) images the spectral components into the Fourier plane, where each component is individually retarded by the different pixels of the LCD. The group delay which a certain pixel introduces is computer-controlled by the voltage applied to the pixel. The whole setup is built symmetrically with respect to the Fourier-plane, so that a second pair of CM and RG reconstructs the laser beam spatially. Interference of the phase-modulated spectral components in space and time yields a shaped temporal laser pulse at the output of the pulse shaper

individual. The fitness, i.e. some measure for the adequacy of the pulse shape with respect to a certain optimization target, is measured in the experiment. Via selection of the best individuals and creation of new pulse shapes being offspring of those best individuals, optimally adapted pulse shapes are found analogous to evolution in biology based on "survival of the fittest". Also other optimization procedures such as the covariance matrix adaption evolutionary strategy (CMA-ES) [95] have been implemented in coherent control [96].

4.3 Isolated Model Systems

In the first part of this review we focus on coherent control of Resonance Enhanced Multiphoton Ionization (REMPI) of isolated model systems in the gas phase using shaped femtosecond laser pulses. We discuss the interaction of well-defined shaped pulses with simple prototype systems such as atoms or dimers in order to allow for a detailed analysis of the underlying physical mechanisms by comparison of experimental results with simulations. In addition, on the femtosecond time scale, atoms or molecules in the gas phase are unaffected by perturbations of the environment such as de-coherence processes due to collisions or interactions with a solvent. Because the coherence of matter waves is the prerequisite for coherent control, we begin our analysis by studying the coherence transfer from light to matter waves on isolated model systems. Specifically, we investigate the transfer of the coherence properties of a two pulse sequence to free electron wave packets. Building on this simple example, the physical mechanism of three-dimensional coherent control of free electron wave packets by polarization-shaped laser pulses is investigated in some detail. A novel experimental tool to measure these three-dimensional photoelectron distributions by tomographic reconstruction is introduced and sample results including the measurement of so-called designer electron wave packets are shown. Then, two examples of adaptive optimization of ionization using polarization-shaped femtosecond laser pulses are presented. The first one demonstrates maximization of the ionization yield in potassium dimers whereas in the second example the projection of a designer free electron wave packet is optimized. The last part of this section is devoted to strong-field control of coherent electron dynamics in atoms and the coupled electron-nuclear dynamics in molecules. Intense laser pulses, entailing non-perturbative interactions of the light field with the atoms or molecules, open new physical mechanisms for efficient coherent control. Prominent strong-field control schemes are presented and discussed with regard to efficient switching of electronic excitation on the ultrafast timescale.

4.3.1 Coherence Transfer from Light to Matter

In order to better understand the physical mechanisms of coherent control and, in particular, coherent control of electron dynamics, it is essential to consider the quantum physical wave nature of the electron. In photo-ionization, for example, the wave picture describes the detached electron as a somehow localized free wave packet consisting of the superposition of many plane waves. Their amplitudes are manipulated by controlling the constructive and destructive quantum mechanical interference by coherent light sources such as ultrafast lasers. Therefore, the transfer of the coherence properties of light to matter waves is crucial to controlling ultrafast coherent dynamics of matter waves. Experimentally, coherence transfer from shaped laser pulses to matter waves was demonstrated on the interference of free electron wave packets by the photo-ionization of atoms with a coherent double pulse sequence [97]. Besides this fundamental aspect of photo-ionization with coherent radiation, interferences in the photoelectron spectrum have become an established tool to characterize ultrashort femtosecond and attosecond laser pulses [98–101].

4.3.1.1 Interference of Free Electron Wave Packets

In the experiments on the interference of free electron wave packets photoelectron kinetic energy spectra were measured to probe the coherence properties of the matter waves. To elucidate the physical picture of the creation, propagation and observation of a coherent free electron wave packet we first consider single pulse ionization and subsequently discuss the transfer of the coherence properties in a double pulse scenario.

Single pulse ionization

In the above experiment, potassium atoms have initially been prepared in the excited state by using the second harmonic ($2h\nu$) of an 800 nm femtosecond laser pulse in order to study one photon ionization (see energy level scheme in Fig. 4.7a). After the ionization from the excited $5p$ state with the fundamental laser radiation at time T_1, a localized electron wave packet with an average kinetic energy of $E_{kin} = 3h\nu - IP$ is launched from the interaction region, where IP denotes the ionization potential. The calculated time evolution of the wave packet is depicted in Fig. 4.7b. Due to the quadratic dispersion relation of matter waves in vacuum $E(k) = \frac{\hbar^2 k^2}{2m}$ the wave packet broadens during its further propagation (T_2 and T_3)—a textbook example of the dispersion of a quantum mechanical matter wave packet (see for example [102, 103]). It turns out that the wave packet evolves into its own momentum distribution [22, 99] as observed with a TOF (time-of-flight) photoelectron spectrometer.

Double pulse ionization

As depicted in Fig. 4.7c, d, the ionization with a coherent double pulse illustrates how a structured electron wave packet is created by the interference of the two partial wave packets. Since the alkali metal potassium possesses only a single outer electron, in the weak-field using infrared pulses at most one electron can be detached from a single atom during the light matter interaction. There is a certain probability amplitude for ionization by the first laser pulse and the same probability amplitude during the second pulse. The corresponding double peaked wave packet at the time T_1 is described by the probability density $|\psi(x, t = T_1)|^2$ shown in Fig. 4.7d. Due to the aforementioned dispersion of matter waves both partial wave packets spread in space and start to overlap. The interference of both partial wave packets—effectively the interference of the electron with itself- is responsible for the subsequent transient interference pattern shown at T_2. At some later time (T_3) a quasi stationary wave packet appears which still broadens in space but does not change its qualitative shape. Again, the wave packet evolves into its momentum distribution which consists of a series of equidistant interference fringes. Those interferences observed in the photoelectron time-of-flight spectrum reveal the transfer of coherence in the double pulse sequence to the electron waves.

4.3.1.2 Analogy to the Double Slit Experiment

The analogy of the interfering matter waves in coherent control to the interference of classical waves in the celebrated Young's double slit experiment has been pointed out already in the early days of coherent control by Brumer and Shapiro [104]. The close analogy of the two interfering free electron wave packets demonstrated in the above experiment to the Young's double-slit experiment is particularly enlightening. In this

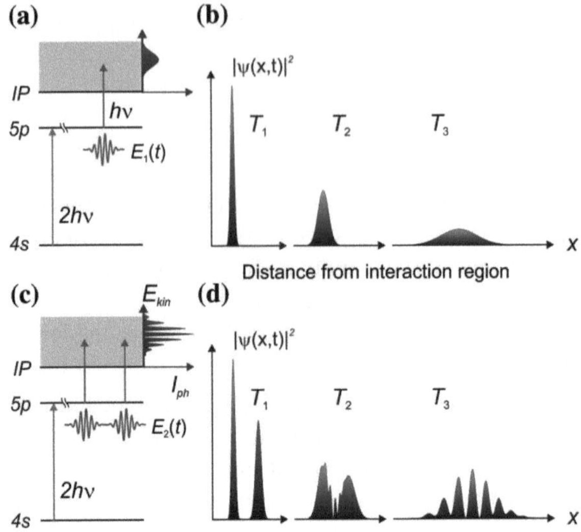

Fig. 4.7 In order to study the coherence transfer from light to matter waves in single photon ionization the initial $5p$ state is prepared by excitation with the second harmonic ($2h\nu$) of an infrared femtosecond laser pulse. **a** Level scheme for the ionization from the potassium $5p$ excited state with a single femtosecond laser pulse ($E_1(t)$) leads to a broad spectrum of photoelectron kinetic energies $E_{kin} = 3h\nu - IP$ (photoelectron spectra above the ionization potential IP). **b** During the time evolution the wave packet broadens in coordinate space as shown for the subsequent time steps $T_1 < T_2 < T_3$. **c** Ionization scheme for double pulse ionization. **d** Due to the coherent light matter interaction, the ionization by a coherent double pulse ($E_2(t)$) leads to an interference of the two coherent matter waves in the continuum. At T_1 both electron wave packets are separated whereas transient interference structures appear at T_2. The long term evolution ($t > T_3$) gives rise to a stationary interference pattern that is observed in the photoelectron spectrum

spirit, the interference of the free electron wave packets is regarded as a "Young's double-slit in the time domain" [97, 98] because the visibility of the interference pattern is complementary to the "which way information" on the photo-ionization process, i.e. the information whether the ionization took place during the first or the second laser pulse. By analyzing the interaction of an intense two pulse sequence in the dressed states basis (see also Sect. 4.3.3 for a discussion of the dressed states), the double slit analogy was extended to strong field ionization [105].

4.3.2 Control by Polarization-Shaped Laser Pulses

By controlling the vectorial properties of the light-matter interaction, polarization shaping gives access to the three-dimensional temporal response of atomic and molecular systems [75]. In this section we start by discussing the physical mechanism of the creation of shaped free electron wave packets by multiphoton ionization with elliptically polarized pulses. This example serves as a prototype for three-dimensional control with polarization-shaped laser pulses [22, 106] which is described in a subsequent section.

4.3.2.1 Physical Mechanisms

In the double pulse experiment, the momentum distribution was shaped by the interference of the two ionization pathways provided by the two pulses [97, 99]. In general, much more versatile manipulation of a quantum system may be attained by controlling the interference of multiple quantum pathways. In order to illustrate the interference of multiple quantum pathways, we consider the 1+2 REMPI of potassium atoms with polarization-shaped 800 nm femtosecond laser pulses. For the analysis, we decompose the REMPI process into an initial step, i.e. control of the amplitudes of superposition states during the interaction via multiple quantum pathways with a time-dependent ellipticity of the pulse (see Sect. 4.2.1.6 on polarization shaping), and the subsequent interference of multiple partial waves during their propagation in the continuum.

Polarization control of multiple pathways

We start by considering the bound and free quantum states characterized by the angular momentum quantum numbers $l = 0, 1, 2, 3 = s, p, d, f$ and $m = -l, ..., l$ ignoring the spin for simplicity [76, 107]. For the absorption of the three required light quanta in a 1+2 REMPI process, three different l-quantum pathways (ladders) are allowed due to the Δl selection rules, i.e. $s \rightarrow p \rightarrow d \rightarrow f$, $s \rightarrow p \rightarrow d \rightarrow p$ and $s \rightarrow p \rightarrow s \rightarrow p$ (see Fig. 4.8). As a consequence, the final continuum states consist of superposition states with f and p character. In multiphoton ionization with linearly polarized laser pulses the $\Delta m = 0$ selection rule applies in addition and therefore the initial $m = 0$ state is exclusively connected with final $m = 0$ states (not shown). In contrast, multiphoton ionization using elliptically polarized light—being a superposition of left and right handed circularly polarized light fields entailing the selection rules $\Delta m = \pm 1$—creates superposition states in the continuum with different values of m. For example, three photon ionization of potassium atoms with elliptically polarized light creates superposition states consisting of four f states with $m = -3, -1, 1, 3$ and two p states with $m = -1, 1$ (see Fig. 4.8). Linear combinations of those states with different amplitudes and phases include for example all electron wave packets which are rotated by an angle θ about the z-axis (see left panels in Fig. 4.10 for linear polarization) and, in addition, the non-cylinder symmetric states (right panels in Fig. 4.10 for elliptical polarization) [76]. Because the absorption of each elliptically polarized photon alters the quantum number by $\Delta m = \pm 1$ there are 18 discernible quantum pathways available to excite a superposition state. Using the notation $l(m)$ for a state with the quantum numbers l and m we can write down the involved pathways explicitly: there are eight pathways for the $s \rightarrow p \rightarrow d \rightarrow f$ ladder, i.e. $s(0) \rightarrow p(-1, 1) \rightarrow d(-2, 0, 2) \rightarrow f(-3, -1, 1, 3)$, six for the $s \rightarrow p \rightarrow d \rightarrow p$ ladder, i.e. $s(0) \rightarrow p(-1, 1) \rightarrow d(-2, 0, 2) \rightarrow$

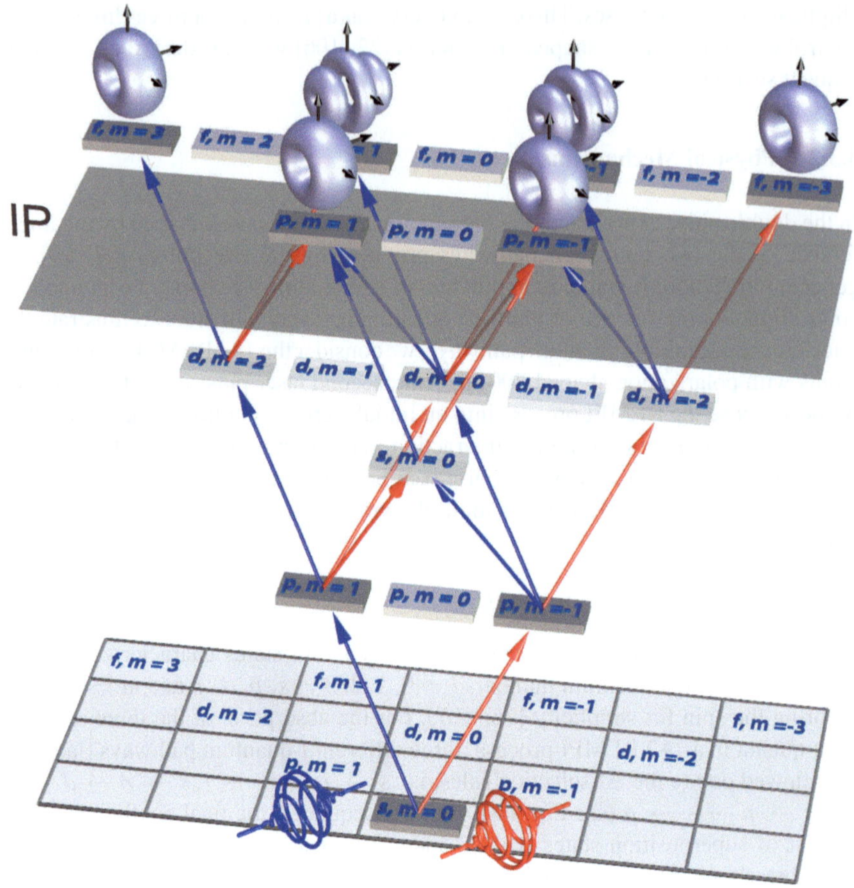

Fig. 4.8 Allowed pathways for multiphoton ionization with elliptically (i.e. a superposition of left- and right-handed circularly) polarized light. The six specific accessible states $f(-3, -1, 1, 3)$ and $p(-1, 1)$ above the ionization potential (IP) interfere to create the designer electron wave packets. Their contributions are determined by the interplay between the time-dependent instantaneous polarization of the laser pulse and the amplitudes and phases of the transition matrix elements. For simplicity, the electron spin is not considered

$p(-1, 1)$ and additional four pathways $s(0) \rightarrow p(-1, 1) \rightarrow s(0) \rightarrow p(-1, 1)$ for the $s \rightarrow p \rightarrow s \rightarrow p$ ladder. The probability for a specific transition is controlled by the corresponding matrix elements and the time-dependent ellipticity of the laser polarization. The amplitudes and phases of a single continuum state (say for instance $f(m = -1)$) are determined by the interference of all allowed pathways (in this case three) for multiphoton ionization. The initial shape of the electron wave packets after the interaction with the pulse is controlled by the amplitudes and phases of all six contributing specific continuum states shown in Fig. 4.8.

Interference in the continuum

In addition to the above mentioned interference effects during the creation of the superposition states in the continuum, the propagation of the electron wave packet will alter the radial shape of the wave packet on its way to the detector. Analogous to the two interfering partial wave packets in the double pulse experiment shown in Fig. 4.7 many different partial wave packets generated by the shaped laser pulse will interfere during their coherent time evolution. For example, wave packets composed of different angular momentum states which are initially separated radially will mix during the propagation giving rise to the complex wave packets observed in the experiment (see Fig. 4.12).

Determination of transition matrix elements

In the above experiments the instantaneous ellipticity of the pulse (see Sect. 4.2.1.6) was used as a control parameter to manipulate the electron wave packet. For the theoretical description of this interaction one needs to have complete knowledge of the involved matrix elements. Because the interference of all those pathways is determined by the different amplitudes and phases of the matrix elements, we can turn the tables, and use the measurement of a shaped free electron wave packet to achieve a complete description of the photo-ionization process in terms of transition amplitudes and phases [108]. In a recent theoretical study incorporating the intra-pulse electronic dynamics a full set of ionization matrix elements for potassium atoms was extracted from extensive experimental data sets on the ionization of potassium atoms with elliptically polarized light [107]. In addition, in that work, three-dimensional photoelectron angular distributions (3dPADs) generated from the extracted matrix elements were compared to experimental, tomographically reconstructed 3dPADs. This technique provides a route to "complete" photo-ionization experiments. Most likely, the technique to determine matrix elements can be refined by employing polarization-shaped pulses which are specifically adapted to this purpose.

4.3.2.2 Tomographic Reconstruction

In this section we will portray a novel technique to measure the three-dimensional shape of free electron wave packets based on a combination of Velocity Map Imaging (VMI) and tomographic reconstruction. Numerous highly differential schemes for the detection of 3dPADs have been developed (see for instance [108] for a topical review) because they contain detailed information on the intra molecular dynamics and the photo-ionization process. A novel application of photoelectron angular distribution measurements is devoted to probing control of atomic and molecular dynamics [56, 109–112].

Velocity Map Imaging (VMI)

An efficient and yet relatively simple method to measure 3dPADs is to project the complete photoelectron distribution onto a two-dimensional detector plane. An

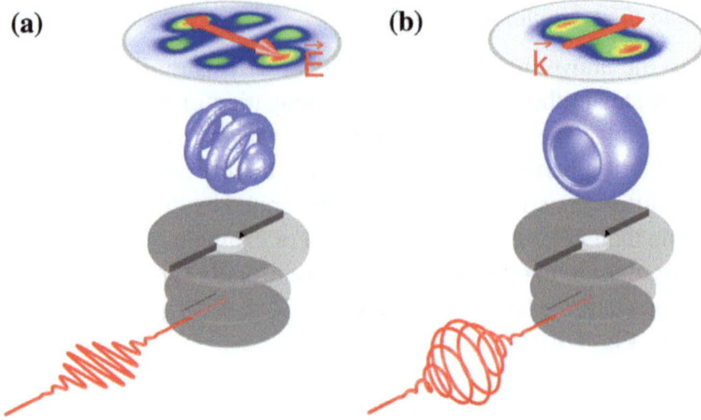

Fig. 4.9 A cylindrically symmetric photoelectron distribution can be reconstructed by the Abel-inversion provided its symmetry axis lies in the detector plane. **a** In case of a linearly polarized laser pulse the 3dPAD is cylindrically symmetric about the polarization vector $\vec{E}$ of the light field and hence $\vec{E}$ needs to be in the detector plane. **b** The PAD is cylindrically symmetric about the $\vec{k}$-vector if the light field is circularly polarized. For the Abel inversion, the $\vec{k}$-vector needs to be parallel to the detector plane

example for this technique is the Velocity Map Imaging (VMI) method [113, 114]. In this scheme photoelectrons created in the interaction region are accelerated in the electric field of a parallel-plate capacitor towards a two-dimensional MCP detector. The detector maps the transversal momentum distribution of the 3dPAD. In the quantum mechanical description, the classical motion of an electron in the spectrometer is replaced by the time evolution of a three-dimensional electron wave packet and thus the measurement of a 3dPAD on the two-dimensional MCP detector is described by the Abel transform of the 3dPAD [114, 115]. If the photoelectron distribution is cylindrically symmetric a single measurement of a two-dimensional distribution is sufficient to reconstruct the 3dPAD via Abel-inversion provided its symmetry axis lies in the detector plane. The two scenarios for which this condition applies are depicted in Fig. 4.9. For ionization with linearly polarized light it is required that the polarization vector $\vec{E}$ of the light field lies within the detector plane (Fig. 4.9a). By contrast, the symmetry axis of a wave packet created with circularly polarized light is the $\vec{k}$-vector. Therefore, in this case the $\vec{k}$-vector needs to be oriented parallel to the detector plane (Fig. 4.9b).

Electron wave packets arising from photo-ionization with linearly polarized light being rotated about the z-axis are rotated out of the $y-z$-plane (see left panels in Fig. 4.10 for linear polarization) and therefore, the Abel-inversion is no longer feasible. In addition, wave packets created with elliptically polarized light are no longer cylindrically symmetric (right panels in Fig. 4.10 for elliptical polarization) violating yet another prerequisite for the Abel-inversion. In general, polarization-shaped pulses will lead to complicated 3dPADs without the required symmetry properties necessitating a novel ansatz for reconstruction.

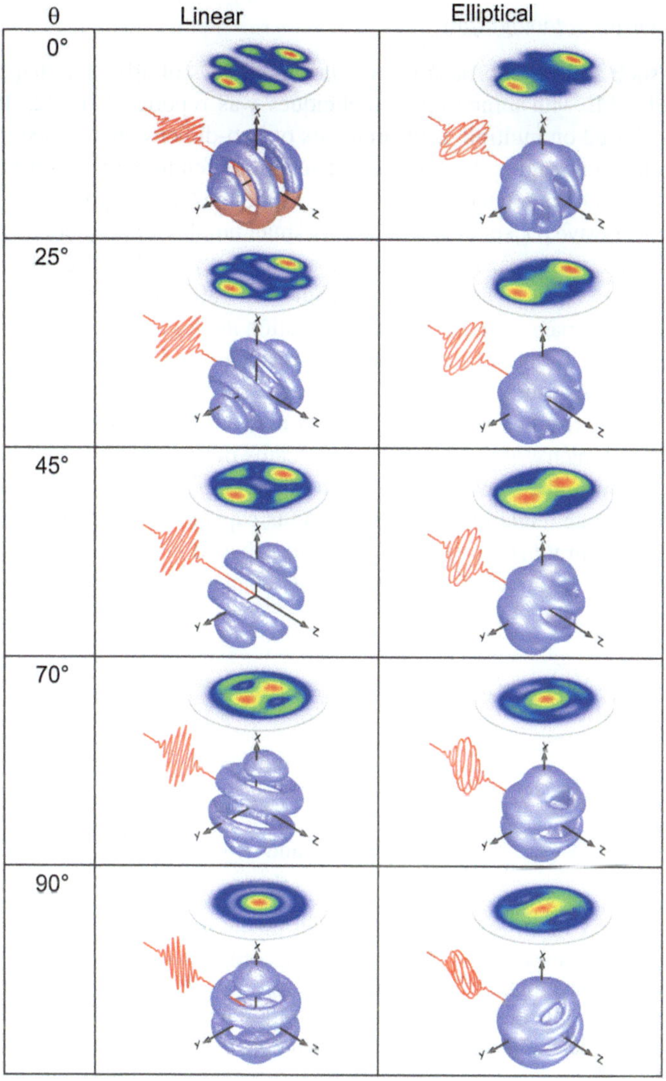

Fig. 4.10 Three-dimensional photoelectron angular distributions (3dPAD) are tomographically reconstructed from measured VMI images. The polarization vector of the ionizing femtosecond laser pulse is rotated about the z-axis by $\theta = 0°, 25°, 45°, 70°$ and $90°$ using a $\lambda/2$ plate. *Left* 3dPADs from three-photon ionization of potassium atoms with a linearly polarized femtosecond laser pulse creates an f-type electron wave packet with little p contribution. The tomographically reconstructed 3dPAD measured at $\theta = 0°$ is compared to the Abel-inverted result (*red surface*), showing excellent agreement [109]. *Right* 3dPADs from elliptical polarization

Three-dimensional tomography

Recently, such a novel approach to measuring 3dPADs of arbitrary shape by combination of VMI with tomographic techniques was reported [21, 22, 106, 109]. The idea is based on multiple measurements of two-dimensional projections of the 3dPAD at different angles. To this end, the incoming femtosecond laser pulse—and hence the 3dPAD—is rotated employing a $\lambda/2$ plate. Depending on the complexity of the electron wave packet only a relatively small number of projections need to be recorded for the reconstruction. In the experiments reported in [22, 109] typically 36 projections to cover an angle interval from 0° to 180° were measured. Some representative VMI images measured for the rotation angles $\theta = 0°, 25°, 45°, 70°$ and 90° are displayed in Fig. 4.10 along with the reconstructed 3dPAD.

Based on the measured data set a tomography algorithm [116] is used to reconstruct the 3dPAD from the measured projections. For the reconstruction no assumptions on the symmetry of the 3dPAD or on details of the ionization mechanism are required. Because a $\lambda/2$ wave plate at an angle of $\theta/2$ rotates any polarization-shaped pulse by an angle of θ this technique is capable of reconstructing electron wave packets from ionization with polarization-shaped laser pulses as well. To see this, we consider the effect of a $\lambda/2$ wave plate at an angle of $\theta/2$ on a polarization-shaped pulse described by (4.49)

$$J_{\lambda/2}(\theta)\begin{pmatrix}\mathcal{E}_{x,mod}(t)\\ \mathcal{E}_{y,mod}(t)\end{pmatrix} = R(\theta/2)\,J_{\lambda/2}R^{-1}(\theta/2)\begin{pmatrix}\mathcal{E}_{x,mod}(t)\\ \mathcal{E}_{y,mod}(t)\end{pmatrix} = R(\theta)\begin{pmatrix}\mathcal{E}_{x,mod}(t)\\ -\mathcal{E}_{y,mod}(t)\end{pmatrix}, \tag{4.69}$$

where

$$R(\theta) = \begin{pmatrix}\cos\theta & -\sin\theta\\ \sin\theta & \cos\theta\end{pmatrix} \text{ and } J_{\lambda/2} = \begin{pmatrix}1 & 0\\ 0 & -1\end{pmatrix} \tag{4.70}$$

denote the matrix of an active rotation about the angle θ and the Jones-matrix for a $\lambda/2$-wave plate, respectively. Note that the y-component changes it sign upon the insertion of the wave plate ($\mathcal{E}_{y,mod}(t) \rightarrow -\mathcal{E}_{y,mod}(t)$). However, since this change-of-sign applies to all rotation angles θ, the tomographic method is unaffected. By insertion of another $\lambda/2$-wave plate at a fixed angle of 0° the change-of-sign could be compensated

$$J(\lambda/2)\,R(\theta/2)\,J(\lambda/2)\,R^{-1}(\theta/2)\begin{pmatrix}\mathcal{E}_{x,mod}(t)\\ \mathcal{E}_{y,mod}(t)\end{pmatrix} = R(\theta)\begin{pmatrix}\mathcal{E}_{x,mod}(t)\\ \mathcal{E}_{y,mod}(t)\end{pmatrix}, \tag{4.71}$$

to yield the rotated modulated field. Figure 4.11 shows the effect of the rotation of a polarization-shaped laser pulse about the propagation axis. Note that in general, if the rotation angle is not a multiple of 90°, the projections look more complicated and thus the underlying spectral modulation function is not easily identified.

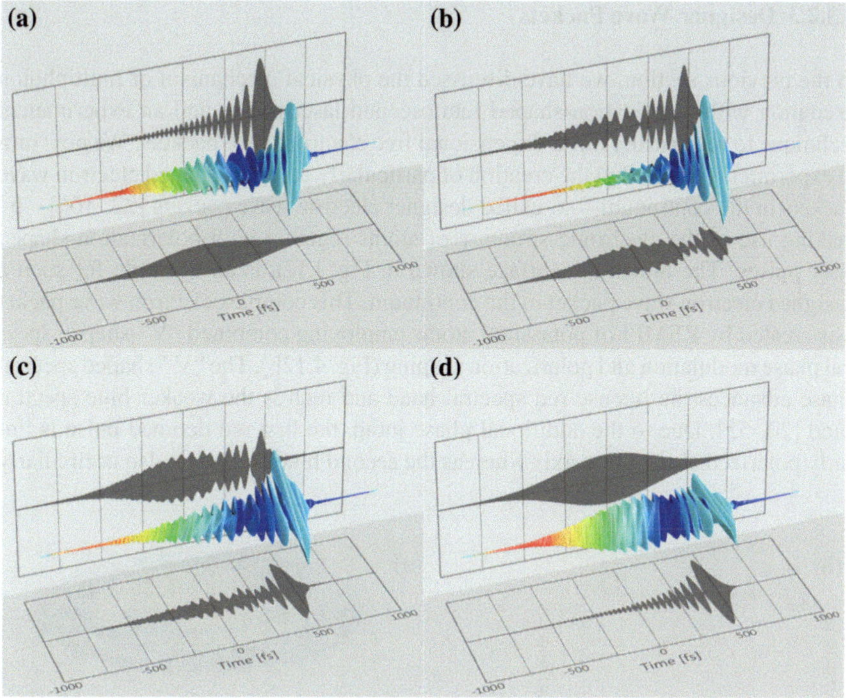

Fig. 4.11 Rotation of a polarization-shaped laser pulse created by GDD and TOD in the two different polarization directions ($\varphi_x(\omega) = \frac{\phi_2}{2!}\omega^2$ and $\varphi_y(\omega) = \phi_1\omega + \frac{\phi_3}{3!}\omega^3$ with $\phi_1 = 500\,\text{fs}$, $\phi_2 = 5000\,\text{fs}^2$ and $\phi_3 = -32,000\,\text{fs}^3$). The rotation angles about the z-axis are 0°, 30°, 60° and 90° from **a** to **d**. As described above, the rotation could be implemented experimentally by using a $\lambda/2$ plate. In general, rotation of the pulse leads to projections (*black shadows*) which appear more complicated

The electron wave packet generated by ionization with a linearly polarized laser pulse was used to validate this procedure [109]. The two iso-surfaces in the left upper panel of Fig. 4.10 at an angle of $\theta = 0°$ show a comparison of the Abel-inverted results (red iso-surface) with the full tomographic reconstruction (bluish iso-surface). The excellent agreement between both techniques confirms the validity of the approach. Lately, this tomographic technique for reconstruction of the three-dimensional momentum distribution has been implemented to study strong-field multiphoton ionization of argon [117], the alignment of naphthalene molecules [112] and the photoelectron angular distributions from rotationally state-selected ammonia [118].

4.3.2.3 Designer Wave Packets

In the previous section, we have discussed the physical mechanism of multiphoton excitation with polarization-shaped femtosecond laser pulses and an experimental technique to reconstruct three-dimensional free electron wave packets. We now turn to experiments devoted to the creation of particularly complex shaped electron wave packets in the continuum—so-called designer electron wave packets [22, 106]—by making use of the electronic structure of atoms together with polarization-shaped laser pulses. The bluish iso-surface shown in Fig. 4.12a is an example for such a designed electron wave packet in the continuum. This complex electron wave packet was created by REMPI of potassium atoms employing combined "V"-shaped spectral phase modulation and polarization shaping (Fig. 4.12b). The "V"-shaped spectral phase advances the intense red spectral band and retards the weaker blue spectral band [20, 55]. Due to the additional phase jump, the first red detuned pulse is linearly polarized along the x-axis whereas the second blue detuned pulse is circularly

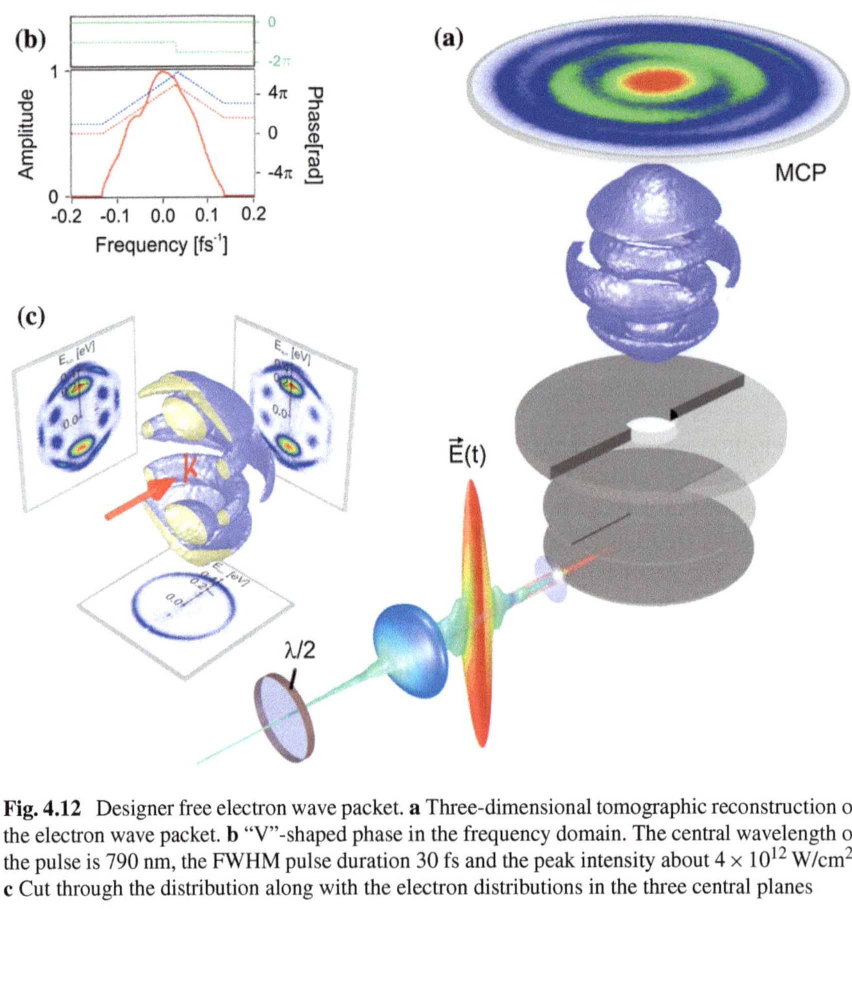

Fig. 4.12 Designer free electron wave packet. **a** Three-dimensional tomographic reconstruction of the electron wave packet. **b** "V"-shaped phase in the frequency domain. The central wavelength of the pulse is 790 nm, the FWHM pulse duration 30 fs and the peak intensity about 4×10^{12} W/cm^2. **c** Cut through the distribution along with the electron distributions in the three central planes

polarized. The cut through the distribution in Fig. 4.12c shows the interior of the complex three-dimensional distribution. In addition, the energy calibrated electron distributions in the three central planes are projected out.

4.3.2.4 Adaptive Optimization by Polarization Shaping

As discussed above, adaptive optimization is a powerful tool to find suitable laser pulses for coherent control. In most applications reported in the literature so far, linearly polarized femtosecond laser pulses were used for the optimization procedure. In this section we discuss two examples of coherent control by adaptive optimization of polarization-shaped laser pulses.

Diatomic Molecules

Soon after the introduction of polarization shaping by Brixner et al. [73] this technique was applied to coherent control of two-photon absorption in atomic rubidium by investigating the atom's fluorescence [74]. The maximization of the ion yield in the REMPI of potassium molecules was the first demonstration of adaptive optimization of molecular dynamics using polarization-shaped femtosecond laser pulses [75]. In that work it was established that polarization-shaped laser pulses increase the ionization yield beyond that obtained with an optimally shaped linearly polarized laser pulse. Figure 4.13a shows comparative optimizations of the K_2^+ yield proving that polarization shaping is superior to phase-only shaping. The inset to Fig. 4.13a shows an example of an optimal laser pulse—shaped in phase and polarization—along with the two orthogonal polarization components represented by black shadows. The physical rationalization of this observation is related to the adaptation of the vectorial electric field to the time evolution of the relevant vectorial transition dipole moments. In particular, the different multiphoton ionization pathways in K_2, involving parallel $\left(A^1\Sigma_u^+ \leftarrow X^1\Sigma_g^+\right)$ and perpendicular $\left(2^1\Pi_g \leftarrow A^1\Sigma_u^+\right)$ dipole transitions (Fig. 4.13b) favor different polarization directions of the exciting laser field during the interaction.

Designer Electron Wave Packets

The two-dimensional projections of the electron wave packets shown in Figs. 4.10 and 4.12 are quite varied and complex. However, recently it was shown experimentally that it is even possible to manipulate the three-dimensional electron wave packet such that an arbitrarily shaped target projection (which is compatible with energy conservation and the symmetry of the photo-ionization process) can be obtained by suitable optimization of the shaped ionization pulse [21]. To this end, the two-dimensional dumbbell-shaped target projection shown in Fig. 4.14 was defined (the boxed target image is labeled with T in Fig. 4.14). During the optimization the

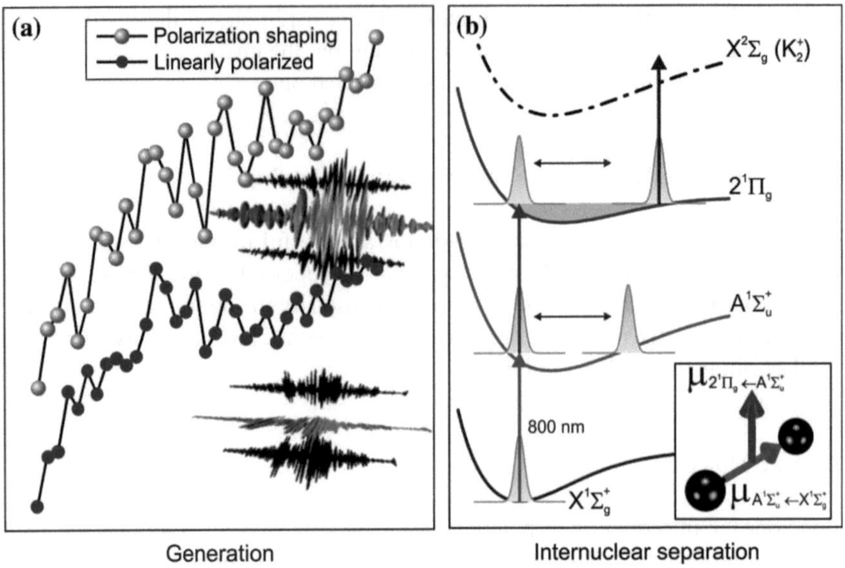

Fig. 4.13 **a** Comparative adaptive optimizations of the K_2^+ yield and an example of an optimal laser pulse—shaped in phase (*lower panel*) and in phase and polarization (*upper panel*). **b** The excitation scheme shows that different multiphoton ionization pathways involve parallel $\left(A^1\sum_u^+ \leftarrow X^1\sum_g^+\right)$ and perpendicular $\left(2^1\Pi_g \leftarrow A^1\sum_u^+\right)$ dipole transitions

target is iteratively approximated by adaption of the spectral phases of the laser pulse. The sequence (1–5) in Fig. 4.14 shows the evolution of the measured projections towards the target (T) during the optimization procedure. While the measurements of the first generation (1) have some similarity with the projection of an f-electron wave packet, see for example Fig. 4.10a for linear polarization, in (2–3) the outer maxima of the dumbbell start to emerge. In the course of the optimization procedure the inner part of the dumbbell is filled (4–5) such that eventually the measured projection converges to the target function. For the optimization, the spectral phase modulation function was parameterized by piecewise linear functions in order to delay different spectral bands with respect to each other and, in addition, a relative phase between both polarization components was applied in order to control the ellipticity of individual spectral bands.

4.3.3 Strong Field Control

The use of intense laser fields is a prerequisite for efficient atomic and molecular excitation. Strong-field interactions beyond the perturbative regime offer a rich variety of laser-induced excitations and enable new physical mechanisms to control the ensuing dynamics. In particular, strong fields alter the potential energy landscape

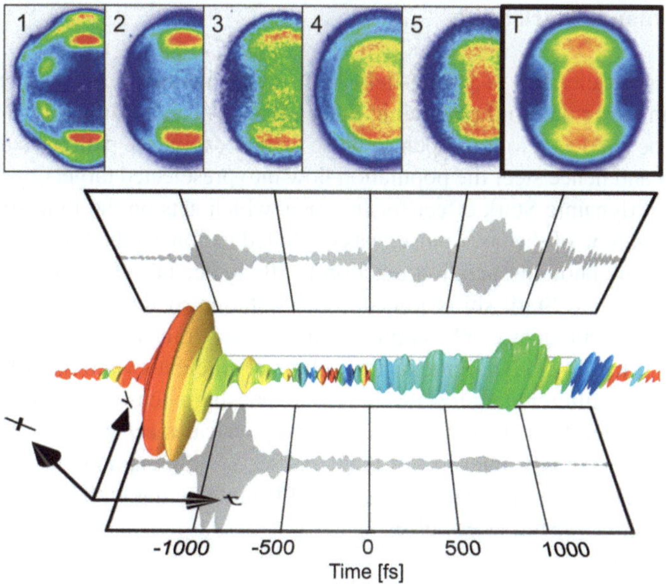

Fig. 4.14 Adaptive optimization of photoelectron angular distribution (PAD) measured by velocity map imaging (VMI) from REMPI of potassium atoms with polarization-shaped femtosecond laser pulses employing an evolutionary algorithm. The pulse parameters are: central wavelength of 790 nm, FWHM pulse duration of 30 fs and peak intensity of about 3×10^{11} W/cm^2

of an atom or a molecule by means of the dynamic (AC) Stark effect. The new light-induced potentials form the stage for non-perturbative quantum dynamics and open new excitation pathways to target channels which are inaccessible in weak laser fields. In this section we focus on strong-field control experiments on model systems which allow the analysis of the underlying physical mechanisms by comparison with simulations. Physical mechanisms to direct the strong-field induced quantum dynamics comprise—in addition to the interference of multiple excitation pathways discussed in Sect. 4.3.2—resonant and off-resonant Dynamic Stark Shifts (DSSs) [56, 119–121], Adiabatic Passage techniques such as RAP (Rapid Adiabatic Passage) and STIRAP (Stimulated Raman Adiabatic Passage) [122–124] and multi-pulse scenarios like Photon Locking (PL) [9, 105, 125–128] being the optical analogue of Spin Locking [129, 130] known from NMR. A common basis for all of these mechanisms is the Selective Population of Dressed States (SPODS) [67, 105]. In the following we will review these techniques and highlight their potential for efficient ultrafast switching of population transfer among different target states.

4.3.3.1 Dynamic Stark Shifts

A general characteristic of strong-field laser-matter interactions are energy shifts of quantum states due to the dynamic Stark effect. Precisely timed dynamic Stark shifts can be used as a tool to open or close certain excitation pathways during the interaction and hence steer the population flow into preselected target channels. The off-resonant dynamic Stark effect for instance, which acts on the time-scale of the intensity envelope of the laser pulse, was exploited to control the branching ratio of a non-adiabatic photodissociation reaction in IBr molecules [121]. In cases where undesired dynamic Stark shifts impede the excitation of the target state ultrafast pulse shaping techniques can be employed to counteract the effect and optimize the transition probability [120]. For example, ultrafast efficient switching of population transfer among different target states in Na atoms using intense chirped laser pulses was demonstrated [56]. It was shown that the physical mechanism governing the control was based on the interplay of RAP (cf. Sect. 4.3.3.3) and DSSs. Figure 4.15a shows the excitation and ionization scheme of Na interacting with intense $\Delta t = 30$ fs FWHM, $\lambda_0 = 795$ nm laser pulses.

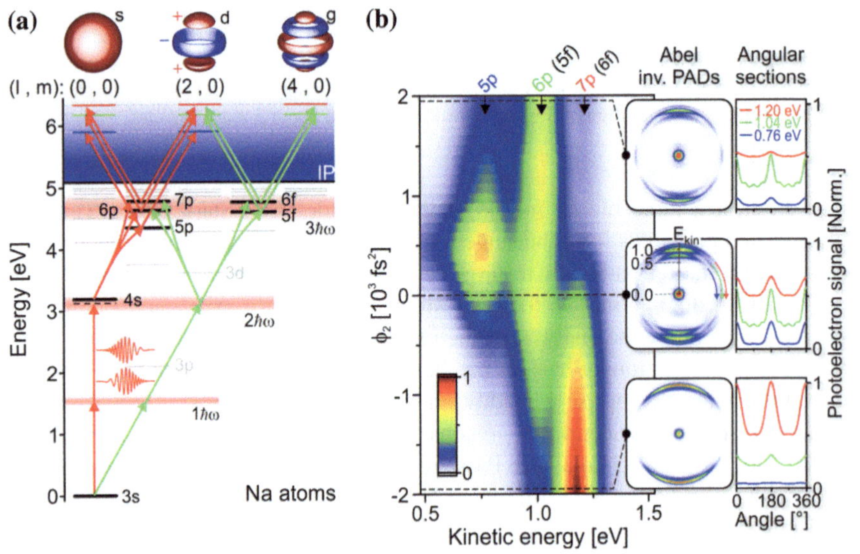

Fig. 4.15 **a** Scheme for simultaneous excitation and ionization of Na atoms by ultrashort chirped laser pulses. *Red horizontal bars* indicate the first, second and third order laser spectrum, respectively. *Red* and *green arrows* indicate the predominant REMPI channels. Photoelectron wave packets from these channels are characterized by distinct symmetries illustrated by the spatial electron density distributions on *top*. **b** Measured photoelectron spectra from strong-field REMPI of Na by intense chirped laser pulses. The *Insets* show Abel inverted PADs for distinct values of ϕ_2 and angular sections at three kinetic energies corresponding to the three main ionization channels around 0.76 eV (5p), 1.0 eV (6p/5f) and 1.2 eV (7p/6f) , respectively. The weak f-type contribution observed around 0.2 eV in the middle PAD has been discussed in terms of two-photon ionization from the transiently populated "virtual" state 3p [56]

Red and green arrows highlight the most relevant REMPI channels. The dominant pathway is (2+1+1) REMPI involving the strongly driven two-photon resonance $4s \leftarrow\leftarrow 3s$ and ionization via the high-lying target states $5p$, $6p$, and $7p$. Photoelectrons from this ionization channel are characterized by a combined s- and d-symmetry and kinetic energies of 0.76 eV ($5p$), 1.0 eV ($6p$) and 1.2 eV ($7p$), respectively. A competing pathway is the (3+1) REMPI process which proceeds via the high-lying target states $5f$ and $6f$. This ionization channel gives rise to photoelectrons wave packets with combined d- and g-symmetry and kinetic energies of 1.0 eV ($5f$) and 1.2 eV ($6f$). Since all excitations are driven coherently by the same laser pulse, photoelectron wave packets launched with the same kinetic energy interfere in the continuum. PADs measured in the experiment unambiguously reflect the symmetries of the photoelectron wave packets and thus provide direct information on the underlying ionization channel. Figure 4.15b shows photoelectron spectra from strong-field excitation and ionization of Na with linearly polarized chirped laser pulses (cf. (4.30) and see Fig. 4.1a) as a function of the spectral chirp parameter ϕ_2 ranging from -2000 to $+2000\,\text{fs}^2$. The intensity of the input pulse was $I_0 = 1 \times 10^{13}$ W/cm^2. For clarity all spectra are presented with energy-resolution. Selected spectra shown in the insets are presented with angular resolution in addition. The experiment demonstrates that a single control parameter, i.e. the chirp, can be sufficient to switch the atomic excitation between different target states. For the bandwidth limited pulse, i.e. $\phi_2 = 0$, three contributions appear in the spectrum with similar intensity. Due to its combined s- and d-symmetry the signal at 1.2 eV is assigned to ionization from target state $7p$. The signal at 1.0 eV exhibits a pronounced g-symmetry and thus originates from target state $5f$. The d-symmetry of the signal at 0.76 eV finally suggests ionization from target state $5p$. For large negative values of ϕ_2 the contribution from the p-channel at 1.2 eV is considerably enhanced whereas the 1.0 eV contribution is strongly suppressed and the 0.76 eV contribution has essentially vanished from the spectrum. This observation indicates the selective excitation of the target state $7p$ by down-chirped laser pulses. Upon inversion of the chirp however, the p-channel is largely suppressed. For large positive values of ϕ_2 the PAD is dominated by the contribution from the f-channel at 1.0 eV, indicating the selective excitation of the target state $5f$ by up-chirped laser pulses. A third prominent region opens up for small positive values of ϕ_2. Here the low energy contribution at 0.76 eV assigned to excitation of and ionization from target state $5p$ is most pronounced. This target state is completely non-resonant with the (third order) laser spectrum and therefore utterly inaccessible in the weak-field regime. Its efficient excitation requires a strong DSS resulting from the interaction with short and intense laser pulses, but is optimized by application of a moderate up-chirp.

In order to gain insights into the physical mechanisms behind the observed population control, quantum dynamics simulations were carried out by solving the Time-Dependent Schrödinger Equation (TDSE). For simplicity the Na atom was modelled by a five-state-system considering only the p-channel. A selection of numerical results is shown in Fig. 4.16a–c. The middle row shows the energy scheme of the five-state-system in a quantum mechanical frame rotating with the instantaneous laser frequency. In this frame, the eigenenergies of the two-photon resonant states

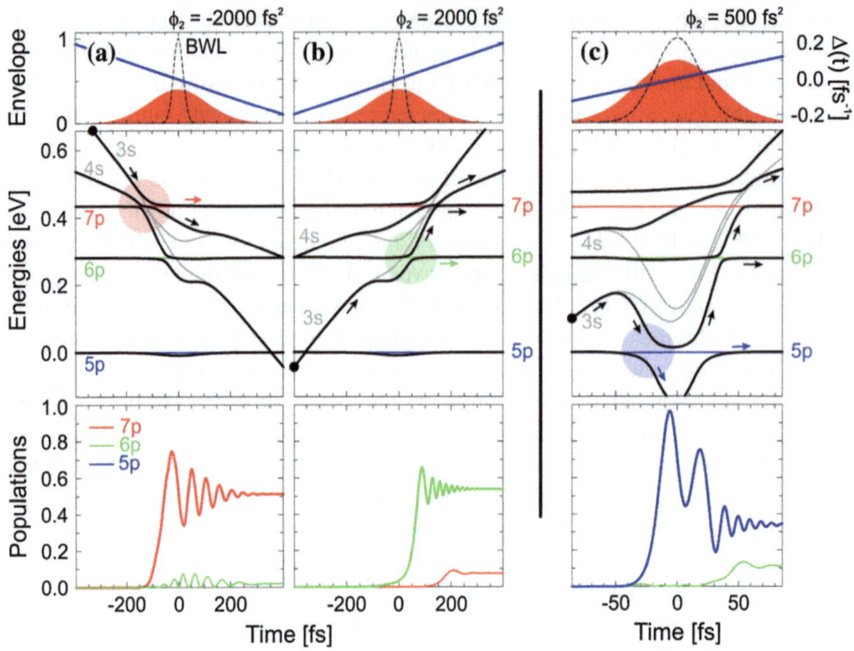

Fig. 4.16 Calculated neutral excitation dynamics of Na atoms interacting with intense chirped Gaussian-shaped laser pulses with chirp parameters of (**a**) $\phi_2 = -2000\text{fs}^2$, (**b**) $\phi_2 = 2000\text{fs}^2$ and (**c**) $\phi_2 = 500\text{fs}^2$. The laser pulses are shown in the top frames, decomposed into their temporal envelope $\mathcal{E}(t)$ (*red background*) and detuning $\Delta(t)$ (*blue line*). In the middle frames *grey* and *coloured lines* display the time-dependent bare states of the Na five-state-system. *Black lines* indicate the adiabatic or dressed atomic states resulting from diagonalization of the system. The *black dots* mark the respective initial conditions while arrows serve to trace the major population flow. The bottom frames show the population dynamics of the relevant target states under the different excitation conditions. Note the different time-axis in (**c**)

$3s$ and $4s$ (grey lines) are chirped due to both the linear detuning $\Delta(t)$ (cf. 4.36) of the pulse and pronounced DSSs induced by the non-resonant intermediate state $3p$. These time-variations give rise to various level crossings (dynamic resonances) with the target states $5p$, $6p$ and $7p$ during the course of the interaction. For example, in the case of down-chirped pulses with $\phi_2 = -2000\,\text{fs}^2$ shown in Fig. 4.16a a three-state bow-tie crossing between states $3s$, $4s$ and $7p$ around $t = -130\,\text{fs}$ (light-red circle) is responsible for the efficient excitation of the target state $7p$. Because the interaction is adiabatic (cf. Sect. 4.3.3.3) the interacting system follows the upper adiabatic state (upper black line) starting at the ground state $3s$ and reaching the upper target state $7p$ after the avoided crossing. For up-chirped pulses with $\phi_2 = +2000\,\text{fs}^2$ shown in Fig. 4.16b the energy scheme is time-reversed. In this case the avoided crossing between ground state $3s$ and target state $6p$ around $t = 50\,\text{fs}$ (light-green circle) is crucial to the excitation. Again adiabatic interaction determines the course of the population flow steering the system into the intermediate target state $6p$. In the third case, moderately up-chirped pulses with $\phi_2 = 500\,\text{fs}^2$ shown in Fig. 4.16c, the laser pulse is much shorter and more intense. This results in stronger and more rapid dynamic Stark chirps creating non-adiabatic, impulsive excitation conditions. The population flow is governed by a Stark chirp induced dynamic resonance between the dressed ground state $3s$ and the target state $5p$ around $t = -25\,\text{fs}$ (light-blue circle). Here adiabaticity is broken leading to an efficient population transfer among the adiabatic states, and eventually into the lower and far off-resonant target state $5p$.

In conclusion, the results presented on strong-field REMPI of Na atoms demonstrate that in strong-field control multiple pathways involving different physical mechanisms are generally at play simultaneously. Here an interplay of adiabatic passage and dynamic Stark shifts induced by a strongly driven two-photon resonance was identified to control the population transfer along different excitation pathways using the chirp as a single control parameter.

4.3.3.2 Selective Population of Dressed States

The essence of strong-field coherent control is the manipulation of dressed state energies and populations [67, 105]. In order to unravel the physical mechanisms underlying dressed state control we start by an analysis of the interaction of intense shaped femtosecond laser pulses with a simple well-defined model system. Two-state atoms serve as an approximation to suitably chosen real systems, which is particularly justified in the case of resonant excitation. The advantage of one-photon resonances—as compared to multi-photon resonances (cf. Sect. 4.3.3.1) or non-resonant excitation—are the strong laser-induced couplings enabling highly efficient excitation. Even at moderate laser intensities non-perturbative dynamics are readily achieved. Moreover, in the resonant case the dynamic Stark effect gives rise to energy splittings rather than unidirectional energy shifts. These so-called Autler-Townes (AT) splittings [131] allow controlled bipolar energy shifts in the order of several 100 meV [67]. As an example, we consider the excitation of K atoms by ultrashort near-infrared laser pulses, as sketched in Fig. 4.17. The laser resonantly couples the states $4s$ and $4p$, and simultaneously ionizes the atom in a non-resonant two-photon process. In the weak-field limit (a) this (1+2) REMPI gives rise to a single peak in the photoelectron spectrum mapping the energy and the population of the $4p$ bare state into the ionization continuum. Since the interaction is perturbative the ground state population remains essentially unaltered and the population transfer to the continuum is negligibly small. In the strong-field regime however, Rabi-cycling in the resonant two-state system causes a splitting of the photoelectron peak into the AT doublet. The origin of this splitting is readily understood in the dressed state picture. In the presence of an intense resonant laser field $\mathcal{E}(t)$ the dressed states of the strongly driven two-state atom repel each other. This repulsion results in an energy splitting of $\hbar\Omega(t) = \mu\mathcal{E}(t)$, where $\Omega(t)$ denotes the Rabi frequency and μ is the electric dipole moment of the transition $4p \leftarrow 4s$. Photoionization maps the *interacting* system, i.e. the dressed states, into the photoelectron spectrum [39]. As a result, two new

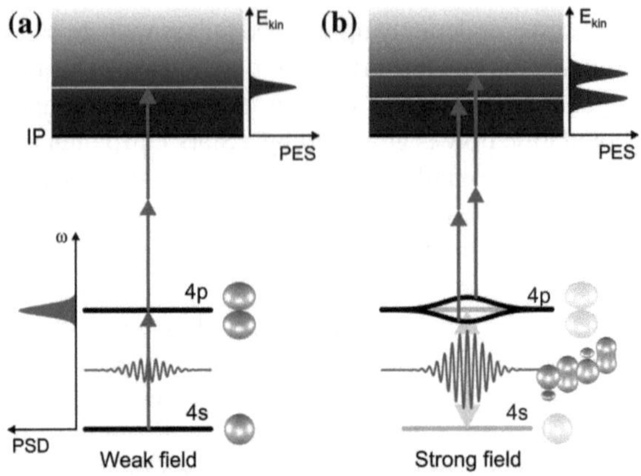

Fig. 4.17 Resonant excitation and simultaneous ionization of a two-state atom (K) by a bandwidth-limited laser pulse. **a** In the weak-field limit the atom is photoionized in a (1+2) REMPI process giving rise to a single peak in the energy-resolved photoelectron spectrum. **b** In the strong-field regime this peak splits into the Autler-Townes (AT) doublet due to Rabi-cycling in the strongly driven neutral system. Here, photoionization maps the dressed states rather than the bare states into the ionization continuum and acts as a sensitive probe for the dressed state energies (AT splitting) and populations (AT contrast)

channels open up in the continuum with energies corresponding to the dressed state energies, and a branching ratio determined by the dressed state populations. Hence, the photoelectrons act as a sensitive probe for the strong-field induced dynamics of dressed states in terms of both energy (AT splitting) and population (AT contrast).

While the manipulation of the dressed state energies is rather straight forward as their splitting follows the temporal envelope of the laser field $\mathcal{E}(t)$, controlling the populations of dressed states is somewhat more sophisticated. The semiclassical spatiotemporal description of the interaction of the atom with an intense resonant laser pulse provides an intuitive physical picture of dressed state population control. The rapidly oscillating laser field $E(t)$ couples to the valence electrons, distorts the spatial electron distribution and induces an oscillating electric dipole moment $\langle\mu\rangle(t)$. Under resonant excitation conditions the induced dipole follows the driving field with a phase shift of $-\frac{\pi}{2}$, in full analogy to a classical driven oscillator. Hence the interaction energy $\langle\mathcal{V}\rangle(t) = \langle\mu\rangle(t) \cdot E(t)$ of the dipole in the external driving field vanishes on the time average. Recalling that quantum mechanically the dressed states are the eigenstates of the interacting system, i.e. of the interaction operator $\boldsymbol{\mathcal{V}}(t) = -\boldsymbol{\mu}\cdot E(t)$, a vanishing interaction energy implies equal population of upper and lower dressed state (no selectivity). In order to accomplish the selective population of a single dressed state, the interaction energy needs to be either maximized (upper dressed state) or minimized (lower dressed state). This requires (i) the preparation of a dipole $\langle\mu\rangle(t)$ oscillating with maximum amplitude and (ii) suitable phase matching of the induced charge oscillation and the driving laser field $E(t)$. In-phase oscillation of laser and dipole leads to a minimization of $\langle \mathcal{V} \rangle(t)$, whereas out-of-phase oscillation of both maximizes $\langle \mathcal{V} \rangle(t)$.

In summary, the key to the control of dressed state populations is the interplay of the optical phase of the laser and the quantum mechanical phase of the induced charge oscillation prepared in a state of maximum coherence. The next sections present two different ways to realize these conditions in order to achieve the Selective Population of Dressed States (SPODS).

4.3.3.3 SPODS via Rapid Adiabatic Passage

In this section we follow up on Sect. 4.3.3.1 and discuss the resonant excitation of two-state atoms by intense chirped laser pulses. Again the K atom is analysed as a prototype of a two-state system interacting with chirped pulses [54]. Figure 4.18 shows energy-resolved photoelectron spectra measured by simultaneous excitation and ionization of K atoms with ultrashort near-infrared laser pulses ($\Delta t = 30\,\text{fs}$, $\lambda_0 = 795\,\text{nm}$, $I_0 = 4 \times 10^{11}\,\text{W/cm}^2$) chirped according to (4.30). The spectra show the AT doublet for different values of the spectral chirp parameter ϕ_2 ranging from -1600 to $1600\,\text{fs}^2$ (see also Fig. 4.1a). For the unmodulated pulse, i.e. $\phi_2 = 0$, the spectrum exhibits both AT components—the slow electrons at $E_{kin} = 0.30\,\text{eV}$ and the fast electrons at $E_{kin} = 0.45\,\text{eV}$—with nearly the same intensity. The resonant bandwidth-limited pulse populates both dressed states in equal measure, achieving no selectivity. For $\phi_2 > 0$ however, the asymmetry of the AT doublet shifts towards the low energetic component. Already at moderate chirp values of around $\phi_2 = 500\,\text{fs}^2$ the high energetic component has vanished completely from the spectrum, indicating the selective population of the lower dressed state by the up-chirped laser pulse. The overall decrease in photoelectron yield is due to the lowered peak intensity of chirped pulses, which impairs the two-photon ionization probability. For down-chirped pulses around $\phi_2 = -500\,\text{fs}^2$ the whole picture is inverted. Here the AT doublet is dominated by the fast photoelectrons whereas the slow electrons are completely suppressed. In this case the upper dressed state is populated selectively.

In order to analyze the physical mechanism the TDSE is solved numerically for a two-state atom interacting with intense resonant chirped laser pulses, including two-photon ionization from the $4p$ state [39]. Figure 4.19 shows calculation results for the case of an up-chirped laser pulse. The pulse is shown in frame (a). Frame (b) displays the population dynamics of the ground state $|s\rangle$ (dashed black line) and excited state $|p\rangle$(dashed-dotted grey line), along with those of the corresponding lower dressed state $|l\rangle$(solid blue line) and upper dressed state $|u\rangle$(solid green line). The induced dipole $\langle \mu \rangle(t)$, i.e. the expectation value of the dipole operator $\boldsymbol{\mu}$, is shown in frame (c) and compared to the driving electric field $E(t)$ (grey line), and finally frame (d) shows the expectation value $\langle \mathcal{V} \rangle(t)$ (black line) of the interaction operator $\mathcal{V}(t) = -\boldsymbol{\mu} \cdot E(t)$ together with the eigenenergies $\varepsilon_l(t)$ of the lower dressed state (blue line) and $\varepsilon_u(t)$ of the upper dressed state (green line). Snapshots of the spatiotemporal electron density dynamics $\varrho\,(\boldsymbol{r},\ t)$ are shown on top of the figure to visualize the relation of the induced electric dipole (blue arrows) and the driving electric field (red arrows).

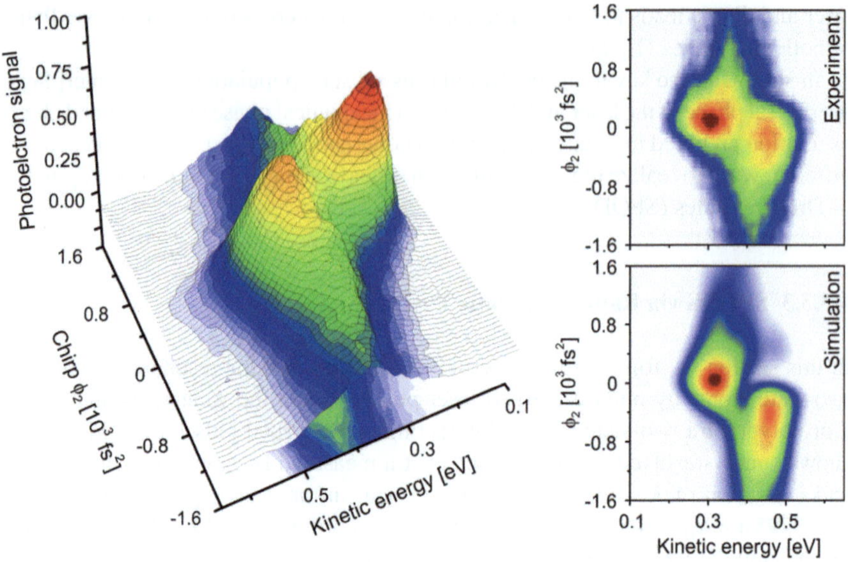

Fig. 4.18 Energy-resolved photoelectron spectra from simultaneous excitation and ionization of K atoms with intense resonant chirped femtosecond laser pulses. The spectra show the AT doublet as a function of spectral chirp parameter ϕ_2. For large positive chirp values (up-chirped pulses) exclusively slow photoelectrons are detected, which indicates the selective population of the lower dressed state in the neutral atom. Upon inversion of the chirp, the photoelectrons switch from the lower to the upper ionization channel. Down-chirped laser pulses promote the fast electrons, indicating the selective population of the upper dressed state

Initially the atom is in the ground state$|s\rangle$ characterized by a spherically symmetric -wave. The dipole moment of this electron distribution vanishes exactly and so does the interaction energy. Therefore, the dressed states are equally populated prior to the interaction. Excitation of the atom gradually launches the coherent charge oscillation which follows the driving field strictly in-phase due to the initial red-detuning of the up-chirped pulse with respect to the atomic resonance. Because the variation of the field in terms of amplitude and instantaneous frequency is sufficiently slow, the dipole follows the field adiabatically so that the initial phase relation is maintained throughout the entire interaction. Thus the phase matching condition is automatically fulfilled. As the dipole gains amplitude the interaction energy is successively lowered and the equal population of dressed states is lifted in favour of the lower dressed state. However, it is not before the dipole oscillation reaches maximum amplitude that full selectivity is attained. This condition is fulfilled around $t = 0$ where the bare states approach the state of maximum electronic coherence. In this time window the interaction energy $\langle \mathcal{V} \rangle (t)$ is minimized and its time average coincides exactly with the energy of the lower dressed state.

According to the semiclassical picture this indicates selective population of the lower dressed state which is in fact verified by the quantum mechanical population

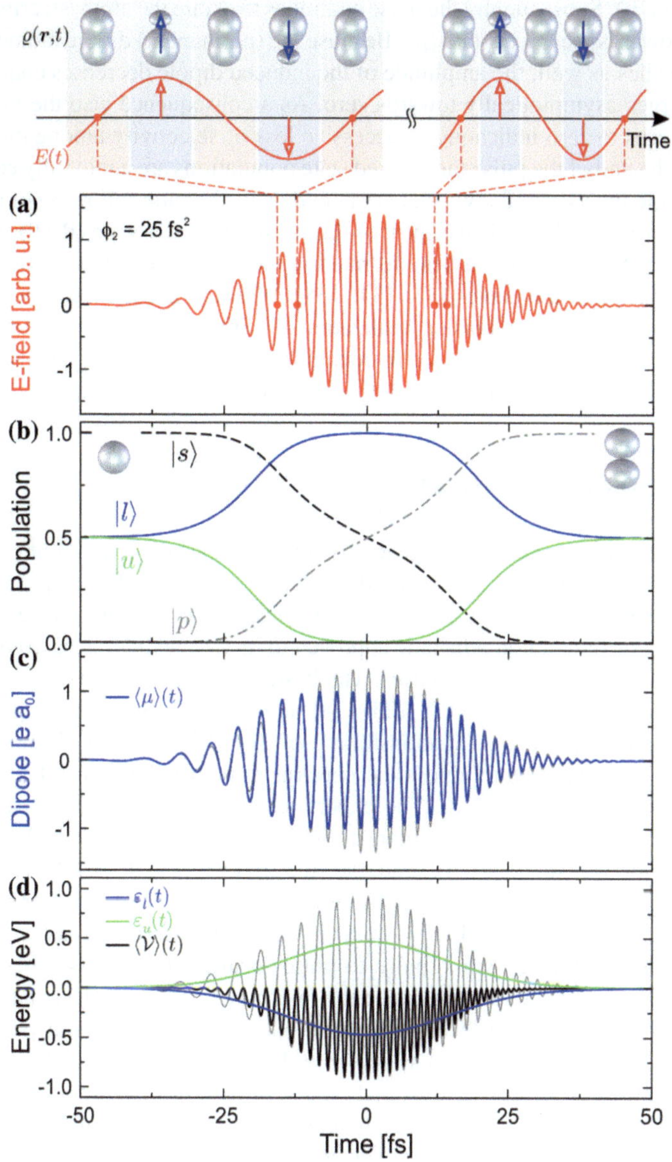

Fig. 4.19 Selective population of the lower dressed state via RAP using a chirped laser pulse. **a** Electric field $E(t)$ of the up-chirped pulse. **b** Calculated population dynamics of the bare states (*dashed lines*) and dressed states (*solid lines*) of the neutral two-state atom. **c** Induced electric dipole moment $\langle\mu\rangle(t)$ (*blue line*) driven by the external laser field (*grey line*). **d** Eigenenergies of the dressed states (*blue line* lower, *green line* upper) and interaction energy $\langle\mathcal{V}\rangle(t)$ of the induced dipole in the driving field (*black line*). The *top* of the figure shows snapshots of the spatiotemporal electron dynamics during selected cycles of the laser field. The coherent charge oscillation (*blue arrows*) is strictly in phase with the optical oscillation (*red arrows*), which is the basis for the selective population of the lower dressed state

dynamics in (b). Subsequently the pulse continues to excite the atom, steering it adiabatically towards the upper state $|p\rangle$. Because the (permanent) dipole moment of the -wave vanishes as well, the amplitude of the induced dipole decreases continuously and converges asymptotically towards zero. As a consequence also the interaction energy returns to zero indicating a successive loss of selectivity among the dressed states. By the end of the pulse the dressed state populations are again fully equalized. The adiabatic inversion of a two-state system due to adiabatic following of the field-induced charge oscillation is the spirit of Rapid Adiabatic Passage (RAP) originally developed in NMR [130] and demonstrated in the optical regime on atoms [9, 54, 123, 132, 133] and molecules [32, 134, 135]. With the phase matching condition 'built-in', RAP entails the transient realization of SPODS during the excitation stage of maximum electronic coherence. To conclude, driving quantum systems adiabatically is one distinct method to control the populations of dressed states.

4.3.3.4 SPODS Via Photon Locking

Adiabatic scenarios such as RAP generally rely on long interaction times and large pulse energies. In practice, it is often desirable to act on the fastest possible time scale in order to outperform intramolecular energy redistribution or decoherence processes. A different, non-adiabatic approach to realize SPODS is based on phase-locked sequences of ultrashort pulses. The basic mechanism behind this approach is Photon Locking (PL) [9, 105, 125–128] analogous to Spin Locking in NMR [129, 130]. An experimental demonstration of SPODS via PL on K atoms using multi-pulse sequences from sinusoidal spectral phase modulation (cf. (4.45) and Fig. 4.2a) is shown in Fig. 4.20. The experimental conditions are the same as in Sect. 4.3.3.3 except for the laser intensity which was set to 7×10^{11} W/cm^2. The sine-amplitude was set to $A = 0.3$ rad, the sine-frequency to $T = 120$ fs.

This parameter combination results essentially in a triple-pulse sequence of an intense main pulse preceded by a much weaker prepulse and succeeded by a likewise weak postpulse, with a pulse-to-pulse separation of 120 fs. The temporal phase of the pulse sequence, and hence the interplay between optical and quantum mechanical phase, is controlled by the sine-phase ϕ of the spectral phase function $\varphi_{SIN}(\omega) = A \sin(\omega T + \phi)$ as shown in Fig. 4.2a. Figure 4.20 shows energy-resolved photoelectron spectra as a function of ϕ varied over two optical periods. As ϕ scans through one optical cycle the photoelectrons switch back and forth between the two ionization channels. For instance at $\phi = \frac{\pi}{2}$ the AT doublet shows predominantly fast electrons at $E_{kin} = 0.50$ eV while slow electrons at $E_{kin} = 0.25$ eV are strongly suppressed.

Here the upper dressed state is populated selectively. In contrast, half an optical cycle later $\phi = \frac{3\pi}{2}$ at the AT doublet is inverted. Here the slow electrons dominate the spectrum while the fast electrons have essentially vanished, indicating selective population of the lower dressed state. The physical mechanism behind the impulsive realization of SPODS was discussed in terms of the bare states [63, 105, 136], the dressed states [63, 67, 105] and the Bloch vector [39]. Here we focus on the

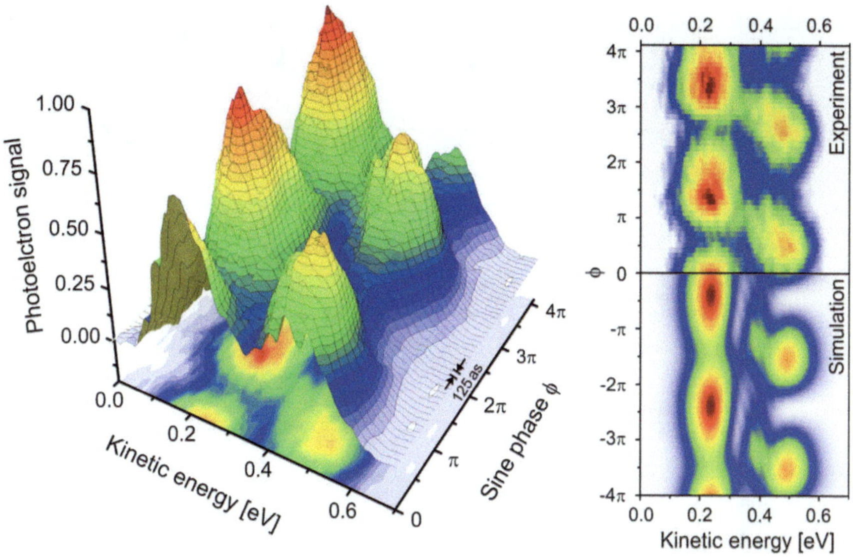

Fig. 4.20 Energy-resolved photoelectron spectra from simultaneous excitation and ionization of K atoms with intense resonant multi-pulse sequences from sinusoidal phase modulation. The spectra show the AT doublet as a function of the sine-phase ϕ which controls the temporal phases of the individual subpulses. As ϕ scans through one optical cycle (2π) the photoelectrons switch back and forth between the two ionization channels. At $\phi = \frac{\pi}{2}$ the spectrum is dominated by fast electrons, indicating the selective population of the upper dressed state. Selective population of the lower dressed state occurs at $\phi = \frac{3\pi}{2}$, i.e. half an optical cycle later, where predominantly slow electrons are detected. The physical mechanism behind the impulsive realization of SPODS via sequences of ultrashort pulses is found to be Photon Locking

interaction of a resonant double pulse sequence with the two-state atom. Figure 4.21 shows simulation results for this case. The assignment of quantities is the same as in Fig. 4.19. The pulse sequence in (a) consists of a weak prepulse followed by an intense main pulse. The prepulse has a pulse area of $\theta = \frac{\pi}{2}$ (see (4.19)) and therefore steers the atom from its groundstate $|s\rangle$ into a state of maximum electronic coherence (see (b)). The oscillating electron density is illustrated on the top left side of the figure. The coherent charge oscillation launched by the prepulse corresponds to a dipole moment $\langle\mu\rangle(t)$ oscillating with maximum amplitude (shown in (c)). Since the dipole is driven on resonance, it follows the external driving field with a phase shift of $-\frac{\pi}{2}$. As a result the interaction energy $\langle\mathcal{V}\rangle(t)$ displayed in (d) vanishes on the time average indicating equal population of the dressed states during the prepulse. Indeed the population dynamics in (b) confirm that no selectivity is obtained throughout the prepulse. The main pulse however, being phase-shifted by $\Delta\zeta = \frac{\pi}{2}$ with respect to the prepulse, oscillates anti-phase with the induced dipole. With its onset the interaction energy is maximized abruptly. Throughout the main pulse its time average coincides with the energy of the upper dressed state indicating the selective population of the upper dressed state. This is confirmed by the dressed state

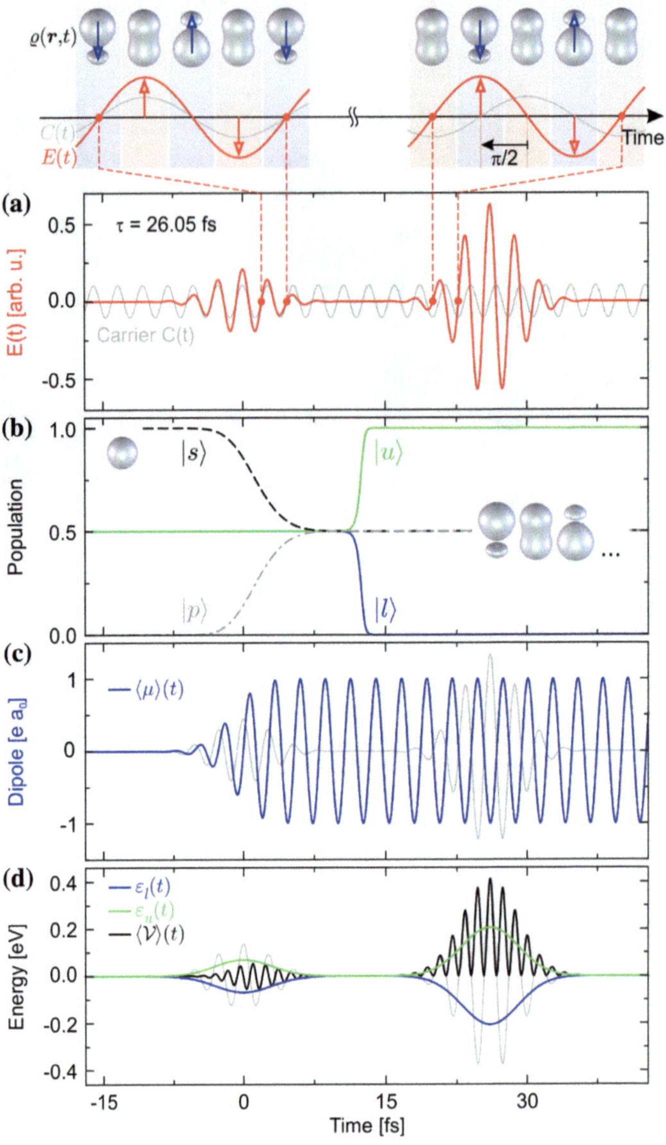

Fig. 4.21 Selective population of the upper dressed state via PL using a sequence of resonant laser pulses. The assignment of the displayed quantities is the same as in Fig. 4.19. The weak prepulse prepares the neutral atom in a state of maximum electronic coherence. Due to the resonant excitation, the induced charge oscillation follows the driving field with a phase shift of $-\frac{\pi}{2}$ as illustrated on the *top left side* of the figure. The intense main pulse, however, is phase-shifted by $\frac{\pi}{2}$ with respect to the prepulse and hence oscillates in anti-phase with the dipole. Therefore the interaction energy $\langle \mathcal{V} \rangle(t)$ is maximized during the main pulse, which is equivalent to the selective population of the *upper* dressed state. Note, that the bare state populations are frozen and the charge oscillation remains unaffected by the interaction with the intense resonant main pulse

population dynamics shown in (b). Remarkably, the bare state populations remain constant, i.e. locked, during the entire interaction with the main pulse—despite the presence of the strong resonant laser field. The phase configuration of field and dipole inhibits further excitation. Instead only the quantum mechanical phase of the two-state system is driven clockwise (not shown) resulting in the described enhancement of the interaction energy. On the other hand, shifting the main pulse by $\Delta\zeta = \frac{\pi}{2}$ with respect to the prepulse leads to in-phase oscillation of field and induced dipole. By this means the interaction energy is minimized and the lower dressed state is populated selectively. Because photoionization is triggered predominantly by the intense main pulse, only the selected dressed state is projected to the continuum resulting in the highly asymmetric AT doublets observed in Fig. 4.20. In conclusion, driving quantum systems by phase-locked sequences of ultrashort laser pulses with carefully adjusted relative phases provides a means for efficient and ultrafast switching between the dressed states.

4.3.3.5 Efficient Control of Concerted Electron-Nuclear Dynamics In Molecules

In this section we extend the principles of SPODS as discussed in the previous sections from atomic prototype systems to the strong-field control of coherent electron dynamics in molecules. The basic physical mechanism behind SPODS realized by shaped femtosecond laser pulses can be summarized as follows: The initial part of the shaped pulse excites a coherent charge oscillation of maximum amplitude which is exploited by a later part of the pulse to either increase or decrease the interaction energy by tailoring the optical phase to the phase of the induced dipole oscillation. In general, the situation in molecules is more complicated than in atoms, due to the coupling between electronic and nuclear degrees of freedom. Vibrational dynamics launched along with the electronic excitation affect both the amplitude and the phase of the induced charge oscillation. A maximum amplitude of the electric dipole oscillation, required for complete selectivity, is obtained only for a maximum overlap of the nuclear wave packets in the ground and excited state. The propagation of nuclear wave packets may thus impede the build-up of the dipole oscillation. Moreover, the wave packet propagation leads to a continuous variation of the electronic resonance, changing the frequency of the electric dipole. This results in an additional phase drift which the laser field has to adapt to in order to maintain a defined phase relation to the dipole. Therefore, simple pulse shapes such as doublepulse sequences or linearly chirped pulses are not expected to be optimal for the efficient control of coupled electron-nuclear dynamics in molecules.

In [69] complex shaped pulses from sinusoidal phase modulation were used for ultrafast switching of electronic excitation between different target channels in a neutral molecule K_2. In this work the potassium molecule was chosen as a model system because it can be treated on a highly accurate level both experimentally and theoretically. The ground state transition $A^1\Sigma_u^+ \leftarrow X^1\Sigma_g^+$ is near-resonant with the spectrum of typical infrared (IR) femtosecond lasers, and the molecule exhibits

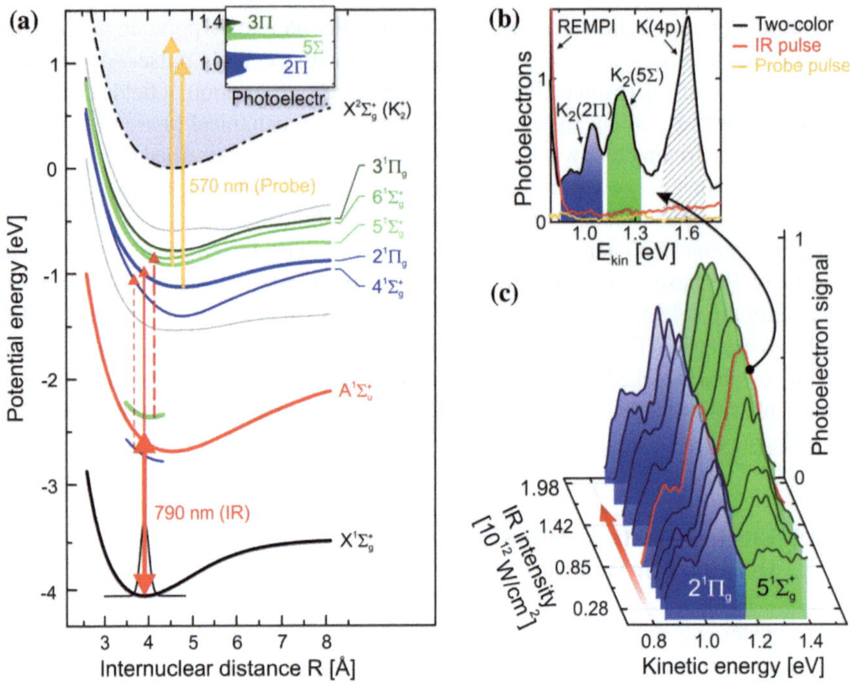

Fig. 4.22 Two-color excitation and ionization of K_2 molecules. **a** Potential energy scheme of K_2 excited by infrared (IR) pump pulses and post-ionized by visible probe pulses. The *short blue* and *green lines* indicate the dressed states of the X-A-subsystem, arising under strong-field excitation of K_2. **b** Two-color photoelectron spectrum recorded with the bandwidthlimited IR pulse. The choice of the probe wavelength ensures background-free detection of molecular signals from the target states $2^1\Pi_g$ and $5^1\Sigma_g^+$. (c) Intensity study of the neutral excitation by the bandwidth limited IR pulse. In the weak-field regime only the lower target state $2^1\Pi_g$ is excited. Above a threshold intensity of 0.5×10^{12} W/cm^2 the upper target state $5^1\Sigma_g^+$ becomes energetically accessible as well and is populated efficiently

a set of high-lying neutral target states for selective excitation. Figure 4.22a shows the potential energy scheme of K_2 excited by $\lambda_0 = 790$ nm laser pulses with an intensity FWHM of 25 fs, and photoionized by a second delayed laser pulse centered at $\lambda_p = 570$ nm. The probe pulse maps the final population of the target states $2^1\Pi_g$ and $5^1\Sigma_g^+$ into the photoelectron spectrum. Its central wavelength was chosen to ensure background free detection of the relevant two-color photoelectron signals (cf. Figure 4.22b). In the weak-field regime only the target state $2^1\Pi_g$ is excited by a resonant two-photon absorption process from the ground state $X^1\Sigma_g^+$ via the intermediate state $A^1\Sigma_u^+$. The corresponding two-color photoelectron spectrum is displayed at the front of Fig. 4.22c. It shows a weak contribution from the $2^1\Pi_g$ state around $E_{kin} = 1.0$ eV and almost no electrons from the $5^1\Sigma_g^+$ state. All spectra in Fig. 4.22c are recorded using a bandwidth-limited IR pulse. Only with increasing IR laser intensity the upper target state $5^1\Sigma_g^+$ becomes accessible as well due to the

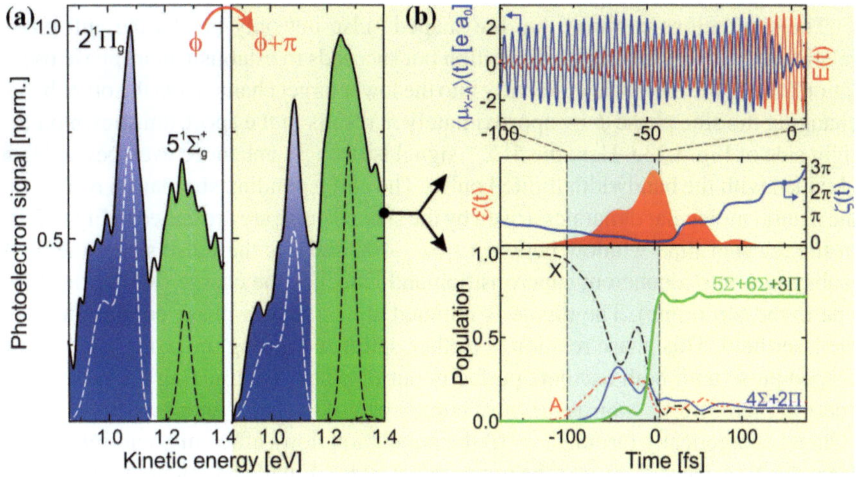

Fig. 4.23 Control of coupled electron-nuclear dynamics in K_2 by sinusoidally phase shaped laser pulses. **a** Two-color spectra indicating efficient and selective population of the lower target state $2^1\Pi_g$ (*left frame*) and upper target state $5^1\Sigma_g^+$ (*right frame*), respectively. The two cases differ only in the applied sine-phases ϕ, which changes by π from *left* to *right*. **b** Simulated quantum dynamics behind the right spectrum in (**a**). The *lower frame* shows the neutral population dynamics of the relevant molecular states. The *middle frame* shows the shaped laser pulse decomposed into its amplitude (*red background*) and phase (*blue line*). The *top frame* illustrates the induced dipole moment (*blue line*) and its phase relation to the real-valued, i.e. fast oscillating laser electric field during the relevant switching time window

increased dressed state energy splitting in the resonant X-A -subsystem. Hence the steep rise of the $5^1\Sigma_g^+$ -contribution around $E_{kin} = 1.25$ eV observed in Fig. 4.22c as the IR intensity grows. At the intensity $I_0 = 8.5 \times 10^{11}$ W/cm^2 the $5^1\Sigma_g^+$ signal already dominates the spectrum indicating more efficient population of the upper target state as compared to the lower target state $2^1\Pi_g$ by the unshaped pulse. The reason for this asymmetry is a slight blue-detuning of the IR laser spectrum with respect to the X-A-resonance. Due to this detuning the induced dipole tends to oscillate in anti-phase with the driving laser field already upon creation, favouring the upper dressed state and thus biasing the system towards the upper target channel. However, by specific tailoring of the IR excitation pulse using, e.g., sinusoidal spectral phase modulation (cf. (4.45)) the influence of the detuning can be compensated effectively. The two-color spectrum shown on the left side of Fig. 4.23a was recorded with the same pulse energy but making use of sinusoidal spectral phase modulation with $A = 0.8$ rad, $T = 45$ fs and $\phi = 1.8$ rad. The $2^1\Pi_g$ contribution is strongly enhanced at the expense of the $5^1\Sigma_g^+$ contribution, indicating the efficient and selective population of the $2^1\Pi_g$ state by the shaped IR pulse. The dashed lines represent calculated photoelectron spectra based on quantum dynamics simulations described in detail in [29].

The simulations confirm, that the shaped pulse not only breaks the anti-phase relation to the induced charge oscillation but succeeds to establish an in-phase oscillation to steer the molecule efficiently into the lower target channel. On the other hand, changing the sine phase ϕ by approximately π results in the spectrum shown on the right side of Fig. 4.23a. Here the $5^1\Sigma_g^+$ signal is strongly enhanced even beyond that obtained with the bandwidth limited pulse. The corresponding simulation results for the neutral molecular dynamics driven by the shaped pulse are presented in Fig. 4.23b. In the relevant time window between $t = -100$ and 0 fs the pulse steers the X-A -subsystem into a coherent superposition and launches the charge oscillation (blue line in the top frame). The dipole is initiated almost in anti-phase configuration to the laser field. This phase relation is further optimized during the rising edge of the shaped pulse until both oscillate perfectly out-of-phase, maximizing the interaction energy. Therefore, as soon as the most intense part of the pulse opens the upper target channel energetically (around $t = 0$) the population flows efficiently and selectively from the X-A-subsystem into the upper target states. Finally a total population yield of 75 % is obtained in the upper target channel while the lower target channel receives only 8 % of population. A detailed analysis of the vibrational dynamics during the build-up of the coherence and the switching of population into the upper target channel reveals an increase of the internuclear distance by 8 % which corresponds to a change of the Bohr frequency in the X-A -subsystem of 100 meV.

In conclusion, specific shaping of the temporal amplitude and phase of an intense ultrashort laser pulse enables efficient control of the coupled electron-nuclear dynamics in molecules. Tailoring the intricate interplay between the driven charge oscillation and the driving laser field, by adapting the optical phase to the induced charge dynamics and introducing directed energy shifts via the field amplitude, provides an avenue to steer the system selectively into predefined target channels that may even be completely inaccessible in the case of weak-field excitation.

4.4 Control of Ionization Processes in Dielectrics

In this last chapter we shortly highlight the extension of experimental control methodologies to ultrafast laser control of incoherent processes with an emphasis on processing of dielectrics on the nanometer scale. Here, primary processes induced by ultrafast laser radiation involve nonlinear electronic excitation where electron-electron collisions at high excitation densities (in the range 10^{21} cm^{-3} for ablation of dielectrics) destroy any coherence imprinted by the light field. In general, the electronic excitation is followed by energy transfer to the lattice and phase transitions that occur on fast (femtosecond to picosecond) but material dependent time scales [137, 138]. Optimal energy coupling with the help of suitably shaped temporal pulse envelopes gives thus the possibility to guide the material response towards user-designed directions, offering extended flexibility for quality material processing [20].

Prototype studies in the above mentioned spirit have been performed on dielectrics, water and metals. For dielectrics a microscope objective was used to obtain a focal

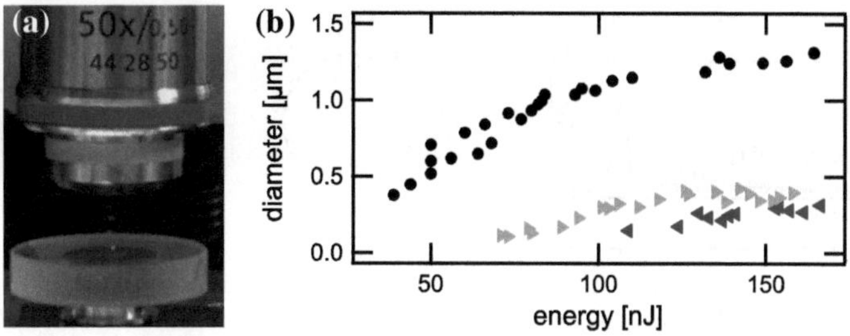

Fig. 4.24 **a** In the materials processing platform, a femtosecond laser pulse is focused with a microscope objective. The Zeiss LD Epiplan 50x/0.5 NA objective delivers a calculated and measured spot diameter of 1.4 μm (1/e^2 value of intensity profile). Due to the nonlinear interaction the laser induced plasma is highly localized. **b** Diameters of ablation structures measured by scanning electron microscopy (SEM) for fused silica for zero, positive and negative TOD modulations (*black circles* for $\phi_3 = 0\,\text{fs}^3$, *red triangles* for $\phi_3 = +6 \times 10^5\,\text{fs}^3$ and *blue triangles* for $\phi_3 = -6 \times 10^5\,\text{fs}^3$)

spot with a calculated and measured diameter of 1.4 μm (see Fig. 4.24a). Different thresholds for material processing with temporally asymmetric pulse shapes were observed (see Fig. 4.24b) which is attributed to control of different ionization processes i.e. multi photon ionization (MPI) and avalanche ionization (AI) [58, 59, 139, 140].

An exemplification of the transient free electron dynamics for unshaped and TOD-shaped pulses used in this experiment is shown in Fig. 4.25. The free electron density n_e is modeled by a single rate equation $\mathrm{d}n_e/\mathrm{d}t = \sigma_k I^k + \alpha n_e I$ implementing MPI with coefficient σ_k and AI with coefficient α. MPI is a k-photon process and is therefore most efficient at high intensities (e.g. unshaped pulses in the middle row of Fig. 4.25). AI on the other hand needs an initial free electron population and time to work efficiently. Negative TOD results in a train of sub-pulses with increasing intensity. Up to the last few sub-pulses the intensity is too low for efficient MPI and the transient free electron density is insufficient for efficient AI (see Fig. 4.25 top). On the other hand positive TOD gives a train of sub-pulses with decreasing intensities which leads to efficient MPI at the beginning and therefore a sufficient initial free electron density which is efficiently amplified by AI (see Fig. 4.25 bottom). Note that for laser pulses with the same energy the free electron density exceeds the critical value for material ablation n_c for a positive TOD but not for negative TOD. This observation has been discussed in a *seed and heat* mechanism based on a refined ionization model employing a multiple rate equation approach [58].

The resulting nanometer scale structures were measured by AFM and SEM and are one order of magnitude below the diffraction limit (see Fig. 4.25). Recently the studies on the dynamics of the free electron plasma created by femtosecond pulses in a thin water jet [141, 142] were extended to a direct observation of the free electron density after excitation with temporally shaped laser pulses by using spectral interference techniques [143, 144]. Employing this technique allowed to directly confirm control of the two ionization mechanisms in the low excitation regime making use of a Drude

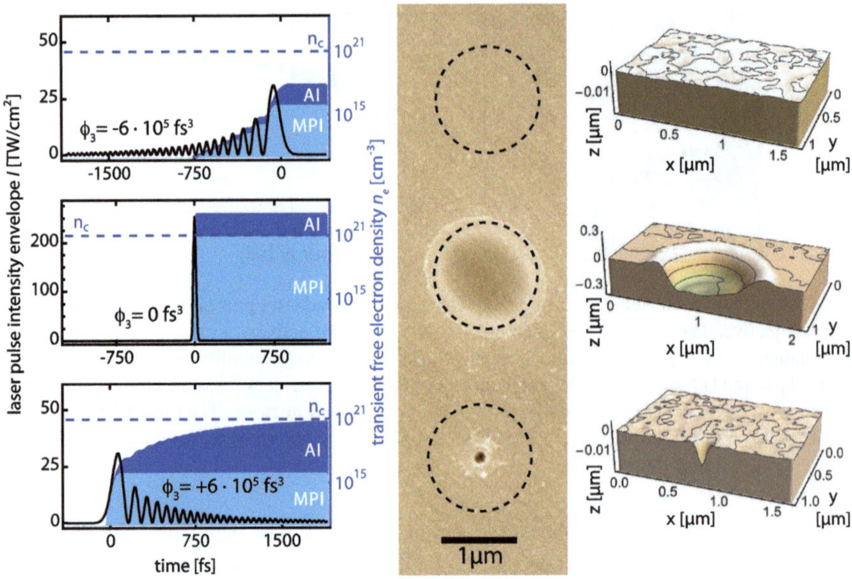

Fig. 4.25 Control of ionization processes in fused silica via asymmetrically shaped femtosecond pulses for three TOD parameters (*top row* negative TOD with $\phi_3 = -6 \times 10^5\,\text{fs}^3$, *middle row* zero TOD with $\phi_3 = 0\,\text{fs}^3$, and *bottom row* positive TOD with $\phi_3 = +6 \times 10^5\,\text{fs}^3$) at the same laser pulse energy. The *left column* shows the temporal profile of the pulse intensity envelope (*solid black line*) and the transient free electron density (*light blue* contribution from multiphoton ionization (MPI) only, *dark blue* contribution from avalanche ionization (AI) in combination with MPI). The critical electron density $n_c = 10^{21}\,\text{cm}^{-3}$ resulting in ablation is marked with a *dashed line*. The *middle column* shows scanning electron microscope (SEM) micrographs including a *dashed circle* for the calculated and measured spot diameter of 1.4 µm (1/e^2 value of intensity profile) and the *right column* shows the corresponding atomic force (AFM) micrographs (note the different z-scaling for the *middle row*)

model and the justified assumption of homogenous excitation within the laser excited volume. The high excitation regime is in the focus of current research.

Regarding the nanoscale structures created with temporally shaped laser pulses in solid dielectrics, the physical picture is still not conclusive [140]. On a first glance the small structures might be attributed to filamentation processes as observed in bulk fused silica under similar excitation conditions [145] as well as on surfaces of dielectrics [146, 147] and reviewed in [148]. However, as filamentation usually needs propagation to occur, the lack of observed propagation structures especially for the TOD shaped pulses may rule out filamentation as the only explanation. While the seed and heat model explains the threshold dependence on the temporally asymmetric pulse shapes, the spatial observations require further extensions to the simulation model. The motivation stems from the simple picture that an initial part of the pulse structure may create free electrons in a spatially very confined region well below the damage threshold via MPI and the remaining pulse exploits AI to reach the critical energy also in a very restricted area. In that simulations rate equations were solved for a Gaussian spatial beam profile taking MPI, AI and recombination into account as a function of various temporal profiles. So far, these simulations did not arrive at a conclusive picture. Currently it is speculated in addition, to what extent nanoplasmonic effects [149] like for example near field effects from a spatially confined region of high electron densities created by a part of the temporally structured laser pulse can also be responsible for part of the observations. Within the context of near field enhancements it has been shown, that by exploiting polarization dependent near field effects [78] nanoscale material processing of dielectrics can be achieved [150–152].

As mentioned above shortly, laser electron excitation with ultrashort laserpulses is followed by electron thermalization on a femtosecond time scale and coupling to the lattice on a femtosecond to picosecond time scale. Finally a free ablation plume is observed on a picosecond to nanosecond time scale that contains spectrochemical information on the ablated material. This laser-induced breakdown spectroscopy (LIBS) is a well-established technique for spectrochemical element analysis in various applications [153, 154]. Although LIBS does not achieve the sensitivity of other methods, as for example mass spectrometry, to date it offers a wide range of different advantages: No special sample preparation is needed and the analysis can be done in air under atmospheric conditions. Usually nanosecond lasers are used in these applications. Femtosecond laser pulses on the other hand offer due to different ablation processes a more precise ablation with less thermal damage and with a higher reproducibility compared to nanosecond pulses [155]. Furthermore the femtosecond laser-induced plasma features a faster temporal decay and a reduced background [156], enabling to work at a higher repetition rate without any gate for data acquisition. On this basis femtosecond laser pulses are suitable to improve the spatial resolution and spectrochemical sensitivity in comparison to ns-LIBS and facilitate fs-LIBS for biomedical applications [157], such as chemical mapping [158] and depth profiling of complex biological systems [156]. In order to minimize the spatial structure and at the same time maximize the spectrochemical sensitivity for fs-LIBS temporal pulse tailoring was exploited for material processing: by applying double pulses an increase in line specific emission was observed, and spatial resolution of few μm and below have been achieved [159].

4.5 Summary and Conclusion

In this review we have presented an overview on recent work on control of ultrafast electron dynamics using shaped femtosecond laser pulses with applications ranging from the manipulation of a single electron in order to create a designer electron wave packet to macroscopic control of electron densities in femtosecond materials processing.

Shaped femtosecond laser pulses are an outstanding tool to manipulate molecular dynamics. Therefore, an introduction to the fundamentals of femtosecond pulse shaping including polarization shaping was presented in view of the applications discussed in this review. Because the physical basis of coherent control is the manipulation of the interferences of matter waves, the coherence transfer from shaped pulses to matter waves was discussed in some detail. As an example for three-dimensional control with polarization-shaped pulses we have studied the creation and detection of designer electron wave packets. It turned out that the underlying physical mechanism is based on the creation of superposition states in the continuum by the interplay between the time-dependent polarization of the laser pulse and the respective complex-valued transition matrix elements as well as the subsequent interference of the partial electron wave packets in the continuum.

Non-perturbative interaction with shaped intense laser fields enables new control scenarios and allows efficient population transfer to the target states. We have considered prototypical examples of strong-field control which highlight specific physical mechanisms of non-perturbative control, such as Dynamic Stark shifts (DSS) and Selective Population of Dressed States (SPODS). The analysis of dressed states populations and energies proved to be particularly suitable to analyze strong-field control. We showed that the concept of SPODS provides a unifying framework to describe adiabatic techniques such as Rapid Adiabatic passage (RAP) and non-adiabatic control enabling ultrafast switching. As a topical example for strong field control we have discussed the manipulation of concerted nuclear-electron dynamics in molecules by shaped intense laser pulses. Tailoring the laser field to the induced charge oscillation in a molecule turned out to be a general theme in coherent control of electron dynamics.

The enduring progress in the refinement of femtosecond laser sources in terms of tunability, pulse-width (associated with increasing bandwidth) and Carrier Envelope Phase (CEP) stabilization opens new perspectives for coherent control of ultrafast dynamics. For instance, due to the decrease of the pulse duration, faster and faster processes such as electron dynamics will come to the fore. In addition, control by the CEP will become increasingly important in this regime—especially for highly non-linear processes. Along with the trend towards shorter pulses goes the broadening of the laser bandwidth promising a much higher degree of attainable control because all molecular degrees of freedom become accessible simultaneously. Furthermore, ultrafast light sources being tunable over a wide spectral band will broaden the range of applications by tailoring the laser central frequency with respect to the application rather than selecting the application based on the available frequencies. Extending recent efforts on the generation and characterization of shaped femtosecond laser pulses in the ultraviolet towards XUV- and X-ray sources should in principle be feasible by combining dispersive elements with absorbing targets. Due to the Kramers-Kronig relation for dielectric functions the insertion of absorbing targets will also include dispersive effects enabling amplitude and phase control. Based on these innovative developments we anticipate a huge variety of novel experiments and applications in coherent control.

Acknowledgments We like to thank Dipl. Phys. Jens Köhler and M.Sc. Dominik Pengel for careful proofreading of the manuscript and Dr. Lars Englert for preparing the Figs. 4.24 and 4.25 for this review. Financial support by DFG via the project WO-848/3-1 and the priority program SPP 1327 is gratefully acknowledged.

References

1. S.A. Rice, M. Zhao, in *Optical Control of Molecular Dynamics* (Wiley, New York, 2000)
2. M. Shapiro, P. Brumer, in *Principles of the Quantum Control of Molecular Processes*, 1st edn. Wiley, Hoboken, 2003)
3. D. Tannor, in *Introduction to Quantum Mechanics: A Time-Dependent Perspective* (Palgrave Macmillan Publishers Limited, Houndmills, Basingstoke, Hampshire, RG21 6XS, England, 2007)
4. D.J. Tannor, S.A. Rice, Adv. Chem. Phys. **70**, 441 (1988)
5. M. Shapiro, P. Brumer, Int. Rev. Phys. Chem. **13**, 187 (1994)
6. T. Baumert, J. Helbing, G. Gerber, in *Advances in Chemical Physics—Photochemistry: Chemical Reactions and Their Control on the Femtosecond Time Scale*, ed. by I. Prigogine, S.A. Rice (Wiley, New York, 1997)
7. H. Rabitz, R. de Vivie-Riedle, M. Motzkus, K. Kompa, Science **288**, 824 (2000)
8. M. Shapiro, P. Brumer, Rep. Prog. Phys. **66**, 859 (2003)
9. D. Goswami, Phys. Rep. **374**, 385 (2003)
10. M. Dantus, V.V. Lozovoy, Chem. Rev. **104**, 1813 (2004)
11. V. Bonacic-Koutecky, R. Mitric, Chem. Rev. **105**, 11 (2005)
12. T. Brixner et al., in *Femtosecond Laser Spectroscopy*, ed. by P. Hannaford (Springer, Berlin, 2005), Chap. 9
13. M. Wollenhaupt, V. Engel, T. Baumert, Ann. Rev. Phys. Chem. **56**, 25 (2005)
14. P. Nuernberger, G. Vogt, T. Brixner, G. Gerber, Phys. Chem. Chem. Phys. **9**, 2470 (2007)
15. J. Werschnik, E.K.U. Gross, J. Phys. B **40**, R175–R211 (2007)
16. W. Wohlleben, T. Buckup, J.L. Herek, M. Motzkus, Chem. Phys. Chem. **6**, 850 (2005)
17. Y. Silberberg, Ann. Rev. Phys. Chem. **60**, 277 (2009)
18. K. Ohmori, Ann. Rev. Phys. Chem. **60**, 487 (2009)
19. C. Brif, R. Chakrabarti, H. Rabitz, New. J. Phys. **12**, 075008 (2010)
20. R. Stoian, M. Wollenhaupt, T. Baumert, I.V. Hertel, in *Laser Precision Microfabrication*, ed. by K. Sugioka, M. Meunier, A. Piqué (Springer, Berlin Heidelberg, 2010), Chap. 5
21. M. Wollenhaupt, T. Baumert, Faraday Discuss. **153**, 9 (2011)
22. M. Wollenhaupt, C. Lux, M. Krug, T. Baumert, Chem. Phys. Chem. **14**, 1341 (2013)
23. S. Thallmair et al. in *Molecular Quantum Dynamics: From Theory to Applications*, ed. by F. Gatti (Springer, Heidelberg, 2014)
24. P. Gaspard, I. Burghardt (eds.), (Wiley, New York, 1997), Chap. 101
25. J.L. Herek, J. Photochem. Photobiol. A **180**, 225 (2006)
26. H. Fielding, M. Shapiro, T. Baumert, J. Phys. B **41**, 070201-1 (2008)
27. H. Rabitz, New J. Phys. **11**, 105030 (2009)
28. H.H. Fielding, M.A. Robb, Phys. Chem. Chem. Phys. **12**, 15569 (2010)
29. H. Braun et al., J. Phys. B **47**, 124015 (2014)
30. M. Wollenhaupt, A. Assion, T. Baumert, in *Springer Handbook of Lasers and Optics*, 2nd edn., ed. F. Träger (Springer, Dordrecht, Heidelberg, London, New York, 2012), Chap. 12
31. A. Monmayrant, S. Weber, B. Chatel, J. Phys. B : At. Mol. Opt. Phys. **43**, 103001-34 (2010)
32. J. Schneider et al., Phys. Chem. Chem. Phys. **13**, 8733 (2011)
33. A. Galler, T. Feurer, Appl. Phys. B. **90**, 427 (2008)
34. B. von Vacano, T. Buckup, M. Motzkus, J. Opt. Soc. Am. B **24**, 1091 (2007)
35. J. Köhler et al., Opt. Express **19**, 11638 (2011)
36. R. Bracewell, in *The Fourier Transform and Its Applications* , 3rd edn. (McGraw-Hill Higher Education, Singapore, 2000)
37. L. Allen, J.H. Eberly, in *Optical Resonance and Two-Level Atoms* , 2nd edn. (Dover Publications, New York, 1987)
38. N. Dudovich, T. Polack, A. Péer, Y. Silberberg, Phys. Rev. Lett. **94**, 083002-4 (2005)
39. M. Wollenhaupt et al., Phys. Rev. A **73**, 063409-15 (2006)
40. T. Bayer, M. Wollenhaupt, T. Baumert, J. Phys. B. **41**, 074007-13 (2008)

41. L. Cohen, in *Time-Frequency Analysis* (Prentice Hall PTR, New Jersey, 1995)
42. E. Sorokin, G. Tempea, T. Brabec, JOSA B **17**, 146 (2000)
43. A. Präkelt, M. Wollenhaupt, C. Sarpe-Tudoran, T. Baumert, Phys. Rev. A **70**, 063407-10 (2004)
44. H.R. Telle et al., Appl. Phys. B **69**, 327 (1999)
45. F.W. Helbing et al., Appl. Phys. B **74**, 35 (2002)
46. A. Apolonski et al., Phys. Rev. Lett. **92**, 073902-4 (2004)
47. G. Sansone et al., Phys. Rev. A **73**, 053408 (2006)
48. M.F. Kling et al., New J. Phys. **10**, 025024-17 (2008)
49. G. Cerullo, A. Baltuska, O.D. Mucke, C. Vozzi, Laser Photonics Rev. **5**, 323 (2011)
50. B. Broers, L.D. Noordam, H.B. van Linden van den Heuvell, Phys. Rev. A **46**, 2749 (1992)
51. C.J. Bardeen, Q. Wang, C.V. Shank, Phys. Rev. Lett. **75**, 3410 (1995)
52. A. Assion et al., Chem. Phys. Lett. **259**, 488 (1996)
53. J. Degert et al., Phys. Rev. Lett. **89**, 203003–203003-4 (2002)
54. M. Wollenhaupt et al., Appl. Phys. B **82**, 183 (2006)
55. P. Nuernberger, Opt. Commun. **282**, 227 (2009)
56. M. Krug et al., New J. Phys. **11**, 105051 (2009)
57. J.D. McMullen, JOSA **67**, 1575 (1977)
58. L. Englert et al., Opt. Express **15**, 17855 (2007)
59. L. Englert et al., Appl. Phys. A **92**, 749 (2008)
60. N.T. Form, B.J. Whitaker, C. Meier, J. Phys. B : At. Mol. Opt. Phys. **41**, 074011 (2008)
61. M. Ruge et al., J. Phys. Chem. C **117**, 11780 (2013)
62. D. Meshulach, Y. Silberberg, Nature **396**, 239 (1998)
63. M. Wollenhaupt et al., J. Mod. Opt. **52**, 2187 (2005)
64. A. Bartelt et al., Phys. Chem. Chem. Phys. **5**, 3610 (2003)
65. V.V. Lozovoy, I. Pastirk, A. Walowicz, M. Dantus, J. Chem. Phys. **118**, 3187 (2003)
66. M. Wollenhaupt, T. Baumert, J. Photochem. Photobiol. A **180**, 248 (2006)
67. M. Wollenhaupt et al., Chem. Phys. Lett. **419**, 184 (2006)
68. J.L. Herek et al., Nature **417**, 533 (2002)
69. T. Bayer et al., Phys. Rev. Lett. **110**, 123003 (2013)
70. N. Dudovich, D. Oron, Y. Silberberg, J. Chem. Phys. **118**, 9208 (2003)
71. J. Voll, R. Vivie-Riedle, New J. Phys. **11**, 105036 (2009)
72. J. Hauer, T. Buckup, M. Motzkus, J. Chem. Phys. **125**, 061101-3 (2006)
73. T. Brixner, G. Gerber, Opt. Lett. **26**, 557 (2001)
74. N. Dudovich, D. Oron, Y. Silberberg, Phys. Rev. Lett. **92**, 103003-4 (2004)
75. T. Brixner et al., Phys. Rev. Lett. **92**, 208301-4 (2004)
76. M. Wollenhaupt et al., Appl. Phys. B **95**, 245 (2009)
77. F. Weise, G. Achazi, A. Lindinger, Phys. Chem.Chem. Phys. **13**, 8621 (2011)
78. M. Aeschlimann et al., Nature **446**, 301 (2007)
79. P. Schön et al., Phys. Rev. A **81**, 013809 (2010)
80. R. Selle et al., Opt. Lett. **33**, 803 (2008)
81. A.M. Weiner, Rev. Sci. Instr. **71**, 1929 (2000)
82. A. Präkelt et al., Rev. Sci. Instr. **74**, 4950 (2003)
83. D.B. Strasfeld, S.-H. Shim, M.T. Zanni, Adv. Chem. Phys. **141**, 1 (2009)
84. T. Hornung, R. Meier, M. Motzkus, Chem. Phys. Lett. **326**, 445 (2000)
85. S. Fechner et al., Opt. Express **15**, 15387 (2007)
86. S. Ruetzel et al., Phys. Chem. Chem. Phys. **13**, 8627 (2011)
87. R.S. Judson, H. Rabitz, Phys. Rev. Lett. **68**, 1500 (1992)
88. T. Baumert et al., Appl. Phys. B **65**, 779 (1997)
89. D. Meshulach, D. Yelin, Y. Silberberg, Opt. Commun. **138**, 345 (1997)
90. C.J. Bardeen et al., Chem. Phys. Lett. **280**, 151 (1997)
91. A. Assion et al., Science **282**, 919 (1998)
92. R.J. Levis, H.A. Rabitz, J. Phys. Chem. A **106**, 6427 (2002)
93. C. Daniel et al., Science **299**, 536 (2003)

94. D. Yelin, D. Meshulach, Y. Silberberg, Opt. Lett. **22**, 1793 (1997)
95. N. Hansen, in *The CMA Evolution Strategy: A Tutorial* (2009)
96. J.W. Wilson et al., Rev. Sci. Instrum. **79**, 033103-5 (2008)
97. M. Wollenhaupt et al., Phys. Rev. Lett. **89**, 173001-4 (2002)
98. F. Lindner et al., Phys. Rev. Lett. **95**, 040401-4 (2005)
99. M. Winter, M. Wollenhaupt, T. Baumert, Opt. Commun. **264**, 285 (2006)
100. P.B. Corkum, F. Krausz, Nat. Phys. **3**, 381 (2007)
101. J. Mauritsson et al., Phys. Rev. Lett. **105**, 053001 (2010)
102. M. Born, Phys. Bl **2**, 49 (1955)
103. C. Cohen-Tannoudji, B. Diu, F. Laloe, in *Quantum Mechanics*, vol. 1. (Wiley, New York, 1977)
104. P. Brumer, M. Shapiro, Ann. Rev. Phys. Chem. **43**, 257 (1992)
105. M. Wollenhaupt et al., J. Opt. B **7**, S270–S276 (2005)
106. M. Wollenhaupt, M. Krug, T. Baumert, Phys. J. **11**, 37 (2012)
107. P. Hockett, M. Wollenhaupt, C. Lux, T. Baumert, Phys. Rev. Lett. **112**, 223001 (2014)
108. K.L. Reid, Mol. Phys. **110**, 131 (2012)
109. M. Wollenhaupt et al., Appl. Phys. B **95**, 647 (2009)
110. D.A. Malik et al., Phys. Rev. A **84**, 043404–043404-5 (2011)
111. A. Vredenborg et al., Chem. Phys. Chem. **12**, 1459 (2011)
112. J. Maurer et al., Phys. Rev. Lett. **109**, 123001 (2012)
113. A.T.J.B. Eppink, D.H. Parker, Rev. Sci. Instr. **68**, 3477 (1997)
114. B.J. Whitaker, in *Imaging in Molecular Dyanmics—Technology and Applications*, (Cambridge University Press, Cambridge, 2003)
115. G.A. Garcia, L. Nahon, I. Powis, Rev. Sci. Instr. **75**, 4989 (2004)
116. A.C. Kak, M. Slaney, in *Principles of Computerized Tomographic Imaging*, (IEEE Press, New York, 1999)
117. C. Smeenk et al., J. Phys. B **42**, 165402 (2009)
118. P. Hockett, M. Staniforth, K.L. Reid, Mol. Phys. **108**, 1045 (2010)
119. T. Frohnmeyer, M. Hofmann, M. Strehle, T. Baumert, Chem. Phys. Lett. **312**, 447 (1999)
120. C. Trallero-Herrero, J.L. Cohen, T. Weinacht, Phys. Rev. Lett. **96**, 063603-4 (2006)
121. B.J. Sussman, D. Townsend, M.Y. Ivanov, A. Stolow, Science **314**, 278 (2006)
122. U. Gaubatz et al., Chem. Phys. Lett. **149**, 463 (1988)
123. N.V. Vitanov, T. Halfmann, B.W. Shore, K. Bergmann, Ann. Rev. Phys. Chem. **52**, 763 (2001)
124. B. W. Shore, Acta Phys. Slovaca **58**, 243 (2008)
125. E.T. Sleva, I.M. Xavier Jr, A.H. Zewail, JOSA B **3**, 483 (1985)
126. Y.S. Bai, A.G. Yodh, T.W. Mossberg, Phys. Rev. Lett. **55**, 1277 (1985)
127. R. Kosloff, A.D. Hammerich, D. Tannor, Phys. Rev. Lett. **69**, 2172 (1992)
128. V.S. Malinovsky, C. Meier, D.J. Tannor, Chem. Phys. **221**, 67 (1997)
129. S.R. Hartmann, E.L. Hahn, Phys. Rev. **128**, 2053 (1962)
130. A. Abragam, in *The Principles of Nuclear Magnetism*, 13th edn. (Clarendon Press, Oxford, 1994) (Reprint)
131. S.H. Autler, C.H. Townes, Phys. Rev. **100**, 703 (1955)
132. P. Balling, D.J. Maas, L.D. Noordam, Phys. Rev. A **50**, 4276 (1994)
133. A.A. Rangelov et al., Phys. Rev. A **72**, 053403-12 (2005)
134. J. Cao, C.J. Bardeen, K.R. Wilson, Phys. Rev. Lett. **80**, 1406 (1998)
135. V.S. Malinovsky, J.L. Krause, Eur. Phys. J. D **14**, 147 (2001)
136. M. Wollenhaupt et al., Phys. Rev. A **68**, 015401-4 (2003)
137. B. Rethfeld, K. Sokolowski-Tinten, D. von der Linde, S.I. Anisimov, Appl. Phys. A **79**, 767 (2004)
138. P. Balling, J. Schou, Rep. Prog. Phys. **76**, 036502 (2013)
139. M. Wollenhaupt, L. Englert, A. Horn, T. Baumert, J. Laser Micro Nanoeng. **4**, 144 (2009)
140. L. Englert et al., J. Laser Appl. **24**, 042002–042002-5 (2012)
141. C. Sarpe-Tudoran et al., Appl Phys Lett **88**, 261109–3 (2006)
142. C. Sarpe et al., New J. Phys. **14**, 075021 (2012)

143. E. Tokunaga, A. Terasaki, T. Kobayashi, Opt. Lett. **17**, 1131 (1992)
144. V.V. Temnov et al., Phys. Rev. Lett. **97**, 237403 (2006)
145. A. Couairon et al., Phys. Rev. B **71**, 125435-11 (2005)
146. Y.V. White et al., Opt. Express **16**, 14411 (2008)
147. B. Delobelle, F. Courvoisier, P. Delobelle, Opt. Lasers Eng. **48**, 616 (2009)
148. A. Couairon, A. Mysyrowicz, Phys. Rep. **441**, 47 (2007)
149. T. Fennel et al., Rev. Mod. Phys. **82**, 1793 (2010)
150. F. Hubenthal et al., Appl. Phys. Lett. **95**, 063101-3 (2009)
151. R. Moraraescu et al., J. Mater. Chem. **21**, 4076 (2011)
152. A.A. Jamali et al., Appl. Phys. A **110**, 743 (2013)
153. J.D. Winefordner et al., J. Anal. At. Spectrom. **19**, 1061 (2004)
154. E.L. Gurevich, R. Hergendorfer, Appl. Spectr. **61**, 233A (2007)
155. J. Cheng et al., Opt. Laser Technol. **46**, 88 (2013)
156. A. Assion et al., Appl. Phys. B **77**, 391 (2003)
157. F. Dausinger, F. Lichtner, H. Lubatschowski (eds.), in *Femtosecond Technology for Technical and Medical Applications, Topics in Applied Physics* (Springer, Berlin Heidelberg, 2004)
158. W. Wessel et al., Eng. Fract. Mech. **77**, 1874 (2010)
159. J. Mildner et al., Appl. Surf. Sci. **302**, 291 (2014)

Part II
Attosecond Pulses for Inducing and Probing Electronic Processes

Chapter 5
XUV Attosecond Photoionization and Related Ultrafast Processes in Diatomic and Large Molecules

Victor Despré, Alexandre Marciniak, Thomas Barillot, Vincent Loriot, Arnaud Rouzée, Marc. J.J. Vrakking and Franck Lépine

Abstract New means of inducing and probing in real-time charge dynamics and energy flow in molecules have evolved with progresses made in laser technology and secondary sources. One significant example is the light source based on high harmonic generation (HHG). Such a source typically allows the production of extreme ultraviolet (XUV) radiations temporally confined in very short pulses down to a few tens of femtoseconds and even attosecond duration. This naturally offers new opportunities to study extremely fast processes in very excited molecular species and to track dynamics down to the ultimate time scale of electronic motion. This chapter summarizes recent results obtained in simple and complex molecular species using these short XUV pulses. It demonstrates the potential impact of pump-probe XUV science in molecular physics, chemistry, plasmonic or astrochemistry. Hints of dynamics below 1 fs in pump-probe experiments are already observed in increasingly large systems showing the way to chemistry on attosecond time scale.

V. Despré · A. Marciniak · T. Barillot · V. Loriot · F. Lépine (✉)
Institut Lumière Matière ILM, CNRS, 10 rue Ada Byron,
69622 Villeurbanne CEDEX, France
e-mail: Franck.lepine@univ-lyon1.fr

V. Despré
e-mail: Victor.despre@univ-lyon1.fr

A. Marciniak
e-mail: Alexandre.marciniak@univ-lyon1.fr

T. Barillot
e-mail: Thomas.barillot@univ-lyon1.fr

V. Loriot
e-mail: Vincent.loriot@univ-lyon1.fr

A. Rouzée · M.J.J. Vrakking
Max-Born Institute (MBI), Max-Born Strasse 2A, 12489 Berlin, Germany
e-mail: Arnaud.Rouzee@mbi-berlin.de

M.J.J. Vrakking
e-mail: Marc.Vrakking@mbi-berlin.de

M. Kitzler and S. Gräfe (eds.), *Ultrafast Dynamics Driven by Intense Light Pulses*, Springer Series on Atomic, Optical, and Plasma Physics 86, DOI 10.1007/978-3-319-20173-3_5

5.1 Introduction

Understanding photo-induced phenomena in molecules require detailed knowledge of all structural and dynamical processes that underlie the complex interplay between all electrons and nuclei in the molecule. Following the advent of widely available femtosecond lasers, new tools and experimental strategies have been developed to observe and control molecules on the picosecond and femtosecond timescale, i.e. the timescale of nuclear motion. Since the beginning of this century, the development of high harmonic generation (HHG) that allows for the production of attosecond XUV pulses enables the investigation of the charge motion in molecules, that typically takes place on the attosecond to few fs time scale [1]. While femtochemistry dealing with resolving nuclear dynamics is now a mature field of research, experiments accessing attosecond timescales are still in their infancy [2]. Attosecond physics experiments in molecules have evolved through two complementary strategies.

The "strong field" approach is based on HHG spectroscopy (HHGS), and uses a molecular target to convert infra-red (IR) or low frequency field, delivered by femtosecond laser source, into energetic extreme ultra-violet (XUV) photons through a highly non-linear process. Because it depends on the initial molecular wavefunction, propagated electron wavepacket and its overlap with the molecular electronic structure at the instant of the recombination, the HHG conversion process encodes structural and dynamical information about the molecule in the properties of the emitted XUV light [3, 4]. Consequently, by studying harmonic spectral, phase and polarization properties it becomes possible to perform spectroscopic measurements using HHGS and to follow in real-time a process that has been initiated by a time-delayed first pump laser pulse. One of the advantages of the HHGS is the possibility to access sub-optical cycle time scale thanks to the time/energy relationship that exists between the instant of electron recombination and energy of the emitted photon. Such timescale is still hardly accessible in a pump-probe approach simply because no experiment using both attosecond pump and probe is currently available. However, benchmark experiments are emerging [5].

The second approach is a pump-probe approach that uses short XUV pulses synthesized by mean of HHG. This pulse is used in a sequence of two pulses to perform pump-probe experiments in a fashion similar to traditional femtochemistry experiments. In femtochemistry, the common approach is to use a first laser pulse to resonantly excite the molecule to a specific electronic state whose dynamics is followed in real time. A typical example is the time resolved investigation of the relaxation dynamics of low excited states accessed by the $\pi -> \pi^*$ transition in the UV. When the pulse duration crosses the femtosecond limit, laws of propagation of electromagnetism waves compel the wavelength in the XUV domain. Because typical molecular ionization thresholds are in the UV, VUV domain, molecular photoionization of inner and outer valence shell is a common process in these experiments with a number of fundamental consequences. Threshold ionization of large molecules corresponds to the removal of an electron from well-defined molecular orbitals. Assuming a non-correlated ground state, the ionization process can be described in terms of single electron mechanism. For higher photon energies, in the VUV/XUV domain, deeper states become accessible and electron correlation and multielectron dynamics becomes increasingly important. Obviously correlation effects are also present in visible or UV photoabsorption but in short pulse XUV physics these processes are inherent to the excitation process leading to the excitation of multiple electrons with a single photon. As a result, photoionization of a complex molecule by a short XUV pulses reveals the multielectronic nature of the molecular wavefunction as demonstrated by Cederbaum et al. [6]. Following the excitation process, charge dynamics can be triggered and relaxation through electronic or vibronic couplings occurs. In the last few years it has been demonstrated that exploring these mechanisms may require attosecond resolution and experiments on few femtosecond timescales show that attosecond dynamics could be soon accessible in experiments on large molecular species [7].

In this book chapter we present examples of recent results utilizing XUV pulses to investigate molecular processes on attosecond and femtosecond timescales. The chosen examples cover several aspects of light-matter interaction, from the initial electron dynamics when the molecule is dressed by a NIR laser pulse, to attosecond charge localization, attosecond photoemission and non-adiabatically driven internal energy relaxation. In these experiments an XUV pulse can be used either to trigger or to probe dynamics in a molecule. We discuss that the first attosecond pump-probe experiment in a molecular target now has been extended to multi-electronic molecules and to larger polyatomic systems. We also discuss predictions for future attosecond experiments in large molecular structures and nanoscale systems such as fullerenes.

5.2 The First Attoseconds of the Light-Matter Interaction: Attosecond Control of Molecular Ionization

When a light pulse interacts with a molecule, the interaction can either be resonant or non-resonant. In the latter case, if the light frequency remains below any resonant transition, the molecule behaves like a driven oscillator and the electrons adiabatically follow the light electric field, creating a time-dependent dipole that vanishes when the pulse ends, i.e. in the absence of resonances there is no population transfer to excited states and this can be regarded as a quasi-static polarization mechanism. This constitutes one of the most fundamental aspects of light-matter interactions and, at the commonly used near-infrared wavelength from available Ti:Sapphire lasers (800 nm), this leads to sub-cycle electron dynamics on the attosecond timescale.

Experimentally, laser-induced polarization of molecules has been extensively exploited within the last decade, for instance in the process of dynamic molecular alignment [8]. Here, the interaction of a molecule with a moderately intense laser pulse forces the most polarizable axis of the molecule into alignment along the laser polarization axis. The interaction of the resulting dipole with the laser electric field induces a torque on the molecule that leads to the observed alignment. Recently

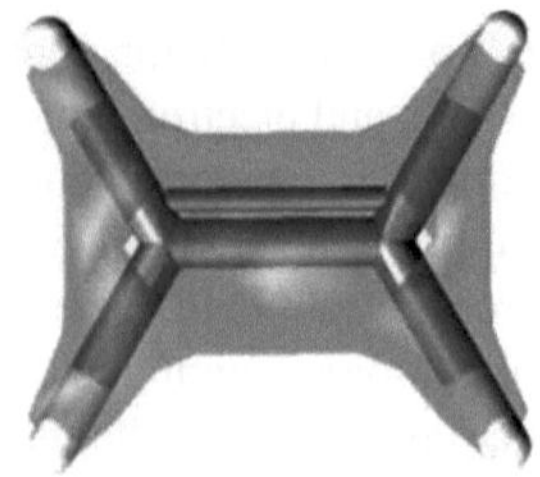

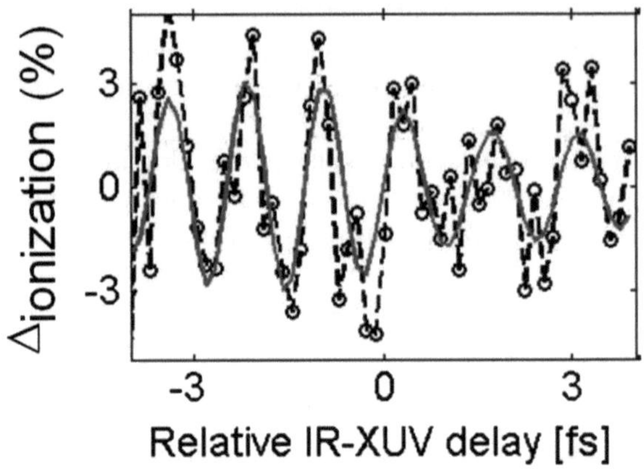

Fig. 5.1 Attosecond molecular ionization control. Attosecond variation of C_2H_4 ionization yield using the combination of an XUV attosecond pulse train and IR pulse. Yield oscillations occur at twice the period of the IR field showing the sub-cycle dependency of the XUV ionization. The IR laser intensity is kept low enough than no molecular ionization is observed from the IR only

we have observed the sub-cycle polarization dynamics in molecules by performing pump-probe experiments on several small to midsize molecules, using the combination of a moderately intense ($\sim 10^{12}$ W/cm^2), NIR femtosecond laser pulse and an XUV attosecond pulse train, overlapping in time and synchronized with attosecond precision [9]. The time-dependent molecular dipole induced by the NIR pulse was probed by attosecond XUV photoionization. The ionization yields of N_2, CO_2 and C_2H_4 were monitored as a function of the XUV-NIR delay. Figure 5.1 shows clear experimental modulations of the ionization yield on the attosecond timescale and shows that the XUV ionization depends on the instantaneous NIR electric field. Because the attosecond pulse train contains one XUV pulse per half NIR cycle, the ionization yield oscillates with a period that is half the period of the NIR driving field. Interestingly, the effect was found to scale with the averaged molecular polarizability and with the molecular size. Therefore, the experiment raises the prospect that attosecond laser pulses can be used to probe the occurrence of time-dependent variations in electronic charge densities in larger molecules. In that respect, it is already striking that two-color XUV-NIR pump-probe experiments are showing dynamical effects in relatively complex systems such as C_2H_4 below 1 fs timescale.

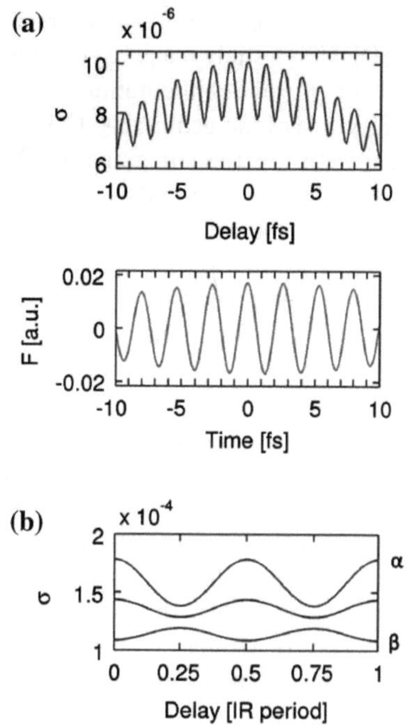

Fig. 5.2 TDDFT calculations of the XUV+IR ionization yield in the case of N_2 molecule. Like in the experiment, ionization attosecond oscillation (**a**) occurs at twice the period of the IR field oscillations (**b**). These oscillations also depend on the molecular alignment with respect to the light polarization (**b**). α corresponds to the molecular axis along and β perpendicular to the laser polarization

Depending on the photon energy of the XUV pulse, dissociative ionization channels can be reached. For instance, in the case of N_2 not only N_2^+ but N^+ ions are observed. As discussed in the next section (see Sects. 5.2 and 5.3) dissociation reactions are also influenced by the IR field but due to couplings of ionic states. In our analysis, we consider the total ionization yield, which means that both ions are following the same oscillations in phase with the NIR electric field (i.e. there is a maximum and a minimum yield in both channels for the same delays) which means that in the present case, the IR field modulates the instantaneous total ionization efficiency of the neutral species.

One of the challenges in attosecond molecular science is to obtain abinitio predictive theoretical description able to describe the observed dynamics. The later experiment is a typical example that contains all ingredients that make a relevant computational approach hardly accessible. First of all, the experiment considers molecular species for which, a priori, interacting many-electrons and nuclei has to be taken into account. Moreover, the molecule interacts with a relatively strong NIR field, preventing a description in terms of linear response theory or perturbation theory. On the contrary, the XUV pulse is weak enough then the molecular response can be treated linearly. However, the photon energy range corresponds to photo-excitation above the ionization threshold meaning that the ionization continuum has to be described. The cumulated constraints of a many-particle system interaction with both non-linear and ionizing laser fields restraints the possible theoretical treatment. We used real time - real space time-dependent density functional theory (TDDFT) to study the pure electron dynamics (Fig. 5.2). Using the exact experimental pump-probe configuration, the time-dependent electron dynamics were computed in a series of molecules: N_2, CO_2, C_2H_4. Calculations were performed in molecules that were fixed in space and with frozen nuclei. The ionization was defined as the loss of charge leaving the discretized calculation box. The calculations were able to reproduce the experimentally observed ionization oscillations as a function of the XUV-NIR pump-probe delay. A closer examination of the time-dependent valence electron distribution shows noticeable changes induced by the NIR light, which corresponds to a modification of the probability to find the electron on one side of the molecule. This was illustrated by plotting the time-dependent Kohn-Sham potential that demonstrates how the molecule is dynamically altered by the laser electric field. When the probe pulse interacts with valence electrons, the XUV field is screened by the light-dressed molecular potential and the photoionization cross sections is modulated due to a modification of the coupling, i.e. dipole matrix element, to the ionization continuum. We note that because the ionization is driven by high energetic photons with a correspondingly weak laser field energetic electrons are removed from the molecules that cannot efficiently be driven back to the ion core neither by the XUV or IR field. The size of the box can therefore be maintained to a limited size. The TDDFT approach used to describe our experiment has been also validated by the algebraic diagrammatic construction-Stieltjes-Lanczos method [10]. Let us notice that polarization effects were also predicted in the case of polar molecules, inducing a phase shift in a streaking trace [11].

This first experiment demonstrates that dynamics occurring on sub-femtosecond timescale can be controlled and observed in a polyatomic molecule using a pump-probe configuration. It is also remarkable that ionization can be controlled on the attosecond timescale in molecular systems such as N_2, CO_2 or C_2H_4 that are electronically so different. Because this observation is not confined to one specific atom or molecule, it makes it a general tool for ultrafast attosecond molecular science. However, to reach an abinitio description of this experiment, a number of questions remain to be elucidated. First of all, the role of the electronic excited states that are populated either by the NIR or the XUV field needs to be investigated. For a NIR laser field, it is clear that Rabi-oscillations would influence the ionization modulation on various timescales that depend on the laser intensity. Similarly, coherent superposition of states will also induce oscillations on a different period than 2ω, determined by the energy splitting between states. These effects will lead to a global decrease of contrast of the attosecond oscillations. In our experiment the XUV spectrum is located several eV above the ionization threshold of the molecules. Considering the XUV photo-excitation, although ionization of the outer valence states is dominating, the population of excited states in the neutral can vary by either absorbing or emitting IR photons. The probability of such processes is weak but certainly plays a role for some of the molecules that we have studied. Disentangling the role of the the NIR and XUV fields requires, for instance, performing molecular frame measurements to identify the contribution of specific orbitals. The experiment would also benefit from the use of electron-ion coincidence techniques in which bound and dissociative channels can be separated. In current state-of-the art experimental set-ups, this remains very challenging, especially considering the weakness of the attosecond oscillation signal. Interestingly the driving field could also be transposed to longer wavelength for which the quasi-static approximation is even more valid, in order to investigate the wavelength dependent polarization effect.

5.3 Photo-Dissociation: Attosecond Control of Dissociation Pathways

In the experiment discussed in Sect. 5.1, the induced dipole resulted from a coupling of electronic states, or quasistatic modification, of the neutral molecule by the NIR laser field. Prior to these experiments, other investigations were performed by our team on molecular hydrogen, revealing a similar coupling taking place in the molecular ion. As in the afore-mentioned experiment, the ionization was performed

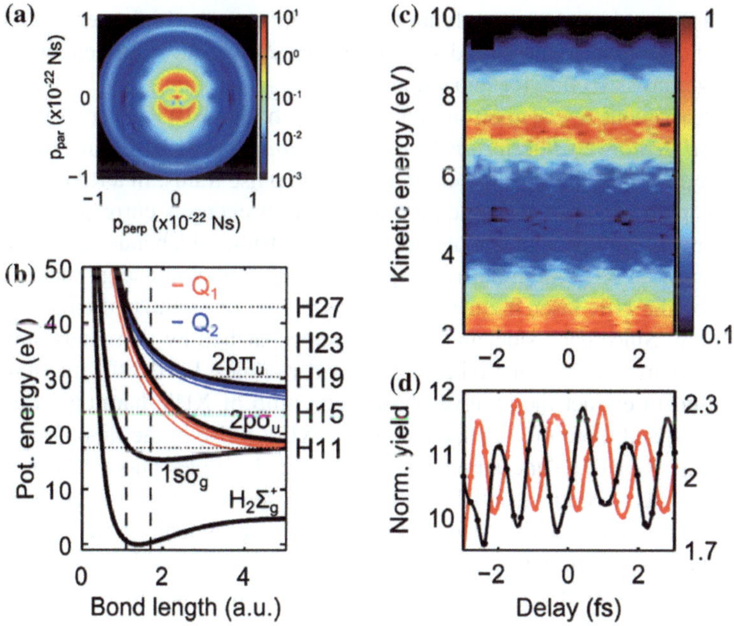

Fig. 5.3 **a** Slice through the 3D momentum distribution of the H^+ fragments formed by dissociative ionization of H_2 molecules by a combined XUV+NIR laser pulse. **b** Relevant potential energy curves of the H_2 molecule and the molecular cation. The position of the harmonics composing the XUV pulse are shown as *horizontal dotted lines* whereas the *vertical dash lines* indicates the Franck Cordon region. **c** Time evolution of the H^+ ions kinetic energy distribution as a function of the time delay between the attosecond pulse train and the NIR laser pulse. **d** Ions yield as a function of the XUV-IR delay for fragments with final kinetic energy centered around 7 and 3 eV

in a combined fields composed by the an attosecond pulse train and a moderately intense (10^{13} W/cm^2), co-propagating NIR field and high energy H^+ fragment ions resulting from dissociative ionization were measured as a function of the XUV-NIR time delay [12].

Dissociative ionization of H_2 at XUV wavelengths is commonly understood in terms of excitation to (i) the $1s\sigma_g$ ground electronic state (producing low energy (E_k <1 eV) H^+ fragments that arise from the Franck-Condon overlap of the neutral ground state with the inner turning point of the $1s\sigma_g$ potential well), (ii) the Q_1 doubly-excited state of the neutral molecule (producing intermediate energy (E_k = 1–5 eV) H^+ fragments that result from auto-ionization in the course of dissociation of the Q_1 state), and (iii) the $2p\sigma_u$ first excited state of the molecular ion. This state is repulsive, and leads to the production of high energy H^+ fragments (E_k>5 eV).

Experimentally, a modulation of the yield of the high energy fragments was observed as a function of the XUV-NIR delay, with a period that, like in the induced polarization experiments, was half the period of the NIR field. Coupling of the $1s\sigma_g$ and the $2p\sigma_u$ ionic states was thought to be responsible for this observation (see Fig. 5.3). As a result of this coupling, the quasi-static laser-dressed eigenstates of the molecular ion in the intense laser field are linear combinations of the field-free $1s\sigma_g$ and $2p\sigma_u$ eigenstates, implying that excitation to the higher energy channel (corresponding to zero laser field strength to the $2p\sigma_u$ state) can benefit from the much larger cross section for ionization to the $1s\sigma_g$ state.

The experiment revealed strong analogies to the RABBITT (*Reconstruction of Attosecond Beating By Interfering Two-photon Transitions* [13]) scheme that is commonly used for the characterization of attosecond pulse trains. In a RABBITT measurement, photoelectron sidebands are measured at energies corresponding to the absorption of an even number of NIR photons. These sidebands result from the absorption of an XUV high-harmonic photon (with a photon energy that is an odd multiple of the NIR laser frequency) and absorption or stimulated emission of an NIR photon. Since each sideband energy can be reached by two interfering pathways (involving the adjacent higher and lower harmonics, RABBITT measurements can be used to extract the relative phase of a comb of XUV harmonic frequency components, and therefore to characterize the XUV pulse in the time-domain. In a "traditional" RABBITT measurement, the NIR field couples photoelectron continuum states that are connected to the same ionic threshold. In contrast, the experiment reported in H_2 can be understood as a RABBITT experiment where the NIR laser couples the $1s\sigma_g$ continuum and the $2p\sigma_u$ continuum, thus providing a more efficient pathway for population of the repulsive, $2p\sigma_u$ electronically excited state than would be available in the absence of the NIR field.

A similar work was performed on the, electronically more complex, O_2 molecule where we have demonstrated that dissociative ionization of O_2 can as well be controlled by the relative delay between an attosecond pulse train and a co-propagating NIR field [14]. Our experiments have revealed a dependence of both the kinetic energy distribution and the angular distribution of the O^+ fragments that are produced on the extreme ultraviolet (XUV) to IR time delay. As in case of the H_2 molecule, an IR-induced coupling between electronic states in the molecular ion is very likely to be responsible of the observed modulation. However, whereas a static picture was sufficient to describe the oscillation observed in the fragment kinetic energy distribution of the H^+ fragments, i.e. the modulation could be well explained in terms of a variation of the ionization probability in one channel with respect to the others due to two path interferences involving two ionization continuums, we cannot rule out in case of O_2 the possible role of additional sequential dissociation pathways that are induced by the IR field after XUV ionization that can as well present sub-cycle oscillation. One has therefore to consider the role of additional coupling between electronic and nuclear degree of freedom by the NIR laser field on the propagation of the coherent superposition of electronic states initially launched by the XUV pulse. Such dynamical theoretical treatment is still nowadays very challenging but with the tremendous development of modern ab-initio quantum chemistry methods allowing for electron correlations, very high electronic excitations and ultrafast nuclear dynamics all to be accurately accounted for, we should be able to fully disentangled the complex dynamics occurring on attosecond and few femtoseconds timescale in molecules having strong multielectronic character exposed to XUV+NIR pulses.

5.4 Attosecond Control of the Charge Localization

The experiment reported in [12] was preceded by another attosecond experiment on hydrogen, in this case involving the combination of an isolated attosecond pulse and a few-cycle NIR pulse [15]. This experiment is considered as the first ever experimental realization of and attosecond pump-probe experiment on molecular target. It addressed elementary charge dynamics in the ${H_2}^+$ molecular ion, i.e. preferred laboratory frame localization of the single remaining electron in the ${H_2}^+$ ion on either (left or right) of the two protons. The electronic wavefunctions of the $1s\sigma_g$ and $2p\sigma_u$ electronic states of ${H_2}^+$ can be viewed as linear combinations of atomic states that have the electron localized on the right or the left proton, i.e.

$$\Psi_g = \frac{1}{\sqrt{2}}(\Psi_{left} + \Psi_{right})$$
$$\Psi_u = \frac{1}{\sqrt{2}}(\Psi_{left} - \Psi_{right})$$

The "+" sign expresses that Ψ_g is a bonding orbital, whereas the "−" sign signifies that Ψ_u is an anti-bonding orbital, consistent with the bound character of the $1s\sigma_g$ potential energy curve and the strongly repulsive nature of the $2p\sigma_u$ potential. These equations can be rewritten to obtain expression for Ψ_{left} and Ψ_{right}:

$$\Psi_{left} = \frac{1}{\sqrt{2}}(\Psi_g + \Psi_u)$$
$$\Psi_{right} = \frac{1}{\sqrt{2}}(\Psi_g - \Psi_u)$$

Hence the breaking of the parity of the electronic wavefunction that accompanies electron localization on the right or left proton requires that the two-color XUV+NIR excitation brings the molecule into a superposition of the $1s\sigma_g$ and $2p\sigma_u$ states. Ionization of the molecule by the XUV attosecond pulse, although capable of reaching both the $1s\sigma_g$ and $2p\sigma_u$ states, cannot do this by itself, since both ionization events are accompanied by distinguishable photoelectrons (in a state with odd angular momentum for ionization producing the $1s\sigma_g$ state, and in a state with even angular momentum for ionization producing the $2p\sigma_u$ state). Note that single-photon XUV ionization can produce a molecular frame asymmetry of the electron, i.e. a correlation or anti-correlation between the direction in which the photoelectron is ejected and the location of the bound electron [16] — these two types of electron localization are not to be confused.

In the pump-probe experiments reported in [15], two mechanisms could be identified for the electron localization under the influence of the XUV+NIR excitation (see Fig. 5.4). In the first of these the NIR acts on the H_2^+ molecular ion as it is dissociating along the $2p\sigma_u$ potential energy curve, and promotes part of the population downward onto the $1s\sigma_g$ potential. This mechanism is very similar to a mechanism that was observed previously in experiments on carrier envelope phase (CEP) control of electron localization, the main difference being that in that case an electron-ion re-collision with a laser-accelerated photoelectron promoted the molecular ion to the $2p\sigma_u$ potential energy curve, instead of the direct XUV photoionization process. The dynamics of electron localization under the influence of the NIR laser field can be well understood within a description using quasi-static laser-dressed potential curves [17, 18]. In fact, it is within such a description that Ψ_{left} and Ψ_{right} appear as eigenstates in the limit that the coupling by the laser exceeds the energy gap between the field-free states. A second mechanism was really unique to the pump-probe experiments reported in [15]. Here, as before, one contribution to the mechanism was XUV ionization launching a wavepacket on the $2p\sigma_u$ potential energy curve. This contribution interfered with a contribution related to the afore-mentioned auto-ionization of the Q_1 state, which produces a continuum vibrational wavepacket on the $1s\sigma_g$ potential. This wavepacket overlaps the continuum vibrational wavepacket emerging from the $2p\sigma_u$ state in energy. Now, the role of the NIR laser field is to act on the photoelectron that accompanies the ionization process. Similar to the NIR-driven formation of sidebands in a RABBITT experiment, the NIR laser can change the energy and angular momentum of the outgoing photoelectron, thereby creating an overlap between the photoelectron wavepackets accompanying the two ionization processes described above, which otherwise would overlap. Thus, intriguingly, this mechanism shows how the correlation between the wavefunction of the outgoing photoelectron

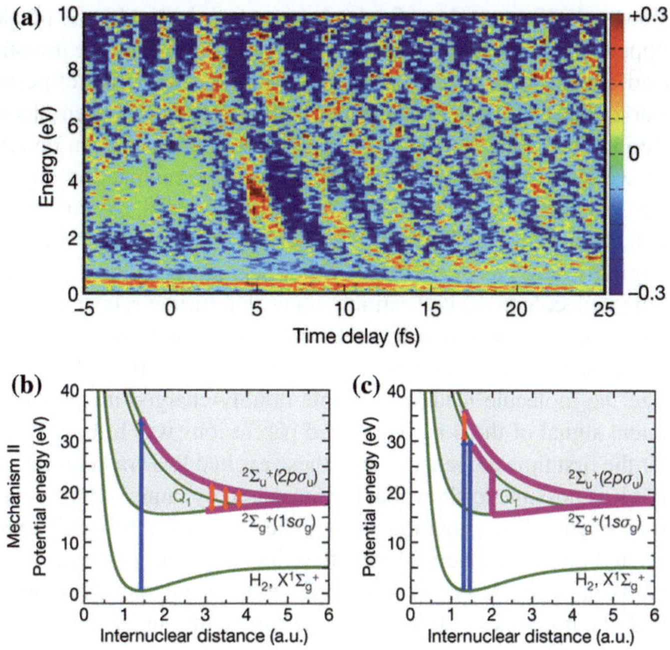

Fig. 5.4 Electron localization has proven itself as a useful observable in some of the early attosecond pump-probe experiments. In the experiment of Sansone et al. [15], electron localization was measured by monitoring the asymmetry in the formation of charged atomic fragments in two-color dissociative ionization $H_2 \rightarrow H^+ + H$ (**a**). The dependence of the normalized asymmetry $A = \frac{P_{left}(E_{kin},\Delta t)-P_{right}(E_{kin},\Delta t)}{P_{left}(E_{kin},\Delta t)+P_{right}(E_{kin},\Delta t)}$ of the proton fragment kinetic energy distribution on the kinetic energy Ekin and the time-delay between the attosecond XUV and few-cycle IR laser pulse Δt allowed to infer two mechanisms for electron localization on the attosecond to few-femtosecond scale, relying, respectively, on coupling of the electronic and nuclear degrees of freedom (**b**) and coupling between the bound and the continuum electron (**c**) (see text for details)

and that of the molecular ion can control the localization of the bound electron in the H_2^+ molecular ion. Clearly, in attosecond experiments one has to be very careful that one does not prematurely seek the explanation of results by looking only at a specific segment of the wave-function, but it is often the total (ion+photoelectron) wavefunction, containing correlations between the photoelectron and the ionic parts that needs to be considered.

5.5 Ultrafast XUV Physics Extended to Large Molecular Species: Case of PAH and Femto-Astrochemistry

One of the factors that prevent coherent control of electron dynamics in polyatomic molecules is the existence of efficient couplings between electron and nuclear degrees of freedom followed by internal vibrational energy redistribution that leads to energy flow and coherence loss. Such sequence is initiated by the breakdown of the Born-Oppenheimer approximation. The possibility to extend the investigation of XUV-induced charge dynamics in large molecular species is therefore intimately linked to our understanding and controls of the interplay between electrons and nuclei. A step in that direction was proposed in a recent experiment on PAH molecules where we have investigated multi-electronic photoionization and non-Born Oppenheimer dynamics for this category of molecules [19]. We have identified a general mechanism where a short XUV pulse ionizes a neutral PAH molecule in a spectral region where electron correlation effects are prominent. The ionization step leaves the molecule in excited cationic states that further relax through very fast non-adiabatic couplings. Within a few tens of femtoseconds the cationic molecule reaches a steady state. This process was probed by a weak NIR probe pulse that can further ionize the molecule leading to stable doubly-charged molecular ions. The time-dependent signal of the doubly charged parent ions was followed in real time revealing for the first time the relaxation of these excited PAH radical species. In this experiment it was possible to show that the measured dynamics was independent on the NIR probe intensity, meaning that clearly the experiment revealed a property of the XUV excited molecules themselves rather than a 2-color process.

Although we have considered a complex situation where both electron correlation and non-adiabatic processes in large molecules are acting in tandem, we have been able to tackle this problem also on the theoretical side. This was possible because the experiment was performed on stable ions where no fragmentation occurred. Consequently, it provides a benchmark example for multi-electronic and non-adiabatic process in XUV excited molecules. The XUV photoinduced mechanism has been investigated by using the combination of a multi-electronic, algebraic diagrammatic construction and non-adiabatic effective Hamiltonian calculation. In the lines of L. Cederbaum's investigations [20], it is especially remarkable that already at photon energy of 10 eV, multielectronic effects cannot be neglected and on the contrary photoexcitation by a single XUV photon involves several electrons. More specifically, the spectral range corresponds to shake-up mechanisms where 1 photon leads to ejection of 1 electron from the molecule and 1 electron is excited, creating 2 holes in the electron density. Another important finding for these excited states is that non-adiabatic relaxations occur on very short timescale. Obviously, vibrational bending or stretching modes involving H atoms can be very fast (for instance C–H bending modes are found at $1500\,\mathrm{cm}^{-1}$). Moreover relaxation can occur on timescales that are inferior to the vibrational period of the considered modes if energy curve crossings occur near the minimum energy of the electronic state. The efficiency and timescales of the electronic relaxation is intimately linked to the exact geometry of the potential energy surfaces, i.e. of the crossing between the multi-dimensional space determined by the vibrational modes, creating seams where the electron wavepacket can propagate from one to another electronic state. The global topology of the seams determines how quickly the molecule relaxes. This implies that only an explicit determination of the electronic excitation and geometry of the potential energy surface allows predicting the relaxation dynamics and therefore designing control strategies. We have demonstrated that the radical excited species relax within a few tens of fs depending on the size of the molecule. In fact the size dependency is very smooth and it appears that even for much larger systems, the relaxation would remain in the few 10s or 100s of fs range. The reason for this is that the relaxation implies slight global distortions of the molecular structure which are similar for all PAH molecules. When the molecular size increases the electron density of states is also increasing but the distribution of excited cations following XUV ionization remains similar. There exists extremely fast relaxation below few femtosecond timescale but the population remains always trapped in specific bottlenecks where the dynamics occurs on a few 10s of fs.

Beyond the fundamental aspects of molecular science and ultrafast dynamics discussed above, these results have direct implications in other fields. One striking implication is the relation with astrochemistry. One of the current mysteries in the investigation of the so-called "molecular universe" [21] remains the identification of absorption and emission bands that have been measured by several telescopes. IR emission bands have allowed to speculate that radical PAH molecules could be responsible for more than 10 % of the carbon in the entire universe, combined with their known reactivity, it is very tempting to consider that radical PAHs are the initial building blocks that lead to increasingly complex aromatic molecules eventually leading to the appearance of life in the universe. In molecular clouds where radical PAHs are assumed to be present, telescopes have also observed remarkable bands that could tentatively be assigned to these molecular species. However, the width of these bands is compatible with extremely short lifetimes of the corresponding excited states, in the range of 100 fs. However, it is generally admitted that low lying excited states have a too weak quantum efficiency to explain these bands, and that higher excited states would decay almost directly faster than predicted by the linewidth (because of the rapid increase of the electronic states density in such complex molecules). In fact, in our experiment it appears that the complex quantum mechanical entangled electron-nuclear dynamics confines the relaxation within a time window of several tens of fs, which is in a very good agreement with linewidth in astrophysical observation. In this context, ultrashort XUV pulses are very versatile tools that can produce excited radical species and follow the subsequent dynamics on very short timescale.

5.6 The Ionization Step: Attosecond Delay in Photoemission in the C_{60} Surface Plasmon Resonance

Although fundamental couplings in the molecule limit attosecond charge dynamics in large systems, theory predicts processes on the attosecond timescale in very large systems. In the following we discuss the C_{60} molecule that is a model system to investigate processes on various timescales [22].

The mechanism described in Sect. 5.1 constitutes a transient process that only occurs during the interaction between the molecule and the NIR pulse. It enables

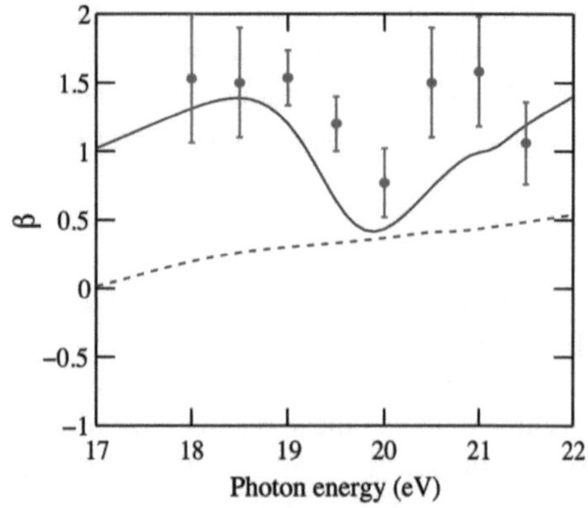

Fig. 5.5 Photon energy dependence of the beta parameter characterizing the photoelectron angular distribution from the HOMO in C_{60} (*blue dots*). TDLDA calculation of the beta parameter as a function of photon energy including correlation effects (*black curve*) or without correlation (*dash line*)

the control of the XUV ionization step with attosecond resolution deciphering the nature of the complex interaction between light electric field and electrons in a molecule. The ionization process, in its simplest form, is certainly the most fundamental mechanism arising on attosecond time scale. As an illustration, the time that an electron takes to reach the ionization continuum has been experimentally investigated in atomic cases [23] and for simple molecules [24], and latter was theoretically investigated for atoms embedded in fullerenes [25]. Quantum mechanically, photoemission delay can be intuitively understood within the basic concepts of scattering theory in terms of phase shifts of the electron wavefunction induced by an external potential. The derivative of this phase shift with respect to the energy can be formally associated to a photoemission time delay through the concept introduced by Wigner [26]. Although, most of the photoionization time delay investigations were carried out in simple atomic system, we have taken a first step towards large molecules by considering the case of the surface plasmon in C_{60} [27]. A surface plasmon resonance (SPR) is an archetype of collectively interacting electrons and therefore it is the ideal case in which microscopic multielectronic effects on photoionization delays can be investigated (Fig. 5.6).

In a recent synchrotron experiment, we have used electron momentum imaging to investigate the photoelectron angular distribution of photoionized C_{60} in the vicinity of its surface plasmon resonance around 20 eV. We observed a variation of the photoelectron angular distribution as a function of the photon energy that is associated to electron correlation effects (see Fig. 5.5). Quantum mechanically, this can be expressed in terms of a phase shift in the outgoing electron wavefunction that is responsible for a variation of the 1-photon radial transition dipole matrix element subsequently modifying the angular distribution.

This phase modification of the electronic wavefunction with photon energy can be interpreted in terms of variation of the photoemission time delay induced by the plasmon resonance. During the photoionization process, electron correlation affects the electron ejection process and depending on the excitation frequency, the plasmon can either act as a potential well or barrier, that either accelerates or decelerates the electron leading either to a decrease or increase of the photoemission time delay (see

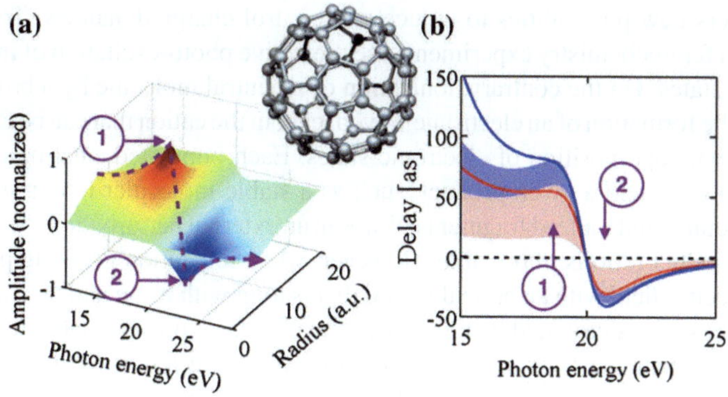

Fig. 5.6 Around the C_{60} *SPR*, the many-electron potential (**a**) acts as a screening potential that modifies both the angular distribution and photoemission time delay (**b**). This can be described semi-classically with a good accuracy using an effective potential

Fig. 5.6). Remarkably, a retardation or acceleration of several tens of attoseconds was predicted, showing not only that the plasmon, and more specifically electron correlation, strongly influences the ionization time delay but also that it does it on a time scale that could be accessible in experiments. TDLDA calculations show that electron correlation indeed determines both the variation of angular distribution and photoemission time-delays. Moreover, an analytical model based on first principles also predicts attosecond variations of the delay around the plasmon resonance in the case of confined electrons in a jellium-like potential. This shows that effect should be observed in other large molecular systems or in metallic nano-sized materials exhibiting a plasmon resonance.

Probing the timescale of the electron ejection gives an accurate measurement of the influence of its surrounding environment and is therefore a sensitive probe of electron correlation. In our investigation of the photoionization of C_{60} molecules, the many-electron interaction clearly influences the ejection dynamics on a timescale that is measurable in current attosecond pump-probe experiment. In the previous example where molecular photoionization was influenced by the time-dependent polarization of the molecule, it is expected that photoemission time-delay will also be affected by the transient dynamics. In general, the measurement of the angular distribution and delay in photoemission could serve as a probe of the electron dynamics in molecules and nano-sized objects in a 4-D (angularly resolved energy/photoemission delay) electron spectroscopy approach.

5.7 Conclusion

We have discussed several aspects of ultrafast mechanisms in molecules induced or probed by short XUV pulses. In these experiments, electron correlation plays a crucial role and large bandwidth required to support the short duration of these pulses offers new possibilities to induce and control charge dynamics. Traditional protocol in femtochemistry experiments uses selective photo-excitation of individual electronic states. On the contrary, ionization of a neutral molecule by a broad pulse results in the formation of an electronic wavepacket in the cation that can be expressed as a coherent superposition of electronic states. Each populated electronic state of the system can yield a given product, such as a stable molecular ion or single and multiple neutral and ionized fragments. Without an external perturbation, control over the final product ratio is impossible. However, when ionization is taking place in a dressed electric field with an optical period that is long with respect to the attosecond XUV pulse, such as provided by a femtosecond NIR laser pulse, one has to consider possible new pathways involving the coupling of the different electronic excited states of the molecules by the additional electric field. In this scheme, the XUV pulse can prepare a coherent superposition of electronic "intermediate" states that can be connected to a final state by the NIR laser field, enabling therefore interferences between pathways involving different intermediate states that are sensitive to the phase between the NIR laser field and the XUV pulse. By controlling the XUV-NIR delay will therefore results in a modulation of the branching ratio of the various products formed. We have shown that such attosecond control can be performed in experiments in the dissociation of several molecules like H_2, N_2, O_2. Subcycle dynamics below 1fs has also been observed experimentally in diatomic molecules and it is remarkable that attosecond dynamics has also been observed in polyatomic systems such as CO_2 and C_2H_4 and these observations allow us to be optimistic that attosecond control of molecular ionization could be performed in increasingly large systems. We have shown that C_{60} can serve as a model system for attosecond investigations with an example of the correlation-induced delay in photoemission at the surface plasmon resonance. Complex systems such as PAH molecules have also been discussed in the context of multielectronic excitation followed by non-adiabatic relaxation. This clearly demonstrates that attosecond and ultrafast XUV molecular physics is just at its infancy and that crossing below 1 fs, there remains an entire world of phenomena to explore in molecules.

References

1. F. Lépine, M.Y. Ivanov, M.J.J. Vrakking, Nat. Photonics **8**, 195–204 (2014)
2. M. Krausz, M. Ivanov, Attosecond physics. Rev. Mod. Phys. **81**, 163–234 (2009)
3. J. Itatani, J. Levesque, D. Zeidler, H. Niikura, H. Pepin, J.C. Kieffer, P.B. Corkum, D.M. Villeneuve, Nature **432**, 867–871 (2004)
4. O. Smirnova, Y. Mairesse, S. Patchkovskii, N. Dudovich, D. Villeneuve, P. Corkum, M.Y. Ivanov, High harmonic interferometry of multi-electron dynamics in molecules. Nature **460**, 972–977 (2009)
5. P.A. Carpeggiani, P. Tzallas, A. Palacios, D. Gray, F. Martín, D. Charalambidis, Phys. Rev. A **89**, 023420 (2014)
6. A.I. Kuleff, L.S. Cederbaum, Dynamics. J. Phys. B **47**, 1240021–12400212 (2014)
7. F. Calegari, D. Ayuso, A. Trabattoni, L. Belshaw, S. De Camillis, S. Anumula, F. Frassetto, L. Poletto, A. Palacios, P. Decleva et al., Science **346**, 336–339 (2014)

8. J.J. Larsen, H. Sakai, C.P. Safvan, I. Wendt-Larsen, H. Stapelfeldt, J. Chem. Phys. **111**, 7774–7781 (1999)
9. C. Neidel, J. Klei, C.H. Yang, A. Rouzée, M.J.J. Vrakking, K. Klünder, M. Miranda, C.L. Arnold, T. Fordell, A. L'Huillier, M. Gisselbrecht, P. Johnsson, M.P. Dinh, E. Suraud, P.G. Reinhard, V. Despré, M.A.L. Marques, F. Lépine, Phys. Rev. Lett. **111**, 033001 (2013)
10. M. Ruberti, R. Yun, K. Gokhberg, S. Kopelke, L.S. Cederbaum, F. Tarantelli, V. Averbukh, J. Chem. Phys. **140**, 184107 (2014)
11. J.C. Baggesen, L.B. Madsen, Phys. Rev. Lett. **104**, 043602 (2010)
12. F. Kelkensberg, W. Siu, J.F. Perez-Torres, F. Morales, G. Gademann, A. Rouzee, P. Johnsson, M. Lucchini, F. Calegari, J.L. Sanz-Vicario, F. Martin, M.J.J. Vrakking, Phys. Rev. Lett. **107**, 043002 (2011)
13. H.G. Muller, Reconstruction of attosecond harmonic beating by interference of two-photon transitions. Appl. Phys. B-Lasers Opt. **74**, S17–S21 (2002)
14. W. Siu, F. Kelkensberg, G. Gademann, A. Rouzée, P. Johnsson, D. Dowek, M. Lucchini, F. Calegari, U. De Giovannini, A. Rubio, R.R. Lucchese, H. Kono, F. Lépine, M.J.J. Vrakking, Attosecond control of dissociative ionization of O_2 molecules. Phys. Rev. A **84**, 063412 (2011)
15. G. Sansone, F. Kelkensberg, J.F. Perez-Torres, F. Morales, M.F. Kling, W. Siu, O. Ghafur, P. Johnsson, M. Swoboda, E. Benedetti, F. Ferrari, F. Lepine, J.L. Sanz-Vicario, S. Zherebtsov, I. Znakovskaya, A. L'Huillier, M.Y. Ivanov, M. Nisoli, F. Martin, M.J.J. Vrakking, Electron localization following attosecond molecular photoionization. Nature **465**, 763–U763 (2010)
16. F. Martin, J. Fernandez, T. Havermeier, L. Foucar, T. Weber, K. Kreidi, M. Schoffler, L. Schmidt, T. Jahnke, O. Jagutzki, A. Czasch, E.P. Benis, T. Osipov, A.L. Landers, A. Belkacem, M.H. Prior, H. Schmidt-Bocking, C.L. Cocke, R. Dorner, Single photon-induced symmetry breaking of H_2 dissociation. Science **315**, 629–633 (2007)
17. P. Dietrich, M.Y. Ivanov, F.A. Ilkov, P.B. Corkum, Two-electron dissociative ionization of H_2 and D_2 in infrared laser fields. Phys. Rev. Lett. **77**, 4150–4153 (1996)
18. F. Kelkensberg, G. Sansone, M.Y. Ivanov, M. Vrakking, A semi-classical model of attosecond electron localization in dissociative ionization of hydrogen. Phys. Chem. Chem. Phys. **13**, 8647–8652 (2011)
19. A. Marciniak, V. Despré, T. Barillot, A. Rouzée, M.C.E. Galbraith, J. Klei, C-H. Yang, C. T. L. Smeenk, V. Loriot, S. Nagaprasad Reddy, A.G.G.M. Tielens, S. Mahapatra, A.I. Kuleff, M.J.J. Vrakking & F. Lépine Femto-Astrochemistry: XUV Multi-electronic excitation followed by ultrafast non-adiabatic relaxation in PAH molecules. Nat. Commun. (in press 2015)
20. J. Breidbach and L. S. Cederbaum J. Chem. Phys. 118, 3983 (2003)
21. A.G.G.M. Tielens, Rev. Mod. Phys. **85**, 1021 (2013)
22. F. Lépine, Multiscale dynamics of C_{60}: from attosecond to statistical physics. J. Phys. B: At. Mol. Opt. Phys topical review. **48**, 122002 (2015)
23. M. Schultze et al., Science **328**, 1658 (2010)
24. S. Haessler et al., Phys. Rev. A **80**, 011404 (2009)
25. G. Dixit, H.S. Chakraborty, M.E. Madjet, Phys. Rev. Lett. **111**, 203003 (2013)
26. E.P. Wigner, Phys. Rev. **98**, 145–147 (1955)
27. T. Barillot, C. Cauchy, P-A. Hervieux, M. Gisselbrecht, S. Canton, Per Johnson, K. Hansen, C. Arnold, S. Sorensen and F. Lépine, Angular asymmetry and attosecond time delay from giant plasmon in C_{60} photoionization. Phys. Rev. A **91**, 033413 (2015)

Chapter 6
Attosecond Electron Spectroscopy in Molecules

Francesca Calegari, Jason Greenwood, Candong Liu, Matteo Lucchini, Maurizio Reduzzi, Giuseppe Sansone, Andrea Trabattoni and Mauro Nisoli

Abstract Electron motion in matter occurs on a temporal scale of a few hundreds of attoseconds, a time scale shorter then an optical period of the visible light. Attosecond science is nowadays a well-established research field and several techniques based on the use of attosecond pulses have been developed to investigate and control the electron dynamics initiated by photo-ionization in atoms, molecules and condensed

F. Calegari (✉) · G. Sansone · M. Nisoli
Institute of Photonics and Nanotechnologies, IFN-CNR,
Piazza L. da Vinci 32, 20133 Milano, Italy
e-mail: francesca.calegari@polimi.it

G. Sansone
e-mail: giuseppe.sansone@polimi.it

M. Nisoli
e-mail: mauro.nisoli@polimi.it

J. Greenwood
Centre for Plasma Physics, School of Maths and Physics,
Queen's University Belfast, BT7 1NN, UK
e-mail: J.Greenwood@qub.ac.uk

C. Liu
State Key Laboratory of High Field Laser Physics, Shanghai Institute of Optics and Fine Mechanics, Chinese Academy of Sciences, Shanghai 201800, China
e-mail: cdliu@siom.ac.cn

M. Lucchini
Department of Physics, ETH Zurich, 8093 Zürich, Switzerland
e-mail: lmatteo@phys.ethz.ch

M. Reduzzi · G. Sansone · A. Trabattoni · M. Nisoli
Department of Physics, Politecnico di Milano, Piazza Leonardo da Vinci 32, 20133 Milano, Italy

M. Reduzzi
e-mail: mauriziobattista.reduzzi@polimi.it

A. Trabattoni
e-mail: andrea.trabattoni@polimi.it

M. Kitzler and S. Gräfe (eds.), *Ultrafast Dynamics Driven by Intense Light Pulses*, Springer Series on Atomic, Optical, and Plasma Physics 86, DOI 10.1007/978-3-319-20173-3_6

matter. In this chapter we will review progress in attosecond technology and in the application of attosecond pulses to the investigation of ultrafast electron processes in molecules.

6.1 Introduction

High-order harmonic generation in gases is now the standard method used for the generation of attosecond pulses, in the form of train of pulses or isolated pulses [1–3]. The main objective of attosecond science is real-time observation and direct control of electronic motion in atoms, molecules, nanostructures and solids. Various techniques have been implemented for applications of sub-femtosecond pulses. Experiments with attosecond temporal resolution present characteristics, which are completely different with respect to the well-established and used femtosecond techniques, where the probe pulse measures the changes of the optical properties of the sample (i.e., absorption or reflection coefficients) induced by the pump pulse. Novel experimental techniques to initiate and probe electron dynamics on an attosecond temporal scale have been introduced in the last decade. In 2002, immediately after the first generation of isolated attosecond pulses, Drescher and coworkers reported on the real-time observation of Auger decay in Krypton [4]: 0.9-fs soft-X-ray pulses have been used for excitation of Krypton atoms and sub-7-fs infrared (IR) pulses for probing electron emission. The IR-induced ionization of shake-up states generated by attosecond excitation has been measured with attosecond resolution, thus leading to real-time observation of electron tunneling in atoms (Neon and Xenon) [5]. Delay in electron photoemission from atoms has been investigated by using both isolated attosecond pulses [6] and trains of attosecond pulses [7, 8]. The attosecond-pump/femtosecond-probe technique has been also extended to condensed matter systems: photoelectron emission from single-crystal tungsten, induced by isolated attosecond pulses, has been probed by the electric field of a 5-fs light pulse, showing a 100-as delay between the emission of photoelectrons that originate from localized core states of the metal, and those that are freed from delocalized conduction-band states [9]. The dynamics of electron localization initiated by attosecond excitation has been studied in H_2 and D_2 molecules [10]. Control of photoionization of D_2 and O_2 molecules has been obtained by employing attosecond pulse trains [11, 40]. More recently, attosecond pulses have been used to photoionize an amino acid, giving rise to a complex and ultrafast electron dynamics [12].

In this chapter we will review progress in attosecond technology and in the application of attosecond pulses to the investigation of ultrafast electron dynamics in molecules. We will concentrate on the generation, characterization and application of isolated attosecond pulses. Section 6.2 is devoted to a very brief review of various techniques implemented for the confinement of the harmonic generation process to a single event, with particular emphasis on the temporal gating schemes. Temporal characterization of isolated attosecond pulses is analysed in Sect. 6.3, where a recently proposed method for the measurement of the carrier-envelope phase of attosecond pulses is described. Application of isolated attosecond pulses to the investigation of electron dynamics in diatomic molecules by using the velocity map imaging spectroscopy is reviewed in Sect. 6.4. Electron dynamics initiated in a biomolecule by attosecond pulses is discussed in Sect. 6.5.

6.2 Temporal Gating Techniques for the Generation of Isolated Attosecond Pulses

Different approaches have been proposed and implemented to confine the harmonic generation process to a single event, leading to the production of isolated attosecond pulses. The first scheme used in 2001 for the generation of isolated attosecond pulses was based on bandpass filtering of the highest energy region of the XUV emission [13, 14]. This technique is based on the fact that the high-energy portion of the XUV spectrum (*cut-off* region) is produced only near the peak of the driving field, thus corresponding to light emitted only within one-half of the laser oscillation period. This generation method requires the use of intense sub-5-fs driving pulses, with stabilized carrier-envelope phase (CEP). Isolated attosecond pulses with a temporal duration down to 80 as have been generated using this method [15] in 2008.

A very effective approach, named polarization gating (PG), is based on the strong sensitivity of the HHG process to the ellipticity of the driving pulse: harmonic generation can be obtained only by using linearly polarized driving pulses. If the polarization of the driving field is temporally modulated from circular to linear and back to circular, the XUV emission is limited to the temporal window where the driving pulse is linearly polarized, as first proposed in 1994 [16]. Two birefringent plates can be used to generate a driving field characterized by almost linear polarization in a temporal window shorter than the separation between two consecutive attosecond pulses, as first proposed by Tcherbakoff et al. [17]. A first multiple-order quarter wave-plate introduces a temporal delay, τ, between the ordinary and extraordinary field components. The second plate is a zero-order quarter wave-plate which transforms the linear polarization into circular and vice-versa so that a short temporal gate of linear polarization can be obtained [18]. By applying this temporal gating scheme, in combination with 5-fs driving pulses with stable CEP, the generation of isolated attosecond pulses at 36-eV photon energy, with a duration of 130 as has been demonstrated in 2006 [19].

The PG technique requires few-cycle CEP-stabilized driving pulses. This is clearly demonstrated in Fig. 6.1, which shows the evolution of the harmonic spectra generated in argon by 5-fs pulses with shaped polarization as a function of the carrier-envelope phase of the driving field. A clear π periodicity of the emitted spectra can be observed, as expected from symmetry considerations. The spectra are composed of well resolved harmonics of the fundamental radiation for certain CEP values, that evolves in a smooth structureless continuum upon a change of $\pi/2$. The measurements indicate that the PG technique can effectively confine the emission both in the

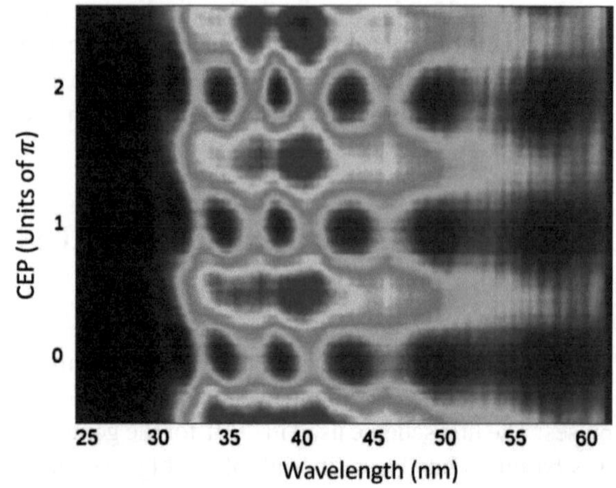

Fig. 6.1 Harmonic spectra generated in argon by 5-fs driving pulses with modulated polarization as a function of the CEP shift

plateau and in the cut-off region, leading to the generation of broadband continua. The generation of XUV continuum corresponds to those CEP values that determine the emission of an attosecond pulse at the center of the gate (around $\epsilon = 0$, where ϵ is the ellipticity of the driving electric field); the rapid variation of the ellipticity strongly reduces the emission of adjacent attosecond pulses one half cycle before or after the main pulse and therefore an XUV continuum is synthesized. Upon changing the CEP by $\pi/2$, two attosecond pulses are generated, symmetrically located with respect to the center of the gate, thus giving rise to a highly modulated spectrum.

It is worth to note that by using the polarization gating technique it is not possible to obtain narrow temporal gates if the duration of the driving pulse is too long while still preserving a high driving intensity in the temporal window of linear polarization. In order to overcome this requirement Chang [20] proposed to add a second harmonic field to the IR driving pulse in the PG scheme resulting in the introduction of the double optical gating (DOG) technique. If the second harmonic is intense enough the time interval between two consecutive attosecond pulses is increased from one-half to one optical cycle, with two main consequences: neutral atom population depletion is reduced and a wider temporal gate can be applied.

A different approach is based on the creation of the required temporal gate by exploiting the laser-induced ionization of the generating medium (ionization gating, IG). If the driving field is intense enough the plasma density rapidly increases on the leading edge of the laser pulse, thus creating a phase mismatch responsible for the suppression of HHG for all later half-cycles. Even if the intensity of the driving field is not enough to create a temporal gate as narrow as required to select a single attosecond pulse, the IG can be used in combination with bandpass filtering to overcome this problem [21, 22]. In 2010, isolated attosecond pulses with temporal duration down to 155 as and an energy on target of 2.1 nJ have been generated using this method [23].

6.3 Streaking Spectroscopy and Carrier-Envelope Phase of Attosecond Pulses

The first temporal characterization of trains and isolated attosecond pulses was demonstrated by measuring the photoelectron spectra released in the continuum by the combination of the extreme ultraviolet (XUV) waveform and a synchronized IR field [14, 24]. In the case of trains of attosecond pulses, the photoelectron spectrum generated by the XUV-only pulses resembles the spectral distribution of the harmonic radiation and it is characterized by discrete photoelectron peaks at energies $E = (2n + 1)\hbar\omega_0 - I_p$, where ω_0 indicates the fundamental frequency of the driving field and I_p is the ionization potential of the target atom. In the presence of an IR field, the photoelectron released by single XUV-photon absorption can either emit or absorb an additional IR photon, resulting in a final kinetic energy given by $E_{2n} = 2n\hbar\omega_0 - I_p$ and $E_{2n+2} = (2n + 2)\hbar\omega_0 - I_p$. These photoelectron peaks are usually referred to as *sidebands* of the main harmonic peak. Two different pathways can contribute to the same sideband of kinetic energy E_{2n}, corresponding to the absorption of a photon of the $2n + 1$ harmonic and emission of an IR photon or to the absorption of a photon of the $2n - 1$ harmonic and absorption of and IR photon. These two paths interfere leading to oscillations in the yield of the photoelectron emission as a function of the delay between the harmonic comb and the IR field oscillations. The relative phase of the harmonics (and therefore the temporal duration of the trains of pulses) can be characterized by determining the relative phase of the sidebands oscillations.

In the case of isolated attosecond pulses, the temporal characterization is still based on the measurement of the photoelectron spectra in the two-color field but the experimental outcome can be more directly interpreted in a classical picture. Once in the continuum, the photoelectron released by the absorption of an XUV photon is accelerated by the electric field of the IR field ending up with an additional final momentum that depends only on the value of the vector potential $\mathbf{A}_{IR}(t)$ at the ionization instant, t_0. By changing the relative delay between the isolated attosecond pulse and the IR field, the photoelectron spectra present modulations in kinetic energy due to the oscillations of the vector potential. This method is known as attosecond streak camera and it was used to derive information about the duration of isolated attosecond pulses [25]. By using the Frequency Resolved Optical Gating for Complete Reconstruction of Attosecond Bursts (FROG-CRAB) algorithm [26], the complete electric field of the IR pulse and the temporal phase and amplitude of the attosecond pulse were completely characterized, demonstrating the generation of single cycle attosecond waveform at 36 eV [19].

The attosecond streak camera combined with the FROG-CRAB algorithm do not provide access to the CEP of the attosecond pulses. In the visible/infrared spectral range, effects of the CEP were first investigated in strong field processes driven by few-cycle laser pulses [27–29]. At the same time experimental techniques for the measurement of the CEP were developed [30]. In the case of few-cycle pulses in the XUV spectral range, CEP effects have been discussed in the photoelectron spectra created by fields with intensity in the range of ($I_{XUV} \geq 10^{14}\,W/cm^2$) in [31, 32], evidencing a CEP-dependent asymmetry along the laser polarization due to the interference between a single and a two-photon ionization pathway. In this unexplored regime, the strong-field electron dynamics is driven by the precise electric field of the attosecond pulse rather than by its intensity profile. In order to relax dramatically the requirement for the peak intensity of the isolated attosecond pulse, recently it was shown that information on the CEP of isolated attosecond pulses can be obtained by applying the attosecond streak camera method using a CEP-stable IR pulse with an intensity higher than the one typically used for the temporal characterization of attosecond waveforms [33]. The attosecond streak camera principle assumes the possibility to neglect contributions of the photo-electron released by the IR field-only through Above Threshold Ionization (ATI), or at least to separate the ATI electron from those released by the absorption of a single XUV photon and then accelerated (streaked) by the IR field. A typical spectrogram simulated in these conditions is shown in Fig. 6.2a. For the considered IR intensity, the photoelectron spectra generated by ATI is negligible and therefore cannot be observed. Upon increasing the IR intensity (see Fig. 6.2b), the amplitude of the energy modulation of the photoelectron spectra increases and a complex structure appears when the photoelectron spectra is shifted towards lower kinetic energies. Finally, for the highest IR intensity considered in Fig. 6.2c, the spectrogram is characterized by a rich modulation structure, which strongly depends on the relative delay. In particular the modulation results to be sensitive to the CEP of the attosecond pulse as shown in Fig. 6.3a, which presents the cut at $\tau = 0$ of the spectrogram shown in Fig. 6.2c for three CEPs of the XUV pulse: ($\varphi_{XUV} = 0, \pi/2$ and π). The modulation presents a regular pattern that shifts linearly upon changing the CEP of the XUV pulse (Fig. 6.3b).

The origin of the modulation and of the CEP-dependence is due to the interference between two photoelectron wave-packets:

(1) The first one is released by ATI of IR-pulse only. As the influence of the XUV field on this process is negligible, the characteristics of the ATI photoelectron spectra does not depend on the relative delay τ between the two pulses. In particular, the photoelectron spectrum emitted by the IR-only extends up to $10U_p$, where U_p is the ponderomotive potential of the IR field. The photoelectrons with energy E between $2U_p$ (5.5 eV) and $10U_p$ (27.3 eV) are due to a rescattering mechanism [34, 35]. Therefore the photoelectron wave-packet contributing to the interference between 11 and 27 eV is a rescattered photoelectron wavepacket.

(2) The second photoelectron wave-packet is released in the continuum by single XUV-photon absorption and then streaked by the synchronized IR field. The energy distribution of this wave-packet strongly depends on the relative delay between the XUV and IR field, accordingly to the simple classical picture described above.

When the two photoelectron wave-packets overlap in energy (in the low energy region for the IR-intensity shown in Fig. 6.2b and over the entire energy range for Fig. 6.2c), the interference between the two amplitudes determines a modulation that depends on the phase difference $\Delta\Phi$ between the two wave-packets:

$$\Delta\Phi = \Phi_{XUV}(E,\tau) + \varphi_{XUV} - \Phi_{IR}(E) \tag{6.1}$$

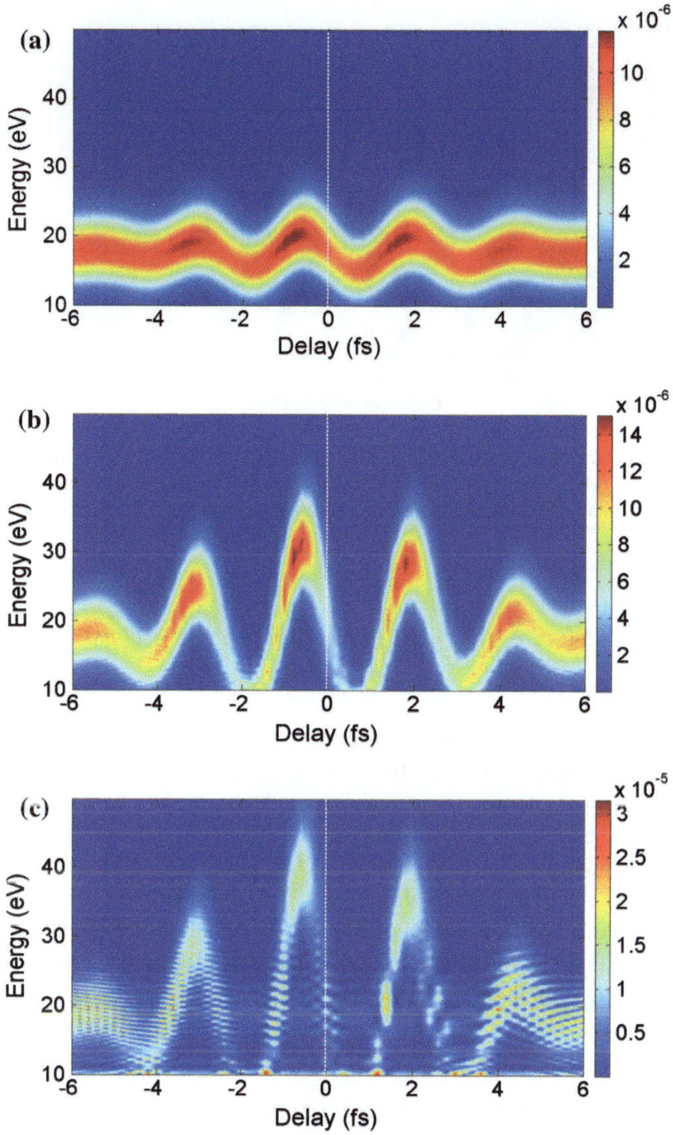

Fig. 6.2 Simulated photoelectron spectra generated by an isolated attosecond XUV pulse ($T_{\mathrm{XUV}} =$ 250 as, $\varphi_{\mathrm{XUV}} = 0$, $\hbar\omega_{\mathrm{XUV}} = 36$ eV, $\mathrm{I}_{\mathrm{XUV}} = 10^{12}$ W/cm^2) and a few-cycle IR field ($T_{\mathrm{IR}} = 5$ fs, $\varphi_{\mathrm{IR}} = 0$, $\hbar\omega_{\mathrm{IR}} = 1.63$ eV) with intensities **a** $\mathrm{I}_{\mathrm{IR}} = 10^{12}$ W/cm^2, **b** $\mathrm{I}_{\mathrm{IR}} = 2 \times 10^{13}$ W/cm^2, and **c** $\mathrm{I}_{\mathrm{IR}} = 4.5 \times 10^{13}$ W/cm^2. The vertical lines indicate the delay $\tau = 0$. Reprinted from [33]

where $\Phi_{\mathrm{XUV}} + \varphi_{\mathrm{XUV}}$ and Φ_{IR} are the phases of the electron wave packets created by the XUV and IR fields, respectively. For the delay $\tau = 0$, this equation can be approximated by:

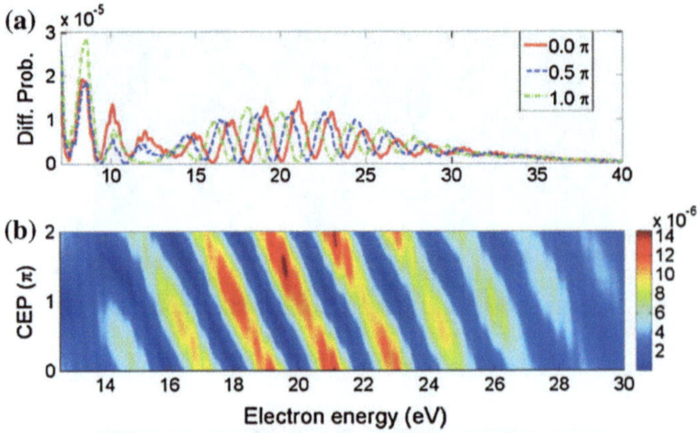

Fig. 6.3 **a** Simulated photoelectron spectra for $\tau = 0$, for the CEPs $\varphi_{XUV} = 0$ (*red solid line*), $\pi/2$ (*blue dashed line*), and π (*green dash-dotted line*). **b** Interference fringes as a function of the electron energy in the CEP range [0; 2π]. Other parameters as in Fig. 6.2c. Reprinted from [33]

$$\Delta\Phi \simeq \frac{\mathbf{p}^2}{2} t_{resc} + \mathbf{p} \int_0^{t_{resc}} \mathbf{A}_{\mathrm{IR}}(t)dt + \varphi_{\mathrm{XUV}} + const. \tag{6.2}$$

where $\mathbf{p}^2/2 = E$ is the final electron energy and t_{resc} is the rescattering time of the most relevant contribution of the rescattered electrons. We have verified that the first term of (6.2) dominates over the second one, leading to a phase difference that depends essentially on the different times the electrons, with final momentum $\mathbf{p}$, enter the continuum ($\tau = 0$ for direct XUV photoionization and $t = t_{resc}$ for the re-scattered electrons). The value of $t_{resc} = 1.8$ fs well matches with the time $\delta t = 1.95$ fs extracted from the photoelectron modulation between 11 and 27 eV. Equation (6.2) also clearly evidences the linear dependence of the position of the interference fringes on the CEP of the XUV pulse.

6.4 Velocity Map Imaging Spectroscopy of Diatomic Molecules

The most conventional approach to perform measurements with attosecond temporal resolution is to use an isolated attosecond pulse in combination with a near infrared femtosecond pulse. Due to the high photon energy of the attosecond pulses (ranging from the extreme ultraviolet to the soft-X spectral region), the excitation of the sample almost inevitably leads to inner valence or core photoelectron emission, followed by a rearrangement of the electronic system. The electron reorganization gives rise to Auger decay, shake-up processes, charge transfer and charge migration in large molecules. Among the different schemes proposed and demonstrated for attosecond pump-probe measurements, in this section we will focus on the velocity map imaging (VMI)technique implemented to study the electron dynamics in small molecules.

Since the first development of the velocity map imaging spectrometer (VMIS) by Eppink and Parker [36], this technique has become a popular tool for measuring angle-resolved ion or electron momentum distributions that result from molecular photo-ionization and subsequent photo-fragmentation [37]. The design of the VMIS has been improved in 2009 by Ghafur [38] and co-workers by integrating the gas injection system in the repeller electrode of the spectrometer. This design is effective at increasing the density of the gas target in the interaction region, thus being particularly suitable for attosecond experiments due to the limited photon flux of the attosecond laser sources.

In 2010 Sansone et al. performed the first pump-probe measurement, using a VMIS, with isolated attosecond pulses to excite H_2 and D_2 molecules [10]. The kinetic energy and angular distribution of H^+ and D^+ ionic fragments were measured as a function of the delay between an isolated attosecond pump pulse (energy range 20–40 eV) and a 6-fs, CEP stable IR probe pulse. From the analysis of the kinetic energies and angular distributions of the photo fragments with respect to the laser polarization, it was possible to identify many contributions including dissociative ionization of the $^2\Sigma_g^+$ and the $^2\Sigma_u^+$ states and autoionization of the Q_1 state, shown in the simplified scheme of Fig. 6.4a. The kinetic energy distribution of the D^+ ions measured as a function of the XUV-pump IR-probe delay is reported in Fig. 6.4b. At kinetic energy <1 eV, a bond softening mechanism [39] induced by the IR probe

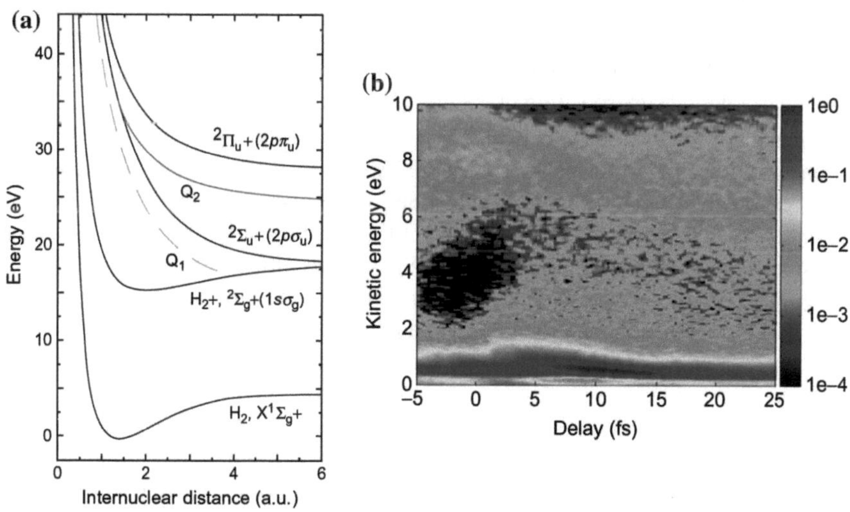

Fig. 6.4 **a** Simplified scheme of the potential energy curves of H_2 and H_2^+. **b** Experimental kinetic energy distributions of D^+ ions as a function of the delay between the attosecond pump pulses and the infrared probe pulses. Reprinted from [10]

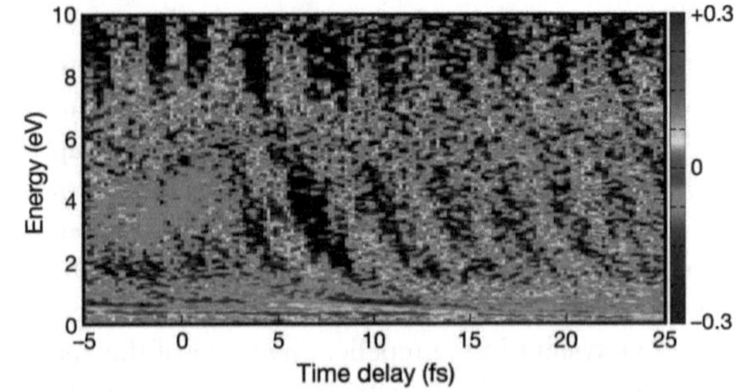

Fig. 6.5 Asymmetry parameter for the formation of D^+ ions after attosecond excitation as a function of the kinetic energy and of the temporal delay between XUV and IR pulses. Reprinted from [10]

pulse has been identified. The increase of the ion signal at kinetic energies around 8 eV has been associated to the IR-induced transition from the Q_1 state to the $^2\Sigma_u^+$ state and to the IR-induced coupling of the $^2\Sigma_g^+$ and $^2\Sigma_u^+$ states both resulting in an increase of cross-section of the $^2\Sigma_u^+$ state.

The experiment revealed that dissociative ionization of H_2 and D_2 molecules by using the combination of isolated XUV pulses and CEP-stable IR pulses inevitably leads to electron localization. This process was identified through the asymmetry parameter, shown in Fig. 6.5, and defined as:

$$A(E_k, \tau) = \frac{N_L(E_k, \tau) - N_R(E_k, \tau)}{N_L(E_k, \tau) + N_R(E_k, \tau)} \tag{6.3}$$

where $N_L(E_k, \tau)$ and $N_R(E_k, \tau)$ are the numbers of ions arriving on the left-hand and right-hand sides of the detector, respectively. As can be seen in Fig. 6.5, in the energy range 2–10 eV the asymmetry parameter exhibits sub-cycle oscillations as a function of the delay between the XUV and the IR pulses. The asymmetry is the result of a coherent superposition of gerade and ungerade states (in this case $^2\Sigma_g^+$ and $^2\Sigma_u^+$), and the relative phase between the two states results in electron localization on one atom ore the other of the molecule. Two mechanisms have been identified to be responsible for this charge localization: in the first mechanism asymmetry is caused by the interference of a wave packet launched directly in the $^2\Sigma_u^+$ state (at zero time delay) and a wave packet in the $^2\Sigma_g^+$ state resulting from autoionization of the Q_1 state. In the second mechanism asymmetry is caused by the interference of a wave packet that is launched in the $^2\Sigma_u^+$ state by direct XUV ionization and a wave packet in the $^2\Sigma_g^+$ state that results from stimulated emission during the dissociation process.

A similar experiment has been performed in 2011 by Kelkensberg et al. by using a moderately intense IR pulse (3×10^{13} W/cm^2) in combination with an attosecond pulse train (APT) [40]. While in the experiment described above the attosecond pulses were used to initiate electron dynamics subsequently probed by the IR field, in this case the IR pulse was used to influence the electronic states that are accessed in photoionization of H_2 by the APTs. The three-dimensional velocity distributions of H^+ and D^+ were measured as a function of the delay between the APT and the

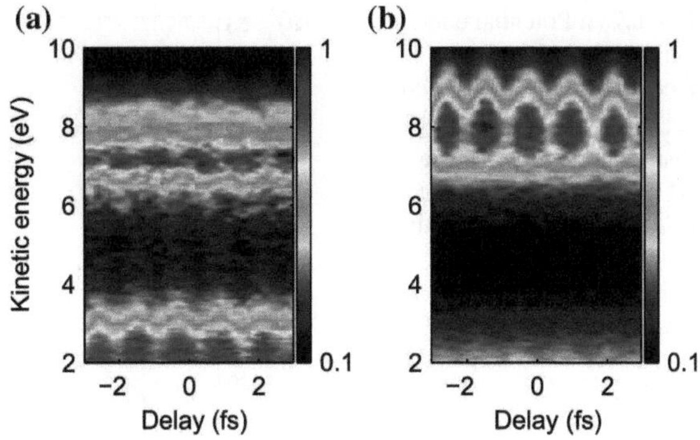

Fig. 6.6 Kinetic energy spectrum of the D^+ fragments as a function of delay between the APT and the IR pulse for fragments ejected at 0° (**a**) and at 90° (**b**). Reprinted from [40]

IR pulse. Also in this case, the ion spectrum is dominated by the channels discussed for the previous experiment: dissociative ionization of the $^2\Sigma_g^+$ and the $^2\Sigma_u^+$ states and autoionization of the Q_1 state to the $^2\Sigma_g^+$ continuum. In Fig. 6.6a, b the kinetic energy spectra for fragments ejected parallel and perpendicular to the laser polarization are shown as a function of the XUV-IR time delay. A clear delay dependence is observed, and different portions of the kinetic energy spectrum oscillate with a period that is half the period of the IR laser. These oscillations have been assigned to the IR-induced coupling between the $^2\Sigma_g^+$ and the $^2\Sigma_u^+$ continua. Theoretical calculations performed in the experimental conditions suggest that, under the influence of the IR pulse, the XUV ionization of the $^2\Sigma_u^+$ state is favored when this state is mixed with the $^2\Sigma_g^+$ state resulting in electron localization and thus in a time-dependent polarization of the molecule.

The use of APTs in combination with IR pulses has been also extended to the study of multi-electrons diatomic molecules such as N_2. In the experiment performed by Lucchini et al. [41], an attosecond pulse train with a spectrum extending up to 55 eV was used to photo ionize N_2 molecules and the subsequent dynamics was probed by a 15-fs IR pulse with a peak intensity of 3–4 × 10^{12} W/cm^2. The kinetic energy and the angular distribution of the N^+ ions resulting from the dissociative ionization and/or Coulomb explosion of the molecule were measured with a VMIs as a function of the XUV-pump IR-probe delay. Figure 6.7a shows a simplified scheme of the potential energy curves of the electronic states of N_2^+ and N_2^{2+}. The XUV photons have enough energy to leave the molecule in a highly excited electronic state of N_2^+ or in a low-excited state of N_2^{2+} (double ionization threshold: 42.88 eV). The kinetic energy spectrum of the N^+ fragments created by the XUV pulse train without (black solid line) and with the IR probe pulse (red dashed line) is reported in Fig. 6.7b. As can be seen from this figure, the presence of the IR probe pulse strongly affects the shape of the kinetic energy spectrum; in particular, the band peaked around 2.3 eV, which can be assigned to autoionization of a superexcited state of of N_2^+ [42], is shifted up to 3.7 eV.

The time-delay dependent kinetic energy spectrum of the N^+ fragments ejected at 0° ± 25° is reported in Fig. 6.8. The band at 3.7 eV created at zero time delay by the presence of the IR probe pulse disappears after 120 fs. This temporal window

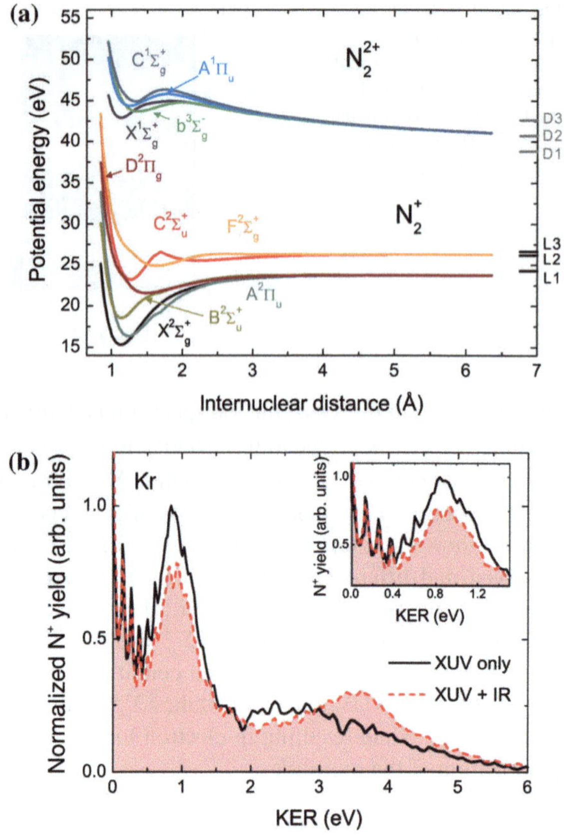

Fig. 6.7 **a** Potential energy curves of the relevant electronic states of N_2^+ and N_2^{2+}. **b** N^+ kinetic energy distribution with only the XUV pump pulse present (*black solid line*) and with both pump and IR probe pulses at zero time delay (*red dashed line* and *shaded area*). Reprinted from [41]

has been assigned to the time required for the auto ionization mechanism to become energetically favorable. Indeed for $\tau < 120$ fs the molecule lacks the energy to autoionize, and the absorption of one or more IR photons allows the transition to the $C^1\Sigma_u^+$ state of N_2^{2+} which gives rise to Coulomb explosion of the molecule. On the other hand, for $\tau > 120$ fs autoionization becomes allowed and the molecule autoionizes. The pump-probe map clearly indicates when the autoionizing state intersects the $C^1\Sigma_u^+$ state and a time-to-internuclear distance mapping of the autoionizing process was obtained.

Finally, it is worth mentioning that the same experimental approach described in the previous paragraphs was recently applied to more complex molecules such as a CO_2 and C_2H_4 [43]. The XUV ionization efficiencies of these molecules were measured as a function of the XUV-pump IR-probe delay, revealing the presence of modulations with the periodicity corresponding to half the period of the IR laser. This modulation of the total yield has been assigned to the time-dependent dipole induced in the molecules by the moderately strong IR field. These encouraging results indicate that the attosecond technology, in combination with VMI-spectroscopy, is now a mature field and ready to be applied to more complex systems.

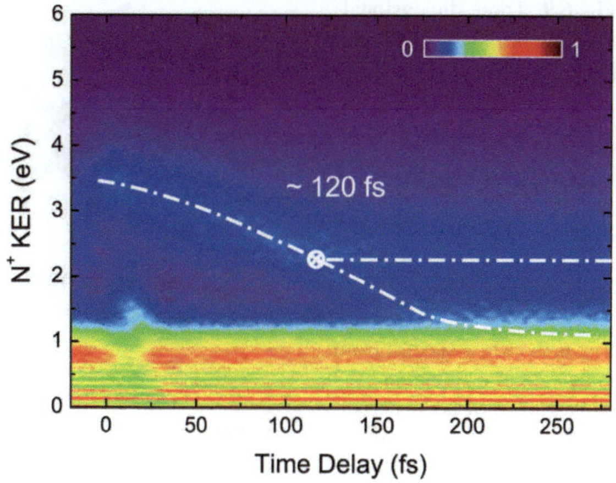

Fig. 6.8 N^+ kinetic energy distribution as function of the delay between the APT and the IR probe pulse for fragments ejected at $0° \pm 25°$. Reprinted from [41]

6.5 Electron Dynamics in Biomolecules

Electron transfer within a single molecule is the fundamental step of many biological processes and chemical reactions. It plays a crucial role in catalysis, DNA damage by ionizing radiation, photosynthesis, photovoltaics, and for switches based on molecular nano-junctions. The investigation of this process is of particular interest and it is the subject of active research [44, 45]. In 2012 Belshaw et al. have performed the first direct measurement of ultrafast electron dynamics in a biomolecular building block, an aromatic amino acid, by using attosecond pulses [12]. An amino acid consists of a central carbon atom (α carbon), linked to an amine ($-NH_2$) group, a carboxylic group ($-COOH$), a hydrogen atom and a side chain, specific for each amino acid. Two amino acids were investigated, phenylalanine and tryptophan, whose molecular structures are shown in Fig. 6.9. In the case of phenylalanine the side chain is a methylene ($-CH_2-$) group terminated by a phenyl ring. In the case of tryptophan an indole group is linked to the methylene group. These amino acids were chosen as model molecules for charge transfer because they each have a radical cation with two charge-acceptor sites at approximately the same binding energy located on the phenyl and amine groups, separated by two singly bonded carbon. Clean plumes of isolated, neutral molecules were produced by a laser induced acoustic desorption (LIAD) technique [46]: a UV desorption laser was focused on the uncoated side of a thin (thickness of 10 μm) tantalum foil, the acoustic wave produced by the UV laser volatilized the sample on the other side of the foil, thus creating a molecular plume in the source region of a Time of Flight (TOF) mass spectrometer. Compared to direct laser desorption methods, LIAD leads to less energy coupling into the molecules, thus giving the possibility to produce fragile neutral molecules with lower internal temperature [46].

Tryptophan and phenylalanine molecules were first irradiated by XUV attosecond pulses produced by high-order harmonic generation in xenon, with a photon energy in

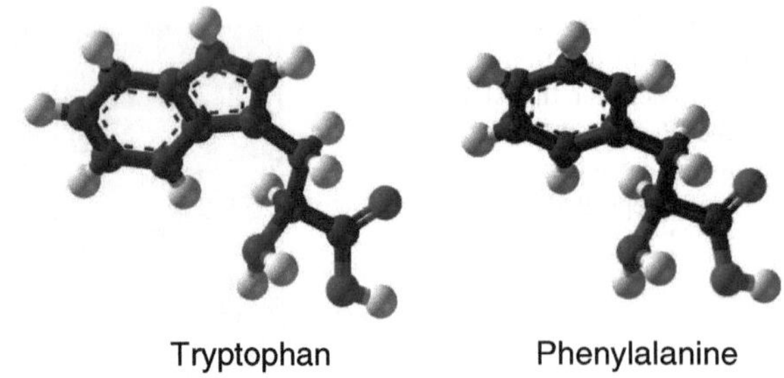

Fig. 6.9 Three-dimensional structure of tryptophan and phenylalanine

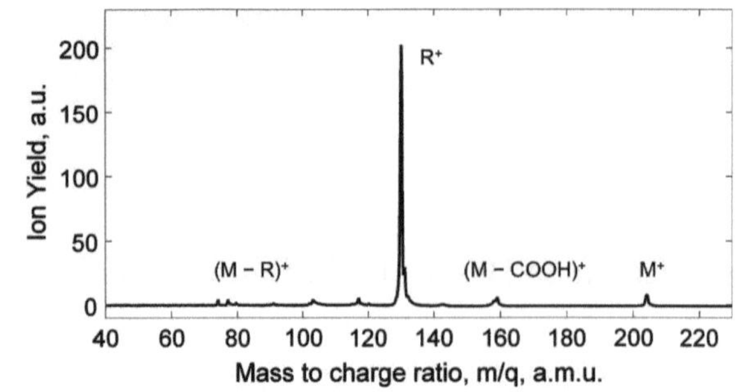

Fig. 6.10 Mass spectra from ionization of tryptophan by XUV pulses. M is the parent ion and R is the side chain group

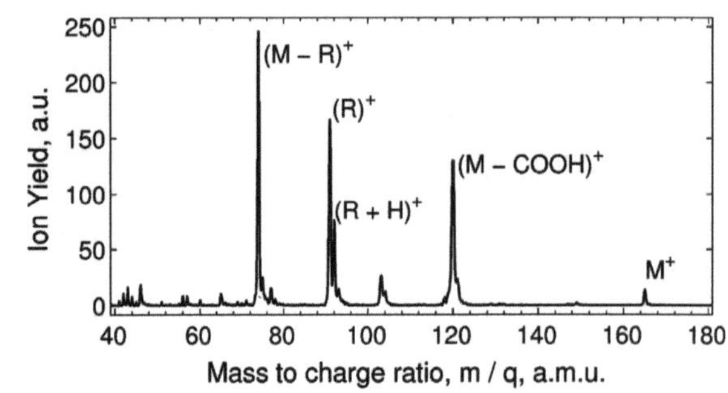

Fig. 6.11 Mass spectra from ionization of phenylalanine by XUV pulses. M is the parent ion and R is the side chain group

the range 16–40 eV. The parent and fragment ions produced were then extracted into the TOF spectrometer for mass analysis. In order to have an estimate of the temporal duration of the XUV pulses used in the experiment, we performed a numerical calculation based on a three-dimensional (3D) propagation model, which takes into account both temporal plasma-induced phase modulation and spatial plasma-lensing effects on the fundamental beam. Due to the temporal gate induced by ionization, the XUV emission is confined within a few recollision events thus giving rise to a double pulse structure with an overall duration of about 1.5 fs.

Figures 6.10 and 6.11 show the measured mass spectra obtained in the case of tryptophan (Try) and phenylalanine (Phe), respectively. In both cases, the main contributions correspond to the parent ion M^+ (204 Da for Try and 165 Da for Phe), loss of the carboxyl group yielding the immonium ion (M–COOH, at 159 Da and 120 Da, respectively), and breakage of the C_α–C_β bond with the charge residing on the

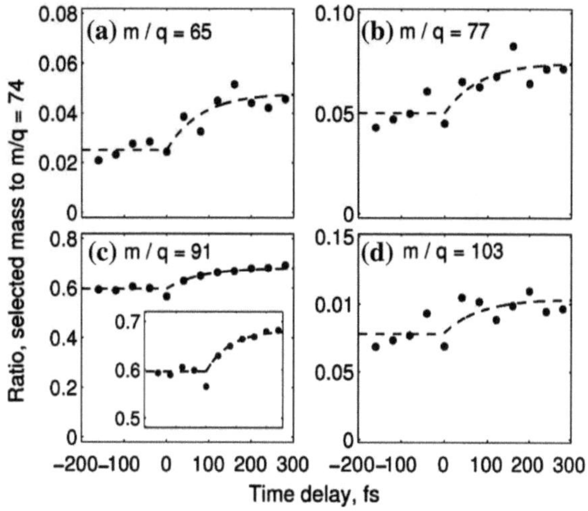

Fig. 6.12 Yields of ions with mass/charge 65 (**a**), 77 (**b**), 91 (**c**) and 103 (**d**) relative to the dominant ion in the spectrum (M–R) as a function of pump (XUV)—probe (VIS/NIR) delay. *Dotted lines* are exponential fits to the data with a time constant of 80 fs. The *inset* in panel (**c**) displays a magnified view for data close to time zero

amine (M–R = 74 Da) or phenyl groups (R = 130 Da for Try, R = 91 for Phe). The XUV pulse is capable of ionizing all valence and some inner shell orbitals, resulting in a wide range of fragment ions.

Then, the temporal evolution of the ionic fragments produced in phenylalanine has been investigated by two-color pump-probe measurements, using attosecond pump pulses and 6-fs visible/near infrared (500–950 nm, VIS/NIR) probe pulses at a variable temporal delay. The pump-probe measurements evidenced two main ultrafast electronic processes. Fragments corresponding to charge residing on the phenyl group (m/q = 65, 77, 91, 103) increase gradually for positive delays, as shown in Fig. 6.12a–d. The time constants for each of these fragments are consistent and give an overall weighted average of about 80 fs. This temporal evolution has been attributed to an internal conversion process into the π_1 state of the phenyl radical cation following initial ionisation of a different orbital by the attosecond pulse [12]. It is known that the π_1 state of the phenyl radical cation absorbs strongly in the visible while the neutral phenyl and neutral and charged amine groups do not. Therefore, increasing population of this state opens up absorption by the VIS/NIR pulse through one or two photon transitions thus enhancing production of fragments corresponding to charge on the ring. A faster process was identified from the yield of immonium dications (m/q = 60), as shown in Fig. 6.13. The experimental data can be fitted by an exponential function with a time constant of about 30 fs, which can be associated to a charge transfer process, where shake-up states can play an important role. It has been suggested that the doubly charged immonium fragment can be a particularly sensitive probe for the dynamics of the charged electron wavepacket coherently generated by the XUV excitation [12].

These measurements provided the first demonstration of an experimental pump-probe scheme capable of studying ultrafast electron transfer, which will allow this phenomenon, and its consequences for a range of biological processes, to be more

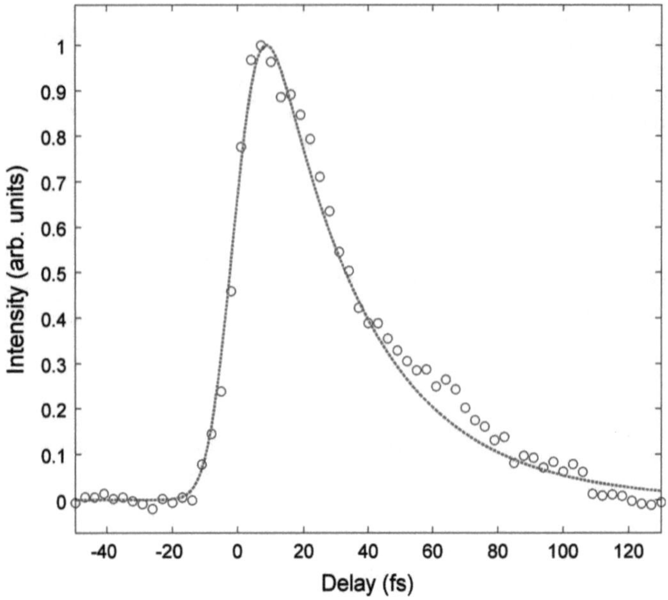

Fig. 6.13 Yield of doubly charged immonium ion (mass/charge = 60) relative to the dominant ion in the spectrum (M–R, mass/charge = 74) as a function of pump-probe delay. The *dotted line* is an exponential fitting curve with a time constant of 30 fs

fully understood. In the future such dynamics could be coherently manipulated with additional ultrashort pulses so that the final destination of the charge could be steered, giving unprecedented quantum control over any subsequent chemical reactivity.

Acknowledgments We would like to acknowledge fruitful collaborations with A. L'Huillier (Lund University), M.J.J. Vrakking (Max Born Institute, Berlin), F. Martín (Universidad Autónoma de Madrid) and M. Hochlaf (Université Paris-Est, Marne-la-Vallée). The research leading to the results presented in this paper has received funding from the European Research Council under the European Community's Seventh Framework Programme (FP7/2007-2013)/ERC grant agreement n. 227355 – ELYCHE, from LASERLAB-EUROPE (grant agreement n. 284464, EC's Seventh Framework Programme), from European COST Action CM1204 XLIC and from the Italian Ministry of Research (FIRB-IDEAS RBID08CRXK).

References

1. F. Krausz, M. Ivanov, Rev. Mod. Phys. **81**, 163 (2009)
2. M. Nisoli, G. Sansone, Prog. Quantum Electron. **33**, 17 (2009)
3. G. Sansone, L. Poletto, M. Nisoli, Nat. Photon. **5**, 655 (2011)
4. M. Drescher, M. Hentschel, R. Kienberger, M. Uiberacker, V. Yakovlev, A. Scrinzi, Th. Westerwalbesloh, U. Kleineberg, U. Heinzmann, F. Krausz, Nature **419**, 803 (2002)

5. M. Uiberacker, Th. Uphues, M. Schultze, A.J. Verhoef, V. Yakovlev, M.F. Kling, J. Rauschenberger, N.M. Kabachnik, H. Schröder, M. Lezius, K.L. Kompa, H.-G. Muller, M.J.J. Vrakking, S. Hendel, U. Kleineberg, U. Heinzmann, M. Drescher, F. Krausz, Nature **336**, 627 (2007)
6. M. Schultze, M. Fiess, N. Karpowicz, J. Gagnon, M. Korbman, M. Hofstetter, S. Neppl, A.L. Cavalieri, Y. Komninos, Th. Mercouris, C.A. Nicolaides, R. Pazourek, S. Nagele, J. Feist, J. Burgdörfer, A.M. Azzeer, R. Ernstorfer, R. Kienberger, U. Kleineberg, E. Goulielmakis, F. Krausz, V.S. Yakovlev, Science **328**, 1658 (2010)
7. K. Klünder, J.M. Dahlström, M. Gisselbrecht, T. Fordell, M. Swoboda, D. Guénot, P. Johnsson, J. Caillat, J. Mauritsson, A. Maquet, R. Taïeb, A. L'Huillier, Phys. Rev. Lett. **106**, 143002 (2011)
8. J.M. Dahlström, D. Guénot, K. Klünder, M. Gisselbrecht, J. Mauritsson, A. L'Huillier, A. Maquet, R. Taïeb, Chem. Phys. **414**, 53 (2013)
9. A.L. Cavalieri, N. Müller, Th. Uphues, V.S. Yakovlev, A. Baltuška, B. Horvath, B. Schmidt, L. Blümel, R. Holzwarth, S. Hendel, M. Drescher, U. Kleineberg, P.M. Echenique, R. Kienberger, F. Krausz, U. Heinzmann, Nature **449**, 1029 (2007)
10. G. Sansone, F. Kelkensberg, J.F. Pérez-Torres, F. Morales, M.F. Kling, W. Siu, O. Ghafur, P. Johnsson, M. Swoboda, E. Benedetti, F. Ferrari, F. Lépine, J.L. Sanz-Vicario, S. Zherebtsov, I. Znakovskaya, A. L'Huillier, M.Yu. Ivanov, M. Nisoli, F. Martín, M.J.J. Vrakking, Nature **465**, 763 (2010)
11. W. Siu, F. Kelkensberg, G. Gademann, A. Rouzée, P. Johnsson, D. Dowek, M. Lucchini, F. Calegari, U. De Giovannini, A. Rubio, R.R. Lucchese, H. Kono, F. Lépine, M.J.J. Vrakking, Phys. Rev. A **84**, 063412 (2011)
12. L. Belshaw, F. Calegari, M.J. Duffy, A. Trabattoni, L. Poletto, M. Nisoli, J. Greenwood, J. Phys. Chem. Lett. **3**, 3751 (2012)
13. I.P. Christov, M.M. Murnane, H. Kapteyn, Phys. Rev. Lett. **78**, 1251 (1997)
14. M. Hentschel, R. Kienberger, C. Spielmann, G.A. Reider, N. Milosevic, T. Brabec, P. Corkum, U. Heinzmann, M. Drescher, F. Krausz, Nature **414**, 509 (2001)
15. E. Goulielmakis, M. Schultze, M. Hofstetter, V.S. Yakovlev, J. Gagnon, M. Uiberacker, A.L. Aquila, E.M. Gullikson, D.T. Attwood, R. Kienberger, F. Krausz, U. Kleineberg, Science **320**, 1614 (2008)
16. P.B. Corkum, N.H. Burnett, M.Y. Ivanov, Opt. Lett. **19**, 1870 (1994)
17. O. Tcherbakoff, E. Mével, D. Descamps, J. Plumridge, E. Constant, Phys. Rev. A **68**, 043804 (2003)
18. I.J. Sola, E. Mével, L. Elouga, E. Constant, V. Strelkov, L. Poletto, P. Villoresi, E. Benedetti, J.-P. Caumes, S. Stagira, C. Vozzi, G. Sansone, M. Nisoli, Nat. Phys. **2**, 319 (2006)
19. G. Sansone, E. Benedetti, F. Calegari, C. Vozzi, L. Avaldi, R. Flammini, L. Poletto, P. Villoresi, C. Altucci, R. Velotta, S. Stagira, S. De Silvestri, M. Nisoli, Science **314**, 443 (2006)
20. Z. Chang, Phys. Rev. A **76**, 051403(R) (2007)
21. I. Thomann, A. Bahabad, X. Liu, R. Trebino, M.M. Murnane, H.C. Kapteyn, Opt. Express **17**, 4611 (2009)
22. M.J. Abel, T. Pfeifer, P.M. Nagel, W. Boutu, M.J. Bell, C.P. Steiner, D.M. Neumark, S.R. Leone, Chem. Phys. **366**, 9 (2009)
23. F. Ferrari, F. Calegari, M. Lucchini, C. Vozzi, S. Stagira, G. Sansone, M. Nisoli, Nat. Phot. **4**, 875 (2010)
24. P.M. Paul, E.S. Toma, P. Breger, G. Mullot, F. Auge, P. Balcou, H.G. Muller, P. Agostini, Science **292**, 1689 (2001)
25. R. Kienberger, E. Goulielmakis, M. Uiberacker, A. Baltuška, V. Yakovlev, F. Bammer, A. Scrinzi, T. Westerwalbesloh, U. Kleineberg, U. Heinzmann, M. Drescher, F. Krausz, Science **427**, 509 (2004)
26. Y. Mairesse, F. Quéré, Phys. Rev. A **71**, 011401 (2005)
27. G.G. Paulus, F. Grasbon, H. Walther, P. Villoresi, M. Nisoli, S. Stagira, E. Priori, S. De Silvestri, Nature **414**, 182 (2001)
28. A. Baltuška, T. Udem, M. Uiberacker, M. Hentschel, E. Goulielmakis, C. Gohle, R. Holzwarth, V.S. Yakoviev, A. Scrinzi, T. Hansch, F. Krausz, Rev. Mod. Phys. **421**, 611 (2003)

29. M. Nisoli, G. Sansone, S. Stagira, S. De Silvestri, C. Vozzi, M. Pascolini, L. Poletto, P. Villoresi, G. Tondello, Phys. Rev. Lett. **91**, (2003)
30. G.G. Paulus, F. Lindner, H. Walther, A. Baltuška, E. Goulielmakis, M. Lezius, F. Krausz, Phys. Rev. Lett. **91**, 253004 (2003)
31. L.Y. Peng, E.A. Pronin, A.F. Starace, New J. Phys. **10**, 025030 (2008)
32. L.Y. Peng, A.F. Starace, Phys. Rev. A **76**, 043401 (2007)
33. C. Liu, M. Reduzzi, A. Trabattoni, A. Sunilkumar, A. Dubrouil, F. Calegari, M. Nisoli, G. Sansone, Phys. Rev. Lett. **111**, 123901 (2013)
34. G.G. Paulus, W. Nicklich, H.L. Xu, P. Lambropoulos, H.J. Walther, Phys. Rev. Lett. **72**, 2851 (1994)
35. G.G. Paulus, W. Becker, W. Nicklich, H. Walther, J. Phys. B **27**, L703 (1994)
36. A.T.J.B. Eppink, D.H. Parker, Rev. Sci. Instrum. **68**, 3477 (1997)
37. M.J.J. Vrakking, Phys. Chem. Chem. Phys. **16**, 2775 (2014)
38. O. Ghafur, W. Siu, P. Johnsson, M.F. Kling, M. Drescher, M.J.J. Vrakking, Rev. Sci. Instrum. **80**, 033110 (2009)
39. P. Bucksbaum, A. Zavriyev, H. Muller, D. Schumacher, Phys. Rev. Lett. **64**, 1883 (1990)
40. F. Kelkensberg, W. Siu, J.F. Pérez-Torres, F. Morales, G. Gademann, A. Rouzée, P. Johnsson, M. Lucchini, F. Calegari, J.L. Sanz-Vicario, F. Martín, M.J.J. Vrakking, Phys. Rev. Lett. **107**, 043002 (2011)
41. M. Lucchini, K. Kim, F. Calegari, F. Kelkensberg, W. Siu, G. Sansone, M.J.J. Vrakking, M. Hochlaf, M. Nisoli, Phys. Rev. A **86**, 043404 (2012)
42. M. Ahmad, P. Lablanquie, F. Penent, J.G. Lambourne, R.I. Hall, J.H.D. Eland, J. Phys. B **39**, 3599 (2006)
43. Ch. Neidel, J. Klei, C.H. Yang, A. Rouzée, M.J.J. Vrakking, K. Klunder, M. Miranda, C.L. Arnold, T. Fordell, A. LHuillier, M. Gisselbrecht, P. Johnsson, M.P. Dinh, E. Suraud, P.G. Reinhard, V. Despré, M.A.L. Marques, F. Lépine, Phys. Rev. Lett. **111**, 033001 (2013)
44. M. Cordes, B. Giese, Chem. Soc. Rev. **38**, 892 (2009)
45. D. Becker, A. Adhikary, M.D. Sevilla, in *Charge Migration, in DNA*, ed. by T. Chakraborty (Springer, New York, 2007), pp. 139–175
46. C.R. Calvert, L. Belshaw, M.J. Duffy, O. Kelly, R.B. King, A.G. Smyth, T.J. Kelly, J.T. Costello, D.J. Timson, W.A. Bryan, T. Kierspel, P. Rice, I.C.E. Turcu, C.M. Cacho, E. Springate, I.D. Williams, J.B. Greenwood, Ohys. Chem. Chem. Phys. **14**, 6289 (2012)

Chapter 7
Controlling Atomic Photoabsorption by Intense Lasers in the Attosecond Time Domain

Xiao-Min Tong and Nobuyuki Toshima

Abstract In this chapter, we introduce the recent developments on the infrared (IR) laser assisted photoionization and photoabsorption experiments and the theories briefly, then present a detailed theoretical method to simulate the IR assisted dynamics by a single attosecond pulse or an attosecond pulse train in the extreme ultraviolet (xuv) regime. The key steps to understand IR assisted atomic photoabsorption processes are (1) the IR laser field modifies an atomic excited and continuum state as a dressed state or Floquet state, (2) the xuv pulse excites the atomic ground state to a Floquet state through different Floquet components. The interference between the transitions to a Floquet state through different components plays the center role to understand the oscillatory structures or control the dynamics of IR assisted photoabsorption processes. Two examples, IR assisted photoionization and photoexcitation of H atoms by xuv pulses, are presented. Existed and possible future applications are also discussed.

7.1 Introduction

Advances of laser technology in the past ten years have opened a door to probe the fast dynamics in femtosecond or even attosecond time domain [1], to advance the femtochemistry [2, 3] towards attochemistry [4]. There are mainly two ways to probe and control the dynamics on a sub-optical cycle time scale. One is to control the wave-form of the intense infrared (IR) laser electric field and in turn to probe and

---

X.-M. Tong (✉) · N. Toshima
Division of Materials Science, Faculty of Pure and Applied Sciences,
University of Tsukuba, 1-1-1 Tennodai, Tsukuba, Ibaraki 305-8577, Japan
e-mail: tong.xiaomin.ga@u.tsukuba.ac.jp

N. Toshima
e-mail: toshima@bk.tsukuba.ac.jp

X.-M. Tong
Center for Computational Sciences, University of Tsukuba,
1-1-1 Tennodai, Tsukuba, Ibaraki 305-8577, Japan

M. Kitzler and S. Gräfe (eds.), *Ultrafast Dynamics Driven by Intense Light Pulses*, Springer Series on Atomic, Optical, and Plasma Physics 86, DOI 10.1007/978-3-319-20173-3_7

control the dynamics of laser material interactions. In this category, one way is to control the carrier-envelop phase (CEP) of an ultrashort IR laser [5–8] to investigate the dynamics of laser material interactions, including atoms [9–11], molecules [12–16], surfaces [17, 18], and solids [19]. The other way is to combine two-color intense lasers by tuning the relative phase between the two to study and control the dynamics of laser material interactions [20–24]. For the second category, one can combine a moderate intense IR laser pulse with a single attosecond pulse (SAP) or an attosecond pulse train (APT). Tuning the arriving time between the two pulses, one has a knob not only to probe but also to control the dynamical processes, like photoabsorption or transparency of materials, in an attosecond time domain. In this category, the moderate IR laser field does not directly affect the ground state properties or in other words, the IR laser pulse itself does not induce excitation or ionization without xuv pulses. Thus the IR field only modifies the structures of the excited and continuum states. The IR field can steer the motion of the excited or continuum electron [25], or from the measured electron, one can get the wave-form of the ultrashort laser [26] or measure the pulse duration of SAPs [27].

In this chapter, we focus on the IR-assisted atomic photoabsorption processes. There are two kinds of experiment: one is to measure the ejected electron [28–30], and the other is to measure the xuv spectra to obtain the apparent photoabsorption cross sections [31–34]. To distinguish the two measurements, we call the first one as photoionization and the second one as photoexcitation or photoabsorption although there are also photoionization processes involved in the second kind of experiment. Since the first kind of experiment can only provide the above-threshold-ionization (ATI) electron information, the photoexcitation is missed in the measurement but the detailed ATI electron momentum distribution can be obtained. For the second kind of experiment, the photoabsorption probability can be measured directly at each xuv photon energy but the detailed ATI momentum distribution is missed. The two kinds of experiment complement with each other. Since the xuv (SAP or APT) pulses are generated by the high-order harmonic generation of a driving IR laser, by merging a second IR laser pulse, one can further study the xuv emission time [35] or relative phase between the APT and the driving IR field [36]. The IR field can also serve as a streaking field, which drives the ionized electron by the xuv pulse to measure the time delay [37] in the photoionization process [38].

Although the applied IR field is moderate intense, the electron IR laser interaction for excited or continuum states is still relatively strong and cannot be treated by a perturbation theory. But we can treat the xuv pulse in a perturbation theory since one-photon absorption is the dominant process. From the theoretical point of view, we need to understand the mechanism of IR assisted atomic photoabsorption qualitatively and explain the available measurements and search for better way to control the dynamics quantitatively. We proposed a mechanism which has explained most of the measurements [28, 29, 35, 36]. In this theoretical frame, atomic structures in an intense IR laser field are described by Floquet states and atoms can be excited or ionized to a Floquet state by an xuv pulse through different Floquet components. The interference of excitation or ionization through different Floquet components results in the oscillation of ionization yields as a function of arriving time of the xuv pulse. Based on the mechanism, we have proposed a way to control atomic transparenc in attosecond time scales by tuning the xuv arriving time. The proposal has been realized in the experiment [39].

To explain each individual measurement and search an effective way to control the IR assisted atomic photoabsorption process, we have to solve the time-dependent Schrödinger equation numerically. For photoionization processes, all the detailed ATI electron information in the outer region are needed so we have to treat the boundary condition carefully [40]. For photoexcitation processes, the xuv photon is absorbed in the inner region, and the processes can be described by apparent photoabsorption cross section. The apparent photoabsorption cross section can be calculated using autocorrelation function or response function [41–44], which does not need the electron information in the outer region. In the following, we first introduce our numerical method, which can treat the two processes in a same way, then present two examples, one for photoionization and one for photoexcitation.

7.2 Theoretical Method

In most cases, only one electron (active electron) involves in the IR assisted atomic photoabsortion and thus the effect of other electrons can be described by a static screening potential [45]. For a many-electron system, a model potential [46] can be used to study the dynamical processes. We choose H atoms as target atoms, but the method can be easily applied to a many-electron atomic system by replacing the electron-nucleus Coulomb interaction with a model potential. All the dynamical information of IR assisted photoabsorption can be obtained by solving the following time-dependent Schrödinger equation (TDSE) (atomic units $\hbar = m = e = 1$ are used unless otherwise stated)

$$i\frac{\partial}{\partial t}\Psi(t) = [H_0 + V_{IR}(t) + V_x(t)]\Psi(t) = H(t)\Psi(t), \tag{7.1}$$

with $\Psi(-\infty) = \psi_g$, the initial ground state wavefunction. Here H_0 is the external field free atomic Hamiltonian, $V_{IR}(t)$ and $V_x(t)$ are the active electron interactions with the IR laser and xuv fields, respectively. In the experiment, we know the IR information, like the IR intensity, pulse duration and so on. We do not know the xuv field intensity, but we know that the xuv field can be treated in a perturbation theory. Thus to solve the above equation, we have to choose the xuv field strength carefully; it is not too strong so that the high-order effects of the xuv field can be neglected and the field strength is not too weak so that the one-photon absorption yield is larger than the multi-photon absorption from the IR laser field. An alternative way is to solve the TDSE in the integral form [11], as

$$\Psi(t) = -i \int_{-\infty}^{t} e^{-i\int_{\tau}^{t} H(t')dt'} [V_{IR}(\tau) + V_x(\tau)] e^{-iH_0\tau} \psi_g d\tau + e^{-iH_0 t}\psi_g. \qquad (7.2)$$

Mathematically the two equations, (7.1) and (7.2), are equivalent. Equation (7.2) looks complicated but it provides better physical insights not only for intense laser problem, but also for collision processes [47–49]. Equation (7.2) shows that the time-dependent wavefunction can be separated into two parts, the dynamically related part (the first term in (7.2)) and dynamically irrelevant part (the second term in (7.2)). Since we focus on the photoabsorption process, a dynamical process, we drop the second term in the following discussion to derive the working equation for IR assisted atomic photoabsorption processes.

7.2.1 Working Equation

Since our goal is to study the intense IR laser assisted photoabsorption, we need to derive the working equations from the general TDSE. The characteristics of IR laser assisted photoabsorption are (1) the xuv field strength is so weak that we can treat it in a first order perturbation method; (2) the IR laser strength is so strong that we cannot treat it in a perturbative way for excited or continuum states; (3) the xuv photoabsorption yield is much larger than the one from the IR laser only. With conditions 1 and 2, we can remove $V_x(t)$ from $H(t)$ in (7.2) and with condition 3, we can remove $V_{IR}(\tau)$ from $[V_{IR}(\tau)+V_x(\tau)]$. Therefore the time-dependent dynamically related wavefunction can be approximated as

$$\Psi_d(t) = -i \int_{-\infty}^{t} e^{-i\int_{\tau}^{t}[H_0+V_{IR}(t')]dt'} V_x(\tau) e^{-iH_0\tau} \psi_g d\tau. \qquad (7.3)$$

This is the working equation to study IR laser assisted atomic photoabsorption processes. To check the validity of the working equation, we have compared the ATI spectra obtained from the first term of (7.2) and (7.3) and confirmed that they are almost identical. Although the numerical efforts to solve (7.2) and (7.3) are the same, (7.3) provides a better physical insight. The physical interpretation of the equation is (1) the electron freely propagates to time τ and (2) the electron is excited or ionized by the xuv field at time τ and then the electron propagates in the combined atomic and IR laser fields. This also implies that the applied moderate intense IR laser field can modify the atomic structures of excited or continuum states since the electron is loosely bound by the atomic potential while the IR laser field cannot perturb the ground state structure since the electron is tightly bound by the atomic potential.

7.2.2 Interpretation of the Working Equation

The transition probability to a final external-field-free state ψ_f can be rewritten as

$$\begin{aligned}P_f &= |\langle\psi_f|\Psi_d(\infty)\rangle|^2 \\ &= |\langle\psi_f| \int_{-\infty}^{\infty} e^{-i\int_{\tau}^{\infty}[H_0+V_{IR}(t')]dt'} V_x(\tau)|e^{-iH_0\tau}\psi_g\rangle d\tau|^2 \\ &= |\int_{-\infty}^{\infty} \langle\Psi_f(\tau)|V_x(\tau)|e^{-iH_0\tau}\psi_g\rangle d\tau|^2\end{aligned} \tag{7.4}$$

with

$$\Psi_f(t) = e^{-i\int_{\infty}^{t}[H_0+V_{IR}(t')]dt'}\psi_f. \tag{7.5}$$

Here $\Psi_f(t)$ in (7.5) stands for a final state ψ_f propagating to time t from time $t = \infty$ inversely, and the transition amplitude from the ground state ψ_g to the final state ψ_f at $t = \infty$ is expressed as the time-integral of the transition amplitude to the state $\Psi_f(\tau)$ at $t = \tau$ in the transformed expression in (7.4). This expression is analogous to the photon-induced transition amplitude in the perturbation theory if we replace $\Psi_f(\tau)$ by $e^{-i\epsilon_f\tau}\psi_f$ with ϵ_f the final sate energy of ψ_f. $\Psi_f(\tau)$ is an IR-laser dressed state, which can be described by the Floquet theorem [50].

In the Floquet theorem, such a dressed state can be expressed as

$$\Psi_f(t) = \sum_n \psi_{f,n}e^{-i(\epsilon_f+n\omega)t} \tag{7.6}$$

for a single color, infinitive long IR pulse. Here $\psi_{f,n}$, a time-independent function, is the Fourier component of the Floquet state . In the IR assisted photoabsorption, the IR pulse duration is much longer than that of the xuv pulse. Thus, the IR assisted photoabsorption can be explained as follows: A dressed state (Floquet state) is formed when an atom is placed into the IR laser field as shown in Fig. 7.1. The energy structure of the Floquet state looks like a multi-slit in the energy domain. The xuv pulse excites or ionizes the atom into a Floquet state through different Floquet components, and the interference between the excitations through different Floquet components results in the oscillation of the absorption probability. From the conservation of parity, even without any simulation, we can predicate that the ionization yields oscillate twice per IR optical cycle as a function of the time delay between the IR and xuv pulses. To understand the detailed measurements, we need

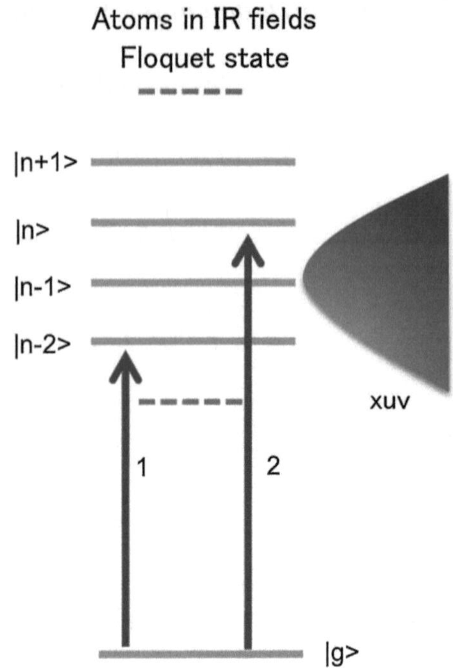

Fig. 7.1 Scheme of the IR assisted photoabsorption by an xuv pulse

to solve the above TDSE numerically. To cope with the two kinds of measurements (to measure the ATI electrons or xuv absorption spectra), we discuss the photoionization and photoexcitation (photoabsorption) processes separately although we solve the same working equation.

7.2.3 Photoionization

For photoionization, we need all the detailed photoelectron information so that we solve (7.3) numerically using the split-operation method in the energy representation [51] and project the continuum wavefunction to Volkov states in the asymptotic region as detailed in [11] to obtain the photoelectron momentum distribution. In this article, we have refined the method in the following two aspects. (1) Instead of describing the laser electron interaction in the length form, we also extended the method to the velocity form to make a cross check of the numerical convergency. (2) Instead of projecting the continuum states onto Volkov states, we project them to atomic Volkov states, or Coulomb Volkov states [52–54]. In principle, the two modifications should not change the results. In practice, those modifications speed up the simulation greatly since we do not need a large space box in the simulation. Most of the details of the numerical methods have been published in our previous works [11, 55, 56].

7.2.4 Photoexcitation (Photoabsorption)

Once we obtained the time-dependent wavefunction in the study of photoionization, we can calculate an autocorrelation function, and then obtain the apparent photoabsorption cross sections. Since we did not present minutely how to derive the apparent photoabsorption cross sections from (7.3) in our previous works, we show the details in this article. Suppose both the IR field and xuv pulse switched on from time $t = -\infty$, the time-dependent wavefunction at time $t \to \infty$, when all the time-dependent fields are concluded, can be written as

$$\begin{aligned}\Psi_d(\infty) &= -i\int_{-\infty}^{\infty} e^{-i\int_t^{\infty}[H_0+V_{IR}(t')]dt'} V_x(t)e^{-iH_0t}\psi_g dt \\ &= -i\int_{-\infty}^{\infty} U(\infty, t)V_x(t)e^{-iH_0t}\psi_g dt,\end{aligned} \tag{7.7}$$

with

$$U(t', t) = e^{-i\int_t^{t'}[H_0+V_{IR}(t'')]dt''}. \tag{7.8}$$

Since the unperturbed initial state component $e^{-iH_0t}\psi_g$ is removed in the wave functions [(7.3) and (7.7)], the total photoabsorption probability when all the pulses are over at time $t \to \infty$ is given by

$$\begin{aligned}P &= \langle\Psi_d(\infty)|\Psi_d(\infty)\rangle \\ &= \int_{-\infty}^{\infty}\int_{-\infty}^{\infty} \langle\psi_g|V_x(t)U(t, t')V_x(t')|\psi_g\rangle e^{i\epsilon_g(t-t')}dtdt' \\ &= \int_{-\infty}^{\infty} E_0^*(\omega)\int_{-\infty}^{\infty}\int_{-\infty}^{\infty} \langle\psi_g\mathbf{r}\cdot\hat{\mathbf{e}}|U(t, t')V_x(t')e^{-i\epsilon_g t'}|\psi_g\rangle e^{i(\epsilon_g+\omega)t}dtdt'd\omega \\ &= \int_{-\infty}^{\infty} E_0^*(\omega)\int_{-\infty}^{\infty} C(t)e^{i(\epsilon_g+\omega)t}dtd\omega,\end{aligned} \tag{7.9}$$

with

$$C(t) = \int_{-\infty}^{\infty} \langle\psi_g\mathbf{r}\cdot\hat{\mathbf{e}}|U(t, t')V_x(t')e^{-i\epsilon_g t'}|\psi_g\rangle dt', \tag{7.10}$$

$$E_x(t) = \int_{-\infty}^{\infty} E_0(\omega)e^{-i\omega t}d\omega, \tag{7.11}$$

and $E_x(t)$ the time-dependent xuv electric field. Here $\hat{\mathbf{e}}$ is the unit vector of the xuv polarization direction and ϵ_g the orbital energy of the ground state and $U(t, t')$ the time propagator from t' to t. Note that $U(t, t')$ depends on both t and t', not only the difference between the two because IR and xuv fields are pulses. The autocorrelation function $C(t)$ is the overlap of $\psi_g \mathbf{r} \cdot \hat{\mathbf{e}}$ with the time-dependent wave function $\Psi_d(t)$ of (7.7). Since ψ_g has a finite distribution in the coordinate space, only the inner part of $\Psi_d(t)$ is required accurately for the overlap integral. The expression of total photoabsorption probability is valid when the depletion of the ground state is negligible small, just as in the present case.

The conventional photoabsorption process can be described by the photoabsorption cross section, which does not depend on the photon field strength. Similarly we can define an apparent photoabsorption cross section to describe IR assisted photoabsorption processes. The photon density of xuv pulse with photon energy ω is

$$J(\omega) = \frac{c}{4\pi\omega} E_0^2(\omega) \tag{7.12}$$

with c the speed of light in vacuum, and the total photoabsorption probability

$$P = \int \sigma(\omega) J(\omega) d\omega. \tag{7.13}$$

Comparing (7.9) and (7.13), the apparent photoabsorption cross section is defined as

$$\begin{aligned} \sigma(\omega) &= \frac{E_0^*(\omega)}{J(\omega)} \int_{-\infty}^{\infty} C(t) e^{i(\epsilon_g + \omega)t} dt \\ &= \frac{4\pi\omega}{E_0(\omega) c} \int_{-\infty}^{\infty} C(t) e^{i(\epsilon_g + \omega)t} dt \end{aligned} \tag{7.14}$$

Equation (7.14) works when both IR and xuv fields are pulsed fields and both are concluded at time $t \to \infty$. If there is no IR laser field or time-dependent field, the above equation goes to the equation we used for photoabsorption in a static field [57]. Here we assume that the xuv field has a phase relation with the applied IR field. If the xuv and IR fields have no phase relation, we need to average over all the phases between the IR and xuv fields from 0 to 2π as we did before in the paper [41], in which the detailed derivations were presented. Note that although the apparent photoabsorption cross section is a function of the xuv photon energy, it also depends on the character of the xuv pulse and arriving time with respect to the IR laser. The information is encoded in the autocorrelation function $C(t)$.

7.3 Results

Although most of the measurements are performed on the rare gas atoms, here we use hydrogen atoms as an example to illustrate the physics. In the dipole approximation, the electron external field can be written as

$$V(\mathbf{r},t) = V_x(t) + V_{IR}(t) = -\mathbf{r}\cdot(\mathbf{E}_{xuv}(t) + \mathbf{E}_{IR}(t)), \tag{7.15}$$

with the IR laser field

$$\mathbf{E}_{IR}(t) = \hat{\mathbf{e}}E_0 e^{-2\ln 2\,(t-t_d)^2/\tau_{IR}^2}\cos(\omega(t-t_d)+\delta), \tag{7.16}$$

with E_0 the IR laser field strength, δ the CEP of the IR laser, τ_{IR} the IR pulse duration of FWHM (full width at half maximum), and t_d the time delay between the IR and xuv pulses. The xuv field

$$\mathbf{E}_{xuv}(t) = \sum_j (-1)^j \mathbf{E}_{SAP}(t - jT_h)e^{-2\ln 2\,t^2/\tau_x^2} \tag{7.17}$$

with T_h the half cycle time of the driving IR used to generate APT and τ_x the APT pulse duration of FWHM. Sum over j represents that the APT contains many SAPs, with its field as

$$\mathbf{E}_{SAP}(t) = \hat{\mathbf{e}}E_x e^{-2\ln 2\,t^2/\tau_{SAP}^2}\cos(\omega_x t). \tag{7.18}$$

Here ω_x is the center energy of the SAP, τ_{SAP} the SAP duration of FWHM, and E_x the field strength of the xuv pulse. $\hat{\mathbf{e}}$ is the unit vector of IR laser polarization direction and we assume that the xuv pulse also polarizes along the same direction. To illustrate the physics of IR assisted dynamics by xuv pulses, we choose the IR laser center energy of $\omega = 1.55$ eV, with the wavelength of 800 nm and the intensity of 10^{13} W/cm^2, the SAP pulse center energy of $\omega_x = 10.0$ eV, which is lower than the ionization threshold of H atoms, the pulse duration $\tau_{SAP} = 0.5$ fs and intensity of 10^{10} W/cm^2. Under such conditions, the multi-photon process by the IR-only is negligible small comparing the single photon process by the xuv pulse.

7.3.1 *IR Assisted Photoionization*

Now, let us first investigate the IR assisted photoionization. Figure 7.2 shows the total photoionization probabilities of IR assisted ionization of H atoms by the xuv pulses from an APT (solid red curve) and an SAP (dashed green curve). When the IR pulse arrives earlier than the APT (SAP) at time delay $t_d = -10.0$ o.c. (optical cycle), the total photoionization yields are very small. For the APT, the ionization

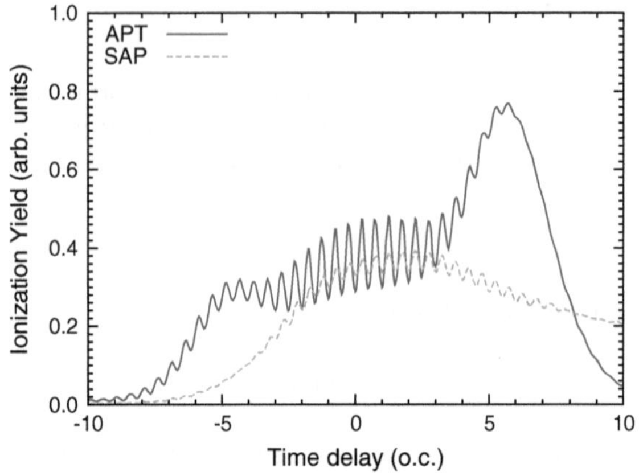

Fig. 7.2 Total ionization yield of IR assisted photoionization of H atoms by an xuv pulse from an APT (*solid red curve*) or an SAP (*dashed green curve*). The APT pulse duration (FWHM) is 5.0 fs and the IR pulse duration is 20 fs

yield increases as the time delay increases and reaches a maximum at about $t_d = 5$ o.c. and then decreases again as the time delay increases further. When the two pulses overlap with each other around $t_d = 0.0$, the yield oscillates twice per IR optical cycle. For the SAP, similarly to that of the APT, the yield also increases as the time delay increases and reaches the maximum at $t_d = 0.0$, and then slowly decreases as the time delay increases further. The oscillation amplitude is about one order smaller than the one by the APT and there is no sharp peak at time delay $t_d = 5.0$ o.c. Those behaviors can be understood qualitatively as follows: Since the xuv center energy is smaller than the ionization potential of H atoms, the xuv-only can mainly excite the atoms with a small ionization yield. If the IR arrives much earlier than the xuv pulse, it does not affect the xuv excitation. If the IR arrives much later than the xuv pulse, it can further ionize the excited states populated by the xuv pulse. This results in the asymmetry of ionization yields with respect to the time delay at $t_d = 0$. When the IR pulse overlaps with the xuv pulse, atomic dressed states (Floquet states) are formed being assisted by the IR field and the atoms can be ionized to dressed states by the xuv pulse. The large oscillation amplitude of the ionization by the APT can be attributed to the narrow energy distribution of the APT in which only a few Floquet states are involved. While for the SAP, which has a broad energy distribution, the atoms can be ionized to many Floquet states with different phases [58]. Therefore, the total ionization yield is less sensitive to the IR laser arriving time.

To confirm the above interpretation, we also plot the excitation probabilities of H atoms by the APT and SAP in Fig. 7.3. The excitation probability by the APT arises first, then reaches the first broad peak at $t_d = -5.0$ o.c., and the second broad peak at $t_d = 0.0$, and then decreases monotonically as the time delay increases further. Since the energy structure of an APT looks like a comb with sharp energy peaks, the xuv energy is off the resonant energy of the transition $1s \rightarrow 2p$ without the IR field, and hence the excitation probability is smaller when the IR arrives much earlier than the APT. When the IR laser overlaps with the APT at $t_d = -5.0$ o.c., the APT can resonantly excite the atoms owing to the ac Stark shift [59], and this results in the

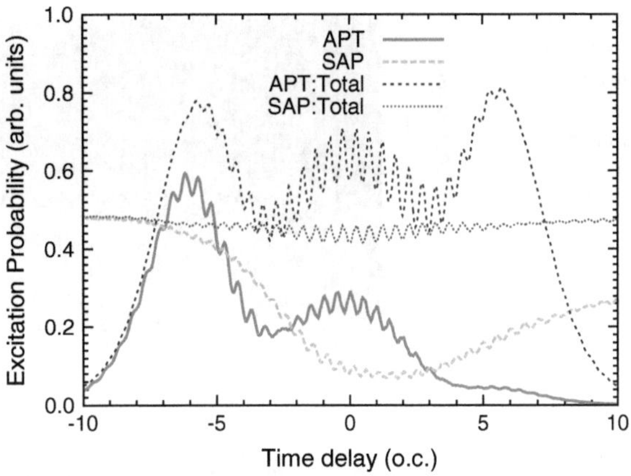

Fig. 7.3 Excitation probabilities of IR assisted photoabsorption of H atoms by an xuv pulse from an APT (*solid red curve*) or an SAP (*dashed green curve*). The total absorption probabilities by the APT (APT:Total, *blue dashed curve*) and the SAP (SAP:Total, *pink dashed curve*) are also presented. The parameters for the IR laser and xuv pulse are the same as the ones used in Fig. 7.2

first broad peak. When t_d increases further to the overlap region at $t_d = 0$, the ac Stark shift may move the $2p$ energy off the resonance again. Therefore the excitation probability decreases as t_d increases and the probability reaches the second broad peak when the two pulses overlaps at $t_d = 0$. As t_d increases further from $t_d = 0$, the ac Stark shift may result in a resonant excitation again. For $t_d > 0$, the IR laser arrives later and the early tail of the IR laser pulse helps the APT to excite resonantly the atoms and the main part of the IR laser ionizes the excited states. Thus, the large ionization peak appears at $t_d = 5.0$ o.c. while the excitation probability decreases oscillating. Since the SAP has a broad energy distribution it can excite the atoms without the IR and thus the excitation probability is large when the IR arrives much earlier than the SAP pulse. As t_d increases and approaches $t_d = 0$, the IR pulse ionize the excited states and the ionization probability increases while the excitation probability decreases. As the time delay crosses the maximum overlap at $t_d = 0$, then the ionization yield decreases and the excitation probability increases again.

We also plot the total absorption probability (ionization + excitation) in Fig. 7.3. For the SAP, the total absorption probabilities are almost constant, insensitive to the arriving time of the IR field. For the APT, the total absorption probabilities are sensitive to the arriving time of the IR pulse. All these are consistent with our interpretation. Interestingly, the total absorption probabilities are almost symmetry with respect to the positive and negative time delay.

To support the above interpretations, let us look at the ATI electron energy spectra as a function of the time delay between the IR and APT pulses in Fig. 7.4. In the negative large time delay region, there are almost no high energy electrons since the direct ionization by the APT is very small. When the IR and APT overlap with each other around $t_d = 0$, there are many sharp ATI peaks. For the peak around the

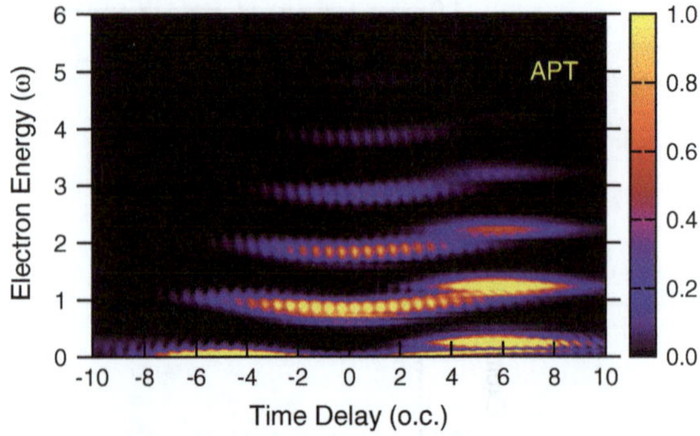

Fig. 7.4 ATI spectra of H atoms by the APT pulse as a function of the time delay between the IR laser pulse and APT. Other parameters for the IR laser and xuv pulse are the same as the ones used in Fig. 7.2

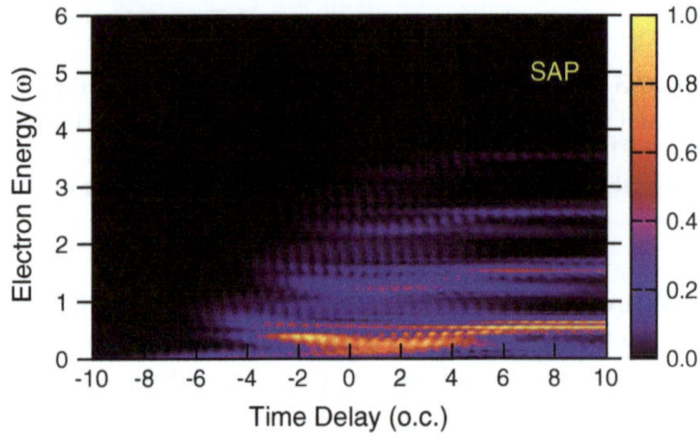

Fig. 7.5 ATI spectra of H atoms by the SAP pulse as a function of the time delay between the IR laser pulse and SAP. Other parameters for the IR laser and xuv pulse are the same as the ones used in Fig. 7.2

electron energy of ω, it bends to the lower energy side due to the ac Stark shift. At this IR intensity, the ac Stark shift is about 0.6 eV, which is larger than the energy needed to shift the 2p state into the resonant energy region by APT. Thus, the resonant excitation appears on both sides around $t_d = -5.0$ and 5.0 o.c. For the first resonant region at $t_d = -5.0$ o.c., the IR laser does not have enough power in the later tail to further ionize the excited states and there are no high energy ATI peaks. For the second resonant regime at $t_d = 5.0$ o.c., the main part of the IR further ionizes the excited states and there are several ATI strips as shown in the figure. Since the excited states are not affected by the ac Stark shift, the ATI peak energies are almost constant against the IR intensity (time delay) and their behaviors look like a Freeman resonance [60]. In the overlap region around $t_d = 0$, the ATI peak yield oscillates twice per o.c. and the oscillation for different ATI peaks are all in phase for the time delay. Therefore, the total ionization probabilities also oscillate significantly.

For the ATI electron spectra from the SAP, as shown in Fig. 7.5, the general pattern is similar to that of the APT pulse. One difference is that we cannot clearly identify the ATI peaks since there are several peaks among one group, e.g. around the electron energy of ω. Since the energy distribution of the SAP is very broad and the SAP pulse can excite the atoms into many possible Floquet states. The phases between the different Floquet states depend on the IR intensity and they may have no phase relation at all [58]. Therefore, the total ionization yield, the sum of ionization from different Floquet states, changes moderately as a function of time delay. If we focus on a specified ATI peak, it oscillates in a similar way as the one by the APT [55].

From above examples, once again, we can say that the IR laser field modifies the atomic excited and continuum structures and the xuv pulse probes the structures. Based on this mechanism, we can control the transparency of materials by tuning the arriving time of the xuv pulse and the character of the xuv pulse as shown in the recent experiment by Ranotovic et al. [39]. Since APT pulses are generated by a driving IR, if we merge a second IR laser with the driving IR, we can extract the emission time of APT with respect to the driving IR field by analyzing the oscillation of ATI spectra as shown in the measurement by Shivaram et al. [35].

7.3.2 IR Assisted Photoexcitation

In the previous section, we analyzed the IR assisted photoionization by monitoring the ejected electron. Since the center xuv energy is lower than the ionization threshold of H atoms, the excitation is the dominant process. Recently, instead of measuring the ejected electrons, one can also measure the apparent photoabsorption cross section. In such experiments [33, 34], apparent photoabsorption cross section defined in (7.14) can be extracted by comparing the xuv energy spectra before and after a gas cell. Figure 7.6 shows the apparent photoabsorption cross sections of H atoms in a 5 fs IR laser field by an SAP. We see there is a very strong absorption line around 10 eV of photon energy, which corresponds to the transition from $1s \rightarrow 2p$. When the IR laser overlaps with the SAP, the strength of the absorption line is reduced significantly and there are several side bands appear. Hence, the IR field redistributes the oscillation strength to different energies and modifies the atomic excited and continuum structures. Unlike the convention photoabsorption cross section in a static field, the Bethe sum rule [61] does not apply to the apparent photoabsorption cross section.

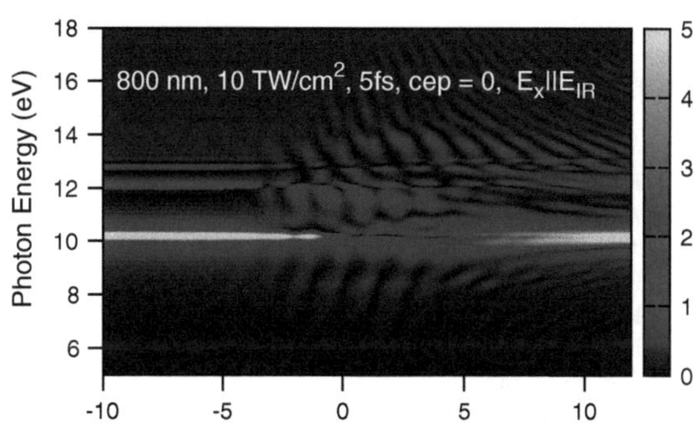

Fig. 7.6 IR assisted apparent photoabsorption cross section as a function of the time delay between the IR and SAP pulses. The IR pulse duration (FWHM) is 5 fs

7.4 Summary

In this chapter we have briefly reviewed the recent developments of the IR assisted photoionization and photoexcitation by an xuv pulse either from the APT or SAP. The IR laser field can assist the xuv photoabsorption by modifying the atomic structure and steering the excited or ionized electron. Since the IR optical cycle is of the order of femtosecond time scale, we can control dynamics in a fraction of one optical cycle or in attosecond time domain. Once we understand the mechanism of the IR assisted atomic photoabsorption , and the roles of the IR laser field, we can search a better way to control the material transparency [39] by tuning the IR laser arriving time as well as the character of the xuv pulse. By streaking the electron wavepacket generated by an SAP, one may image atomic or molecular structures [62]. Differently from the imaging in use of middle IR lasers [63], this method gives more freedom to control the electron wavepacket. Since the xuv has a broad energy distribution and it may excite a material into many excited states, tuning the time delay between the pump (xuv pulse) and the probe (IR laser pulse), one may resolve the detailed excited structures with an energy resolution much better than the limitation from the uncertainty principle of either the xuv pulse or IR pulse as shown in recent experiments [4, 64]. In this article we focused on the xuv photon energy bellow the ionization threshold. For the xuv photon energy well above the ionization threshold, the ejected electron by the xuv pulse can be streaked into the momentum space in which both the xuv and streaking IR fields information are encoded. From the streaking momentum distribution , one can obtain the xuv pulse information, the wave-form of the IR field or fundamental information of the time delay in the photoionization processes. Therefor, the IR assisted xuv photoabsorption has opened a door for us to *understand* the electron dynamics, *control* material properties and *image* molecular structures in ultrashort (femtosecond or even attosecond) time domain.

Acknowledgments This works was supported by a Grand-in-Aid for Scientific Research (C24540421) from the Japan Society for the Promotion of Science and part of the simulations was supported by HA-PACS Project for advanced interdisciplinary computational sciences by exa-scale computing technology.

References

1. F. Krausz, M. Ivanov, Rev. Mod. Phys. **81**, 163 (2009)
2. A.H. Zewail, Science **242**, 1645 (1988)
3. A.H. Zewail, J. Phys. Chem. A **104**, 5660 (2000)
4. P. Ranitovic, C.W. Hogle, P. Rivire, A. Palacios, X.M. Tong, N. Toshima, A. Gonzlez-Castrillo, L. Martin, F. Martn, M.M. Murnane, H. Kapteyn, Proc. Natl. Acad. Sci. U.S.A. **111**, 912 (2014)
5. D.J. Jones, S.A. Diddams, J.K. Ranka, A. Stentz, R.S. Windeler, J.L. Hall, S.T. Cundiff, Science **288**, 635 (2000)
6. S.T. Cundiff, J. Ye, Rev. Mod. Phys. **75**, 325 (2003)
7. G. Sansone, C. Vozzi, S. Stagira, M. Pascolini, L. Poletto, P. Villoresi, G. Tondello, S.D. Silvestri, M. Nisoli, Phys. Rev. Lett. **92**, 113904 (2004)

8. T. Wittmann, B. Horvath, W. Helml, M.G. Schätzel, X. Gu, A.L. Cavalieri, G.G. Paulus, R. Kienberger, Nature Physics **5**, 357 (2009)
9. G.G. Paulus, F. Grasbon, H. Walther, P. Villoresi, M. Nisoli, S. Stagira, E. Priori, S.D. Silvestri, Nature **414**, 182 (2001)
10. G.G. Paulus, F. Lindner, H. Walther, A. Baltuška, E. Goulielmakis, M. Lezius, F. Krausz, Phys. Rev. Lett. **91**, 253004 (2003)
11. X.M. Tong, K. Hino, N. Toshima, Phys. Rev. A **74**, 031405 (2006)
12. V. Roudnev, B.D. Esry, I. Ben-Itzhak, Phys. Rev. Lett. **93**, 163601 (2004)
13. M.F. Kling, C. Siedschlag, A.J. Verhoef, J.I. Khan, M. Schultze, T. Uphues, Y. Ni, M. Uiberacker, M. Drescher, F. Krausz, M.J.J. Vrakking, Science **312**, 246 (2006)
14. X.M. Tong, C.D. Lin, J. Phys. B **40**, 641 (2007)
15. X.M. Tong, C.D. Lin, Phys. Rev. Lett. **98**, 123002 (2007)
16. M. Kremer, B. Fischer, B. Feuerstein, V.L.B. de Jesus, V. Sharma, C. Hofrichter, A. Rudenko, U. Thumm, C.D. Schröter, R. Moshammer, J. Ullrich, Phys. Rev. Lett. **103**, 213003 (2009)
17. C. Lemell, X.M. Tong, F. Krausz, J. Burgdorfer, Phys. Rev. Lett. **90**, 076403 (2003)
18. A. Apolonski, P. Dombi, G.G. Paulus, M. Kakehata, R. Holzwarth, T. Udem, C. Lemell, K. Torizuka, J. Burgdorfer, T.W. Hansch, F. Krausz, Phys. Rev. Lett. **92**, 073902 (2004)
19. A. Schiffrin, T. Paasch-Colberg, N. Karpowicz, V. Apalkov, D. Gerster, S. Mhlbrandt, M. Korbman, J. Reichert, M. Schultze, S. Holzner, J.V. Barth, R. Kienberger, R. Ernstorfer, V.S. Yakovlev, M.I. Stockman, F. Krausz, Nature **493**, 70 (2013)
20. H.S. Nguyen, A.D. Bandrauk, C.A. Ullrich, Phys. Rev. A **69**, 063415 (2004)
21. M. Muramatsu, M. Hita, S. Minemoto, H. Sakai, Phys. Rev. A **79**, 011403 (2009)
22. H. Li, D. Ray, S. De, I. Znakovskaya, W. Cao, G. Laurent, Z. Wang, M.F. Kling, A.T. Le, C.L. Cocke, Phys. Rev. A **84**, 043429 (2011)
23. X. Xie, S. Roither, D. Kartashov, E. Persson, D.G. Arbó, L. Zhang, S. Gräfe, M.S. Schöffler, J. Burgdörfer, A. Baltuška, M. Kitzler, Phys. Rev. Lett. **108**, 193004 (2012)
24. J. Wu, A. Vredenborg, L.P.H. Schmidt, T. Jahnke, A. Czasch, R. Dörner, Phys. Rev. A **87**, 023406 (2013)
25. R. Kienberger, M. Hentschel, M. Uiberacker, C. Spielmann, M. Kitzler, A. Scrinzi, M. Wieland, T. Westerwalbesloh, U. Kleineberg, U. Heinzmann, M. Drescher, F. Krausz, Science **297**, 1144 (2002)
26. E. Goulielmakis, M. Uiberacker, R. Kienberger, A. Baltuska, V. Yakovlev, A. Scrinzi, T. Westerwalbesloh, U. Kleineberg, U. Heinzmann, M. Drescher, F. Krausz, Science **305**, 1267 (2004)
27. G. Sansone, E. Benedetti, F. Calegari, C. Vozzi, L. Avaldi, R. Flammini, L. Poletto, P. Villoresi, C. Altucci, R. Velotta, S. Stagira, S. De Silvestri, M. Nisoli, Science **314**, 443 (2006)
28. P. Johnsson, J. Mauritsson, T. Remetter, A. L'Huillier, K.J. Schafer, Phys. Rev. Lett. **99**, 233001 (2007)
29. P. Ranitovic, X.M. Tong, B. Gramkow, S. De, B. DePaola, K.P. Singh, W. Cao, M. Magrakvelidze, D. Ray, I. Bocharova, H. Mashiko, A. Sandhu, E. Gagnon, M.M. Murnane, H. Kapteyn, I. Litvinyuk, C.L. Cocke, New J. Phys. **12**, 013008 (2010)
30. C. Neidel, J. Klei, C.H. Yang, A. Rouzée, M.J.J. Vrakking, K. Klünder, M. Miranda, C.L. Arnold, T. Fordell, A. L'Huillier, M. Gisselbrecht, P. Johnsson, M.P. Dinh, E. Suraud, P.G. Reinhard, V. Despré, M.A.L. Marques, F. Lépine, Phys. Rev. Lett. **111**, 033001 (2013)
31. E. Goulielmakis, Z.H. Loh, A. Wirth, R. Santra, N. Rohringer, V.S. Yakovlev, S. Zherebtsov, T. Pfeifer, A.M. Azzeer, M.F. Kling et al., Nature **466**, 739 (2010)
32. H. Wang, M. Chini, S. Chen, C.H. Zhang, F. He, Y. Cheng, Y. Wu, U. Thumm, Z. Chang, Phys. Rev. Lett. **105**, 143002 (2010)
33. M. Chini, X.W. Wang, Y. Cheng, Y. Wu, D. Zhao, D.A. Telnov, S.I. Chu, Z.H. Chang, Sci. Rep. **3**, 1105 (2013)
34. X. Wang, M. Chini, Y. Cheng, Y. Wu, X.M. Tong, Z. Chang, Phys. Rev. A **87**, 063413 (2013)
35. N. Shivaram, H. Timmers, X.M. Tong, A. Sandhu, Phys. Rev. A **85**, 051802 (2012)
36. N. Shivaram, H. Timmers, X.M. Tong, A. Sandhu, Phys. Rev. Lett. **108**, 193002 (2012)
37. E.P. Wigner, Phys. Rev. **98**, 145 (1955)

38. M. Schultze, M. Fiess, N. Karpowicz, J. Gagnon, M. Korbman, M. Hofstetter, S. Neppl, A.L. Cavalieri, Y. Komninos, T. Mercouris, C.A. Nicolaides, R. Pazourek, S. Nagele, J. Feist, J. Burgdoerfer, A.M. Azzeer, R. Ernstorfer, R. Kienberger, U. Kleineberg, E. Goulielmakis, F. Krausz, V.S. Yakovlev, Science **328**, 1658 (2010)
39. P. Ranitovic, X.M. Tong, C.W. Hogle, X. Zhou, Y. Liu, N. Toshima, M.M. Murnane, H.C. Kapteyn, Phys. Rev. Lett. **106**, 193008 (2011)
40. X.M. Tong, C.D. Lin, Phys. Rev. A **73**, 042716 (2006)
41. T. Shirahama, X.M. Tong, K.I. Hino, N. Toshima, Phys. Rev. A **80**, 043414 (2009)
42. R. Santra, V.S. Yakovlev, T. Pfeifer, Z.H. Loh, Phys. Rev. A **83**, 033405 (2011)
43. S. Chen, M.J. Bell, A.R. Beck, H. Mashiko, M. Wu, A.N. Pfeiffer, M.B. Gaarde, D.M. Neumark, S.R. Leone, K.J. Schafer, Phys. Rev. A **86**, 063408 (2012)
44. M. Murakami, S.I. Chu, Phys. Rev. A **88**, 043428 (2013)
45. X.M. Tong, S.I. Chu, Phys. Rev. A **55**, 3406 (1997)
46. X.M. Tong, C.D. Lin, J. Phys. B **38**, 2593 (2005)
47. X.M. Tong, K. Hino, N. Toshima, Phys. Rev. Lett. **97**, 243202 (2006)
48. X.M. Tong, K. Hino, N. Toshima, Phys. Rev. Lett. **101**, 163201 (2008)
49. X.M. Tong, N. Nakamura, S. Ohtani, T. Watanabe, N. Toshima, Phys. Rev. A **80**, 042502 (2009)
50. S.I. Chu, D.A. Telnov, Phys. Rep. **390**, 1 (2004)
51. X.M. Tong, S.I. Chu, Chem. Phys. **217**, 119 (1997)
52. J. Bauer, J. Phys. B **34**, 1343 (2001)
53. D.G. Arbó, J.E. Miraglia, M.S. Gravielle, K. Schiessl, E. Persson, J. Burgdörfer, Phys. Rev. A **77**, 013401 (2008)
54. K.L. Ishikawa, K. Schiessl, E. Persson, J. Burgdörfer, Phys. Rev. A **79**, 033411 (2009)
55. X.M. Tong, P. Ronitovic, C.L. Cocke, N. Toshima, Phys. Rev. A **81**, 021404 (2010)
56. X.M. Tong, P. Ranitovic, D.D. Hickstein, M.M. Murnane, H.C. Kapteyn, N. Toshima, Phys. Rev. A **88**, 013410 (2013)
57. X.M. Tong, S.I. Chu, Phys. Rev. A **61**, 031401 (2000)
58. X.M. Tong, N. Toshima, Phys. Rev. A **81**, 043429 (2010)
59. F. He, C. Ruiz, A. Becker, U. Thumm, J. Phys. B **44**, 211001 (2011)
60. R.R. Freeman, P.H. Bucksbaum, H. Milchberg, S. Darack, D. Schumacher, M.E. Geusic, Phys. Rev. Lett. **59**, 1092 (1987)
61. J.G. Wang, T. Kato, I. Murakami, Phys. Rev. A **60**, 3750 (1999)
62. M.H. Xu, L.Y. Peng, Z. Zhang, Q. Gong, X.M. Tong, E.A. Pronin, A.F. Starace, Phys. Rev. Lett. **107**, 183001 (2011)
63. D.D. Hickstein, P. Ranitovic, S. Witte, X.M. Tong, Y. Huismans, P. Arpin, X. Zhou, K.E. Keister, C.W. Hogle, B. Zhang, C. Ding, P. Johnsson, N. Toshima, M.J.J. Vrakking, M.M. Murnane, H.C. Kapteyn, Phys. Rev. Lett. **109**, 073004 (2012)
64. W. Cao, G. Laurent, S. De, M. Schöffler, T. Jahnke, A.S. Alnaser, I.A. Bocharova, C. Stuck, D. Ray, M.F. Kling, I. Ben-Itzhak, T. Weber, A.L. Landers, A. Belkacem, R. Dörner, A.E. Orel, T.N. Rescigno, C.L. Cocke, Phys. Rev. A **84**, 053406 (2011)

Chapter 8
Photoionization Time Delays

J. Marcus Dahlström, Morgane Vacher, Alfred Maquet, Jérémie Caillat and Stefan Haessler

Abstract The material presented in this chapter is based on important advances realized in "attophysics" which make feasible to follow the motion of electrons in atoms and molecules with attosecond-level time resolution. In this context, time-delays have been recently determined in the process of photoionization by extreme-ultra-violet (XUV) pulses and the question of the significance of these measured delays arises. As we shall outline here, numerical experiments show that they are intimately related to the structure of the ionized species' continuous spectrum. Another point addressed here is that, in experiments, the measurements have the common

J.M. Dahlström
Department of Physics, Stockholm University, AlbaNova University Center,
10691 Stockholm, Sweden
e-mail: marcus.dahlstrom@fysik.su.se

J.M. Dahlström
Max-Planck Institute for the Physics of Complex Systems,
Noethnitzerstr. 38, 01187 Dresden, Germany

M. Vacher
Department of Chemistry, Imperial College London, SW7 2AZ London, United Kingdom
e-mail: m.vacher@imperial.ac.uk

A. Maquet · J. Caillat
UPMC, UMR 7614, Laboratoire de Chimie Physique - Matière et Rayonnement 11,
rue Pierre et Marie Curie, 75231 Paris Cedex 05, France
e-mail: alfred.maquet@upmc.fr

A. Maquet · J. Caillat
CNRS, UMR 7614, Laboratoire de Chimie Physique - Matière et Rayonnement 11,
rue Pierre et Marie Curie, 75231 Paris Cedex 05, France
e-mail: jeremie.caillat@upmc.fr

S. Haessler (✉)
Photonics Institute, Vienna University of Technology,
Gußhausstrße 27/387, 1040 Vienna, Austria
e-mail: stefan.haessler@tuwien.ac.at

S. Haessler
LOA, ENSTA ParisTech, CNRS, Ecole Polytechnique, Université Paris-Saclay,
828 bd des Maréchaux, 91761 Palaiseau Cedex, France

M. Kitzler and S. Gräfe (eds.), *Ultrafast Dynamics Driven by Intense Light Pulses*, Springer Series on Atomic, Optical, and Plasma Physics 86, DOI 10.1007/978-3-319-20173-3_8

characteristic to be performed in the presence of an auxiliary infra-red (IR) field, used to "clock" the timing of the process. This implies to adapt the theory treatment to handle such "two-color" photoionization processes. We review a systematic analysis of these features that are characteristic of this class of electronic transitions, when viewed in the time domain.

8.1 Introduction

"Attophysics" has emerged in the 2000s with the advent of a new generation of radiation sources delivering attosecond pulses of Extreme Ultra-Violet (XUV) radiation via the High-order Harmonic Generation (HHG) process [1–3]. These new sources allow the photoionization process to be considered as taking place at a well defined time within a brief time window with attosecond resolution. This opens up the possibility to achieve the real-time probing and control of electron (or hole) dynamics on a sub-femtosecond time scales [4, 5]. It is in this very active context that several different experiments have evidenced attosecond time-delays that are associated with the photoionization process [6–12]. Complementary interpretations of these experiments have been reviewed within the frameworks of both the time-independent [13] and the time-dependent picture [14].

This set of results has been obtained thanks to the advances realized in both the attosecond metrology and generation schemes of XUV pulses through HHG. One has led to the generation of isolated pulses with duration down to 80 as [15], while the other has resulted in the emission of attosecond pulse trains, with bursts of XUV radiation having durations down to 63 as [16]. When recombining such XUV pulses with an infrared (IR) laser radiation, one can realize two-color pump-probe experiments, with sub-femtosecond temporal resolution. One important outcome has been to study the photoionization of atoms in the time domain, thus evidencing intrinsic time-delays associated with the process [6, 9–11]. As we shall show, the physical significance of the delays so measured depends on the electronic structure of the excited species, including the influence of resonant states. Regarding the input of attophysics in this latter field, c.f. [17–19].

Before to proceed, we wish to make clear the distinction existing between the attosecond scale relevant to intra-atomic electronic transitions and the much longer time scale governing the detection of the wave packet created in the ionization process. When the atom is irradiated by an attosecond XUV pulse, the width of the energy content of the photoionized wave packet is governed by the time-energy uncertainty principle. While traveling over a *macroscopic* distance towards the detector, the motion of the maximum of the density probability is essentially classical. In parallel, all energy components of the electron wave packet (EWP) will spread so that the spatio-temporal structure also reaches the macroscopic realm. This allows for direct detection of the electron energy distribution as a function of time-of-flight to the detector. The important point is that, although the electronic transition is taking place on an attosecond scale, in the presence of the IR clock a signature of the time-development of the process can be extracted from the photoelectron spectrum.

The organization of the chapter is as follows: In Sect. 8.2, we will follow the wave packet evolution using time-dependent simulations and extract the delays due to the microscopic interactions that have occurred in the target. The results are directly compared with theoretical delays obtained from time-independent scattering theory. In Sect. 8.3 we provide an interpretation of the microscopic delays in photoionization using a formalism designed to compute two-photon (XUV-IR) matrix elements. In Sect. 8.4 we give a review of experimental work on photoionization delays using two-color (XUV-IR) fields. Finally, in Sect. 8.5 we present our conclusions.

8.2 Phase-Shifts and Time-Delays

8.2.1 Formal Definition of a Photoionization Delay

The characterization of the dynamics of a photoelectron wave packet commonly refers to the notion of scattering time-delay developed in the context of quantum collision theory [20, 21]. A wave packet scattered by a potential experiences a delay as compared to free motion. Wigner and his collaborators [20, 21] first pointed out that this delay can be characterized through the spectral derivative of a phase shift η:

$$\tau = \hbar \frac{\partial \eta}{\partial E} \tag{8.1}$$

where E represents the particle's energy.[1] A stationary-phase derivation shows that this delay, evaluated at the energy of maximum amplitude, represents the delay induced by the potential on the traveling particle, compared to the time it takes a reference particle (typically a free particle) with the same energy to cover the same distance: In (8.1), it is implicit that η is the phase difference (or phase *shift*) between the scattered and the reference waves.[2] It must be clear also that the reference has to be chosen to match the scattering wave in the incoming region (usually far from the interaction region). Moreover, since a collision treated quantum mechanically is described with the help of wave packets, its dynamics can never be *completely* represented by a unique delay: The group delays are representative quantities in principle only for near Fourier-limited, ideally Gaussian distributions.

Extending Wigner's formalism to photoionization processes is straightforward if one admits that a photoelectron experiences a "half collision" as it is ejected, while remaining under the influence of the ionic core [13]. Further, a group delay can be assigned to the spectral phase that the photoelectron wave packet accumulates

[1]Other formal definitions can be invoked, see [22] for a comprehensive review.

[2]The stationary-phase approximation is discussed in e.g. [23].

during and *upon* the transition. This comes in addition to the imprint of the driving electric field phase exploited in standard XUV pulse characterization techniques such as RABBIT [24, 25] or FROG-CRAB [26].

8.2.1.1 Scattering Delay

In the case of one-photon ionization, the only contribution to the phase of the EWP is associated with the real-valued continuum wave function *selected* by the transition [27]. Just as for collisions, that phase is actually a *phase shift*, the definition of which comes with the choice of a reference.

When taking the free-particle as a reference, the overall scattering phase associated with photoionization decomposes as the sum of a short-range and a long-range contributions,

$$\eta = \eta_{\mathrm{SR}} + \eta_{\mathrm{LR}} \tag{8.2}$$

The long-range term, η_{LR}, is due to the asymptotic Coulomb tail associated with any atomic or molecular ionic core potential and it depends logarithmically on the electron position. The short-range term, η_{SR}, contains a signature of the detailed interactions between the released electron and the remaining ones which constitute the ionic core, as well as the multi-center nuclear structure in the case of molecules.

According to (8.2), the group delay associated with scattering upon photoionization reads as the sum

$$\underbrace{\frac{\partial \eta}{\partial E}}_{\tau_{\mathrm{sca}}} = \underbrace{\frac{\partial \eta_{\mathrm{SR}}}{\partial E}}_{\tau_{\mathrm{SR}}} + \underbrace{\frac{\partial \eta_{\mathrm{LR}}}{\partial E}}_{\tau_{\mathrm{LR}}} \tag{8.3}$$

of a long-range delay τ_{LR} and a short-range delay τ_{SR}. While τ_{LR} is *everywhere* modified by the infinite-range influence of the Coulomb potential, τ_{SR} converges after some finite distance representative of the size of the atom or the molecule, i.e. beyond the influence of the short-range component of the potential. As already mentioned, the implicit reference in (8.2) and (8.3) is the free motion.

Note that for practical reasons, scattering phase shifts and Wigner-like delays are usually assigned to angular momentum *components* of the complete EWP expanded on a partial wave basis. Partial-wave decompositions are essential in atomic calculations as they allow the use of angular momentum theory. Formal links between the Wigner delay, the lifetime matrix, Q, and the scattering matrix, S, were first derived by Smith [21]. In the present context, such constructions provide a way to separate the total wave packet into partial wave packets in different ionic channels. The dynamics of each of these can be properly characterized by a simple Wigner-like group delay.

Besides, when choosing specific representations of the continuum, the scattering *phase shift* sometimes manifests as a contribution to the phase of a complex valued wave-function. This occurs for example when treating the continuum with conventional scattering waves, initially developed in the S-matrix formalism to represent a full collision. This phase then shows up as the argument of the associated (otherwise real-valued) *transition matrix element*: One indeed often refers to the scattering phase shift as the "phase of the dipole matrix element". Similarly, it has been shown recently that the one-photon dipole matrix element becomes complex if correlation effects are included perturbatively on a basis of uncorrelated single particle function products [28].

8.2.1.2 Beyond 1-Photon Transitions

The fact that the real-valued dipole operator itself does not bring any contribution to the phase of the released EWP reflects the instantaneous nature of *single* photon absorption (or its annihilation in the quantum-electrodynamics formalism). However, in multiphoton transitions associated with above-threshold ionization (ATI), i.e. when intermediate states lie in the continuous spectrum, the transition operator (typically the Green operator in a 2-photon transition) becomes complex, thus bringing an additional contribution to the spectral phase of the released wave packet.

Following the line of reasoning developed for the scattering delay, this additional phase can be used in the time domain to introduce a "transition delay" τ_{tra} accounting for the temporal shaping of the EWP as it builds up *during the transition* [29]. In particular, it can be interpreted as a transition *duration* when the energy transferred by one of the photons coincides with the excitation energy towards a resonant state possessing a significant lifetime τ_{R}: τ_{tra} is then related to τ_{R} and represents the time spent by the system in this intermediate resonance before transiting toward the final state.[3] Experimentally, accessing this "transition phase" requires an interferometric setup with two paths leading coherently to the same final continuum state. These requirements are met by the RABBIT technique [31], a method initially designed to characterize attosecond XUV pulse trains using an IR pulse as an external clock [24, 25, 32, 33]. This technique was thus revisited in [7, 8, 29, 34] to investigate XUV-IR 2-color photoionization in the time-domain, where two distinct 2-photon transitions probe each other.

Furthermore, it was shown that the phase of the transition matrix element in two-photon XUV-IR ionization through an intermediate continuum is related to the short-range *scattering phase shift* (η_{SR}) of the *intermediate state* [10, 13, 35, 36]. It was verified numerically for hydrogen that the intermediate scattering phase could be recovered with high fidelity if a universal phase shift, called the *continuum–continuum phase*, was subtracted from the two-photon matrix element [36]. This reinterpretation of RABBIT measurements [10], which states that the one-photon ionization delays can be experimentally obtained, has now been verified by diagrammatic methods for both neon and argon atoms [37]. In this reinterpretation, the IR pulse is seen as a probe used to characterize the dynamics of XUV-driven 1-photon

[3] See e.g. [30] for a comprehensive derivation and exploitation of transitions delays in the particular context of resonant X-ray Raman scattering.

ionization.[4] The RABBIT setup shares fundamental similarities with the streaking technique [14, 36, 38], which (i) also consists of HHG-based XUV-pump IR-probe schemes, (ii) was initially conceived to characterize XUV pulses (using a FROG-CRAB analysis), and (iii) had already been revisited to access photoionization delays [9]. The main difference lies in the pulse properties: XUV pulse train and weak IR field in RABBIT *versus* broad, single XUV pulse and moderate IR field for streaking. Section 8.3 of the present chapter is dedicated in more details to the specific subject of probing one-photon ionization delays using dressing IR fields, but first we turn to time-domain simulations of microscopic delays and compare the results with scattering theory.

8.2.2 Ionization Dynamics in Numerical Experiments

The physical relevance of the scattering delay as defined above can be addressed in simulations of photoionization, where the actual dynamics observed in numerical experiments is compared to the delays evaluated from energy-dependent scattering phase shifts, using Wigner's formalism. To this end, we used simple models of atoms allowing extensive numerical experiments where the formal delays are compared to the exact (numerical) solution of the time-dependent Schrödinger equation (TDSE).

8.2.2.1 The Model Atoms

Our generic model consists in a single active electron (with position x) interacting with an effective core through a symmetric potential in one dimension. We will consider three different potentials defining three different systems: A, B and C. Atom A is built on a simple soft-Coulomb potential,[5]

$$V(x) = -\frac{1}{\sqrt{x^2 + a^2}} \tag{8.4}$$

where the regularizing parameter a is the main knob to control the energies of the bound states. The potentials for atoms B and C read

$$V(x) = -\frac{1}{\sqrt{x^2 + a^2}} + G(x) \tag{8.5}$$

where $G(x)$ represents two symmetric hyper-gaussian barriers, see Fig. 8.1. In such models, the short-range barriers are adjusted to design shape resonances, accounting

[4] Note that the two reinterpretations of RABBIT in terms of transition *or* scattering delays are not contradictory, but rather complementary.

[5] The equations are displayed in atomic units: $\hbar = m = e = 1/(4\pi\epsilon_0) = 1$.

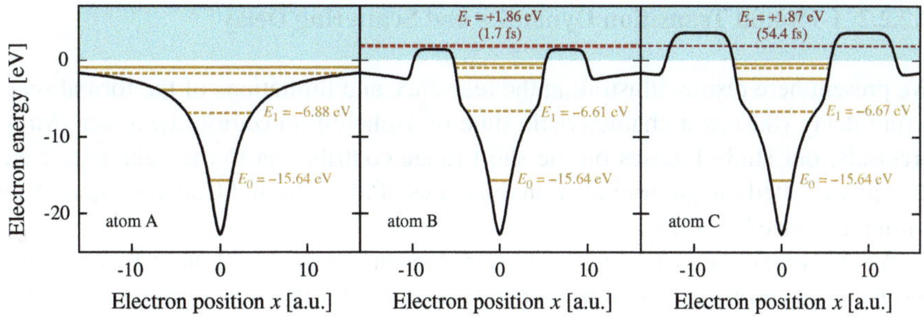

Fig. 8.1 Model potentials used in our simulations. The potential for model atom A (*left*) consists in a simple soft-Coulomb term. The potentials for model atoms B (*center*) and C (*right*) contain additional barriers inducing structures in the continuum. *Horizontal lines* represent discrete energy levels: bound states ($E < 0$, *yellow*) and resonances ($E > 0$, *red*). The *line styles* indicate the parity of each state: even (*full*) or odd (*dashed*). The energy (and life-time) of the states relevant for our simulations are also indicated

qualitatively for the mean-field interaction between the active electron and the frozen cores. The spectral positions of the shape resonances depend mainly on the barrier separation, and their widths on the height and width of the barriers.

Note that because we use symmetric potentials, the discrete bound state wave-functions are alternately even and odd (starting with an even wave-function for the ground state) whereas degenerate even and odd continuum states are found for any positive energy. This property is of first importance in the simulations since a dipole transition between two such states is allowed only if their parity is opposite, allowing for a direct treatment of the continuum in terms of real-valued wave-functions [27, 29, 39], with defined parities. In the following, we denote $\phi_{E,p}(x)$ the continuum eigenfunction associated with the positive energy E and parity p ($+1$ and -1 for an even and odd parity respectively). The potentials were adjusted to create a relatively broad, *odd* resonance near threshold in atom B, and a narrower one, also odd, in atom C (see Fig. 8.1).

The time evolution of the system's wave-function $\psi(x,t)$ follows the time-dependent Schrödinger equation:

$$i\frac{\partial}{\partial t}\psi(x,t) = \left[-\frac{1}{2}\frac{\partial^2}{\partial x^2} + iA(t)\frac{\partial}{\partial x} + V(x)\right]\psi(x,t), \tag{8.6}$$

where the interaction with light is treated in the velocity gauge ($A \cdot p$) representation of the dipole approximation. We solved it numerically using the Crank-Nicolson algorithm [40]. The temporal dependence of the vector potential $A(t)$ is defined by a carrier frequency ω_0 and a generic $\sin^2$-envelope centered at $t = 0$, with a duration of 100 cycles FWHM. In the range of considered photon energies (15.8 eV $< \omega_0 <$ 18.8 eV), it represents pulse durations of ~25 fs, which lies between the lifetimes of the resonances in atom B (1.7 fs) and in atom C (54.4 fs). The XUV pulse intensities were kept low enough so that non-linear effects could be safely neglected.

8.2.2.2 1-Photon Transition Dynamics and Scattering Delay

We present here results illustrating the relevance and limitations of the formal scattering delay (8.3) as a characteristic time of 1-photon ionization dynamics. More precisely, our study focuses on the short-range contribution to the delays, τ_{SR}: we thus investigated the photoionization dynamics of atoms B and C, taking atom A as a reference (see Fig. 8.1).

We first detail the procedure for a near resonant ionization of atom B, where the resonance is short-lived enough to induce negligible distortions on the bell-shaped temporal and spectral profiles, but where the delay is large enough to be visualized on the wave packet temporal profile. We present afterwards the results obtained near a sharper resonance and in a smooth continuum.

Near a broad resonance

Our analysis consists in merging temporal and spectral information retrieved from photoelectron wave packets. Figure 8.2a displays the spectral profile $\sigma(E)$ of the wave packet created by ionizing atom B initially in its ground state, with a pulse of central energy $\omega_0 = 17.55$ eV. The spectrum was retrieved by using the spectral analysis technique described in [41]. The wave packet is centered close to the resonant

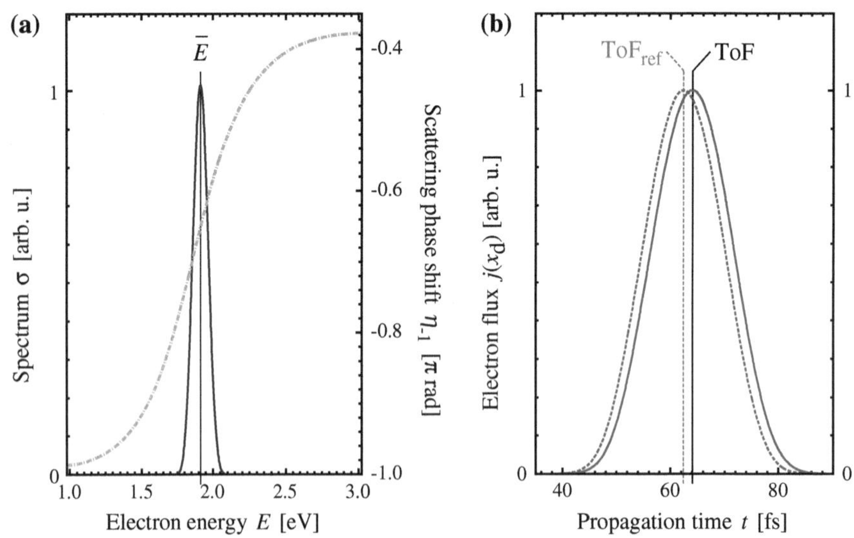

Fig. 8.2 Model atom B (ground state) @ 17.55 eV— **a** *Full blue curve* electron spectrum (the *vertical line* indicates the mean energy under the peak $\bar{E}$); *dash-dotted green curve* phase shift of the odd scattering state (atom A as a reference); **b** *full red curve* electron flux $j(x_d)$ evaluated at $x_d = 1000$ a.u. versus propagation time; *dashed red curve* $j(x_d)$ evaluated at the same distance when photoionizing the reference atom A with a pulse adjusted to obtain the same average energy $\bar{E}$; *vertical lines* average times under the peaks, TOF (*full*) and TOF$_{ref}$ (*dashed*) respectively. For a better readability, the spectrum and flux data were normalized to 1 at their maximum values

energy (E_R =1.87 eV), within its width (Γ_R = 0.383 eV). The associated mean photoelectron energy computed as

$$\bar{E} = \frac{1}{P_{\text{ion}}} \int_{\text{peak}} E \times \sigma(E)\,\mathrm{d}E \tag{8.7}$$

where P_{ion} is the total ionization probability, is equal to 1.86 eV. From the temporal point of view, we characterized the wave packet dynamics by computing the electron flux as a function of time at a position x_{d} = 1000 a.u, beyond the short-range part of the potential, as

$$j(x_{\text{d}}, t) = \text{Im}\left\{[\psi(x_{\text{d}}, t)]^{\star}\psi'(x_{\text{d}}, t)\right\} \tag{8.8}$$

where ψ' is the first order derivative of ψ with respect to x and * stands for complex conjugate. From the flux, shown in Fig. 8.2b, we defined a numerical "time-of-flight" as the average arrival time at the virtual detector,

$$\text{TOF}(x_{\text{d}}) = \frac{1}{P_{\text{ion}}} \int_{\text{peak}} t \times j(x_{\text{d}}, t)\,\mathrm{d}t. \tag{8.9}$$

In the considered case, it amounts to 64.04 fs. That average value is indeed close to the time it would take a free particle with a kinetic energy of 1.86 eV to travel the distance of 1000 a.u (65.42 fs), and the difference evidences the influence of the potential on the electron motion, all along the way to the detector.

To reveal the role of the short-range barriers on the release dynamics, and more specifically on the TOF value value, we simulated the photoionization of the reference atom A (with the interaction potential restricted to the soft Coulomb term) with a central pulse frequency finely adjusted to create a wave packet with the same average energy $\bar{E}$ = 1.86 eV. The flux computed at x_{d} for this reference transition presents a temporal profile similar to the near resonant transition in atom B (also displayed in Fig. 8.2b). It is however clearly shifted towards shorter times: The associated average time-of-flight, TOF_{ref}, is equal to 62.40 fs. The time-of-flight difference

$$\Delta\text{TOF} = \text{TOF}(x_{\text{d}}) - \text{TOF}_{\text{ref}}(x_{\text{d}}) \tag{8.10}$$

is equal to 1.64 fs. It is a numerical evaluation of the delay induced on the released wave packet by the short-range barriers in atom B, seen from a time-dependent perspective, to be compared to the formal scattering delay.

We thus numerically computed the real-valued continuum eigenfunctions selected by the transition [39] $\phi_{E,p}(x)$ for atom B and $\phi^{\text{ref}}_{E,p}(x)$ for the reference-atom A and finally determined their relative phase shift $\eta_p(E)$. The phase shift $\eta_{-1}(E)$ computed in the energy range covered by the wave packet is shown as a dashed curve in Fig. 8.2a. It undergoes a smooth $\simeq 0.6\pi$ jump spread over the resonance width. Its spectral derivative evaluated at the average energy $\bar{E}$, i.e. the formal scattering delay τ_{sca}, is equal to 1.65 fs, which is remarkably close to the value of ΔTOF.

The procedure was then repeated for a set of pulse frequencies near the resonance. Figure 8.4b compares the values of ΔTOF (circles) obtained in the time-dependent treatment (8.8–8.10), to the group delays τ_{sca} (full line) derived from the scattering phase shifts and evaluated at the mean wave packet energies $\bar{E}$. The two sets of data follow a typical bell-shaped curve, with a maximum at resonance reaching the resonance lifetime $\Gamma_R^{-1} = 1.7$ fs. They are in very good agreement with the most noticeable discrepancy at resonance ($\bar{E} \simeq 1.87$ eV), i.e. where the wave packet distorsion is expected to be maximum.

Near a narrow resonance

The situation is different when the transition reaches the vicinity of a resonance narrower than the pulse spectral width. A representative case is given by the resonant ionization of atom C taken in its ground state.

In the wave packet launched by a $\hbar\omega_0 = 17.55$ eV pulse, the resonance now manifests as a sharp, prominent, peak distorting the otherwise smooth, bell-shaped spectral profile, as displayed in Fig. 8.3a. Furthermore, the electron flux shown in Fig. 8.3b (still computed at $x_d = 1000$ a.u.), now displays an asymmetric tail towards

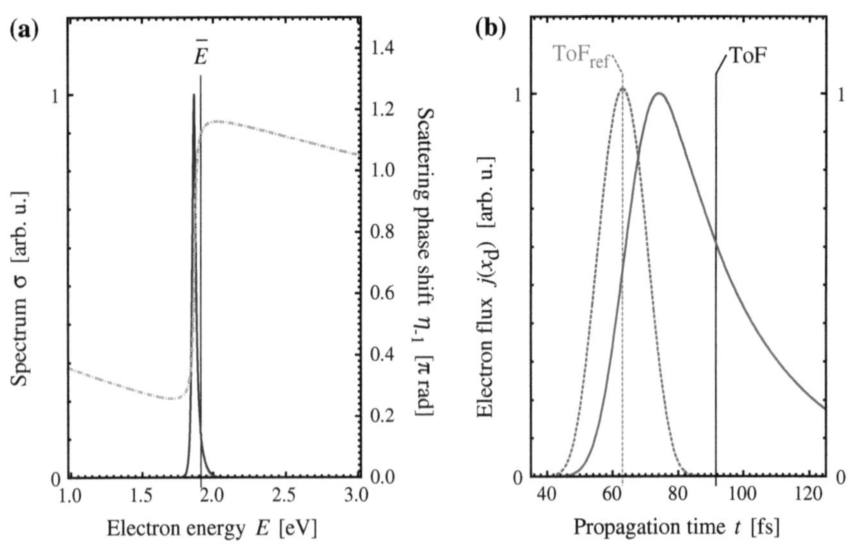

Fig. 8.3 Model atom C (ground state) @ 17.55 eV— **a** *Full blue curve* electron spectrum (the *vertical line* indicates the mean energy under the peak $\bar{E}$); *dash-dotted green curve* phase shift of the odd scattering state (atom A as a reference); **b** *full red curve* electron flux $j(x_d)$ evaluated at $x_d = 1000$ a.u. versus propagation time; *dashed red curve* $j(x_d)$ evaluated at the same distance when photoionizing the reference atom A with a pulse adjusted to obtain the same average energy $\bar{E}$; *vertical lines* average times under the peaks, TOF (*full*) and TOF$_{ref}$ (*dashed*) respectively. For a better readability, the spectrum and flux data were normalized to 1 at their maximum values

larger times—a direct signature of the resonance's exponential decay. At this point, it is already clear that such a structured wave packet cannot be accurately characterized by a single energy (e.g. $\bar{E}$), nor its elongated temporal profile by a single time (e.g. TOF). Taking again atom A as a reference provides a time of flight difference $\Delta\text{TOF} = 28.46$ fs, while the formal delay derived from the scattering phase shift $\eta_{-1}(E)$ (shown in Fig. 8.3a) at the average energy $\bar{E}$ is above 50 fs.

We again repeated the simulations by changing the central frequency, and reported the delays in Fig. 8.4c. Although both sets of data display the same peaked shape centered at the resonant energy $E_R = 1.86$ eV, they present clear discrepencies. In particular, the maximum value reached by ΔTOF is approximately twice smaller than the actual resonance lifetime and the ΔTOF peak is broadened by the light pulse width, as compared to the τ_{sca} peak.

In a smooth continuum

The formal scattering delay, validated before as representative of the ionization dynamics near a *broad* resonance, is expected to be just as relevant in a smooth continuum. To illustrate this, we simulated photoionizations of atom B, now initially considered in its *first excited state* with odd symmetry. The pulse central frequency was varied to end up in the same electron energy region, between 1 and 3 eV. Selection rules are such that now, the odd resonance is transparent to one-photon transitions, for parity reasons.

We followed the same time-energy analysis as for the two previous cases. The delays, computed as the time-of-flight difference ΔTOF on the one hand and as the phase shift derivative τ_{sca} on the other hand, are compared in Fig. 8.4a. The delay evolves monotonically, and the two sets of data are indeed in very good agreement. The results presented above illustrate the validity of the formal scattering delay as representative of the *average* one-photon ionization dynamics, at least when the created wave packets are unstructured enough both in the energy and time domains. We mention that the significance and relevance of the formal transition delay can be evidenced in a similar manner in two-photon ionization simulations [29, 34].

8.3 Analysis of Two-Photon XUV +IR Ionization

A real-world time-of-flight spectrometer for detecting photoelectrons operates on the basis of separating different energy (or momentum) components through macroscopic propagation [42, 43]. Indeed, this constitutes *a frequency domain measurement* and information about microscopic time-delays cannot be resolved. However, if the photoionization process is modulated by an IR laser field, it is possible to encode some temporal information into the frequency domain by construction of a so-called *spectrogram* over photoelectron kinetic energy (or momentum) and sub-cycle delay between XUV and IR fields [26, 31, 44, 45]. This is done by repeating the experiment as a function of the delay between XUV and IR fields. Still, timing information about the one-photon process is not directly recorded and further theoretical calibration is required, as we shall briefly discuss in this section.

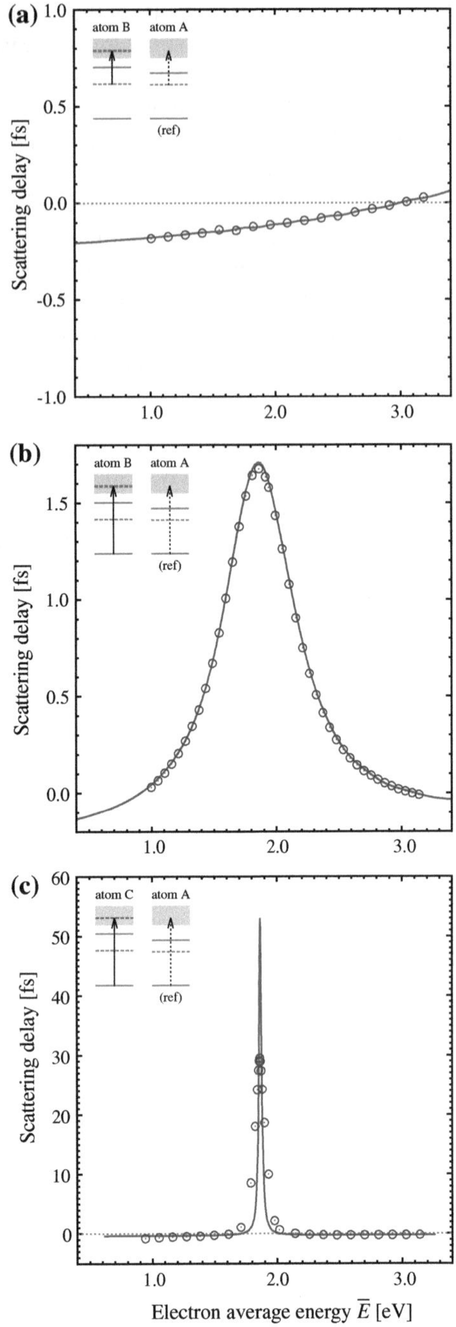

Fig. 8.4 Scattering delays evaluated as a time-of flight difference (*circles*) and as a formal group delay (*full curves*), as a function of the mean photoelectron energy $\bar{E}$ (see text). **a** *Smooth continuum*: Ionization of atom B compared to ionization of atom A, both from their first excited state (the resonance is transparent to the transition, for parity reasons); **b** *broad resonance*: ionization of atom B compared to ionization of atom A, both from their ground state; **c** *sharp resonance*: ionization of atom C compared to ionization of atom A, both from their ground state

In the presence of an IR laser field, the photoelectron released upon absorption of one XUV photon can further absorb and emit IR photons, thereby changing its energy. Provided that the IR field is weak, the energy of the electron will change as the result of the exchange of only *one* IR photon at most, so that the overall process amounts to a two-photon transition. These measurements are based on the interference of several multiphoton pathways that lead to the same final continuum state. The contribution from the final state scattering phase cancels out and the so-called *transition phase* through the different pathways will determine the observed delay in the measurement.

In the following, we will restrict the analysis to ATI in a smooth continuum where the photoelectron interacts perturbatively with both the XUV and IR fields through second-order perturbation theory. The two-photon process induced by the linearly polarized field, $E(\omega)$, is expressed as an S-matrix:

$$S_{fi}^{(2)} = \frac{1}{2\pi i} \int d\omega' \, M_{fi}^{(2)}(\omega') \, E(\omega') \, E(\omega''), \tag{8.11}$$

where the two-photon matrix element from the initial state $|i\rangle$ (with energy ϵ_i) and to the final state $|f\rangle$ (with energy ϵ_f) is

$$M_{fi}^{(2)}(\omega') = \lim_{\varepsilon \to 0^+} \sum\!\!\!\!\!\!\int_n \frac{\langle f|z|n\rangle\langle n|z|i\rangle}{(\epsilon_i + \omega' - \epsilon_n + i\varepsilon)}, \tag{8.12}$$

with ω' being the energy of the first absorbed photon and where the integral-sum runs over intermediate states n (with energy ϵ_n). The integral over ω' in (8.11) can be seen as a sum over all *quantum paths* that lead from the initial state to the final state with the energy-conservation requirement: $\omega' + \omega'' = \epsilon_f - \epsilon_i$. In our two-color case, the field has two different components at XUV and IR frequencies for the pump and probe fields, respectively. The dominant contribution to (8.11) can then be identified as the process with absorption of one XUV photon followed by exchange of one IR photon. If both components have a broad bandwidth then there is a broad range of photon-pairs available. However, the effects of this convolution are negligible if one of the fields is quasi-monochromatic. We will assume that the IR field is quasi-monochromatic with frequency $\omega_{\mathrm{IR}} = \omega$ and with a bandwidth Δ_{IR} that is narrower than the bandwidth of the XUV field with frequencies $\Omega_>$ and $\Omega_<$. The two dominant quantum paths become:

$$\begin{aligned} S_{fi}^{(a)} &\approx \frac{1}{2\pi i} M_{fi}^{(2)}(\Omega_<) E(\Omega_<) E(\omega) \Delta_{\mathrm{IR}} \\ S_{fi}^{(e)} &\approx \frac{1}{2\pi i} M_{fi}^{(2)}(\Omega_>) E(\Omega_>) E^*(\omega) \Delta_{\mathrm{IR}}, \end{aligned} \tag{8.13}$$

corresponding to absorption (a) and emission (e) of a laser photon from the fields as depicted in Fig. 8.5. Note that the intermediate XUV photon energy is different in the two paths to ensure global energy conservation. For this class of multiphoton

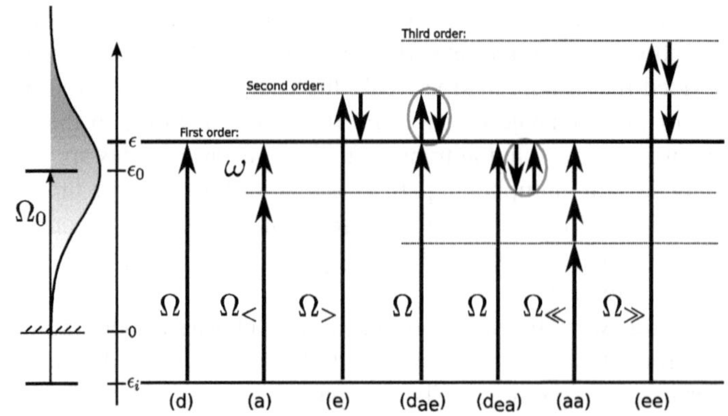

Fig. 8.5 Photon-diagrams for laser-assisted photoionization. Single XUV-photon absorption contributes with a direct path *(d)* to the final state. The dominant two-photon processes involve the absorption or emission of a laser photon indicated by paths *(a)* and *(e)*. Higher-order processes involve the exchange of more laser photons. Figure adapted from [35]

processes numerically exact computations can only be performed in one-electron (hydrogen-like) and two-electron (helium-like) systems. In single-ionization events the wave packet can be described as a single excited electron at a large radial distance far from the ionic core. Within the single-active electron approximation, it is feasible to derive a convenient approximation method designed to evaluate the time delay induced by the IR-driven transition in the continuum. It has been demonstrated that the main results from this analysis also hold for the delay in photoionization from many-electron atomic systems [37, 46].

8.3.1 Asymptotic Approximation for ATI Transition Amplitudes

In order to obtain an estimate for the *phase* of the two-photon matrix element, we will rely on an approximation, which utilizes the asymptotic continuum forms of the final state and of the first-order perturbed wavefunction. Specializing to the case of a hydrogenic system with nuclear charge Z, the S-matrix in (8.11) can be evaluated approximately, as shown in [36]. One notices that the phase of the two-photon matrix element contains only phase-shifts that are governed by the angular momentum λ of the intermediate state, i.e. a state that can be reached via *single*-photon ionization. More precisely, for a given transition channel (characterized by the angular momenta of the intermediate and final state: $\ell_i \to \lambda \to L$), the phase of the matrix element reduces to:

$$\begin{aligned}\arg[S_{fi}^{(L,\lambda,m_i)}] = \pi + \arg[Y_{L,m_i}(\hat{k})] + \phi_< + \phi \\ - \frac{\pi\lambda}{2} + \eta_\lambda(\kappa) + \phi_{cc}(k,\kappa,Z),\end{aligned} \tag{8.14}$$

where $\phi_<$ and ϕ are the phases of the XUV field, $\Omega_<$, and of the IR laser, ω, respectively; and where the *continuum–continuum phase* is approximated by

$$\phi_{cc}(k, \kappa, Z) \equiv \arg \left\{ \frac{(2\kappa)^{iZ/\kappa}}{(2k)^{iZ/k}} \frac{\Gamma[2 + iZ(1/\kappa - 1/k)] + \gamma(k, \kappa)}{(\kappa - k)^{iZ(1/\kappa - 1/k)}} \right\}, \tag{8.15}$$

where k and κ are the final and intermediate momentum, respectively; and Z is the residual charge of the ion. The first term in (8.15), which only includes the effects of long-range phase distortion due to the ionic potential, corresponds to the label (P) in Fig. 8.6a. The additional term, $\gamma(k, \kappa)$ in (8.15), is a correction term that arises from amplitude effects due to the ionic potential,

$$\gamma(k, \kappa) = iZ \frac{(\kappa - k)(\kappa^2 + k^2)}{2\kappa^2 k^2} \Gamma[1 + iZ(1/\kappa - 1/k)], \tag{8.16}$$

which is shown with the label (P+A) in Fig. 8.6a. The result presented as (P+A′) in Fig. 8.6 is a fit to the exact calculation for hydrogen by Taïeb [36]. We stress that the final state scattering phase, $\eta_L(k)$, cancels out and that it does not enter in (8.14). In the next section we explain how the "delay in photoionization" can be constructed using two-photon matrix elements.

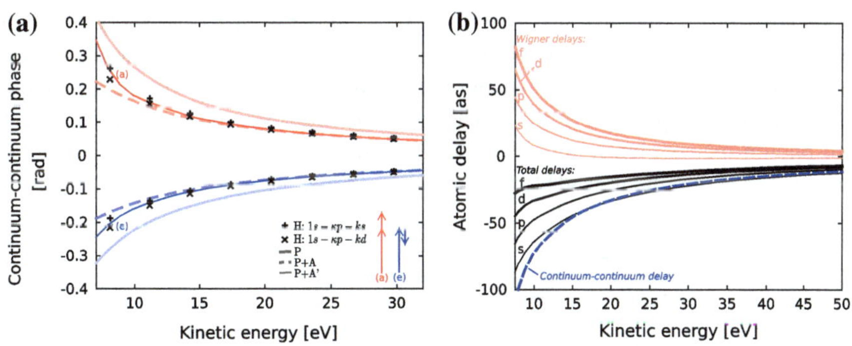

Fig. 8.6 **a** Continuum–continuum phases: ATI phase-shifts for absorption (*a*) of one 800 nm laser photon; and for stimulated emission (*e*) of one laser photon, in a Coulomb potential with $Z = 1$. The asymptotic approximation (P) provides the correct qualitative behaviour, while the long-range amplitude corrections (P+A and P+A′) lead to quantitative agreement, at high enough energy, with the exact calculations in hydrogen (+ and ×). **b** Ionization delays: the Wigner delays for hydrogen with a photoelectron of s, p, d or f-character (*red*, *upper curves*) plus the universal continuum–continuum delay for the laser-probing process with $Z = 1$ and $\hbar\omega = 1.55$ eV (*blue*, *dashed curve*), yields the total delay in laser-assisted photoionization (*black*, *lower curves*). Figure adapted from [35]

8.3.2 Extracting Time-Delay Information from Laser-Assisted Photoionization Signals

Having established the asymptotic approximation for the complex amplitudes of ATI processes, we now turn to the *probability* for the emission of a photoelectron with energy $\epsilon_k = \epsilon_i + \Omega$, as depicted in Fig. 8.5. The probability is given by the square of the sum of the amplitudes:

$$P_k \approx |S_d + S_a + S_e|^2 = |S_d|^2 + |S_a|^2 + |S_e|^2 + 2\Re\left\{S_d^*(S_a + S_e) + S_a^* S_e\right\}, \tag{8.17}$$

where d, a and e label the paths associated with the lowest-order processes: (d), (a) and (e) in Fig. 8.5. The total probability depends on the relative phase of all individual quantum paths. In experiments, the phase of the two-photon amplitudes labelled (a) and (e) can be controlled by changing the sub-cycle delay, τ, between the probe field and the attosecond pulses. More precisely, one controls the relative phase of the IR field as compared to the XUV field, $\phi \equiv \omega\tau$ in (8.14). The probe-phase dependence is $S_a \propto E(\omega) \propto \exp[i\omega\tau]$ and $S_e \propto E^*(\omega) \propto \exp[-i\omega\tau]$. This implies that the cross-terms in (8.17) vary differently as a function of τ: (d)–(a) and (d)–(e) are modulated with periodicity $\omega\tau$ associated with the exchange of only one IR photon [47]; while the cross-term (a)–(e) is modulated with periodicity $2\omega\tau$ due to the two IR photons involved [31, 48]. The maximal probability for photoemission in a RABITT sideband[6] occurs when the amplitudes associated with paths (a) and (e) are in phase, arg$[S_a]$ = arg$[S_e]$, which, using the explicit phases of the relevant two-photon matrix elements in (8.14) and assuming that one intermediate angular channel (λ) is dominant, leads to:

$$\tau = \overbrace{\frac{\phi_{\Omega_>} - \phi_{\Omega_<}}{2\omega_r}}^{\tau_{GD}} + \underbrace{\frac{\eta_\lambda(\kappa_>) - \eta_\lambda(\kappa_<)}{2\omega_r}}_{\tau_\lambda} + \underbrace{\frac{\phi_{cc}(k,\kappa_>) - \phi_{cc}(k,\kappa_<)}{2\omega_r}}_{\tau_{cc}}, \tag{8.18}$$

where the momenta $\kappa_<$ and $\kappa_>$ correspond to absorption of an XUV photon $\Omega_<$ and $\Omega_>$, respectively. In (8.18), we observe that the XUV-IR delay, τ, that maximizes the yield is a sum of three delays:

- τ_{GD}: the group delay of the XUV field is relative to the subcycle oscillations of the IR probe field.

[6] It has been shown theoretically that the delay of the photoelectron spectrogram in an attosecond streak-camera experiment, as compared to the time variation of the IR vector potential, is equal to the shift of the corresponding RABITT sidebands, c.f. [14, 36, 38].

- τ_λ: the Wigner delay is the "delay" in single-photon ionization (must be deduced from the difference with the delay in a reference system, e.g. hydrogen).
- τ_{cc}: the continuum–continuum delay is a measurement-induced delay from the interplay of the IR field and the long-range Coulomb potential.

The sum: $\tau_\lambda + \tau_{cc} = \tau_\theta$ constitutes the so-called "atomic delays". Its variation in terms of the kinetic energy of the photoelectron are shown in Fig. 8.6b for the case of hydrogen the angular momentum, λ. A similar separation of the time-delays as shown in (8.18) was reported by the group of Burgdörfer for streaking of single attosecond pulses by solving numerically the time-dependent Schrödinger equation [49, 50]. Interestingly, they also showed that τ_{cc} (there called 'Coulomb-laser coupling') could be calculated from a classical ensemble of electron trajectories.

The delays presented in (8.18) are calculated from the *finite-difference* approximations to the actual derivatives, $\tau_{GD} = \partial\phi_\Omega/\partial\Omega$ and $\tau_\lambda = \partial\eta_{\kappa,\lambda}/\partial\Omega$. For these approximations to be valid, we must require that the spectral phases vary slowly, $|\Delta\phi_\Omega| \equiv |\phi_{\Omega>} - \phi_{\Omega_<}| \ll 2\pi$, with a small phase variation over the energy range spanned by two laser photons, $\Delta\Omega = 2\omega$. In this sense, "slow" reactions, such as autoionization of a resonance discussed in Sect. 8.2.2, require a more detailed evaluation of the spectral integral in (8.11) [34, 51]. In this context, we wish to mention the work that has been made also using the soft-photon approximation (SPA) where Volkov solutions are used as approximate continuum states [52–54]. The advantage of this approximation is that it allows for analytical solutions to non-perturbative IR interactions with the photoelectron, however, because the electron–ion Coulomb interaction is omitted neither the scattering phases nor the continuum–continuum phases will be recovered by SPA. Prior to these successful demonstrations of the separability of the τ_λ and τ_{cc} in (8.18), the influence of the short-range scattering phase-shifts was identified by Yakovlev et al. [55] and the importance of the long-range Coulomb tail was discussed by Zhang and Thumm [56].

The effects of electron–ion interaction for laser-assisted XUV ionization has also been studied by Smirnova and coworkers using the Eikonal-Volkov Approximation (EVA) [57]. With EVA it was possible to provide accurate results for the photoionization delays from hydrogen [58]. More recently, (8.18) has been validated, not only for hydrogen, but also for noble-gas atoms by diagrammatic methods [37]; effective two-electron models [50]; and for the H_2^+ molecule [59].

8.3.2.1 The Delay Paradox

Equation (8.18) opens up for the possibility to measure (i) $\tau_{\rm GD}$ of the attosecond pulse *or* (ii) τ_λ of the photoelectron. The former requires exact knowledge of $\tau_\theta = \tau_\lambda + \tau_{cc}$, which then must be subtracted from the experimental data; while τ_λ requires exact knowledge of $\tau_{\rm GD}$ and compensation of the universal τ_{cc}. The above statements contain a paradox because our ability to measure one of these quantities requires exact knowledge of the other.

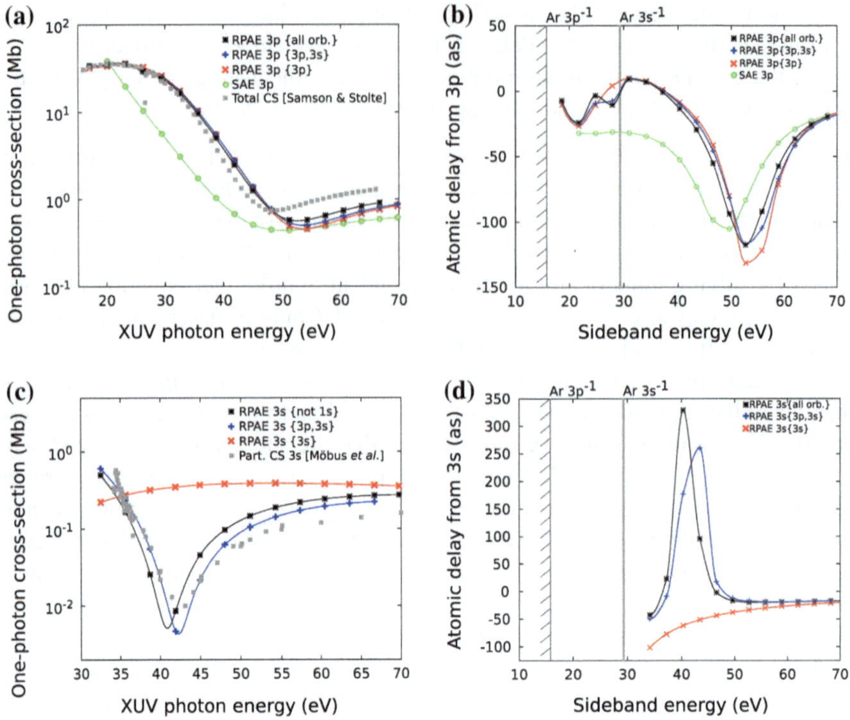

Fig. 8.7 **a** Theoretical partial photoionization cross-section for Ar$3p^{-1}$ compared to the total photoionization cross-section measurements by Samson and Stolte [60]. **b** Atomic delay for Ar$3p^{-1}$ observed along the polarization axis of the fields. **c** Theoretical partial photoionization cross-section for Ar$3s^{-1}$ compared to partial cross-section measurements by Möbus et al. [61]. **d** Atomic delay for Ar$3s^{-1}$ observed along the polarization axis of the fields. The correlated orbitals in the RPAE model are listed in the curly brackets. The IR photon is 1.55 eV. SAE calculations are performed using the potential proposed by Muller [62]. Figure adapted from [63]

The first observation of attosecond pulse structures in 2001 [24] relied on subtraction of τ_θ computed for argon within the single-active electron approximation (SAE) using the potential proposed by Muller [62]. In Fig. 8.7a we show the cross-section for ionization from the $3p$ subshell in argon by absorption of one XUV photon for both the Muller potential ($\circ$) and using the Random-Phase Approximation with Exchange (RPAE[7]) for many-body screening effects ($+$, $\times$, $*$) [63]. Each model reproduces a Cooper minimum in the photoionization cross section [64] in qualitative agreement with the experimental total cross-section determined by Samson and Stolte [60]. The RPAE result, which includes ground-state correlation, is in excellent agreement with the measurement close to the ionization threshold. At higher energies the total cross-section differs from the partial RPAE cross-section due to opening of other ionization channels. In Fig. 8.7b we show the corresponding atomic delays (com-

[7]For details about the RPAE theory see [28].

puted from two-photon matrix elements) for argon within the SAE (∘) and for RPAE screened XUV photon (+, ×, ∗). The SAE agrees qualitatively with the correlated calculations, but it can hardly be used for calibration of the experimental data on the attosecond time scale.

In Fig. 8.7c, d we show the partial cross-section and atomic delays for photoionization from the $3s$ subshell of argon, respectively [63]. Inter-orbital correlation effects (+, ∗) give rise to a minimum in the partial cross-section, as first pointed out by Amusia [65], and to a large *positive* peak in the atomic delay [12] that is not present for models which only include intra-shell correlation (×). In Fig. 8.7c the experimental measurements of the partial photoionization cross section by Möbus and co-workers [61] is in qualitative agreement with the RPAE result (+, ∗).[8] By subtraction of the data in Fig. 8.7b, d, an atomic delay difference of ∼25 as is found at sideband 24 (∼37.2 eV) between the $3p$ and $3s$ subshells. Unfortunately, this value does not agree well with the experimental measurements of ∼100 as performed at Lund University [10, 12].

While the long-range phase, $\eta_{\mathrm{LR}}(x)$ in (8.2), is known from early quantum scattering theory [66], the accurate determination of the short-range phase, η_{SR}, presents a numerical challenge for complex systems due to correlation effects [67]. Multi-electron screening effects on the delay in photoionization were first studied using the state-specific expansion approach (SSEA) [9] and by the RPAE for absorption of a single XUV photon [68, 69]. Further works have included the multi-configurational Hartree-Fock method (MCHF) [51] and the time-dependent local density approximation (TDLDA) [70].

It was soon argued that the IR probe field should affect the photoionization delay measurement due to coupling to the long-range Coulomb potential [10, 36, 38, 49, 58]. Correlation effects in two-color fields (XUV +IR) have been benchmarked for helium [46]. Approximate methods for larger atoms include the time-dependent R-matrix [71], effective two-electron models [72], diagrammatic perturbation theory [37, 63] and application of the B-spline R-matrix (BSRM) for inclusion of shake-up processes [73].

Despite all this theoretical activity, the experimentally measured neon delay of 21 ± 5 as between the $2p$ and $2s$ subshell at an XUV photon energy of ∼100 eV [9] remains larger than the simulated values. The experimentally measured argon delay of ∼100 as between the $3p$ and $3s$ subshell at an XUV excitation energy of ∼37 eV [10, 12] has been best reproduced by MCHF [51]. A further study of the correlation effects by diagrammatic methods has evidenced that the sign of the $3s$ delay peak depends on the details of the correlation model [63], but further work is needed to pin-point important interactions beyond the RPAE that include virtual shake-up (two holes and one electron) and knock-out (one hole and two electrons) processes.

We predict that in the coming years, the interpretation of attosecond experiments will offer a new testing ground for many-body calculations. Future experiments will provide interesting opportunities to test various theoretical methods for light-induced

[8] Surprisingly, the inclusion of correlation with the L-shell (∗) brings the calculation further away from the experimental measurements.

electron-electron interactions in connection to measurements of phases and delays in complex atomic and molecular systems. In the next section we provide an overview of how the paradox of attosecond delays was overcome experimentally by performing relative delay measurements from different initial states of the target.

8.4 Review of Experimental Delay Measurements

The first experimental observations of photoionization time delays were made with the same experimental setups as those used for the characterization of attosecond pulses, but with one important modification: In these *relative* delay experiments, *two* photoelectron spectrograms were recorded simultaneously from two different initial states of the target [6, 9–12]. Because it was the same attosecond pulses that triggered ionization from both initial states, the unknown exact shape of the attosecond pulses canceled out and the paradox of delays was circumvented. In these measurements photoelectron 'replicas' of the attosecond pulses are generated at different average kinetic energies corresponding to each initial state, say A and B. These replicas were then characterized by using the IR-field as an ultrafast phase-gate so that the relative atomic delay between the two states could be observed: $\Delta\tau_\theta(A, B) = \tau_\theta(A) - \tau_\theta(B)$. Alternatively, if the stability of the attosecond pulses is good enough, the relative atomic delay can be studied sequentially in different target systems, thus revealing differences in the atomic delays associated with the two systems [7].

The experimental work on photoionization time-delays has been carried out on many-electron systems: condensed matter [6], noble gas atoms [9, 10, 74] and molecules [7].

In the following, we will proceed with a brief overview of the current experimental efforts, adapted and extended from the corresponding section out of [13]. We will only review work on time-delay and phase measurements using attosecond pulse trains (APT) where we have taken an active part.

8.4.1 *Atomic-Delay Measurements Using Attosecond Pulse Trains*

In an experiment conducted by Klünder and co-workers using an APT with a photon energy of ~35 eV [10], *two* RABITT scans were recorded simultaneously by ionizing both the $3p$ and $3s$ orbitals of argon. A schematic illustration of the experiment is shown in Fig. 8.8. As explained above, the unknown temporal structure of the attosecond pulses is subtracted without ever knowing the exact shape. The result is a relative delay of ~100 as between the ionization from the $3p$ and $3s$ subshells at the same ionizing photon energy. A large part of this measured delay was attributed to τ_{cc}, see Fig. 8.6b, due to the different final energies of the photoelectrons originating

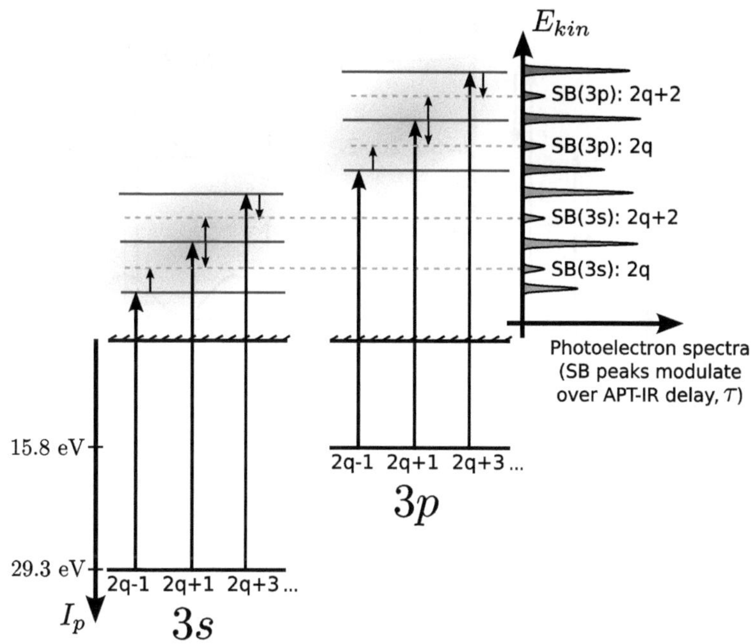

Fig. 8.8 *Delay experiment for the* $3s$ *and* $3p$ *shells in Ar using APT:* The APT is composed of three odd harmonics, $2q + 1$, that ionize (pump) electrons from either initial shell, $3s$ or $3p$, into the continuum. After absorbing one harmonic photon, the photoelectron can either absorb or emit a laser probe photon in order to reach an even number sideband (SB) state, labelled by $2q$. The SB probability oscillates with the APT-IR delay, τ, due to interference between the two quantum paths. Information about the attosecond timing is found in the relative modulation offset between the same sideband numbers from different initial states [10]. Figure adapted from [35]

from the two different subshells. In this work, the high-order harmonic comb was first passed through a thin chrome foil acting as a band-pass to select four odd harmonics (21–27) of the 800-nm driving laser. In this way, only three sidebands, SB: 22, 24, 26, were produced when ionizing argon atoms from either orbital. This truncation of the harmonic comb was made to prevent different sidebands from different orbitals from overlapping in photoelectron kinetic energy.

The analysis of the experiment [10] was first carried out within the SAE approximation using Hartree-Fock phase-shifts from the literature [75]. As shown in Fig. 8.7c, d, the photoelectrons from the *inner* subshell, $3s$, couple strongly with the other electrons, and the corresponding dynamics is greatly altered due to correlation. This important effect was identified by Kheifets by considering RPAE effects in the absorption of a single XUV-photon [69].

The influence of a narrower atomic resonance has been observed experimentally by Swoboda and co-workers [8]. Figure 8.9 illustrates the principle of the experiment. Here, an APT has the particular advantage that the discrete harmonics to which it corresponds in the spectral domain can be tuned towards a specific energy region

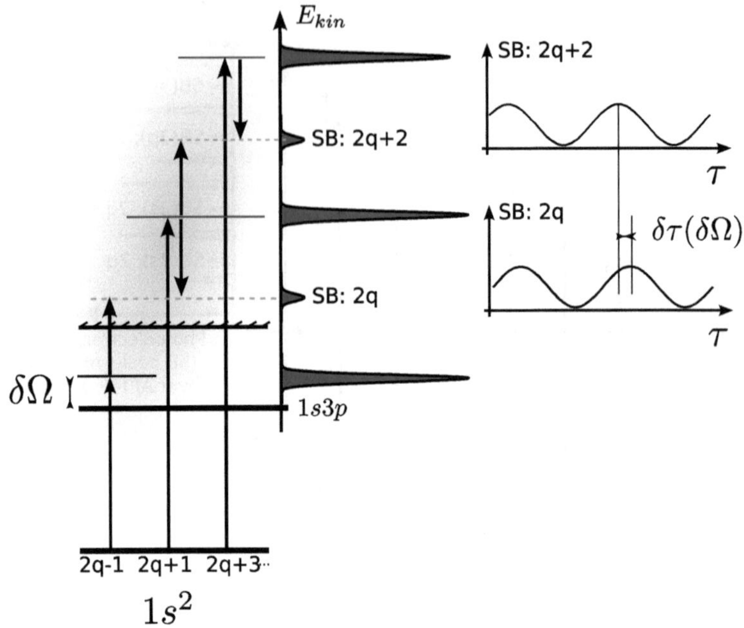

Fig. 8.9 *Phase induced by the* $1s3p$ *state in He using APT:* The phase in photoionization, $\delta\tau$, depends on the detuning, $\delta\Omega$, between the harmonic, $(2q-1)\omega = 15\omega$, and the bound state, $1s3p$. The effect arises from the π-jump that occurs in the two-photon amplitude, when passing the resonance. The higher-lying sideband, SB:$2q+2$, acts as an important "reference clock" in this experiment, which is mostly insensitive to small frequency changes of the harmonic fields [8]. Figure adapted from [35]

of interest. The frequency of the harmonics from the HHG process can be tuned naturally by changing the fundamental (driving) laser pulse frequency, $\omega + \delta\omega$, so that the high-order harmonic frequencies increase or decrease by $\delta\Omega = (2q+1)\delta\omega$. It was found that the phase of the modulation of the lowest-lying sideband in Helium depended critically on the frequency of the harmonics.

The observed modulation-shifts were attributed to a phase jump occurring when a below-threshold harmonic, $15\omega < I_p^{(He)}$, was scanned over the intermediate "resonant" *bound* state, $1s3p\ ^1P_1$. By tuning the harmonic over the sharp resonance, an abrupt π-shift is expected. The experimental π-shift is "smoothed" due to the finite duration of the laser probe field and of the APT ($\sim$30 fs), which is much shorter than the resonance lifetime. In order to fully resolve the abrupt rise of the delay due to pure atomic effects, one needs harmonics that are narrower in frequency than the atomic resonance, i.e. an APT which is longer that the resonance lifetime.

A similar situation was observed by Haessler and co-workers with a short-lived complex continuum resonance in nitrogen molecules [7, 34] covered by harmonic 11 of the 800-nm pump laser. Its lifetime was more comparable to the duration of the APT, implying that the observed modulation-shifts contain some information about the resonance lifetime. The experiment, very similar to the scheme shown in Fig. 8.9 but with the resonance situated just *above* the ionization threshold, compared ionization from the HOMO and the HOMO-1 of nitrogen (corresponding to ions in their ground or first excited state, respectively), with additional resolution of vibrational states of the created ions, enabled by the relatively narrowband harmonics. As an additional "external" reference, the corresponding measurements were also made ionizing the $3p$-orbitals of argon atoms, where the harmonic instead directly creates a photoelectron, $11\omega > I_p^{(Ar)}$, and where resonances do not contribute considerably.

The resonant channel involved in the nitrogen ionization starts by promoting an electron from the HOMO-2 to a state of auto-ionizing character from where it is ejected when its corresponding hole is filled by an electron from the HOMO. At the resonant XUV-photon energy, ionization to the ionic ground state is thus dominated by this resonant channel—at least in excited vibrational states, which are less efficiently populated by direct ionization from the HOMO. In the experiment, a large 0.9π-modulation-shift of sideband 12 was observed for a single detuning $\delta\Omega$. The accompanying theoretical study, including numerical TOF simulations like those in Sect. 8.2, showed that the derivative of the modulation phase shift with respect to the *detuning* would give a physically representative Wigner-like delay for the two-photon XUV +IR ionization channel involving the resonance, reaching values of several femtoseconds. Interestingly, it is close to the lifetime of the resonance, suggesting a chronology (the electron spends a few femtoseconds in the intermediate resonance before auto-ionization and interaction with the IR field) in an a-priori non-sequential two-photon ATI transition. Note that this delay for the two-photon transition naturally includes the continuum–continuum delay, which is here not to be considered a "measurement-induced artefact" but part of the studied ionization dynamics.

Recently, Månsson and co-workers studied non-sequential double-ionization of xenon atoms [74]. They compare, at a total absorbed energy of 40.1 eV (sideband 26), the reference, single ionization of the xenon $5p$ state, with the non-sequential double-ionization channel leading to the $^1D_2$ excited state of doubly-charged xenon ions, measured in parallel by detecting the two released photo-electrons in coincidence. This experimental work is important because it represents a way to study the *phase* of two-electron wave packets.

8.5 Conclusions

In this chapter, we have illustrated how subtle short-range distorsions in atomic and molecular potentials can lead to variations in photoionization delays on the attosecond and femtosecond time scale. Such delays are nowadays measurable in experiments using two-color (XUV-IR) laser fields. The time-domain perspective opened by such studies provides possibly more intuitive physical pictures as compared to the analytical derivations of stationary state phase shifts. Thus, exploring the time domain brings a complementary view from the traditional frequency domain that prevails in standard spectroscopy.

Acknowledgments We acknowledge that elements of this chapter are reproduced from [13, 63].

References

1. P. Agostini, L.F. DiMauro, Rep. Progr. Phys. **67**(6), 813 (2004)
2. P.B. Corkum, F. Krausz, Nat. Phys. **3**(6), 381 (2007)
3. F. Krausz, M. Ivanov, Rev. Mod. Phys. **81**(1), 163 (2009)
4. F. Lepine, M.Y. Ivanov, M.J.J. Vrakking, Nat. Phot. **8**(3), 195 (2014)
5. S.R. Leone, C.W. McCurdy, J. Burgdorfer, L.S. Cederbaum, Z. Chang, N. Dudovich, J. Feist, C.H. Greene, M. Ivanov, R. Kienberger, U. Keller, M.F. Kling, Z. Loh, T. Pfeifer, A.N. Pfeiffer, R. Santra, A. Schafer, K. Stolow, U. Thumm, M.J.J. Vrakking, Nat. Phot. **8**(3), 162 (2014)
6. A. Cavalieri, N. Müller, T. Uphues, V. Yakovlev, A. Baltuška, B. Horvath, B. Schmidt, L. Blümel, R. Holzwarth, S. Hendel et al., Nature **449**(7165), 1029 (2007)
7. S. Haessler, B. Fabre, J. Higuet, J. Caillat, T. Ruchon, P. Breger, B. Carré, E. Constant, A. Maquet, E. Mével et al., Phys. Rev. A **80**(1), 011404 (2009)
8. M. Swoboda, T. Fordell, K. Klünder, J. Dahlström, M. Miranda, C. Buth, K. Schafer, J. Mauritsson, A. L'Huillier, M. Gisselbrecht, Phys. Rev. Lett. **104**(10), 103003 (2010)
9. M. Schultze, M. Fieß, N. Karpowicz, J. Gagnon, M. Korbman, M. Hofstetter, S. Neppl, A. Cavalieri, Y. Komninos, T. Mercouris et al., Science **328**(5986), 1658 (2010)
10. K. Klünder, J. Dahlström, M. Gisselbrecht, T. Fordell, M. Swoboda, D. Guénot, P. Johnsson, J. Caillat, J. Mauritsson, A. Maquet et al., Phys. Rev. Lett. **106**(14), 143002 (2011)
11. S. Neppl, R. Ernstorfer, E.M. Bothschafter, A.L. Cavalieri, D. Menzel, J.V. Barth, F. Krausz, R. Kienberger, P. Feulner, Phys. Rev. Lett. **109**, 087401 (2012)
12. D. Guénot, K. Klünder, C. Arnold, D. Kroon, J. Dahlström, M. Miranda, T. Fordell, M. Gisselbrecht, P. Johnsson, J. Mauritsson et al., Phys. Rev. A **85**(5), 053424 (2012)
13. J. Dahlström, A. L'Huillier, A. Maquet, J. Phys. B **45**(18), 183001 (2012)
14. R. Pazourek, S. Nagele, J. Burgdörfer, Faraday Discuss. **163**, 353 (2013)
15. E. Goulielmakis, M. Schultze, M. Hofstetter, V. Yakovlev, J. Gagnon, M. Uiberacker, A. Aquila, E. Gullikson, D. Attwood, R. Kienberger et al., Science **320**(5883), 1614 (2008)
16. D.H. Ko, K.T. Kim, J. Park, J.H. Lee, C.H. Nam, New J. Phys. **12**(6), 063008 (2010)
17. M. Drescher, M. Hentschel, R. Kienberger, M. Uiberacker, V. Yakovlev, A. Scrinzi, T. Westerwalbesloh, U. Kleineberg, U. Heinzmann, F. Krausz, Nature **419**(6909), 803 (2002)
18. L. Argenti, C. Ott, T. Pfeifer, F. Martín, J. Phys.: Conference Series **488**(3), 032030 (2014)
19. C. Ott, A. Kaldun, P. Raith, K. Meyer, M. Laux, J. Evers, C.H. Keitel, C.H. Greene, T. Pfeifer, Science **340**(6133), 716 (2013)
20. E.P. Wigner, Phys. Rev. **98**, 145 (1955)
21. F.T. Smith, Phys. Rev. **118**(1), 349 (1960)
22. C.A. de Carvalho, H.M. Nussenzveig, Phys. Rep. **364**(2), 83 (2002)
23. D.J. Tannor, *Introduction to Quantum Mechanics—A Time-Dependent Perspective* (University Science Books, Sausalito CA, 2007)
24. P.M. Paul, E.S. Toma, P. Breger, G. Mullot, F. Augé, P. Balcou, H.G. Muller, P. Agostini, Science **292**(5522), 1689 (2001)
25. Y. Mairesse, A. de Bohan, L.J. Frasinski, H. Merdji, L.C. Dinu, P. Monchicourt, P. Breger, M. Kovacev, R. Taïeb, B. Carré, H.G. Muller, P. Agostini, P. Salières, Science **302**(5650), 1540 (2003)
26. Y. Mairesse, F. Quéré, Phys. Rev. A **71**(1), 011401 (2005)
27. H. Park, R.N. Zare, J. Chem. Phys. **104**, 4554 (1996)

28. M. Amusia, *Atomic Photoeffect* (Plenum Press, New York, 1990)
29. M. Vacher, A. Maquet, R. Taïeb, J. Caillat (in preparation)
30. F. Gel'mukhanov, H. Ågren, Phys. Rep. **312**, 87–330 (1999)
31. H.G. Muller, Appl. Phys. B **74**(1), s17 (2002)
32. W. Boutu, S. Haessler, H. Merdji, P. Breger, G. Waters, M. Stankiewics, L.J. Frasinski, R. Taïeb, J. Caillat, A. Maquet, P. Monchicourt, B. Carré, P. Salières, Nat. Phys. **4**, 545 (2008)
33. S. Haessler, J. Caillat, W. Boutu, C. Giovanetti-Teixeira, T. Ruchon, T. Auguste, Z. Diveki, P. Breger, A. Maquet, B. Carré et al., Nat. Phys. **6**(3), 200 (2010)
34. J. Caillat, A. Maquet, S. Haessler, B. Fabre, T. Ruchon, P. Salières, Y. Mairesse, R. Taïeb, Phys. Rev. Lett. **106**(9), 093002 (2011)
35. J. Dahlström, A. L'Huillier, J. Mauritsson, J. Phys. B **44**(9), 095602 (2011)
36. J.M. Dahlström, D. Guénot, K. Klünder, M. Gisselbrecht, J. Mauritsson, A. L'Huillier, R. Taïeb, Chem. Phys. **414**, 53 (2013)
37. J.M. Dahlström, T. Carette, E. Lindroth, Phys. Rev. A **86**, 061402 (2012)
38. C.H. Zhang, U. Thumm, Phys. Rev. A **82**(4), 043405 (2010)
39. R. Gaillac, M. Vacher, A. Maquet, R. Taïeb, J. Caillat, Attosecond photoemission dynamics encoded in real-valued continuum wave-functions, submitted (2015)
40. W.H. Press, S.A. Teukolsky, W.T. Vetterling, B.P. Flannery, *Numerical Recipes: The Art of Scientific Computing* (Cambridge University Press, Cambridge, 2007)
41. K.C. Kulander, K.J. Schafer, J.L. Krause, *Atoms in Intense Laser Fields* (Academic Press Inc, United States, 1992)
42. P. Kruit, F. Read, J. Phys. E: Sci. Instrum. **16**(4), 313 (1983)
43. A.T. Eppink, D.H. Parker, Rev. Sci. Instrum. **68**(9), 3477 (1997)
44. J. Itatani, F. Quéré, G.L. Yudin, M.Y. Ivanov, F. Krausz, P.B. Corkum, Phys. Rev. Lett. **88**(17), 173903 (2002)
45. F. Quéré, Y. Mairesse, J. Itatani, J. Mod. Opt. **52**(2), 339 (2005)
46. R. Pazourek, J. Feist, S. Nagele, J. Burgdörfer, Phys. Rev. Lett. **108**, 163001 (2012)
47. M. Chini, S. Gilbertson, S.D. Khan, Z. Chang, Opt. Expr. **18**(12), 13006 (2010)
48. V. Véniard, R. Taïeb, A. Maquet, Phys. Rev. A **54**(1), 721 (1996)
49. S. Nagele, R. Pazourek, J. Feist, K. Doblhoff-Dier, C. Lemell, K. Tőkési, J. Burgdörfer, J. Phys. B **44**(8), 081001 (2011)
50. S. Nagele, R. Pazourek, J. Feist, J. Burgdörfer, Phys. Rev. A **85**(3), 033401 (2012)
51. T. Carette, J.M. Dahlström, L. Argenti, E. Lindroth, Phys. Rev. A **87**, 023420 (2013)
52. L.B. Madsen, Am. J. Phys. **73**(1), 57 (2005)
53. A. Maquet, R. Taïeb, J. Mod. Opt. **54**(13–15), 1847 (2007)
54. A.J. Galn, L. Argenti, F. Martín, New J. Phys. **15**(11), 113009 (2013)
55. V.S. Yakovlev, J. Gagnon, N. Karpowicz, F. Krausz, Phys. Rev. Lett. **105**(7), 073001 (2010)
56. C.H. Zhang, U. Thumm, Phys. Rev. A **84**(3), 033401 (2011)
57. O. Smirnova, M. Spanner, M.Y. Ivanov, J. Phys. B **39**(13), S323 (2006)
58. M. Ivanov, O. Smirnova, Phys. Rev. Lett. **107**(21), 213605 (2011)
59. Q.C. Ning, L.Y. Peng, S.N. Song, W.C. Jiang, S. Nagele, R. Pazourek, J. Burgdörfer, Q. Gong, Phys. Rev. A **90**, 013423 (2014)
60. J. Samson, W. Stolte, J. Electron. Spectrosc. Relat. Phenom. **123**(23), 265 (2002)
61. B. Möbus, B. Magel, K.H. Schartner, B. Langer, U. Becker, M. Wildberger, H. Schmoranzer, Phys. Rev. A **47**, 3888 (1993)
62. H.G. Muller, Phys. Rev. A **60**, 1341 (1999)
63. J.M. Dahlström, E. Lindroth, J. Phys. B **47**(12), 124012 (2014)
64. J.W. Cooper, Phys. Rev. **128**(2), 681 (1962)
65. M. Amusia, V. Ivanov, N. Cherepkov, L. Chernysheva, Phys. Lett. A **40**(5), 361 (1972)
66. L.D. Landau, E.M. Lifshitz, *Quantum Mechanics (Non-Relativistic Theory)* (Pergamon Press, Oxford, 1965)
67. K. Chadan, P.C. Sabatier, *Inverse Problems in Quantum Scattering Theory* (Springer, New York, 1989)
68. A. Kheifets, I. Ivanov, Phys. Rev. Lett. **105**(23), 233002 (2010)

69. A.S. Kheifets, Phys. Rev. A **87**, 063404 (2013)
70. G. Dixit, H.S. Chakraborty, M.E.A. Madjet, Phys. Rev. Lett. **111**, 203003 (2013)
71. L. Moore, M. Lysaght, J. Parker, H. van der Hart, K. Taylor, Phys. Rev. A **84**(6), 061404 (2011)
72. S. Nagele, R. Pazourek, J. Feist, J. Burgdörfer, Phys. Rev. A **85**, 033401 (2012)
73. J. Feist, O. Zatsarinny, S. Nagele, R. Pazourek, J. Burgdörfer, X. Guan, K. Bartschat, B.I. Schneider, Phys. Rev. A **89**, 033417 (2014)
74. E.P. Mansson, D. Guenot, C.L. Arnold, D. Kroon, S. Kasper, J.M. Dahlstrom, E. Lindroth, A.S. Kheifets, A. L'Huillier, S.L. Sorensen, M. Gisselbrecht, Nature Phys. **10**(3), 207 (2014)
75. D.J. Kennedy, S.T. Manson, Phys. Rev. A **5**(1), 227 (1972)

Part III
Surfaces, Nanostructures and Solids in Strong Laser Fields

Chapter 9
Ultrafast Nanoplasmonic Photoemission

Péter Dombi

Abstract The interaction of ultrashort laser pulses with plasmonic nanostructures enables the extreme, unprecedented localization of electromagnetic fields in both space and time. Ultrashort plasmonic fields can be bound to the closest nanoscale vicinity of nanoparticles and thin films and this way, new aspects of fundamental photoemission and other electron phenomena can be demonstrated. Here, we will review these phenomena including strong-field nanoplasmonic photoemission and keV electron acceleration on the nanoscale.

9.1 Introduction

9.1.1 Introduction to Surface Plasmon Enhanced Electron Phenomena

Intense ultrashort pulses of laser light have been exploited in a number of fundamental interaction studies involving strong electromagnetic fields and atoms/molecules. These interactions laid the foundations of, for example, femtochemistry and attophysics, representing prominent examples in this field. For all of these cases, the femtosecond laser beam needs to be focused onto a target where the minimum achievable interaction volume is limited by diffraction. When ultrashort pulses impinge on atomic and/or molecular samples, this volume is typically some cubic microns. This also caps the maximum achievable on-target intensity. If a free-space laser beam could be confined to well below the wavelength scale, the benefit would be an increased spatial resolution in applications, moreover, enhanced electromagnetic fields at nanoscale interaction volumes could be achieved. A solution for this challenge is readily offered by surface plasmons induced by field-driven charge oscillations in metallic nanolayers or nanostructures. The advantage offered by these

P. Dombi (✉)
MTA "Lendület" Ultrafast Nanooptics Group, Wigner Research Centre for Physics, Konkoly-Thege M. út 29-33, Budapest 1121, Hungary
e-mail: dombi.peter@wigner.mta.hu

M. Kitzler and S. Gräfe (eds.), *Ultrafast Dynamics Driven by Intense Light Pulses*, Springer Series on Atomic, Optical, and Plasma Physics 86, DOI 10.1007/978-3-319-20173-3_9

configurations is twofold: (i) they offer the sub-wavelength confinement of the electromagnetic field and (ii) this way substantial enhancement of the nano-localized electric field can be achieved with respect to that of a focused laser pulse.

Several fundamental physical phenomena show new aspects when the interaction is taking place in nano-localized fields. Here, we will review the recently discovered new facets of electron emission from metals with the help of ultrashort laser pulses and plasmonic nanostructures. The advancement of femtosecond laser technology together with increased research on nanoemitters enabled these fundamental discoveries. We will also show some routes towards the application of these unique, ultrafast photoemission phenomena.

9.1.2 Surface Plasmons

It was observed early that the efficiency of many chemical and physical processes involving laser-surface interactions can be significantly enhanced by the nanoscale roughness of the metal surface involved. These processes and applications include Raman scattering [1], plasmonic biosensors [2], and some nonlinear optical phenomena such as surface-harmonic generation [3, 4] and optical rectification [5]. A rough surface that can provide such a signal enhancement is a random, nanostructured, metallic surface which facilitates an electric field enhancement and surface plasmon coupling. It is both metal films and the nanoscale voids and protrusions of these materials that can support plasmons. Therefore, it is important to introduce basic phenomena related to the generation of these electromagnetic fields.

Surface plasmons are charge-density oscillations of conduction band electrons found in thin metal films or nanoparticles (Fig. 9.1). For metal films, these oscillations can support propagating waves on the metal film (propagating surface plasmons, (PSPs), whereas for nanoparticles the oscillations typically remain localized to the nanoparticle vicinity (localized surface plasmons, LSPs). Surface plasmons are well known to confine electromagnetic energy to nanoscale spatial domains [6] and thus they can possess higher electric field strengths than that of the laser responsible for their generation in both cases. This opens the possibility of using ultra-compact, table-top, low-power lasers for inducing strong-field light-matter interactions in highly confined surface plasmon fields.

The simplest means of producing PSPs is by way of the Kretschmann-Raether coupling geometry (Fig. 9.1). A thin metal film (ca. 40–70 nm thick) coats the hypotenuse face of a right-angle prism. After satisfying the momentum-matching resonance condition at a certain angle of incidence, the incoming laser field becomes evanescently confined to a nanoscale surface layer (some 200–300 nm in width) at the surface of the thin film. The electromagnetic near field at the metal-air or metal-vacuum boundary decays exponentially away from the interface [6]. Even though lateral dimensions of the PSP on the surface correspond to a diffraction-limited focal spot size, the remaining one-dimensional field nanoconfinement is sufficient to induce both surface plasmon enhanced photoemission and electron acceleration. These phenomena

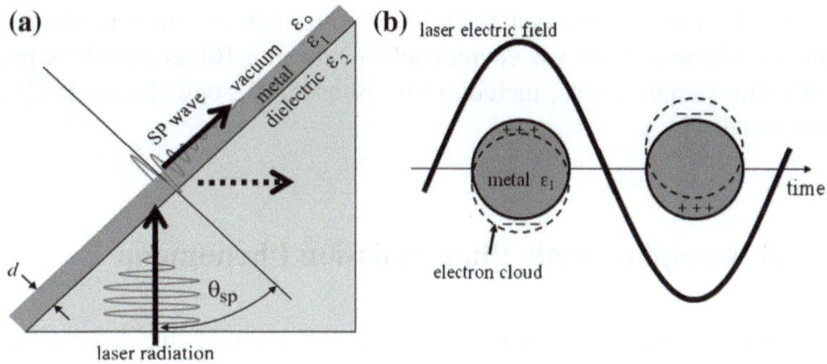

Fig. 9.1 Surface plasmon coupling of free-space electromagnetic waves in **a** Kretschmann-Raether configuration at a specific resonant incidence angle θ_{SP} and with **b** metal nanoparticles where the electron cloud can be seen moving collectively under the influence of the laser field. The generated surface bound electromagnetic wave in (**a**) is called a propagating surface plasmon (PSP), whereas in (**b**) localized surface plasmon oscillation (LSP) is induced. *Arrows* in (**a**) indicate the flow of energy. In case of perfect PSP coupling, the reflected beam from the metal surface indicated by the *dashed arrow* disappears

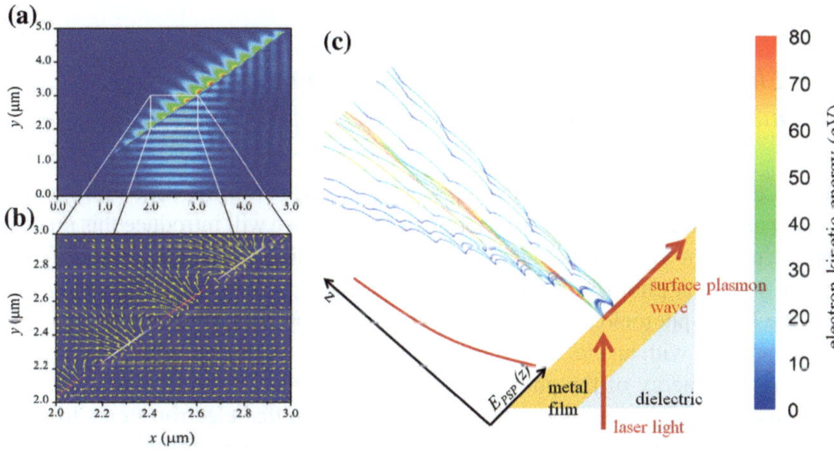

Fig. 9.2 **a** Electric field distribution of a propagating surface plasmon wave on a metal nanolayer in the Kretschmann-Raether coupling geometry. **b** Depicts electric field vectors. *Source* [16]. **c** Illustrates the trajectories of photoemitted electrons in the PSP field and the ponderomotive, cycle-by-cycle acceleration process in the inhomogeneous field for some sample electron trajectories

will be detailed below. Typical electric field distribution of a PSP is visualised in Fig. 9.2a, b.

The other form of surface plasmon generation uses localized charge oscillations in metal nanoparticles induced through the application of an external optical field, so-called localized surface plasmons (LSPs, see Fig. 9.1b). Certain metal nanoparticles can sustain LSPs, which exhibit a pronounced resonance behavior coming together with strongly localized, evanescent fields [7, 8]. The combination of the sharp optical resonance condition and sub-wavelength field localization [6] renders these particles ideal for various applications, including (bio)sensors [9], optoelectronics [10, 11] and photovoltaics [12].

9.2 Novel Nanoplasmonic Photoemission Phenomena

The nanoconfinement property of plasmonic fields has remarkable consequences, as already observed for some applications [1–4]. Moreover, enhancements in photoelectron emission quantum efficiency by more than three orders of magnitude have been observed in the presence of PSP excitation from noble metal films using the Kretschmann-Raether configuration [14, 15] compared to non-plasmonic photoemission yields from metal surfaces. Other intriguing photoemission phenomena both in PSP and LSP fields will be detailed below after reviewing basic photoemission processes upon laser-solid interaction.

9.2.1 Linear Versus Nonlinear Photoemission and Photocurrents

Electron emission from a metal can take place in various forms depending on the intensity and wavelength of the interacting light. Here, we will introduce this process from the point-of-view of its linearity. In the next section we will analyse different types of nonlinear photoemission phenomena, as well.

In the linear photoelectron emission (in short, photoemission) regime, a single photon interacts with an electron in the metal to excite it to the vacuum niveau. Provided that the energy of the photon is higher than the work function of the metal, the photoemission process is linear, since one photoemitted electron stems from an incident photon. Since typical metal work functions are a few electronvolts, linear photoemission can be induced by ultraviolet photons, and the number of photoelectrons is directly proportional to the intensity of the incoming light.

On the other hand, for intense optical fields in the visible spectrum (when the energy of each individual photon is lower than the work function of the metal), nonlinear photoemission can still take place where the joint action of multiple photons is needed to set one electron free. One example is when nonlinear photoemission occurs in the form of the simultaneous, non-sequential absorption of several photons by the metal film. In these cases, the number of photoelectrons no longer scales linearly with the intensity of the light source. Some types of these phenomena will be reviewed below, after introducing basic scale parameters of these interaction processes.

Photocurrents from metal surfaces can be generated with or without the presence of surface plasmons. In the presence of plasmons, their electric field (and not that of the laser pulse) must to be considered when describing the photoemission currents.

9.2.2 Scale Parameters in Photoemission Processes

Simple equation of motion considerations yield that the average oscillation energy of a free electron in the field of an infinite electromagnetic plane wave is

$$U_p = \frac{e^2 E_l^2}{4m\omega^2} \tag{9.1}$$

where the electron charge and rest mass are denoted by e and m, respectively. ω is the angular frequency, and the amplitude of the electric field strength of the laser light is given by E_l.

Fields that are inhomogeneous in space change this picture only slightly. For example, in case of a laser pulse which is focused, U_p can be generalized by taking the spatial dependence into account: $U_p \rightarrow U_p(\boldsymbol{r})$ following the spatial variation of the electric field amplitude, $E_l(\boldsymbol{r})$. The generalized $U_p(\boldsymbol{r})$ quantity is commonly called the ponderomotive potential, as the cycle-by-cycle motion of an electron in such an oscillating field follows the gradient of $U_p(\boldsymbol{r})$, (see Fig. 9.2c). The ponderomotive potential determines the cycle-by-cycle motion and gradual energy gain of a free electron in a spatially inhomogeneous laser field. In addition, this quantity also proved to be very useful for the distinction between different regimes of nonlinear photoemission from metals (as well as photoionization from atoms).

The analysis of Keldysh yielded a famous perturbation parameter, γ that is an efficient scale parameter to describe bound-free transitions (e.g., photoemission, tunneling) in a general way that are induced by laser fields. This adiabaticity parameter is defined as

$$\gamma^2 = \frac{W}{2U_p} = \left(\frac{\omega\sqrt{2mW}}{eE_l}\right)^2 \tag{9.2}$$

where W is the work function of the metal. It can be shown that photoemission from metals and photoionization mechanisms are distinctively different for $\gamma >> 1$ (low laser intensities) and $\gamma << 1$ (high laser intensities), to be discussed below.

9.2.3 Mechanisms of Photoemission and Related Phenomena

The distinction between photoemission phenomena depends on the extent of distortion of the metal surface potential by the external laser field. Here, we will classify these processes in three different categories, making a distinction between (i)

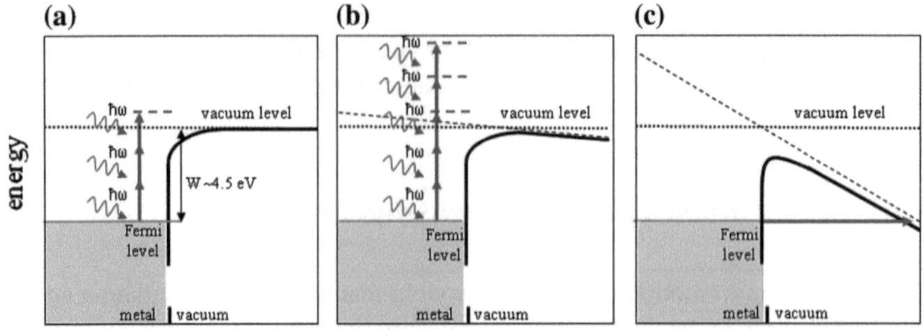

Fig. 9.3 Mechanisms of photoelectron emission for different laser intensities. **a** For low laser intensity, the simultaneous action of multiple photons can set an electron free. **b** By increasing the laser intensity, the metal surface potential becomes slightly distorted by the contribution of the laser field (*dashed line*). The dominant photoemission mechanism is still the absorption of multiple photons but in this case, more photons than the necessary minimum can interact with the electron. This phenomenon resulting in higher final electron kinetic energies is called above threshold photoemission. **c** When the laser field is strong enough to distort the potential substantially, direct tunnelling of an electron from the Fermi level is possible. This phenomenon is called tunnelling or strong-field photoemission

multi-photon-induced photoemission, (ii) above threshold photoemission and (iii) tunneling or strong-field emission, schematically illustrated in Fig. 9.3.

9.2.3.1 Multi-photon Induced Nanoplasmonic Photoemission and Applications

When the value of the Keldysh parameter is $\gamma >> 1$, meaning low intensities (i.e., where the field of the laser pulse is not sufficient to significantly distort the binding potential), multi-photon-induced processes dominate at visible wavelengths (Fig. 9.3a). These nonlinear processes can be described by a perturbative approach since the incident laser field represents only a perturbation term in the Hamiltonian describing the system. As such, the interaction of light with matter is predominantly non-adiabatic and it is governed by the evolution of the amplitude of the laser field, or, in other words, the intensity envelope of the laser pulse. In the case of metals (4–6 eV work function), multi-photon-induced photoemission typically takes place with visible or near-infrared light (1–3 eV photon energy) with 2–5 photons needed to generate one emitted electron. The energy balance of this n-photon process can be described by

$$n\hbar\omega = E_{\text{kin}} + W \tag{9.3}$$

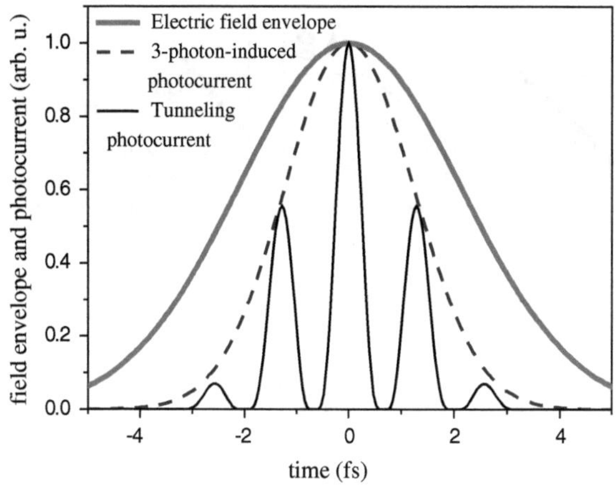

Fig. 9.4 Examples of electron emission temporal profiles for a few-femtosecond laser pulse with a duration of 3.5 fs (intensity full width at half maximum (FWHM)). The *thick solid curve* depicts the field envelope evolution. The *dashed curve* is the photocurrent temporal distribution in case of a three-photon-induced photoemission according to formula (9.4). The *thin solid curve* is the photocurrent profile in case of tunneling electron emission from the surface, determined by the Fowler-Nordheim equation (for further details, see text)

with E_{kin} being the kinetic energy of the photoemitted electron, W representing the work function to be overcome by the simultaneous action of n photons with $\hbar\omega$ energy each. In this case, we can assume that the evolution of the surface photocurrent $j(t)$ is proportional to the nth power of the intensity of the laser field:

$$j(t) \propto I^n(t) \tag{9.4}$$

This relationship yields a very good approximation of the temporal profile for multi-photon-induced photoemission, as confirmed by a full quantum mechanical description of the process [17]. In this case, it is the instantaneous amplitude of the field oscillation which determines the emission probability (illustrated in Fig. 9.4), for a Gaussian laser pulse intensity profile, $I(t)$, the electron emission curve, $j(t)$, has a full width at half maximum (FWHM) that is $\sqrt{n}$ times shorter than the FWHM of the original $I(t)$ curve. Measuring the photocurrent as a function of laser intensity and plotting this curve on a double logarithmic scale enables easy determination of the exponent n by measuring the slope of this curve.

The inherent field enhancement of surface plasmons enables the generation of significantly enhanced photocurrents from plasmonic samples. Since it is the surface plasmon field that matters for determining the total photocurrent, (9.4) suggests huge photocurrent enhancement even for moderate field enhancement factors. Systematic studies with plasmonic Au, Ag, Cu, and Al thin films revealed photoemission yield enhancement factors of $\times 50 \ldots \times 3500$ (or field enhancements of between

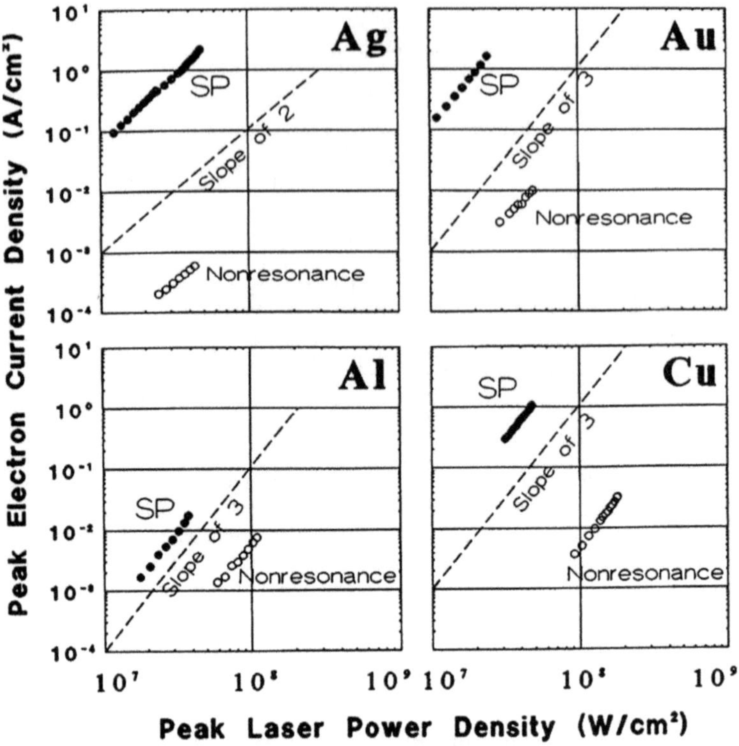

Fig. 9.5 The enhancement of PSP-induced multi-photon photoemission yield as a function of the intensity of the incident laser beam for four different surfaces plotted on double logarithmic scales. The slope of each linear fit equals the nonlinearity of the photoemission process. The lower data sets marked as "nonresonance" depict photoelectron yield from the same metal film without PSP coupling but with a similar illumination geometry. The substantial increase of the PSP-enhanced photoelectron yield is clearly illustrated with the *upper curves* plotted with *solid symbols* and marked with "SP". *Source* [15]

$\times 2$ and $\times 8$) [15] as depicted in Fig. 9.5. The curves show the intensity dependence of the photoelectron yield on double logarithmic scales. Therefore, the slope of each linear fit equals the degree of nonlinearity of the photoemission process. It can be seen (since there is no deviation from the linear fits) that multi-photon-induced emission is the only contributing mechanism to the photoemission process. To estimate the magnitude of plasmonic field enhancement, the PSP-induced photoelectron yield was compared to nonlinear photoemission without PSP coupling in the same film.

The fact that the electron yield is much higher with PSP coupling than with direct surface illumination without PSP coupling underlines a very important feature of the process, namely that it is the field of surface plasmons that induces the observed photocurrent.

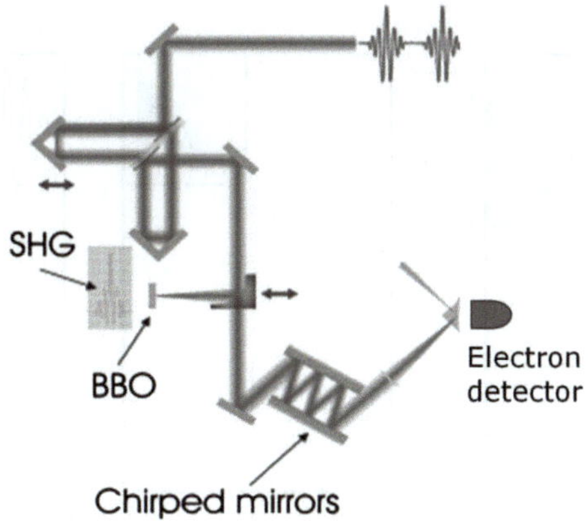

Fig. 9.6 Experimental scheme for time-resolved measurement of ultrashort surface plasmon wavepackets. Laser pulses with 5 fs duration are entering the setup with 5 nJ pulse energy. After passing through an interferometer they can be either characterized by a beta-barium-borate (*BBO*) crystal by second harmonic generation (*SHG*) autocorrelation. Alternatively, by removing the parabolic focusing mirror, the interferometer output is focused onto a prism to generate PSPs. Pulse propagation in the prism material was pre-compensated for by introducing broadband chirped mirrors. Nonlinear autocorrelation of the PSP wavepacket is detected in vacuum by recording the plasmonic photoemission current as a function of the delay in the interferometer. *Source* [19]

The nonlinearity of plasmonic photoemission offers an additional benefit for the investigation of dynamical processes of PSPs. Because of the rather high, third or fourth order nonlinearity, autocorrelation-type measurements can be carried out with the help of photoemitted electrons. This provides valuable information on the dynamics in the metal [18], moreover, if the multiphoton photoemission process is induced by surface plasmon fields, such nonlinear dynamic measurements can reveal temporal information on PSP wavepackets [19]. Such a scheme is illustrated in Fig. 9.6.

The plasmonic sample is placed after an interferometer where the delay between the arms can be changed. Plasmonic photocurrent is measured as a function of this delay. This way, the setup acts a higher order autocorrelator, the nonlinearity of which is dependent on the order of the multi-photon-induced electron emission process. It is, however, the plasmonic field which induces photoemission, not that of the laser pulse. This way, temporal characterization of ultrashort PSP wavepackets can be performed, similarly to the autocorrelation characterization of an ultrashort laser pulse.

This concept was recently implemented experimentally [19] and results are depicted in Fig. 9.7. The photon order in the photoemission process was found to be 4 and it is indeed a fourth order autocorrelation function of a short pulse that is measured (Fig. 9.7a) with the method described above. This already indicates the few-cycle nature of the PSP wavepacket. Further analysis and comparison with calculated autocorrelation functions of the independently characterized laser pulse (Fig. 9.7b) resulted in a retrieved PSP wavepacket length of 6.5 fs [19]. It also has to be noted that such a broadband plasmonic wavepacket, of course, heavily spreads

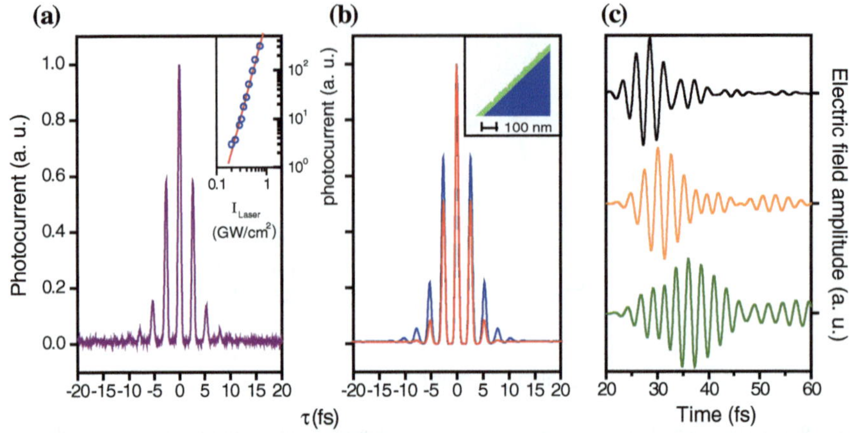

Fig. 9.7 **a** Measured interferometric time-resolved signal of PSP-enhanced photoemission induced by few-cycle pulses. The reconstructed *blue curve* in **b** is calculated using the simulation of PSP coupling on the actual, sputter-deposited metal film with 5-fs exciting laser pulses (see text for details and also the *inset* of **b** for the illustration of a part of the modeling geometry with the prism material in *blue* and the sputter-deposited silver film in *green*). The *red curve* in **b** is the calculated fourth-order autocorrelation function of the SPP-exciting, 5-fs laser pulse, plotted for reference. The *inset* of (**a**) illustrates the fourth-order dependence of the total PSP-induced, free-space photocurrent on a double logarithmic scale. The slope of the fit is $n = 4.05 \pm 0.11$. The *curves* in (**c**) are typical, computed few-cycle SPP waveforms at different locations on the Ag surface exhibiting a 2.8 nm rms roughness. (*Source* [19])

upon propagation, the measured duration is only valid upon launching the PSP pulse. Independent, first-principles simulations of the whole experimental scheme (including a slight surface roughness of the metal film involved) was performed by solving Maxwell's equations for this problem. Results of this also confirm that the electric field waveforms at all points of the rough surface is indeed a few-cycle oscillation, hinting at the generation of few-cycle plasmonic wavepackets. They propagate along the line where PSPs are generated. These calculated waveforms at some representative parts of the surface are depicted in Fig. 9.7c.

With these experiments not only PSP field enhancement could be quantified with simple experimental methods, but time-resolved information could be gained on PSP dynamics. Multi-photon induced photoelectrons proved to be helpful tools in achieving both of these goals.

9.2.3.2 Above Threshold Photoemission with Nanoemitters

At slightly higher intensities than that needed for multi-photon-induced photoemission, it is also possible that a larger number of photons are absorbed than the minimum required. Figure 9.3b illustrates this case of above-threshold photoemission. This process, the likelihood of which decreases with the increasing number of absorbed photons, is termed as "above-threshold photoemission" [20–24]. Above-threshold photoemission (ATP) bears close analogy to the well-known above-threshold ionization process taking place in atomic systems, discovered in 1979 [25].

In the decades since the discovery of above-threshold ionization, several efforts were made to demonstrate its metal surface analogy, ATP. Direct evidence of this process was not delivered since collective electron excitation, space-charge spectrum broadening and other adverse effects are known to smear electron spectra of extended surfaces illuminated by extended beams.

Eventually, one had to use nanoemitters to demonstrate such effects. Electrochemically etched metal nanotips enabled several fundamental discoveries in recent years [26–28]. The obvious demonstration of ATP for the first time was one of these [29]. Even though such nanotips do not typically exhibit plasmonic resonances, it is worth introducing this experiment here.

The experimental scheme and the results are depicted in Fig. 9.8. The illumination geometry of the sharp tungsten nanotip with a 8–10 nm radius of curvature is shown in Fig. 9.8a together with the whole experimental setup in Fig. 9.8b.

Measurement of the kinetic energy spectrum of the electrons emitted from a single tip nanoemitter show features with clear signs of ATP. Steps in the electron spectra in Fig. 9.8c are separated by exactly the photon energy (~1.5 eV). Moreover, the probability of the absorption of an additional photon decreases exponentially with the photon number, as expected from a perturbative high-order process. This is also evidenced by Fig. 9.8c. By increasing the intensity, higher-order processes become more likely, as intuition also suggests for the ATP case. Up to 9th order ATP can be

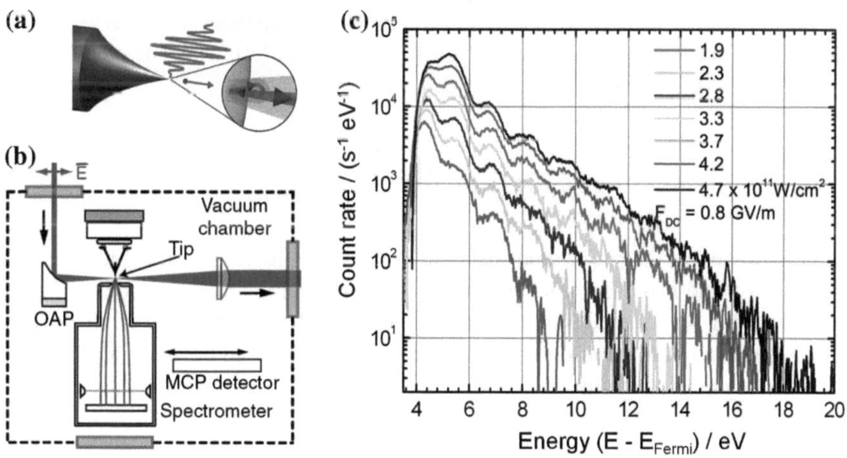

Fig. 9.8 **a** Sharp nanotip illuminated with ultrashort laser pulses. **b** Shows the experimental setup in which the kinetic energy spectrum of photoemitted electrons from a single nanotip can be measured. *OAP* Off-axis parabolic mirror, *MCP* Microchannel plate. Measured electron spectra for different incident laser intensities are shown in (**c**). Clear steps separated by the photon energy (~1.5 eV) can be observed in these spectra with the number of steps becoming higher with increasing laser intensity. This delivers proof for the above threshold photoemission process. *Source* [29]

observed for the highest intensities in this experiment, representing a clear advantage of the nanoemitter geometry over all previously used experimental schemes. The only disadvantage posed by non-plasmonic nanotips is the reduced field enhancement factor up to some 6–8 times based solely on the tip effect. Together with the limited set of shapes that can be realized with a nanotip, it is expected that future applications will rather be driven based on plasmonic nanoparticles.

9.2.3.3 Introduction to Strong-Field (Tunnelling) Photoemission

At higher interacting laser intensities than that needed for multi-photon-induced processes (including ATP), strong-field photoemission can also take place. This is characterized by the scheme in Fig. 9.3c with the electron being able to tunnel out from the metal. As opposed to the previous cases, where the photocurrent is typically governed by the field amplitude of the laser beam, here, the interaction is determined by the instantaneous field strength of the laser pulse (or, equivalently, of the surface plasmon field). Photocurrent generated in this way follows the field evolution of the pulse envelope instantaneously, without any cycle-averaging effects and without any delay. Therefore, this process is also termed as adiabatic emission. This interaction type is a well-known phenomenon that occurs within the context of photoionization of atoms in intense laser fields, representing the first step, for example, in a high harmonic generation process.

Field emission electron currents can be approximated by more complex formulae than multi-photon-induced emission. Approaches to deliver a closed-form analytic expression are based on various quantum mechanical tunnelling considerations. The formula most commonly applied in the literature to metals (for both static and oscillating laser fields) is the Fowler-Nordheim equation [30, 31]. The electric field dependence of the tunneling current is described by

$$j(t) \propto \frac{e^3 E_l(t)^2}{8\pi h W t^2(w)} \exp\left(-\frac{8\pi\sqrt{2m} W^{3/2}}{3he\,|E_l(t)|} v(w) \right) \tag{9.5}$$

where $E_l(t)$ denotes the laser field strength, e and m the electron charge and rest mass respectively, and h is Planck's constant. W is work function of the metal, $v(w)$ is a slowly varying function taking into account the image force of the tunneling electron with $0.4 < v(w) < 0.8$ and the value of the function $t(w)$ can be taken as $t(w) \approx 1$ for tunneling emission with

$$w = e^{3/2}\sqrt{E_l/4\pi\varepsilon_0}/W \tag{9.6}$$

The rapidly varying, characteristic form of the $j(t)$ curve following the field cycles is illustrated in Fig. 9.4. Field emission/tunneling occurs primarily at those instants when the field strength reaches its maximum value, notably, for photoemission from metals excited with visible wavelengths. Experimental investigation of pure field emission (i.e., $\gamma << 1$) is not easily realizable, since the damage threshold of bulk metal surfaces and thin films is around 10^{13} W/cm^2, which is very close to the intensity where the $\gamma \sim 1$ condition is met.

In order to investigate these processes experimentally, a practical workaround is necessary. The use of far-infrared sources has proved to be suitable for this purpose where the $\gamma \sim 1$ condition can be met at much lower intensities [31]. In addition, plasmonic field enhancement on metallic surfaces can be exploited in the visible spectral region so that the $\gamma << 1$ condition can be satisfied for metal films without damage. This latter method is also more advantageous due to the lack of few-cycle, ultrashort, laser sources in the far-infrared domain.

9.2.3.4 Strong-Field Photoemission Enhanced in Nanoplasmonic Environments

Photoemission from metals can be particularly well enhanced by plasmonic nanoparticles and the corresponding LSPs (see Fig. 9.1b). The difference, with respect to the already introduced PSPs is manyfold: (A) LSP electromagnetic fields are confined not only along one spatial dimension, but in all three. Plasmonic fields are concentrated to nanoscale volumes as small as some thousand nm^3, and they are typically localized at the sharpest features of the nanoparticles (see Fig. 9.9). (B) As a result, higher field enhancement factors can be achieved. In this case the field enhancement can be several hundreds, whereas, for propagating surface plasmons it is typically not more than twenty. This way, strong-field interactions can be induced at extremely low laser intensities. (C) The spatial distribution of these electromagnetic fields can be flexibly tailored by the nanoparticle geometry (for this, compare Fig. 9.9a, b). (D) This way, the photoemission and electron acceleration processes, as well as spatial electron distributions can also be controlled to a high degree. In order to demonstrate these potentials, we will introduce recent experimental results on this topic here [32].

The concept of our experiment is schematically illustrated in Fig. 9.10 and relies on the electron beam lithographic fabrication of plasmonic gold nanoparticles on glass substrates, illuminated by linearly polarized femtosecond laser pulses (95–110 fs duration with a central wavelength of 805 nm). These laser pulses were delivered by a long-cavity Ti:sapphire oscillator with some 200 nJ pulse energy [33–35]. After coupling the pulses into LSP oscillations of the nanoparticle sample placed in vacuum, strong-field photoemission and photoacceleration processes were investigated by time-of-flight electron spectrometry. A core element of the experimental concept was the fabrication of resonant and off-resonant nanorods with respect to the given laser wavelength, as well as of resonant bowtie nanostructures with significantly higher plasmonic field enhancement than for the nanorods. This way, both plasmonic resonance and field enhancement effects can be effectively correlated with photoemission spectra by switching between the different sample types.

Different sets of nanoparticle arrays were fabricated with 0.01 mm^2 surface area each, comprised of approximately 80,000 rod-type nanoparticles on a conductive, transparent substrate. The individual nanorods had a sufficient distance so that

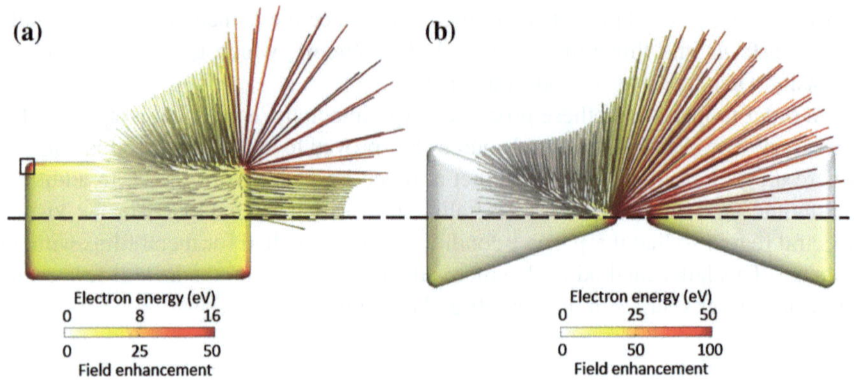

Fig. 9.9 Resonant surface plasmon generation and electron acceleration at metal nanoparticles. **a** The *colour-coded surface* in three quadrants of the nanoparticle shows the absolute value of the electric field enhancement on the nanoparticle surface. The *upper part* (above the *dashed line*) depicts the electron trajectories for one quadrant of the nanoparticles, with colour coding according to the final kinetic energies of the photoemitted and plasmonically accelerated electrons. **b** Same as panel (**a**) but for a *bowtie-shaped* nanoparticle with higher field enhancement factors. *Source* [32]

Fig. 9.10 Experimental scheme for the nanoplasmonic photoemission experiment from metal nanoparticles. Fs laser pulses are incident on a glass substrate on which arrays of metal nanoparticles were fabricated with electron beam lithography. Photoemitted electrons (*white arrows*) are analyzed by a time-of-flight spectrometer, the input opening of which is also depicted. *Source* [32]

they were not coupled electromagnetically. The nanorod height and width measured 40 and 87 nm, respectively, and the length was 120, 152 and 183 nm for three different arrays, as shown by Fig. 9.11b–d with scanning electron microscope (SEM) images. Measured optical resonances of the different nanoparticle geometries are in Fig. 9.11a. Resonances are peaked between 732 and 877 nm, representing blue-shifted, on-resonance and red-shifted nanorods, respectively, with respect to the laser wavelength indicated with the dashed box.

Upon femtosecond illumination, electrons are generated from the nanoparticles. The spectra of these electron beams are shown in Fig. 9.11f. Remarkably, high-energy electrons with kinetic energies up to 19 eV were observed, being well above the photon energy of 1.55 eV corresponding to the laser wavelength. This indicates the presence of ponderomotive electron acceleration effects.

To confirm this hypothesis, we also carried out intensity scans of the input laser beam and recorded multiple spectra this way for all nanoparticle types involved. Results of this are shown in Fig. 9.12. It is clear from Fig. 9.12e that the cutoffs of the electron spectra scale approximately linearly with the intensity of the exciting pulses and that the highest photoemission yields and highest kinetic energies are observed from plasmonically resonant nanoparticles. They are then followed by the red-shifted and blue-shifted nanorods in this respect.

These observations are further corroborated by additional experiments performed with resonant coupled nanotriangle pairs (placed in a bowtie geometry). These closely spaced nanoparticle dimers are known to exhibit a much stronger field enhancement in the gap region between the nanotriangles. The gaps of the bowties are 20 nm wide with only a slight ± 2 nm variation throughout the array. The results of the bowtie measurements, with the same experimental conditions as for the nanorods, together with the particle geometries and optical resonance properties are shown in Figs. 9.11 and 9.12 with green curves.

It is clearly visible that for the bowtie case, significantly lower incident laser intensities are sufficient to generate the same photoelectron yield and electron energies than for the nanorods. Carrying out experiments with intensities above 25 GW/cm^2 were made impossible by optical damage of the nanoparticles, also indicating higher electric field enhancement than in the case of nanorods. Similarly to nanorods, the linear scaling of the cutoff electron energies can be observed here. This scaling resembles the linear scaling of the ponderomotive energy with the intensity given in (9.1). The reason for this phenomenon will be introduced in the next sections.

9.2.4 Electron Acceleration Phenomena in Plasmonic Fields

As previously shown, strong laser fields can change the fundamental nature of the interaction of light with matter. It was also argued that strong fields need not necessarily be induced by intense, amplified laser pulses, PSP/LSP field enhancement can readily compensate for a substantial amount in focused laser intensity.

As already introduced, after an electron is set free by photoemission, it can undergo ponderomotive-type, cycle-by-cycle acceleration along the gradient of sharply decaying, nanolocalized electromagnetic fields. Figure 9.2c shows such a process. Here, we will analyze and show examples of this phenomenon.

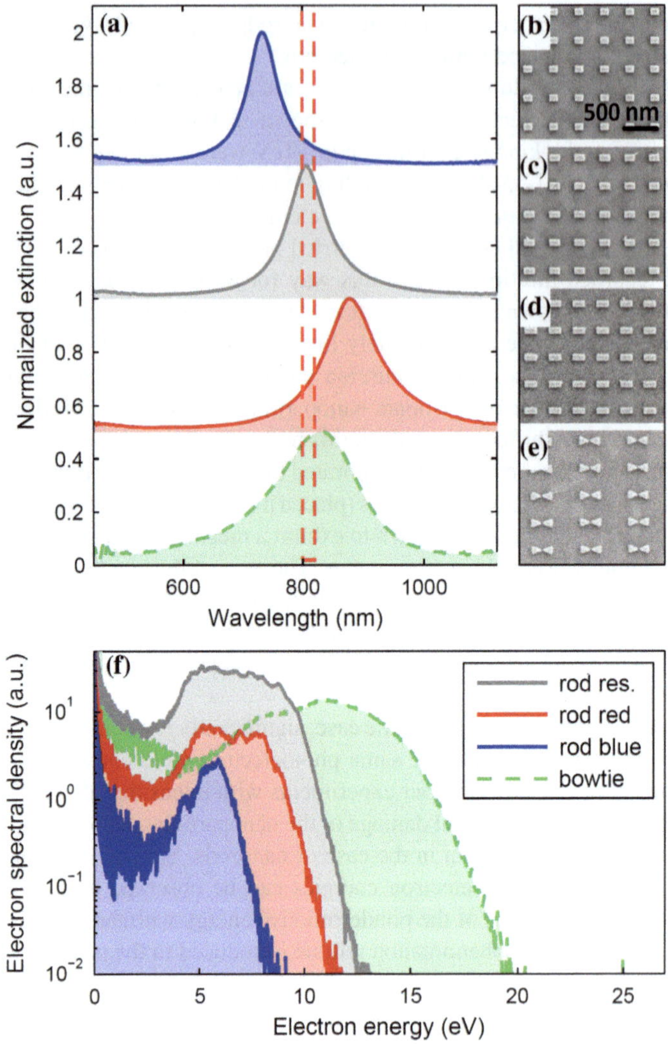

Fig. 9.11 Measured optical spectra and electron kinetic energy distributions for nanorods and bowtie nanoparticles. **a** Measured extinction spectra for nanorods with dimensions of **b** $120 \times 87 \times 40\,\text{nm}^3$, **c** $152 \times 87 \times 40\,\text{nm}^3$, and **d** $183 \times 87 \times 40\,\text{nm}^3$, which are *blue-shifted*, in resonance, and *red-shifted* with respect to the excitation bandwidth centered at $\lambda_{exc} = 805\,\text{nm}$ (see *dashed box*). Spectra are offset for clarity. The *bottom curve* in panel (**a**) depicts the spectrum for a *bowtie structure* shown in (**e**) with 90 nm width, 40 nm height, and 260 nm length (20 nm gap). **f** Electron spectra for different particle geometries and for a laser peak intensity of $25.1\,\text{GW/cm}^2$. The data below 3 eV are of limited validity due to instrumental restrictions of the time-of-flight spectrometer. *Source* [32]

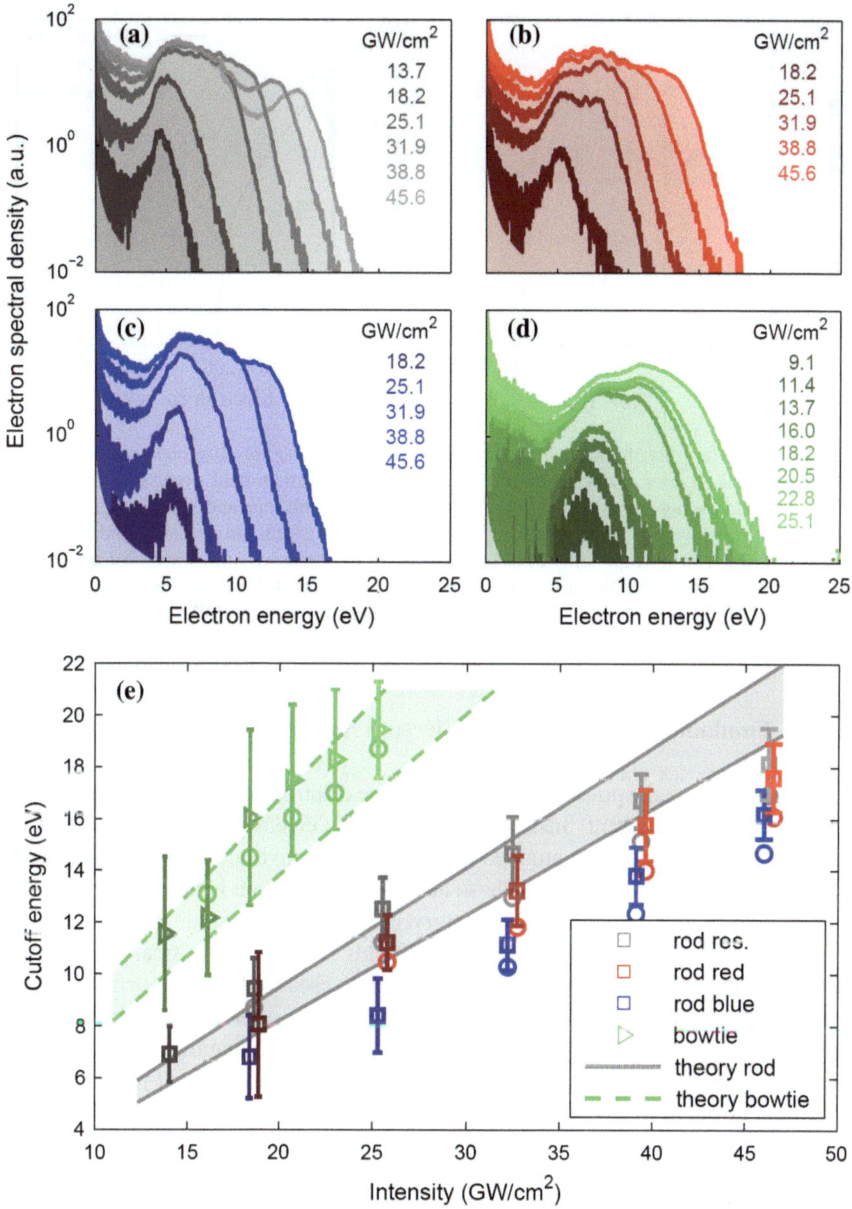

Fig. 9.12 Electron spectra as a function of peak intensity for **a** resonant rod, **b** *red-shifted* rod, **c** *blue-shifted* rod, and **d** resonant bowtie structure. **e** Cutoff energies of the electron spectra, as a function of laser intensity. The *solid* and *dashed lines* show simulation results for the resonant rod and bowtie nanoparticle, respectively. The influence of different cutoff angles θ_{cut} = 6–10° (accounting for the acceptance cone of the electron spectrometer) on the cutoff energies is indicated by the *shaded areas*. *Source* [32]

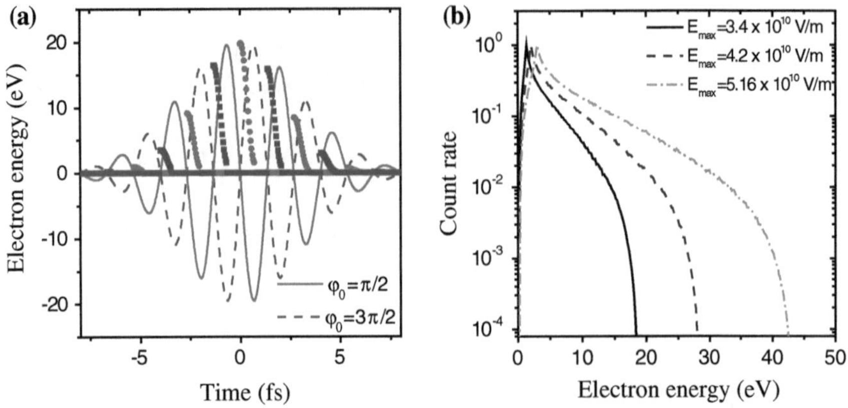

Fig. 9.13 **a** Surface-plasmon-accelerated electron energy as a function of the photoemission instant of the electrons (*scatter plots*). The electric field (with a maximum field strength of 3.4×10^{10} V/m) of the plasmon-generating 5-fs laser pulse is illustrated with *solid* and *dashed lines*. The pulse has either a "positive sine" (*solid line*) or "negative sine" waveform (*dashed line*) under the same pulse envelope. The corresponding electron energies for the "positive sine" waveform are depicted as *circles*, whereas for the "negative sine" waveform, as *squares*. **b** Electron spectra calculated for such pulses for different peak intensities of the laser pulse driving the PSP wave. *Courtesy* Péter Rácz

9.2.4.1 Fundamentals of Plasmonic Electron Acceleration

Now let us analyze the ponderomotive plasmonic electron acceleration process along a laser pulse, for different "instants of birth" of the electrons. It is obvious that the final electron kinetic energy attained in the ponderomotive acceleration process will be very highly dependent on the phase at which the electron is photoemitted and at which it starts to run a classical trajectory in the PSP/LSP field. We analyzed the final energy of electron trajectories in PSP fields by solving classical equations of motions in the field. We disregarded electrons that rescatter from the surface or those where the Lorentz force drives them toward the surface right at their "birth instant". Figure 9.13a illustrates the temporal distribution of the remaining electrons' final kinetic energies as a function of the electron photoemission instant in a maximum PSP field strength of 3.4×10^{10} V/m generated from a 5-fs-long optical pulse at 800 nm central wavelength. In this analysis, these representative, emitted electrons are collected from the central spot of the illuminated area on the metal film.

Overall, the corresponding electron kinetic energy distribution presented in Fig. 9.13 exhibits a close similarity to that of atomic electrons accelerated by an ionizing laser field [36]. However, contrary to atomic electrons, only $\sim$1/4 of all emission instants contribute significantly to the acceleration process in PSP field acceleration. This is due to the electron recollision and reabsorption processes. Macroscopically observable emission distributions and electron kinetic energy spectra can be numerically calculated after the assessment of a large number of trajectories with appropriate sampling representing the whole illuminated surface and the full PSP pulse. In case of flat metal-dielectric surfaces, this method delivers a very good approximation [37]. Figure 9.13b shows full electron spectra for electron acceleration in PSP fields calculated in the above manner. Important features such as the rapid decay after the low-energy peak and the linear scaling of the spectral cutoffs with intensity (as expected from the pondermotive scaling in (9.1)) is clearly visible. These findings also highlight the mechanisms behind the spectra observed in case of LSP excitation of nanoparticles e.g. in Fig. 9.12e. In the LSP case, the contributing phenomena are completely analogous.

9.2.4.2 Experimental Results on Plasmonic Electron Acceleration

Apart from the enhancement of photoemission yield, PSP/LSP fields can also accelerate the electrons that are set free from the surface, thanks to the ponderomotive electron acceleration mechanism. Examples of LSP fields accelerating electrons up to 20 eV kinetic energy have been shown in Sect. 2.3.4. It was found, however, that PSP fields are more efficient when it comes to accelerating electrons to keV kinetic energies on the nanoscale.

The strength, duration, and spatial extent of the PSP evanescent plasmonic field determine the final kinetic energy gained by the electrons. Spectrally resolved measurements of PSP photoemission delivered experimental confirmation of this powerful particle acceleration mechanism several years ago [38–41]. The special feature of these electron spectra is the scaling of cutoff energies already explained above.

The first experiment to demonstrate that energetic electrons can be generated using surface plasmon waves was performed by Zawadzka et al. [38, 39]. They used a Kretschmann PSP coupling configuration and laser pulse durations of between 100 and 150 fs. They demonstrated PSP-enhanced electron spectra reaching 400 eV at a laser intensity of 40 TW/cm^2. Instead of the Kretschmann PSP coupling method, Kupersztych et al. employed a gold-coated diffraction grating to channel laser pulses into PSP waves. With this method and using 60-fs-long pulses with an 8 GW/cm^2 focused intensity, they also demonstrated PSP electron acceleration [40]. However, the highest electron energy obtained was $\sim$50 eV, indicating that grating coupling and PSP enhancement are much weaker compared to the Kretschmann coupling configuration.

Some 10 years ago, Irvine et al. [41] reported an even more remarkable electron energy gain by accelerating electrons in PSP fields up to 400 eV energy using merely 1.5 nJ pulse energy (focused to 1.8 GW/cm^2) delivered by a simple Ti:sapphire oscillator. However, most interesting is the fact that the PSP-enhanced electron spectrum became quasi-monoenergetic peaking at 300 eV with a FWHM of only $\sim$80 eV.

Further experiments were also performed by spectroscopic characterization of ultrafast plasmonic electron bunches generated and accelerated by few-cycle laser pulses [19]. This is a particularly intriguing scheme as the driver pulse duration (5 fs) was comparable to the surface plasmon lifetime. Figure 9.6 also illustrates the experimental setup used for this experiment with the only difference that the

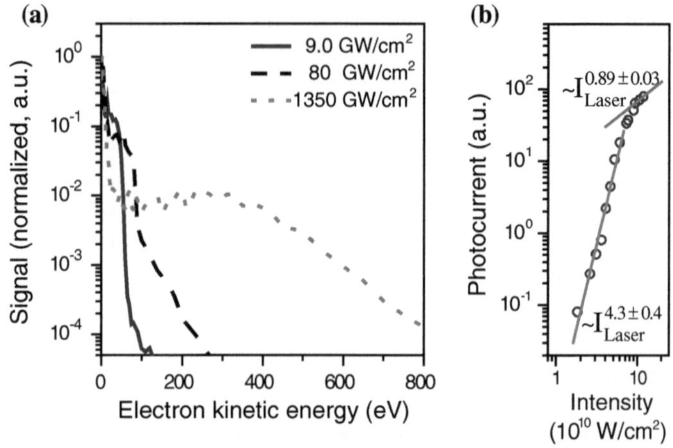

Fig. 9.14 **a** PSP-induced electron acceleration spectra at different incident laser intensities measured in a setup similar to that in Fig. 9.6. **b** Integrated photocurrent as a function of intensity. It shows that within the intensity range of interest in these measurements, the dominant emission mechanism changed from multi-photon-induced photoemission to tunneling emission at around 60 GW/cm^2 focused intensity

interferometer is omitted and few-cycle pulses hit the prism surface after dispersion pre-compensation with chirped mirrors. The electron detector is transformed into a retarding field spectrum analyzer by placing a grid between the prism surface and the electron multiplier tube. By changing the electron retarding voltage on the grid and measuring the signal as a function of this voltage, the integral of electron spectra can be measured easily.

This way, it was demonstrated that plasmonic-field electron acceleration can be an efficient electron energy gain technique even with few-cycle excitation. Figure 9.14 depicts the measured energy distributions at different laser intensities (measured at the surface without accounting for PSP field enhancement) where the highest focused intensity of 1.35×10^{12} W/cm^2 is slightly below the damage threshold intensity of the Ag film ($\sim 1.5 \times 10^{12}$ W/cm^2, as measured). Remarkably, energy levels approaching 1 keV can be reached within the short-lived evanescent PSP field.

9.2.5 Surface Plasmon Induced Electron Acceleration in the Mid-infrared

We have seen that in the strong-field regime, in which $\gamma \leq 1$, the ponderomotive energy becomes comparable to the electron binding energy. This regime requires typical focused laser intensities of 10^{14} W/cm^2 at 800 nm central wavelength (e.g. in case of a laser-atom interaction, without plasmonic field enhancement). Here,

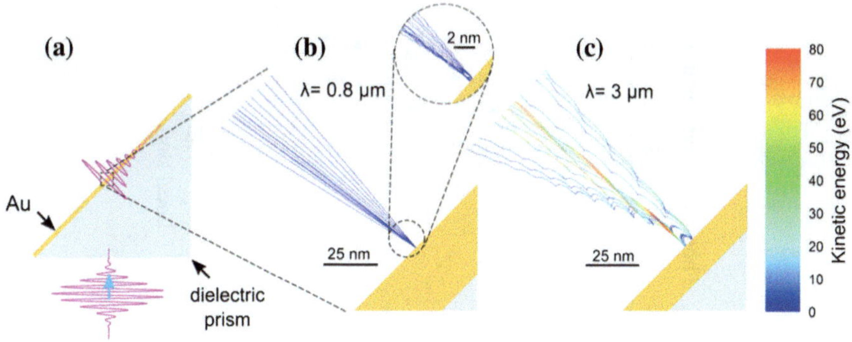

Fig. 9.15 Illustration of the concept of strong-field photoemission and electron acceleration in nanolocalized surface plasmon fields generated on thin gold films by focusing 9-cycle laser pulses in the Kretschmann-Raether coupling geometry (**a**). The advantage of using a long-wavelength mid-infrared femtosecond source is evident by depicting electron trajectories for surface plasmon excitation at (**b**) 800 nm central wavelength, 4 GW/cm^2 focused intensity and 24-fs FWHM pulses (~9 optical cycles) and (**c**) 3 μm central wavelength, 4 GW/cm^2 focused intensity and 90-fs FWHM pulses (~9 optical cycles). It can be seen that both the electron quiver amplitudes and the achievable kinetic energies are substantially increased in the long-wavelength case, in accordance with fundamental, ponderomotive scaling laws. *Source* [42]

we have shown how it is possible to achieve strong-field interactions at orders of magnitude lower focused intensities (40–60 GW/cm^2) by making use of PSP/LSP field enhancement. Thus, the fundamental question arises that what is the lowest intensity to attain strong-field phenomena and the corresponding electron kinetics.

We will show here that the transition between multi-photon induced (perturbative) and strong-field light-solid interactions can be shifted to unprecedentedly low focused intensities of below 1 GW/cm^2 with the help of femtosecond mid-infrared pulses and strong nanoplasmonic field confinement. Thanks to both of these effects, electron energies can be generated in a nanoscale plasmonic acceleration process that exceed the photon energy of the PSP generating pulse by almost two orders of magnitude. Accessing the strong-field regime at unprecedentedly low intensities is enabled by the usage of a mid-infrared femtosecond source. This way one can exploit the $1/\lambda$ scaling of the Keldysh-parameter and the λ^2 scaling of the ponderomotive energy of the electrons. The concept and the corresponding scalings are illustrated in Fig. 9.15 showing numerical, semi-classical simulations of the plasmonic photoemission and electron acceleration phenomenon involving PSPs on a metal thin film.

For these experiments, we used a mid-infrared optical parametric chirped pulse amplifier (OPCPA) source [43] delivering 90 fs pulses at a central wavelength of 3.1 μm. These pulses were loosely focused onto the hypotenuse face of a right-angle CaF_2prism where they were efficiently coupled to PSPs on a 15 nm thick film of gold. Plasmonic photocurrent and surface plasmon enhanced photoemission spectra were then measured with a retarding grid analyzer followed by an electron multiplier.

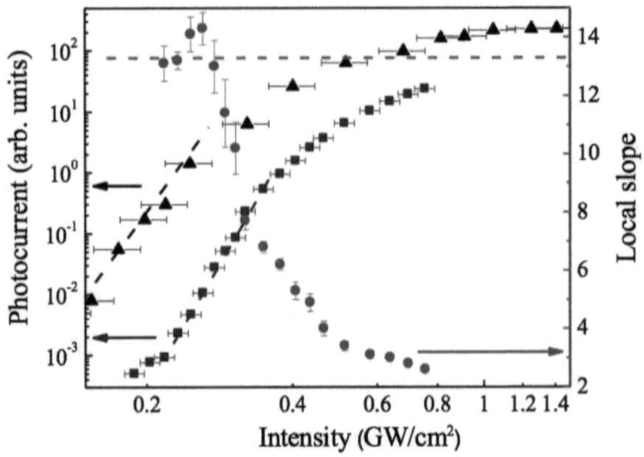

Fig. 9.16 Focused laser intensity dependence of the total plasmonic photocurrent for two independent exemplary scans (offset for clarity, *black triangles* and *red squares*). Spot sizes (FWHM) of 870 μm (*black triangles*) and 1400 μm (*red squares*) were used and the pulse energy was varied as described in the main text. The slopes of the fits to the initial sections are 12.3 ± 1.8 and 13.1 ± 0.6, respectively. The local slope of the second curve is also plotted (*green circles*) to illustrate the gradual transition between multi-photon-induced and tunneling (strong-field) photoemission at very low laser intensities. The electron multiplier gain of each measurement was set such that for the maximum intensity the signal did not show a saturated trace on an oscilloscope. After that, quantitative voltage signal was acquired with a lock-in amplifier. *Source* [42]

Two independent measurements of the intensity dependence of the plasmonic photocurrent are depicted in Fig. 9.16 on a double logarithmic scale. We controlled the intensity by varying the pump power of the last amplifier stage in the OPCPA. This procedure did not lead to modifications of the temporal profile of the pulse. The curves in Fig. 9.16 show that the total photocurrent first scales highly nonlinearly with the intensity (according to a ~13th power law), as expected. Here, multi-photon induced photoemission with a photon energy of 0.4 eV takes place and the work function is ~5.1 eV for polycrystalline gold. The power-law scaling behaviour substantially changes at around 0.6 GW/cm^2 focused intensity where the local slopes of the curve start to become reduced. This is a well-known signature of the transition to tunnelling, however, it takes place here at unprecedentedly low intensities, suggesting strong field enhancement of PSPs.

To determine the magnitude of the field enhancement responsible for this tunneling transition, one can also measure the electron spectra for a number of incident laser intensities and evaluate the spectral cutoffs, as shown in Fig. 9.17a. One can see that whereas the ponderomotive energy of electrons according to (9.1) is only between 1.3 and 3.1 meV for the incident laser intensities, orders of magnitude higher electron energies can be achieved in the PSP field. Analyzing maximum kinetic energies of the spectra, i.e. the cutoffs, we find cutoff values of up to 47 eV.

With the same setup, independent measurements of the maximum (cutoff) kinetic energy of the electrons as a function of intensity were carried out, without resolving electron spectra. This could be achieved by increasing the retardation field and

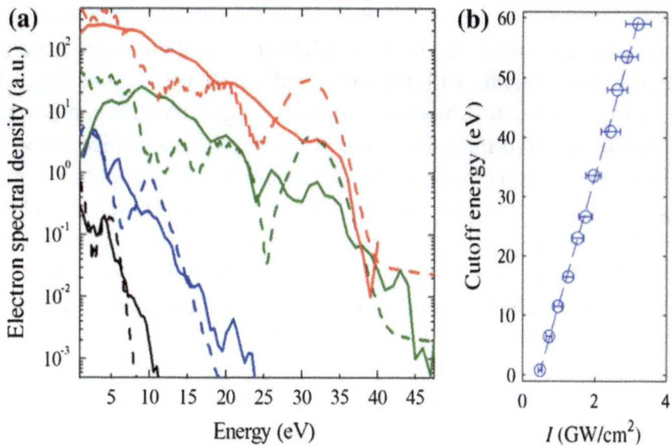

Fig. 9.17 **a** Plasmonic photoemission electron spectra for intensities $1.4\,\text{GW/cm}^2$ (*black*), $1.7\,\text{GW/cm}^2$ (*blue*), $2.8\,\text{GW/cm}^2$ (*green*) and $3.5\,\text{GW/cm}^2$ (*red*) in the strong-field photoemission regime. The evaluated plasmonic field enhancement factors are between 28 and 34. *Dashed lines* show the spectra calculated from the quantum mechanical model involving a plasmonic field enhancement factor. Note that the respective spectra are offset along the y axis for clarity. **b** Maximum kinetic energy of plasmonically accelerated photoelectrons as a function of focused laser intensity as an independent measurement. For a given intensity we increased the retardation field until the noise level was reached, being equivalent to the signal without any beam incident on the prism. The linear dependence of cutoff on intensity confirms the classical ponderomotive scaling law. All measurements were carried out with a spot size (FWHM) of 870 μm and the pulse energy was varied as described in the main text. *Source* [42]

monitoring the voltage where the electron multiplier signal reached the noise level. These results (shown in Fig. 9.17b) confirm that the electron acceleration process in the enhanced PSP field is governed by the classical ponderomotive scaling law. According to (9.1), the maximum kinetic energy of electrons should be proportional to $I\lambda^2$ and this linear scaling with the intensity is observed clearly in the same intensity range as that of Fig. 9.16. Since the acceleration process is independent of the photoemission mechanism, a linear intensity dependence is observed irrespective of the multiphoton or strong-field nature of the emission process.

The experimentally determined cutoff values in Fig. 9.17a allow one to determine the maximum field enhancement factor at the plasmonic thin film in the following way. It is known that the highest energy electrons undergo rescattering on the surface after a fraction of the light oscillation cycle [44–46]. Provided that the emission and rescattering events take place at the most favorable phases of the oscillating electromagnetic field, the electron can gain a kinetic energy amounting to as much as ten times the ponderomotive potential [45, 46]. Thus, by evaluating the cutoffs of the measured spectra, we can determine the maximum value of the ponderomotive potential within the focal spot. The cutoff scaling law is universally valid irrespective of the medium, i.e. atom, surface, etc., and thus it is applicable here, too. Since rescattering takes place within a half cycle of the field and this initial phase of electron motion takes place within the closest nanoscale proximity of the surface (see e.g. Fig. 9.15), one can measure the highest local field enhancement factor experimentally.

Carrying out this analysis with the measured cutoff values (see Fig. 9.17a), one finds an increase of the cutoff by three orders of magnitude compared to the values expected from the focused intensities (see (9.1)) and the mentioned $10 \times U_p$ cutoff law. This shows that a high effective field enhancement factor is present, provided by the plasmonic thin film. The observed cutoff increase corresponds to field enhancement factors between 28 and 37 for the analysed spectra. It has to be noted that the maximum field enhancement factor of a perfectly flat gold film with 15 nm thickness is 19 at this wavelength. We attribute the somewhat higher measured values to additional surface roughness of the metal film [47].

In order to confirm these conclusions from a quantum mechanical point-of-view, one can also carry out a theoretical analysis of the strong-field light-matter interaction the results of which are shown in Fig. 9.17a with dashed lines. The theoretical model used has already been described and employed for the calculation of the electron photoemission from metal nanotips [48]. To match the cutoff energies of a given experimental electron spectrum, one can use the experimental intensity and vary the field enhancement factor in the simulation, representing the only fit parameter in this case. The parameters used for the plots in Fig. 9.17a agree to within a factor of two with the experimentally determined enhancement factors. Best-match modeled electron spectra in Fig. 9.17a yielded field enhancement factors of between 60 and 70, representing higher enhancement values than those gained by the cutoff evaluation of measured spectra. This is, however, a satisfactory agreement on the magnitude of the field enhancement in this particular configuration taking into account the limitations of the quantum mechanical model used.

In conclusion, it was demonstrated that with the help of PSP field enhancement and long excitation wavelengths, one can achieve strong-field light-matter interaction at extremely low incident laser intensities, opening a new paradigm in strong-field physics. High electron energies exceeding the ponderomotive energy in the incident light field by several orders of magnitude were achieved in an all-plasmonic electron acceleration scheme within the closest nanoscale proximity of the metal film.

9.3 Conclusions and Outlook

The generation of surface plasmons at metal-vacuum interfaces (PSPs) and in nanoparticles (LSPs) is known to confine the electromagnetic energy to subwavelength spatial dimensions either along one (PSPs) or along all three (LSPs) spatial coordinates. This electric field confinement, along with the high field enhancement, result in a number of intriguing phenomena including strong-field interactions at unprecedentedly low intensities (some GW/cm^2) and electron acceleration gradients exceeding a few GeV/m. The generation of high-energy, ultrashort electron bunches by using low-intensity, compact lasers can open the doorway for novel time-resolved investigations, as well. The high plasmonic field confinement allows access to strong-field interactions, which, until recently, has been the terrain of multi-TW lasers.

The detailed investigation of ultrafast plasmonic electron acceleration in PSP fields brought important discoveries in recent years which involved a large number of novel features such as all-optical, keV electron acceleration on the nanoscale or strong-field plasmonic photoemission. In addition, photoemitted and photoaccelerated electrons can be used to characterize few-cycle surface plasmon wavepackets. The findings discussed here, supported with experimental observations, focused on the ultrafast physics phenomena involved in the photoemission mechanisms of electrons in strong nanoplasmonic fields and acceleration of freely propagation electrons on the nanoscale. By exploiting PSP/LSP field enhancement, plasmon-induced tunneling emission from the metal surface at low laser intensity was demonstrated.

Controlling localized surface plasmon fields via both the shape of the optical pulse and the nanostructure geometry provide an additional method of optical control of the interaction processes involved. By tailoring the nanostructures' interaction with the laser, ultrafast electron processes can be exploited for a number of applications including extreme ultraviolet light generation [49] and the construction of nanostructured plasmonic photocathodes [50, 51]. The investigation of ultrafast strong-field plasmonic phenomena can be viewed as a natural route that will ultimately take us to the observation of many fundamental processes in nature involving biochemical reactions, molecular, lattice and collective electron dynamics, as well as their coherent control via shaping electric field evolution and distribution on the nanoscale.

Acknowledgments The author acknowledges support from the "Lendület" program of the Hungarian Academy of Sciences, the National R&D Office (OTKA project 109257) and the Partner Group Program of the Max Planck Society.

References

1. K. Kneipp et al., Single molecule detection using surface-enhanced Raman scattering (SERS) Phys. Rev. Lett. **78**, 1667–1670 (1997)
2. S. Lal, S. Link, N.J. Halas, Nano-optics from sensing to waveguiding. Nat. Photon. **1**, 641–648 (2007)
3. H.J. Simon, D.E. Mitchell, J.G. Watson, Optical second-harmonic generation with surface plasmons in silver films. Phys. Rev. Lett. **33**, 1531–1534 (1974)
4. J.C. Quail, J.G. Rako, H.J. Simon, R.T. Deck, Optical second-harmonic generation with long-range surface plasmons. Phys. Rev. Lett. **50**, 1987–1990 (1983)
5. D. Polyushkin, I. Márton, P. Rácz, P. Dombi, E. Hendry, W.L. Barnes, Mechanisms of THz generation from silver nanoparticle and nanohole arrays illuminated by 100 fs pulses of infrared light. Phys. Rev. B **89**, 125–426 (2014)
6. H. Raether, *Surface Plasmons on Smooth and Rough Surfaces and on Gratings* (Springer-Verlag, Berlin, 1988)
7. J.A. Schuller et al., Plasmonics for extreme light concentration and manipulation. Nat. Mater. **9**, 193–204 (2010)
8. H. Atwater, The promise of plasmonics. Sci. Am. **296**, 56–63 (2007)
9. S.A. Maier, *Plasmonics: Fundamentals and Applications* (Springer, Berlin, 2007)

10. J.N. Anker, W.P. Hall, O. Lyandres, N.C. Shah, J. Zhao, R.P. Van Duyne, Biosensing with plasmonic nanosensors. Nat. Mater. **7**, 442–453 (2008)
11. D. Koller, A. Hohenau, H. Ditlbacher, N. Galler, F. Reil, F.R. Aussenegg, A. Leitner, E. List, J.R. Krenn, Organic plasmon-emitting diode. Nat. Photon. **2**, 684–687 (2008)
12. A.L. Falk, F.H.L. Koppens, C.L. Yu, K. Kang, N. de Leon Snapp, A.V. Akimov, M.-H. Jo, M.D. Lukin, H. Park, Near-field electrical detection of optical plasmons and single-plasmon sources. Nat. Phys. **5**, 475–479 (2009)
13. V.E. Ferry, L.A. Sweatlock, D. Pacifici, H.A. Atwater, Plasmonic nanostructure design for efficient light coupling into solar cells. Nano Lett. **8**, 4391–4397 (2008)
14. T. Tsang, T. Srinivasan-Rao, J. Fischer, Surface-plasmon-enhanced multiphoton photoelectric emission from thin silver films. Opt. Lett. **15**, 866–868 (1990)
15. T. Tsang, T. Srinivasan-Rao, J. Fischer, Surface-plasmon field-enhanced multiphoton photoelectric emission from metal films. Phys. Rev. B **43**, 8870–8878 (1991)
16. S.E. Irvine, Laser-field femtosecond electron pulse generation using surface plasmons. Ph. D. Thesis, University of Alberta, Canada (2006)
17. C. Lemell, X.-M. Tong, F. Krausz, J. Burgdörfer, Electron emission from metal surfaces by ultrashort pulses: determination of the carrier-envelope phase. Phys. Rev. Lett. **90**, 076403 (2003)
18. H. Petek, S. Ogawa, Femtosecond time-resolved two-photon photoemission studies of electron dynamics in metals. Prog. Surf. Sci. **56**, 239–310 (1997)
19. P. Dombi et al., Observation of few-cycle, strong-field phenomena in surface plasmon fields. Opt. Express **18**, 24206 (2010)
20. W.S. Fann, R. Storz, J. Bokor, Observation of above-threshold multiphoton photoelectric emission from image potential surface states. Phys. Rev. B **44**, 10980 (1991)
21. G. Farkas et al., Above-threshold multiphoton photoelectric effect of a gold surface. Opt. Eng. **32**, 2476 (1993)
22. M. Aeschlimann et al., Observation of surface enhanced multiphoton photoemission from metal surfaces in the short pulse limit. J. Chem. Phys. **102**, 8606 (1995)
23. F. Bisio et al., Mechanisms of high-order perturbative photoemission from Cu(001). Phys. Rev. Lett. **96**, 087601 (2006)
24. F. Banfi et al., Experimental evidence of above-threshold photoemission in solids. Phys. Rev. Lett. **94**, 037601 (2005)
25. P. Agostini et al., Free-free transitions following six-photon ionization of xenon atoms. Phys. Rev. Lett. **42**, 1127 (1979)
26. P. Hommelhoff, Y. Sortais, A. Aghajani-Talesh, M.A. Kasevich, Field emission tip as a nanometer source of free electron femtosecond pulses. Phys. Rev. Lett. **96**, 077401 (2006)
27. R. Bormann et al., Tip-enhanced strong-field photoemission. Phys. Rev. Lett. **105**, 147601 (2010)
28. B. Piglosiewicz et al., Carrier-envelope phase effects on the strong-field photoemission of electrons from metallic nanostructures. Nat. Phot. **8**, 37–42 (2014)
29. M. Schenk, M. Krüger, P. Hommelhoff, Strong-field above-threshold photoemission from sharp metal tips. Phys. Rev. Lett. **105**, 257601 (2010)
30. V.T. Binh, N. Garcia, S.T. Purcell, Electron field emission from atom-sources: fabrication, properties, and applications of nanotips. Adv. Imaging Electron Phys. **95**, 63 (1996)
31. G. Farkas, S.L. Chin, P. Galarneau, F. Yergeau, A new type of intense CO_2 laser induced electron emission from a gold surface. Opt. Comm. **48**, 275 (1983)
32. P. Dombi et al., Ultrafast strong-field photoemission from plasmonic nanoparticles. Nano Lett. **13**, 674–678 (2013)
33. P. Dombi, P. Antal, Invetisgation of a 200-nJ chirped-pulse Ti: sapphire oscillator for white light generation. Laser Phys. Lett. **4**, 538 (2007)
34. P. Dombi, P. Antal, J. Fekete, R. Szipöcs, Z. Várallyay, Chirped-pulse supercontinuum generation with a long-cavity Ti: sapphire oscillator. Appl. Phys. B **88**, 379 (2007)
35. J. Fekete, P. Rácz, P. Dombi, Compression of long-cavity Ti:sapphire oscillator pulses with large-mode-area photonic crystal fibers. Appl. Phys. B **111**, 415–418 (2013)

36. G. Reider, XUV attosecond pulses: generation and measurement. J. Phys. D **37**, R37–R48 (2004)
37. P. Dombi, P. Rácz, B. Bódi, Surface-plasmon enhanced electron acceleration with few-cycle laser pulses. Laser Part. Beams **27**, 291 (2009)
38. J. Zawadzka, D. Jaroszynski, J.J. Carey, K. Wynne, Evanescent-wave acceleration of femtosecond electron bunches. Nucl. Instr. Meth. Phys. Res. A **445**, 324–328 (2000)
39. J. Zawadzka, D. Jaroszynski, J.J. Carey, K. Wynne, Evanescent-wave acceleration of ultrashort electron pulses. Appl. Phys. Lett. **79**, 2130–2132 (2001)
40. J. Kupersztych, P. Monchicourt, M. Raynaud, Ponderomotive acceleration of photoelectrons in surface-plasmon-assisted multiphoton photoelectric emission. Phys. Rev. Lett. **86**, 5180–5183 (2001)
41. S.E. Irvine, A. Dechant, A.Y. Elezzabi, Generation of 0.4-keV femtosecond electron pulses using impulsively excited surface plasmons. Phys. Rev. Lett. **93**, 184801 (2004)
42. S.M. Teichmann, P. Rácz, M. Ciappina, J.A. Pérez-Hernandez, A. Thai, J. Fekete, L. Veisz, J. Biegert, P. Dombi, Strong-field plasmonic photoemission in the mid-IR $< 1\mathrm{GW/cm}^2$ at intensity, Sci. Rep. **5**, 7584 (2015)
43. O. Chalus, A. Thai, P.K. Bates, J. Biegert, Six-cycle mid-infrared source with 3.8 μJ at 100 kHz. Opt. Lett. **35**, 3204–3206 (2010)
44. C. Farkas, C. Tóth, A. Köházi-Kis, Above-threshold multiphoton photoelectric effect of a gold surface. Opt. Eng. **32**, 2476–2480 (1993)
45. G.G. Paulus, W. Becker, W. Nicklich, H. Walther, Rescattering effects in above-threshold ionization: a classical model. J. Phys. B: At. Mol. Opt. Phys. **27**, L703–L708 (1994)
46. P. Rácz, S.E. Irvine, M. Lenner, A. Mitrofanov, A. Baltuska, A.Y. Elezzabi, P. Dombi, Strong-field plasmonic electron acceleration with few-cycle, phase-stabilized laser pulses. Appl. Phys. Lett. **98**, 111116 (2011)
47. P. Rácz, S.E. Irvine, M. Lenner, A. Mitrofanov, A. Baltuska, A.Y. Elezzabi, P. Dombi, Strong-field plasmonic electron acceleration with few-cycle, phase-stabilized laser pulses. Appl. Phys. Lett. **98**, 111116 (2011)
48. M. Krüger, M. Schenk, M. Föster, P. Hommelhoff, Attosecond physics in photoemission from a metal nanotip. J. Phys. B **45**, 074006 (2012)
49. I. Park, K. Seungchul, C. Joonhee, D.H. Lee, Y.J. Kim, M.F. Kling, M.I. Stockman, S.W. Kim, Plasmonic generation of ultrashort extreme-ultraviolet light pulses. Nat. Photon. **5**, 677 (2011)
50. W.S. Graves, F.X. Kartner, D.E. Moncton, P. Piot, Intense superradiant X rays from a compact source using a nanocathode array and emittance exchange. Phys. Rev. Lett. **108**, 263904 (2012)
51. R.K. Li, H. To, G. Andonian, J. Feng, A. Polyakov, C.M. Scoby, K. Thompson, W. Wan, H.A. Padmore, P. Musumeci, Surface-plasmon resonance-enhanced multiphoton emission of high-brightness electron beams from a nanostructured copper cathode. Phys. Rev. Lett. **110**, 074801 (2013)

Chapter 10
Highly Nonlinear and Ultrafast Optical Phenomena in Metallic Nanostructures

L. Wimmer, M. Sivis, G. Herink, S.V. Yalunin, K.E. Echternkamp and C. Ropers

Abstract This Chapter presents recent findings on nonlinear ionization and photoemission processes at metallic nanostructures. A particular emphasis is placed on processes which—due to the localized excitation in optical near-fields—exhibit different properties and scalings from their counterparts in the gas phase or at planar surfaces. The Chapter is structured in two parts. The first part discusses various regimes in highly nonlinear photoelectron emission from metallic nanotips, including field-driven photoemission at near- and mid-infrared frequencies, and the control of localized photoemission using intense terahertz transients. In the second part, multiphoton and strong-field ionization of atomic gases in plasmonic antennas and waveguides is presented. It is demonstrated that local ionization is enabled by optical field enhancements in various types of structures. At the same time, fundamental physical limitations preventing efficient high-harmonic generation in nanostructures are discussed.

L. Wimmer · M. Sivis (✉) · G. Herink · S.V. Yalunin · K.E. Echternkamp · C. Ropers
IV. Physical Institute—Solids and Nanostructures, University of Göttingen,
Friedrich-Hund-Platz 1, 37077 Göttingen, Germany
e-mail: msivis@gwdg.de

L. Wimmer
e-mail: wimmer@ph4.physik.uni-goettingen.de

G. Herink
e-mail: herink@ph4.physik.uni-goettingen.de

S.V. Yalunin
e-mail: yalunin@ph4.physik.uni-goettingen.de

K.E. Echternkamp
e-mail: echternkamp@ph4.physik.uni-goettingen.de

C. Ropers
e-mail: cropers@gwdg.de

M. Kitzler and S. Gräfe (eds.), *Ultrafast Dynamics Driven by Intense Light Pulses*, Springer Series on Atomic, Optical, and Plasma Physics 86, DOI 10.1007/978-3-319-20173-3_10

10.1 Introduction

Throughout recent years, significant progress was made in the study of highly nonlinear, ultrafast ionization and photoemission processes in nanoscale matter by employing the enhancement of ultrashort laser pulses in tailored metallic nanostructures such as sharp needles, nanoparticles, resonant antennas or waveguides. Two important optical features of these metallic nanostructures are their capability to localize electromagnetic radiation at sub-wavelength scales and, at the same time, to enhance incident light intensities in nanometric hot spots by several orders of magnitude. The associated sub-wavelength localization allows for a detailed analysis and deeper understanding of ultrafast ionization processes driven by highly inhomogeneous fields.

In this context, nonlinear photoemission from metallic and dielectric structures [6–12] and nanostructure-enhanced atomic gas ionization causing extreme-ultraviolet (EUV) light generation [13–19] were shown to exhibit scalings strongly deviating from bulk behavior. In this Chapter, we discuss a set of our recent studies in this field on photoelectron emission and atomic ionization in nanostructures [11, 12, 15, 17, 18].

In the first Section, experiments on nonlinear photoemission from sharp metallic tips are presented. The emission and acceleration of photoelectrons with ultrashort, intense laser pulses are examined for a variety of experimental conditions. Ultrashort near- and mid-infrared laser pulses are used to induce photoemission, and electron kinetic energy spectra are analyzed. In particular, a novel regime of strong-field photoemission is found, in which the electrons are accelerated in the rapidly decaying near-field within a fraction of the optical half-cycle. Furthermore, streaking-type two-color experiments are carried out by additionally coupling delayed single-cycle terahertz (THz) transients onto the tip, enabling a control of the photoemission current and nanoscale electron trajectories.

The second Section covers recent findings about EUV light generation in plasmonic nanostructures, induced by the plasmonic enhancement of atomic gas ionization. Specifically, the feasibility of coherent high-harmonic generation in resonant nanoantennas and tapered hollow waveguides, as reported previously [13, 14], is critically revisited [15, 17, 18].

10.2 Photoelectron Dynamics at Sharp Metal Nanotips

In this Section, we investigate the effects of field-enhancement and localization on photoelectron emission in optical near-fields. Sharp metal nanotips are employed to locally enhance the incident optical fields at the tip apex. The excitation wavelengths in our experiments span from near-infrared (NIR) to terahertz (THz) frequencies. As demonstrated in the following, strong-field excitation at long wavelengths facilitates access to sub-cycle electron dynamics in spatially confined electric fields [11, 20].

10.2.1 Nonlinear Photoemission

Conventional photoemission, involving photon energies exceeding the electron binding energy (or work function) Φ, exhibits photoemission yields proportional to the incident light intensity on a surface [21]. In contrast, nonlinear photoemission is governed by the simultaneous absorption of multiple photons of an energy below the work function, leading to the emission of a single electron in the perturbative process of multiphoton photoemission. For higher intensities and at long wavelengths, in a different conceptual picture, the optical electric field can bend the surface potential to a barrier which allows for quantum-mechanical tunneling of electrons (optical field emission). The transition between the multiphoton photoemission and optical field emission regimes is characterized by the dimensionless *Keldysh parameter*, i.e., the square root of the ratio of the work function and the ponderomotive energy U_p [22]:

$$\gamma = \sqrt{\frac{\Phi}{2U_p}}. \tag{10.1}$$

Here, the *ponderomotive energy* $U_p = \frac{e^2F^2}{4m\omega^2}$ is the average kinetic energy of an electron with mass m and charge e oscillating in an electric field of amplitude F with the angular frequency ω. For moderate intensities or high frequencies, the ponderomotive energy is negligible compared to the metal work function ($\gamma \gg 1$), and the photoemission process can be described by *multiphoton photoemission*, as depicted in Fig. 10.1a. At metallic nanotips, multiphoton photoemission was reported, e.g., in [6, 8, 23]. In this regime, the photocurrent J scales with a power law in the intensity I, corresponding to the photon number n: $J \propto I^n$.

Under such conditions ($\gamma > 1$), multiphoton transitions of the lowest possible photon order n_0 dominate [24–26]. At ponderomotive energies comparable to the photon energy (for higher intensities or lower frequencies), the absorption of excess photons ($n > n_0$) results in additional spectral peaks at multiples of the photon energy (*Above threshold photoemission* [27, 28]), which was observed for metal nanotips in [7]. For increasing intensity, the growing kinetic energy contained in the quiver motion of the final state of the electron after photoemission (Volkov state) eventually leads to the closing of photoemission channels of low order [29].

At even higher intensities and/or longer wavelengths ($\gamma \ll 1$), the emission dynamics can be described by a field-driven instead of a photon-driven process. In this strong-field regime, the emission is viewed as optically-induced tunneling through a potential barrier, adiabatically following the applied electric field (see Fig. 10.1b). The actual barrier height is lowered to an effective work function Φ_{eff} due to the Schottky effect (see, e.g. [30, 31]). Electrons are liberated via tunneling through the potential barrier around the peak of the laser field (cf. Fig. 10.1b). In this adiabatic regime, the current density j is described using the *Fowler-Nordheim* equation from static tunneling [32], which in its simplest version takes the form:

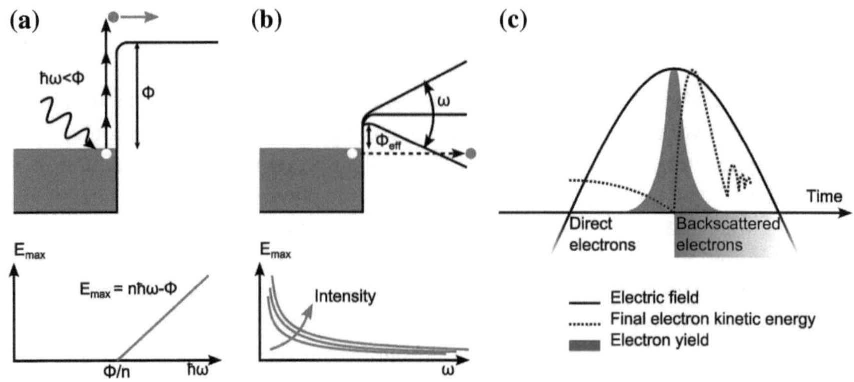

Fig. 10.1 Nonlinear photoemission processes. **a** Multiphoton photoemission (*top*) leads to a slightly modified photoelectric law for the electron kinetic energies (*bottom*). **b** Optically-induced tunneling (*top*) and subsequent field acceleration exhibits drastically different kinetic energy scalings. **c** Components of the two-step photoemission model for one optical half-cycle: the photocurrent (*gray area*) increases nonlinearly with the sinusoidal electric force (*solid line*). For direct (not rescattered) electrons, the kinetic energy gain (*dashed line*) in the electric field reaches a maximum value of $2\,U_p$. Rescattered electrons reach maximum energies up to $10\,U_p$

$$j(F) \propto \Theta(-F) \cdot F^2 \cdot \exp\left(-\frac{4}{3}\sqrt{\frac{2m}{\hbar^2}}\frac{\Phi^{3/2}}{e^1 F^1}\right). \tag{10.2}$$

Here, $\hbar$ is planck's constant, and $\Theta(-F)$ is the Heaviside step function. Subsequent to photoemission, the electrons are exposed to the unscreened optical electric field, which accelerates them to kinetic energies determined by the ponderomotive energy [33]. The kinetic energy distributions resulting from this process can be calculated within a modified simple man's model, as previously employed to describe scalings in strong-field atomic ionization and high-harmonic generation [34, 35]. In a first step, electrons are generated by Fowler-Nordheim tunneling, and in a second step, their respective kinetic energy is calculated via classical propagation of point charges in the optical field. The resulting electron energy depends on the phase of the electromagnetic wave at the moment of photoemission (see Fig. 10.1c), and overall spectra are a result of weighting with the final kinetic energy after acceleration the instantaneous current density. Due to the rectifying nature of a metal-vacuum interface, only half-cycles driving electrons out of the surface contribute to the emission. For electrons emitted early in the optical half-cycle (increasing electrical force on electrons), the final kinetic energy reaches a maximum value of $2\,U_p$, while photoelectrons generated after the peak of the force curve are accelerated back, with a fraction elastically rescattered at the metal surface and reaching final kinetic energies of up to $10\,U_p$ in the case of spatially homogeneous fields [34, 36, 37].

10.2.2 Sub-cycle Electron Dynamics in Highly Localized Electric Fields

The field-driven photoemission in localized fields, e.g., induced at the apex of metallic nanotips, exhibits characteristic electron dynamics, which may strongly deviate from the situation in diffraction-limited optical foci. In a far-field optical focus, photoemitted electrons interact with the entire remainder of the oscillatory field after the instance of emission. In contrast, this electron quiver motion may be suppressed in confined near-fields, as the typical spatial extension of the enhanced near-field (in the range of few to few tens of nanometers) can approach the dimensions of an electron's excursion path.

Generally, the kinetic energy change ΔE of a photoemitted electron in a space- and time-dependent electric field F is described by the integral of the force acting along its trajectory x(t), starting at the time of emission t_e:

$$\Delta E = -\int_{t_e}^{\infty} eF(x(t), t)dt. \tag{10.3}$$

Figure 10.2 schematically depicts electron trajectories for a sequence of increasing confinement in the driving field. When the photoelectrons are accelerated out of the high-field region by multiple cycles of a pulse with long pulse duration, as depicted in Fig. 10.2b, the electrons experience a focal intensity gradient which leads to a ponderomotive force. As a result, the electrons gain an energy offset of 1 U_p in addition to the kinetic energy distribution in spatially homogeneous fields (compare Sect. 10.2.1) [38].

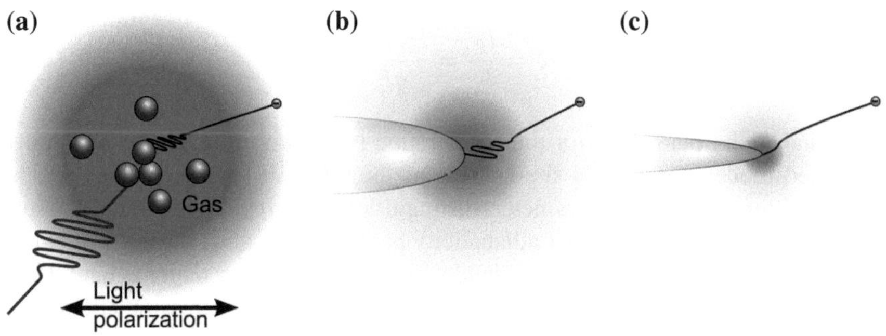

Fig. 10.2 Different regimes in the field-driven acceleration of photoelectrons. **a** In the quiver regime, the optical pulse ends, before the electron has significantly moved out of the optical focus. **b** For more confined fields, the effective interaction time of the electron with the field is limited to few optical cycles. **c** In the sub-cycle regime, the electron escapes the highly confined field within less than one optical half-cycle

In photoemission experiments with very short pulses and diffraction limited foci, the quiver amplitude is much smaller than the decay length of the field, and the photoelectrons do not experience the spatial inhomogeneity during the pulse duration (cf. Fig. 10.2a). Thus, their momentum change Δp is given by the temporal integral of the time-dependent electric field, which is proportional to the vector potential A [39]:

$$\Delta p = -\int_{t_e}^{\infty} eF(t)dt = eA(t_e). \tag{10.4}$$

In this case, the photoelectron energy is shifted compared to the long pulse limit, a phenomenon first observed in above threshold ionization spectra from atoms [40].

Photoemission in highly localized, sub-wavelength fields allows for a third regime, in which electrons rapidly leave the driving field, potentially within a fraction of an optical half-cycle (see Fig. 10.2c). In the extreme, quasi-stationary limit of this *sub-cycle* acceleration regime, the electron kinetic energy reflects the momentary electrical field at the instant of emission:

$$\Delta E = -\int_{x_e}^{\infty} eF(x_1 t_e)dx, \tag{10.5}$$

Where x_e is the position of emission. Sub-cycle electron dynamics can be achieved at long wavelengths, high intensities (corresponding to high ponderomotive energies) and very short decay lengths of the field. The quiver regime (see (10.4)) and the quasi-static regime (see (10.5)) represent two extreme cases of electron propagation in temporally and spatially varying electric fields. The transition between these scenarios can be characterized by a dimensionless *spatial adiabaticity parameter* δ, which compares the corresponding length scales, namely the quiver amplitude $l_q = \frac{eF}{m\omega^2}$ of the photoelectrons and the decay length l_F of the electrical field [11]:

$$\delta = \frac{l_F}{l_q} = \frac{l_F m\omega^2}{eF}. \tag{10.6}$$

This adiabaticity parameter for the acceleration (step two in the above-mentioned simplified model) takes on a similar role as the Keldysh parameter does for the emission (step one), as both parameters distinguish between cycle-integrated and sub-cycle dynamics. For low spatial adiabaticity parameters $\delta \ll 1$, the corresponding quiver amplitude in a homogeneous field would be much larger than the field decay length, leading to sub-cycle acceleration. Similar to sub-cycle emission conditions, this acceleration regime can be reached at high intensities and long wavelengths. However, the Keldysh parameter has a different frequency dependency as the δ-parameter and also depends on the work function of the material, while the spatial adiabaticity parameter is a function of the field decay length. Thus, both parameters are independently adjustable, and strong-field dynamics at nanostructures facilitate ultrafast emission adiabatically following the electric field at $\gamma \ll 1$, with quasi-instantaneous acceleration of the charges at $\delta \ll 1$.

10.2.3 Photoemission from Gold Nanotips Induced by Near- and Mid-infrared Femtosecond Pulses

In order to study strong-field electron dynamics in highly confined fields, we analyze photoemission from single metal nanotips in the near- and mid-infrared range, investigating the effects of field-driven electron dynamics on the photoelectron energy spectra [11]. The experimental setup is depicted in Fig. 10.3a. A Ti:Sapphire laser amplifier system provides 30-fs pulses at 800 nm wavelength, and tunable optical pulses from the near- to the mid-infrared are produced via optical parametric amplification and difference frequency generation, resulting in an accessible wavelength range between 0.8 and 10 µm. This broad tuning range enables a detailed study of the transition from ponderomotive electron acceleration to field-driven electron dynamics. The laser pulses are focused onto an electrochemically etched gold nanotips with radii of curvature of few tens of nanometers. Photoemission is induced in the enhanced field at the tip apex for polarization parallel to the tip axis. The kinetic energy spectra of the photoelectrons are measured using a multichannel-plate (MCP) detector combined with a retarding field analyzer.

Figure 10.3b shows examplary spectra for increasing laser intensities at a fixed excitation wavelength of 3.8 µm for two tips with different radii of curvature. The tips are moderately biased to −40 V to optimize the detection efficiency. In Fig. 10.3b, the additional bias potential is subtracted from the final kinetic energy. For growing intensities, we observe a strong increase in the maximum electron kinetic energy up to several hundred electron volts. Corresponding to the absorption of about one thousand photon energies by a single electron, it is clear that a description in terms of multiphoton orders is no longer practical for this field-driven process.

10.2.3.1 Wavelength-Scaling of the Near-Field Electron Acceleration

We investigate the wavelength-dependence of the photoelectron dynamics by recording kinetic energy spectra for varying excitation wavelengths at sub-100 fs pulse durations as measured with interferometric photocurrent autocorrelations at the nanotip (see Fig. 10.4b). In these measurements, an identical pair of pulses is interfered on the tip, and the photoemission yield is recorded as a function of the relative pulse delay.

Figure 10.4a displays a set of electron energy spectra for excitation wavelengths between 0.8 and 8.0 µm at constant emitted charge per pulse. In adiabatic strong-

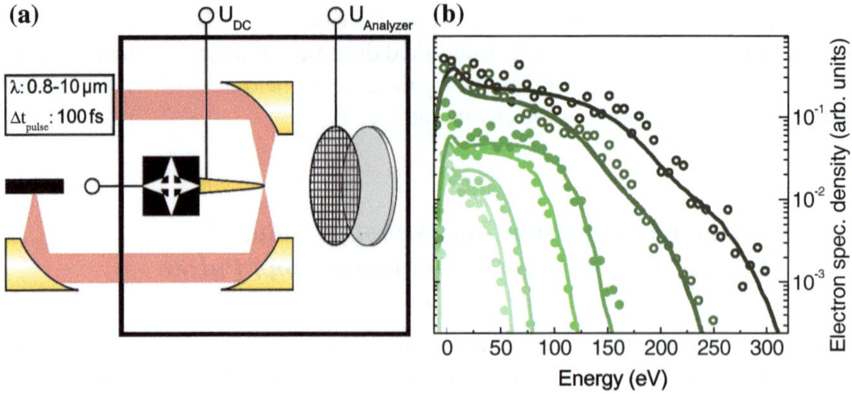

Fig. 10.3 **a** Scheme of the experimental setup. **b** Electron energy spectra for increasing laser intensity recorded at 3.8 μm (adapted from [11]). The measured spectra from two tips with different radii (tip radii: 12 nm (*solid circles*) and 22 nm (*open circles*)) were reproduced by a modified two-step model (*solid lines*)

field emission, the photocurrent is determined by the field strength, and a constant emitted charge at comparable pulse durations is an indication for nearly constant local electric field strengths. In comparing these measurements, we find a saturation in the maximum kinetic energy for long wavelengths, see Fig. 10.4a—in contrast to the quadratic increase of kinetic energy as expected from pure ponderomotive energy scalings (compare Sect. 10.2.1). This observation can be directly attributed to sub-cycle electron acceleration in the optical near-field. According to (10.6), the quiver amplitude of electrons in a homogeneous electromagnetic field increases with the wavelength due to the longer acceleration periods. In the sub-cycle regime (see Sect. 10.2.2), the quiver amplitude begins to exceed the decay length of the near-field at the tip apex, which in a first approximation can be taken as a weak function of wavelength. The resulting saturation of the maximum kinetic energy indicates electron dynamics driven by the momentary electric field.

For a quantitative analysis of the wavelength-scaling, we evaluate the spectral high-energy cutoff, represented by the energy exceed by the fastest 1 % of the electron distribution. Figure 10.5a shows the cutoff energies as a function of wavelength for two different intensities. At higher intensity, the transition to sub-cycle acceleration dynamics is found to be steeper as a result of the increased quiver amplitude for high intensities. The observations are reproduced by a modified two-step model for the calculation of kinetic energy distributions, and the numerical cutoff energies are presented in Fig. 10.5a. The gray shaded area is formed by the cutoff energies of direct (lower curve) and of elastically rescattered (upper curve, 100 % rescattering probability) electrons. The experimental data are found within this corridor.

In homogenous fields, the cutoff energy of rescattered electrons typically exceeds the cutoff of direct electron by a factor of 5 (compare Sect. 10.2.1). Here, we find that with increasing wavelength, the cutoff energies of direct and rescattered elec-

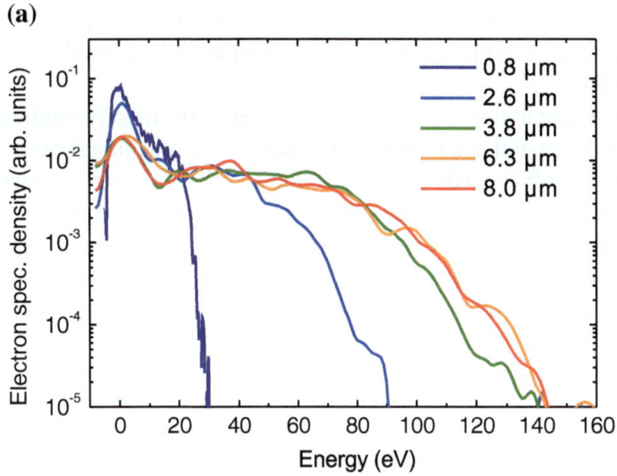

Fig. 10.4 **a** Electron energy spectra for varying wavelengths. The cutoff energy saturates for wavelengths above 3.8 μm. **b** Photocurrent autocorrelation measurements for several wavelengths (adapted from [11])

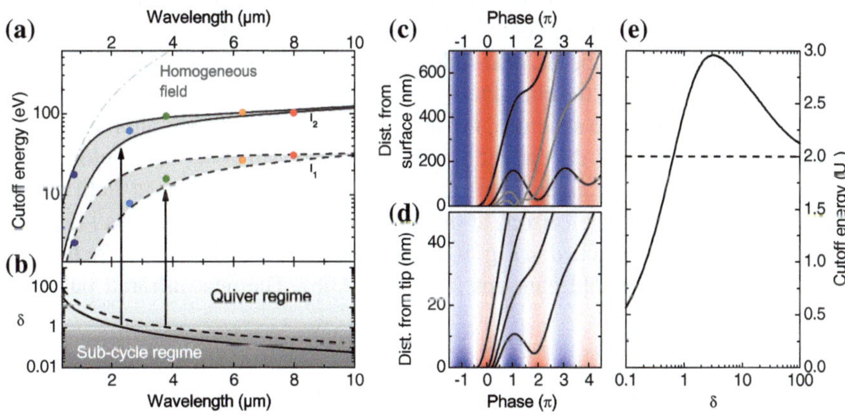

Fig. 10.5 High energy cutoff in a spatially decaying electric field. **a** Experimental cutoff energies for varying wavelength at two different local intensities ($I_1 = 5.4$ TWcm^{-2} and $I_2 = 40$ TWcm^{-2}). The *gray areas* are limited by simulations of the cutoff energy for complete rescattering (*upper bound*) and without rescattering (*lower bound*). **b** δ-Parameter associated to the experimental data. **c** Electron trajectories in a spatially homogeneous field. **d** Electron trajectories in an electric field with a decay length in the order of the electron quiver amplitude. **e** Simulated maximum energy of the directly emitted photoelectrons as a function of the δ-parameter (*solid black line*). *Dashed line* Homogeneous case ($\delta \to \infty$) shown for comparison (**a–d** adapted from [11])

trons converge as a result of reduced back-acceleration towards the surface in the sub-cycle regime. Figure 10.5c, d illustrate the characteristic electron trajectories in spatially homogeneous and highly inhomogeneous fields, respectively. Whereas in

Fig. 10.5c, a large fraction of trajectories shows backscattering at the metal surface, the same emission times result in trajectories for which the photoelectrons directly leave the field in Fig. 10.5d.

Figure 10.5e presents a calculation of the maximum energy of directly emitted electrons as a function of the spatial adiabaticity parameter δ. Approaching the transition at $\delta = 1$, the energy cutoff exceeds $2\,U_p$ due to the additional ponderomotive energy from the gradient force in the near-field. In the limit of sub-cycle field-driven acceleration ($\delta \ll 1$), the energy cutoff decreases below $2\,U_p$ and is finally proportional to the electric field [20]. This scaling may be employed, e.g., to directly map the time-dependent electric near-field at a nanostructure in a streaking scheme, as demonstrated below.

10.2.4 Nanostructure Streaking with Ultrashort THz Pulses

In order to phase-resolve field-driven electron dynamics, we conduct streaking-type experiments with phase-stable single-cycle THz transients and femtosecond NIR pulses. All-optical streaking allows for the mapping of temporal information on, e.g., the kinetic energy of photoelectrons [41, 42]. Numerous schemes and applications for nanostructure-based near-field streaking have been theoretically studied (i.e., [43–46]), discussing the characterization of plasmonic near-fields with attosecond XUV pulses. In these schemes, the attosecond pulses generate photoelectrons, which experience a phase-dependent acceleration in the streaking near-field.

Here, we transfer this concept to longer wavelengths and present near-field streaking of NIR-induced photoelectrons at the nanotip using THz frequencies [12]. The experimental setup is sketched in Fig. 10.6. Both NIR and the THz pulses are focused collinearly onto single gold nanotips (see Fig. 10.6b). The near-infrared pulses of

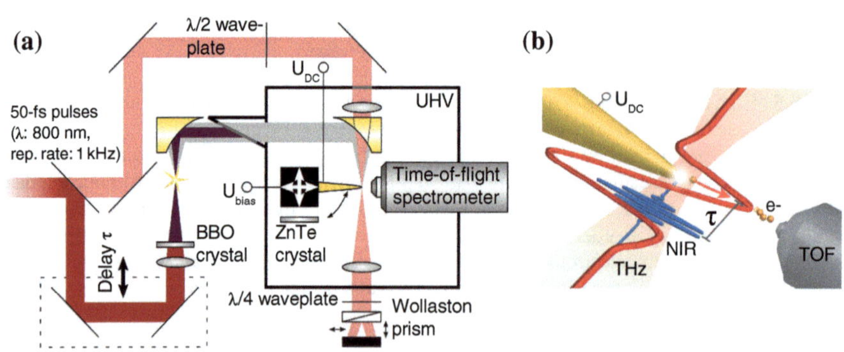

Fig. 10.6 **a** Experimental setup for a streaking experiment at a single nanotip using NIR and THz pulses. **b** Close-up of the streaking scheme (adapted from [12])

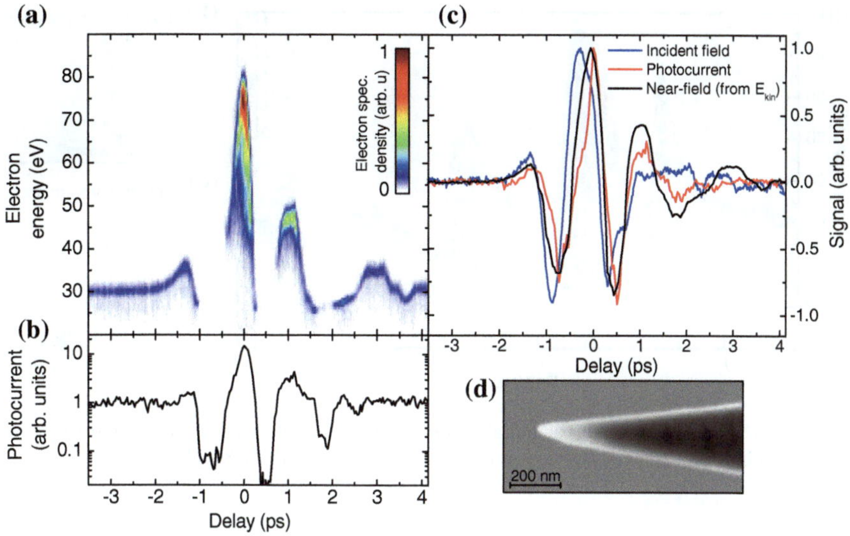

Fig. 10.7 Near-field streaking at a very sharp nanotip (radius of curvature 10 nm) (adapted from [12]). **a** Spectrogram. **b** Photocurrent extracted from spectrogram (**a**). **c** Comparison between the incident electric field and assembled waveforms from the delay-dependent photocurrent and electron kinetic energy. **d** Scanning electron microscope image of the nanotip employed in the measurements

50-fs duration and 800 nm centre wavelength generate photoelectrons, which are subsequently streaked in the tip-enhanced THz electric field.

The THz transients are produced with the AC-bias method in a light-induced air plasma [47, 48]. This method provides phase-stable THz pulses and the control of pulse energy and carrier-envelope-phase. The air plasma is generated with infrared pulses of up to 2 mJ pulse energy, a small fraction of which is frequency-doubled and overlapped with the fundamental beam. The photoelectron spectra are detected with a time-of-flight electron spectrometer or a retarding field analyzer. A series of kinetic energy spectra, recorded as a function of the relative time delay between both pulses, forms a spectrogram as shown, for example, in Figs. 10.7a and 10.8a. The tip is moderately biased to draw the photoelectrons to the detector. We characterize the incident THz transient in-situ at the position of the nanotip using electro-optic sampling in a ZnTe crystal.

10.2.4.1 Experimental Observation of Field-Driven Streaking

A streaking spectrogram at a very sharp gold tip with a radius of curvature of 10 nm is displayed in Fig. 10.7. This spectrogram shows the impact of the THz transient on the electron kinetic energy: The THz electric field accelerates the electrons by 50 eV, resulting in the final energy of 80 eV (bias tip voltage: −30 V). At large

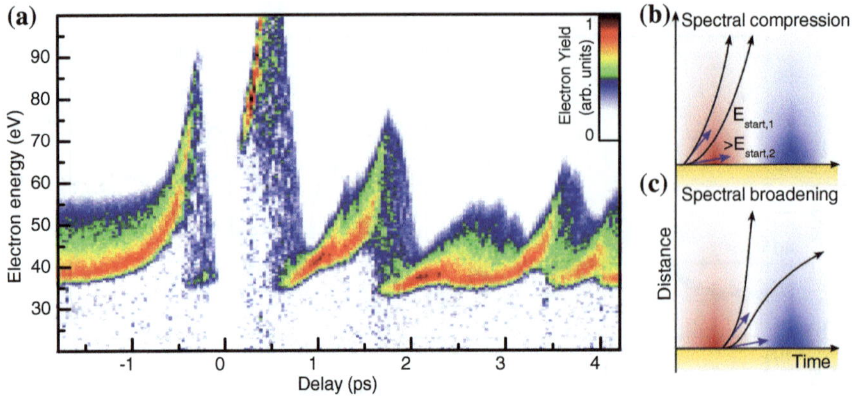

Fig. 10.8 **a** Spectrogram with broad initial energy spectrum and an effective decay length of around 40 nm. **b** Scheme for spectral compression and **c** broadening (adapted from [12])

negative delays, the 800 nm pulse precedes the THz transient, and the photoelectrons are not affected by the THz field. Thus, the electron energy is determined only by the static bias voltage. Due to the rectifying effect of the field emission geometry, the photocurrent is suppressed in the half-cycles of the THz electric force pointing towards the tip.

The spectrally integrated photocurrent as a function of the delay is plotted in Fig. 10.7b and shows the current suppression and enhancement depending on the phase of the THz transient, with the enhancement induced by the impact of the THz field on the effective work function (Schottky effect). The peak of the photocurrent is in-phase with the maximum of the kinetic energy. This observation represents the limiting case of acceleration by a momentary electric field. Electrons emitted at the peak field are quasi-statically accelerated ($\delta \ll 1$) and leave the enhanced near-field with minimal temporal integration.

Due to the suppression of the photocurrent in half-cycles of negative electric force, the spectrogram contains only information about the accelerating half-cycles of the THz near-field. Thus, the THz transient with opposite polarity is measured in addition, in order to characterize the entire continuous THz waveform. In the experiment, the polarity of the THz transient can be inverted by a rotation of the BBO crystal to obtain the complementary spectrogram [12]. The energy expectation value of the assembled waveform[1] is shown in Fig. 10.7c. This curve (black) represents the local THz waveform. A comparison with the incident electric field, characterized via electro-optic sampling (blue), reveals the time-domain response of the nanostructure. We find subtle waveform modulations and a slight phase shift. These effects can be attributed to the electromagnetic response of the nanotip and are reproduced in an antenna circuit model [49, 50].

[1]The bias potential is subtracted and the spectrograms of opposing polarity are combined on a common energy scale.

10.2.4.2 Controlling Electron Trajectories in Time-Varying Near-Fields

At the transition to the sub-cycle acceleration regime, the electron dynamics sensitively depend on the initial electron energy and the temporal slope of the electric field. Figure 10.8a shows a spectrogram at a nanotip with a decay length longer than in the previous spectrograms, and with a broader initial energy spectrum induced by the NIR pulse. Here, the width of the kinetic energy spectra is modulated by the slope of the THz field. We observe that streaking by a temporally decreasing electric field leads to a broadening of the kinetic energy spectra, whereas an increasing streaking field causes spectral compression.

The initial energy of the photoelectrons at the instant of emission determines the interaction time with the streaking field. Electrons with low kinetic energy are accelerated over longer time intervals as compared to faster electrons. Therefore, slow electrons are more sensitive to the temporal evolution of the streaking field. For increasing electric fields, initially slower electrons gain more kinetic energy than their faster counterparts. As a result, the kinetic energy distribution can be compressed, cf. Fig. 10.8b. Accordingly, decreasing streaking fields (see Fig. 10.8c) result in a broadening of the initial spectra.

The dependence of electron trajectories on the slope of the streaking field enables the control of electron spectra and electron pulse durations via the relative delay between NIR and THz pulses. The presented spectral compression could be employed for the generation of optimized electron spectra for ultrafast optical-pump/electron-probe experiments.

10.2.4.3 Modeling THz Streaking at a Nanotip

The streaking spectrograms are simulated by classical particle propagation in the temporally and spatially varying THz streaking field as shown in Fig. 10.9. Due to the different pulse durations of NIR and THz transients, quasi-instantaneously generated electron spectra are propagated in the THz field as described in Sect. 10.2.3. The electron trajectories in the temporally- and spatially-varying streaking near-field are numerically integrated for different initial electron velocities. This model enables a detailed reproduction of the main features of the spectrograms, especially the modulation of the spectral width.

The basic principle of such spectral reshaping can be analytically described with a model presented in the following. We derive the final kinetic energy E_{final} of a photoelectron propagating in an exponentially decaying near-field $F(x,t) = F_0(t)\exp(-x/l_F)$ (field decay length l_F) and in the linear temporal slope of the streaking transient. The final energy E_{final} for a electron with the initial energy E_0 at $x_e = 0$ is given by the path integral over the spatially varying time-dependent field:

$$E_{final} = E_0 - \int_0^\infty eF(t,x)dx. \tag{10.7}$$

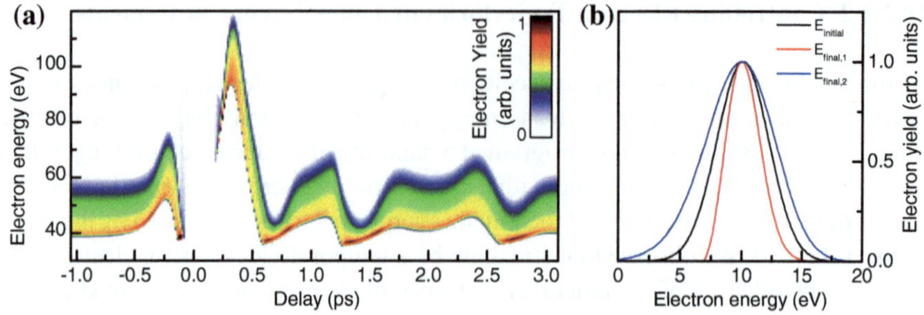

Fig. 10.9 **a** Numerical Simulation of a spectrogram with broad initial energy distribution and a with a decay length of around 40 nm (adapted from [12]). **b** Spectral reshaping computed with analytical model. Increasing and decreasing field strength lead to compression (*red*, $E_{final,1}$) and broadening (*blue*, $E_{final,2}$) of the initial energy spectrum (*black*). The spectra are shifted to a common peak energy for comparison

For the integration, we parametrize the time as a function of space:

$$t(x) = t_e + \int_0^x \frac{dx'}{v(x')}, \tag{10.8}$$

where t_e is the time of photoemission. In the first order, the velocity $v(x)$ is determined by the kinetic energy in the static electric field $F(x) = F_0(t_e)\exp(-x/l_F)$:

$$v(x) = \sqrt{\frac{2}{m}\left[E_0 - el_F F_0(t_e) + el_F F_0(t_e)\exp\left(-\frac{x}{l_F}\right)\right]}. \tag{10.9}$$

The temporal evolution of the electric field is introduced by the Taylor expansion of linear order facilitating the integration:

$$F(t,x) = \left(F_0(t_e) + (t - t_e)\cdot \left.\frac{\partial F_0}{\partial t}\right|_{t=t_e}\right)\exp\left(-\frac{x}{l_F}\right). \tag{10.10}$$

This expression enables the analytical solution of the integral (10.7), yielding the following term for the final kinetic energy:

$$E_{final} = E_0 - eF_0(t_e)l_F - e\left.\frac{\partial F}{\partial t}\right|_{t=t_e}\frac{\sqrt{2m}l_F^2}{\sqrt{E_0 - eF_0(t_e)l_F} + \sqrt{E_0}}. \tag{10.11}$$

The average velocity of the electron in the near-field is $v = \left(\sqrt{E_0 - eF_0(t_e)l_F} + \sqrt{E_0}\right)/\sqrt{2m}$. The energy gain depends on the initial energy and the width of an energy distribution is—under the present assumptions—proportional to the slope of the THz transient, as illustrated in Fig. 10.9.

In this section, we discussed strong-field photoemission in the vicinity of nanostructures and identified sub-cycle electron dynamics at near- and mid-infrared as well as THz frequencies. The extreme case of field-driven electron acceleration is characterized by a spatial adiabaticity parameter $\delta \ll 1$. The field confinement affects all relevant scalings, including wavelength- and intensity-dependent cutoff behaviors and the spectral reshaping of photoelectron spectra in streaking configurations. Some of these features may be employed in the ultrafast and high-resolution characterization of optical near-fields, while others, such as the THz trajectory control, may be utilized in the temporal compression of ultrashort electron pulses for time-resolved electron imaging and diffraction.

10.3 Extreme-Ultraviolet Light Generation in Plasmonic Nanostructures

In recent years, nanoscale plasmonic light confinement in metallic nanostructures has been extensively used to study numerous linear and nonlinear optical phenomena in strongly-enhanced, inhomogeneous fields [51–53]. Specifically, high intensity enhancements in tailored structures, such as sharp needles, resonant antennas or waveguides, have enabled the investigation of highly nonlinear effects, including strong-field photoemission and acceleration (see Sect. 10.2), in nanostructure-based implementations.

Another prominent example for this development is the generation of extreme-ultraviolet (EUV) light in gaseous media by utilizing bow-tie antennas and tapered hollow waveguides [13–19]. Figure 10.10 schematically illustrates the concept of such approaches. In particular, ultrashort low-energy laser pulses are enhanced in nanometric hot spots leading to orders of magnitude higher local intensities, either between the triangular tips (see Fig. 10.10a) or at the very end of the conical taper

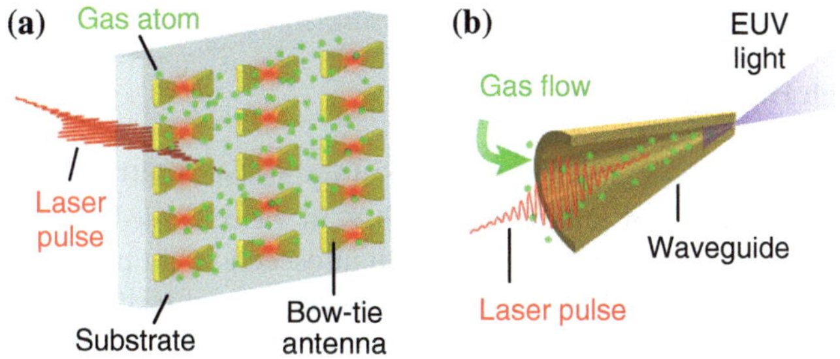

Fig. 10.10 Schematic illustration of plasmon-enhanced gas excitation and EUV light generation by using **a** bow-tie nanoantennas and **b** tapered hollow waveguides

(see Fig. 10.10b). Noble gas atoms can be injected into these hot spots and get excited in the strong near-fields, resulting in the emission of EUV photons.

The very first reports about a successful realization of these experimental schemes [13, 14] suggested that the produced radiation originates from high-order harmonic generation (HHG) [34, 54, 55] and, therefore, raised high expectations that such compact sources could be harnessed for an eased and efficient generation of attosecond EUV pulses [56–61]. If correct, this scheme would enable the integration of attosecond science into the domain of nonlinear nano-optics, with potential benefits for ultrafast spectroscopy [62, 63] or coherent nanoscale photonic imaging [64].

However, despite particular efforts to reproduce plasmon-enhanced HHG [15, 17–19], an independent experimental verification of this effects was not reported up to date. Recently, our studies revealed that the concept of HHG in nanostructures is fundamentally flawed [15, 17, 18]. Specifically, the tiny nanostructure volume displaying field enhancement is found to prohibit an efficient generation of coherent radiation. Whereas local intensities high enough for the nonlinear excitation and ionization of atoms are reached, we could show that the dominant radiation process is not in fact HHG, but incoherent atomic and ionic fluorescence.

In the following, we present the main results of our comprehensive study on EUV light generation in plasmonic nanostructures. Our experiments and spectral analysis clearly evidence that the EUV emission seen under the reported conditions is predominantly caused by incoherent fluorescence stemming from multiphoton and strong-field gas excitation and ionization.

10.3.1 Strong-Field EUV Light Generation from Gas Atoms

Currently, the key approach to obtain access to EUV and soft X-ray wavelengths on laboratory scales is the conversion of intense visible or infrared laser radiation in gaseous media. Two distinct excitation processes are responsible for the generation of EUV light in gas atoms. In particular, ultrashort high-energy laser pulses can cause the emission of highly-energetic photons via nonlinear excitation of gas atoms accompanied by either high-order harmonic generation or fluorescence.

10.3.1.1 High Harmonic Generation

High-order harmonic generation—first realized in the late 1980' [54, 55]—is the frequency up-conversion of intense laser pulses in gaseous media and has paved the way to lab-scale sources enabling coherent attosecond light pulses at EUV and soft-X-ray wavelengths. In short, the physical mechanism of HHG in a single atom picture can be summarized in a semi-classical simple man's model [34]: (1) A linearly polarized, ultrashort (pico- to femtosecond) laser pulse with sufficient peak intensity (typically in excess of 10^{13} W/cm^2 for visible or infrared wavelengths) ionizes a gas atom, creating a quasi-free electron. A part of the electrons' wavefunction stays in the remaining ion. (2) The electron is accelerated in the oscillating strong laser field (with frequency ω_0 and intensity I), gaining additional ponderomotive energy $U_p \propto I/\omega_0^2$. (3) The electron returns to its parent ion and causes the emission of high-energetic photons upon re-collision. As a result of this coherent process only radiation at odd multiples (harmonics) of the driving frequencies ω_0 is generated in each half cycle of the laser pulse [65]. The HHG process retains the optical and temporal properties of the driving laser pulses and, thus, has the potential to enable EUV attosecond pulses [66, 67].

Beyond these unique characteristics, it has to be noted that the HHG process is rather inefficient due to a low re-collision probability of the electrons with their parent ions [68]. In general, sufficient photon fluxes are only obtainable by considering the coherent buildup of the harmonic radiation over a large propagation distance in a mesoscopic (tens to hundreds of micrometers or even millimeters) generation volume. Specifically, the high harmonic output scales quadratically with the pressure-length product [55]. However, such a coherent radiation buildup is only obtainable under ideal—phase-matched—conditions that are influenced by propagation effects like dephasing between the driving and the generated radiation or absorption and defocusing [65].

To date, state-of-the-art experimental concepts for efficient HHG typically rely on exposing gas filled capillaries or hollow waveguides to ultrashort, intense laser pulses, either by employing chirped-pulse amplification [69, 70] or enhancement cavities [71, 72] at kHz and MHz repetition rates, respectively.

10.3.1.2 EUV Fluorescence

In addition to high harmonic radiation, the excitation of gas atoms in intense light fields can also cause fluorescent light emission due to electronic transitions stemming from multiphoton absorption, strong-field ionization and plasma formation [73–75]. The characteristic energy of a fluorescent photon is given by the transitions between different atomic or ionic states. At EUV wavelengths, the photon energies are on the order of several tens of electron volts (eV), which implies that, in the case of excitation with infrared light frequencies (photon energies below 2 eV) multiple photon absorption is required. Hence, the process exhibits an intensity threshold of about 10^{13} W/cm^2, which corresponds to the intensities required for HHG.

In contrast to HHG, the emission characteristics of fluorescence are governed by the intrinsically incoherent nature of the generation process with long spontaneous lifetimes of the excited energy levels. Therefore, the properties of the driving laser pulses are lost, and the fluorescent radiation is temporally and spatially incoherent.

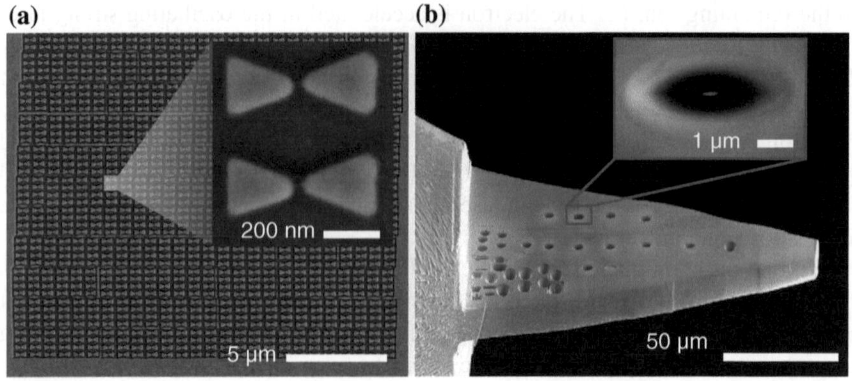

Fig. 10.11 Scanning electron microscope images of the used nanostructures. **a** An array of bow-tie nanoantennas fabricated on the smooth gold film of 100 nm thickness deposited on a sapphire plate. *Inset* shows a close-up view. **b** Tapered waveguides milled into a 10 μm thick gold plateau at the end of a conically etched gold wire. *Inset* shows a top-view of an individual waveguide with elliptical cross section

10.3.2 Experimental Methods

The experiments conducted in this study are based on the excitation of rare gas atoms by using plasmonic enhancement of ultrashort low-energy laser pulses in tailored metallic nanostructures (see Fig. 10.10). The particular nanostructures used are resonant bow-tie antennas on sapphire substrates and tapered hollow waveguides milled into gold bulk supports (cf. scanning electron microscope images in Fig. 10.11). Both types of structures are fabricated via focused ion beam (FIB) lithography. Compared to other nano-fabrication techniques, such as electron beam lithography, the used FIB milling represents a convenient method to produce prototype-like structures with very high control during the fabrication process. Therefore, by using the FIB technique and starting with high quality supports, i.e. smooth gold films on sapphire for the antennas and smooth gold platforms for the waveguides, it is possible to optimize the quality of the structures with respect to a high field-enhancement [56] and thermal stability upon laser irradiation [76].

The experimental setup used to investigate the EUV light generation in the nanostructures is schematically depicted in Fig. 10.12. It consists of a vacuum generation chamber and a flat-field EUV spectrometer. Optical illumination is provided with a 78 MHz Ti:sapphire laser oscillator delivering 8-fs, nJ pulses at a centre wavelength of 800 nm. The generated EUV radiation is collected within a solid angle of $\pm 1.2°$ and spectrally resolved with a flat-field spectrometer. Detection of the EUV photons is realized with an imaging microchannel-plate (MCP) detector in combination with a CCD camera. The inset of Fig. 10.12a shows an exemplary flat-field image for a measurement with xenon gas.

The excitation schemes for the measurements using arrays of bow-tie antennas and single tapered hollow waveguides are depicted in Fig. 10.12b, c, respectively.

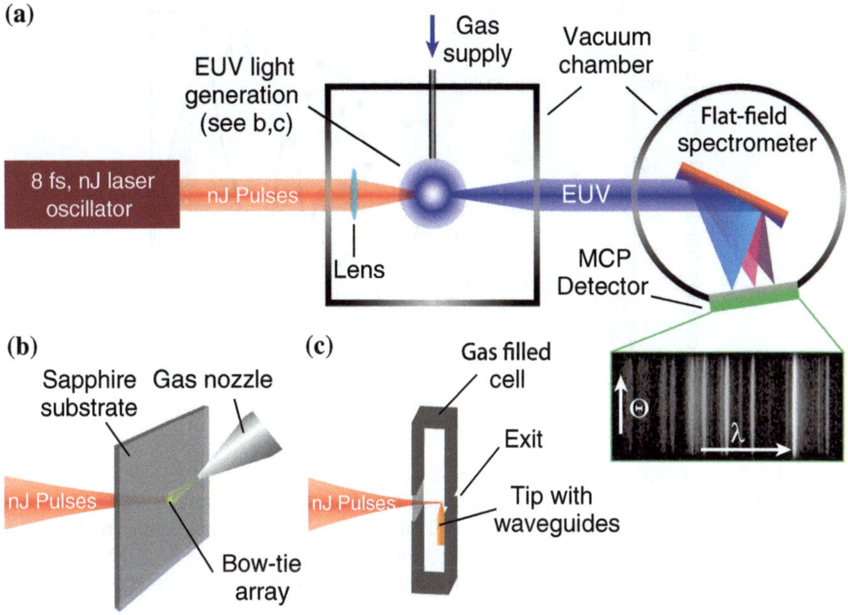

Fig. 10.12 Experimental setup for plasmon-enhanced EUV light generation. **a** Schematic of the vacuum setup consisting of a generation chamber and an EUV flat-field spectrometer. The spectra (xenon fluorescence spectrum in the *inset*) are detected with a microchannel-plate detector (MCP) in combination with a CCD camera. **b**, **c** Excitation schemes for the measurements using **b** resonant bow-tie antennas and **c** tapered waveguides

In both cases, the low-energy pulses are tightly focused on the respective specimen, enabling incident intensities in excess of 0.1 TW/cm^2. Noble gas atoms are exposed to the plasmon-enhanced near-fields of the particular nanostructures and are efficiently excited (see Fig. 10.10). In the measurements with the bow-tie antennas, the gas atoms are supplied via a nozzle facing the structures, whereas the hollow waveguides are placed in a purged gas cell with a small exit aperture (200 μm) for the collection of the generated EUV radiation. Backing gas pressures up to 500 mbar were applied in both scenarios.

10.3.3 Results and Discussion

Figure 10.13 shows an exemplary plasmon-enhanced EUV spectrum (solid black curve) obtained from argon using an array of bow-tie antennas (structure *iv* in Fig. 10.14a). The most prominent spectral features (higher spectral resolution is provided in the second grating diffraction order; red curve) are identified to be fluorescence from neutral and singly ionized argon atoms, with a signal-to-noise ratio

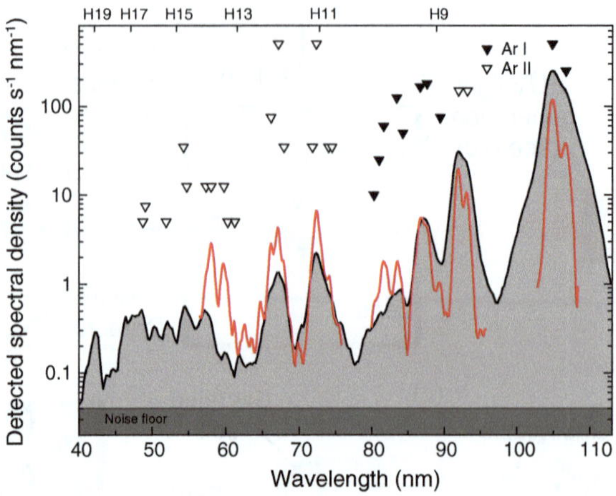

Fig. 10.13 Extreme-ultraviolet spectrum from argon gas atoms excited via plasmonic field-enhancement in an array of bow-tie nanoantennas (see Fig. 10.14a structure *iv*). *Solid black/red curves* represent spectra taken in the frist/second diffraction orders of the used spectrometer grating. Fluorescence line positions for neutral/singly ionized argon atoms are indicated by *filled/open triangles*. *Upper scale* indicates the positions for expected high-harmonic orders (H9–H19). Graph adapted from [15]

larger than 10^3. The exact wavelength positions of expected atomic and ionic fluorescence lines are indicated by the triangles [77]. Although conditions for strong-field ionization are clearly reached, evidenced by significant fluorescence signals from singly ionized atoms, no coherent HHG signals were detectable at harmonic orders (H9-H19 wavelengths indicated in the upper x-axis of Fig. 10.13) of the driving laser frequency.

Figure 10.14 shows an extended set of similar measurements for argon and xenon gas, using several bow-tie geometries (see Fig. 10.14a) with different plasmonic resonances. The resonances are determined by analyzing dark-field scattering spectra, which are shown in Fig. 10.14b for the shortest and longest bow-tie antennas. In agreement with the previous measurements, all spectra shown in Fig. 10.14c, d exclusively originate from fluorescent emission without any signature of coherent high harmonic radiation. Due to the characteristic energy transitions of xenon, the spectral features shown in Fig. 10.14d are completely different compared to those of the argon spectra. Whereas the overall spectral shape is qualitatively comparable for measurements with the same gas species, the strength of the fluorescence signal is governed by the plasmon resonances and the structural quality of the different bow-tie antennas.

Further experiments with tapered hollow waveguides instead of bow-tie antennas yield results which also indicate the dominance of fluorescent over high-harmonic emission in plasmon-enhanced implementations. Figure 10.15 presents the EUV spectra from argon, xenon and neon gas, which were all obtained by using the waveguide structure shown in the inset of Fig. 10.11b. Notably, here, the recorded sig-

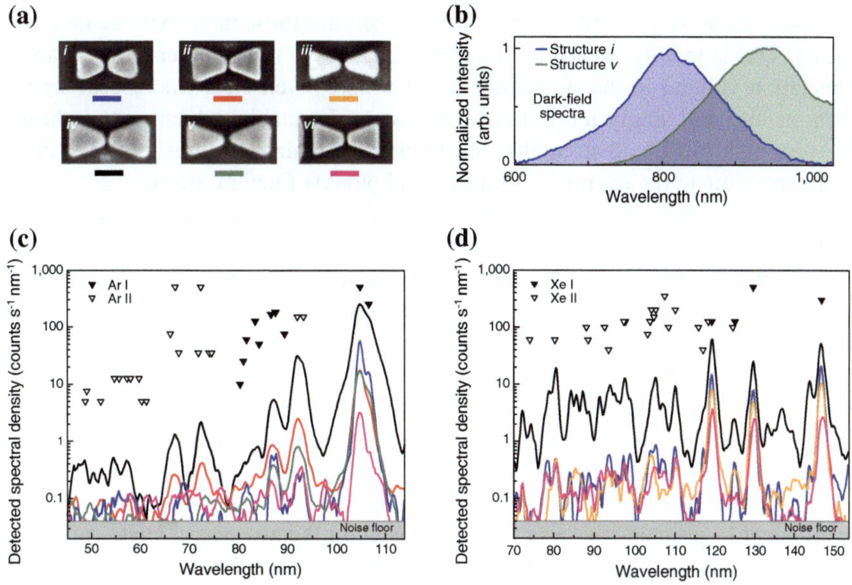

Fig. 10.14 Plasmon-enhanced EUV spectra from argon and xenon with bow-tie antennas. **a** Scanning electron microscope images of individual bow-tie antennas. *Color-coded scale bars* 200 nm. **b** Dark-field scattering spectra of two antenna arrays with different bow-tie arm lengths. **c, d** EUV spectra recorded with different bow-tie antennas (indicated by the *color code* see scale bars in **a**) for **c** argon and **d** xenon gas. Graphs and images adapted from [17]

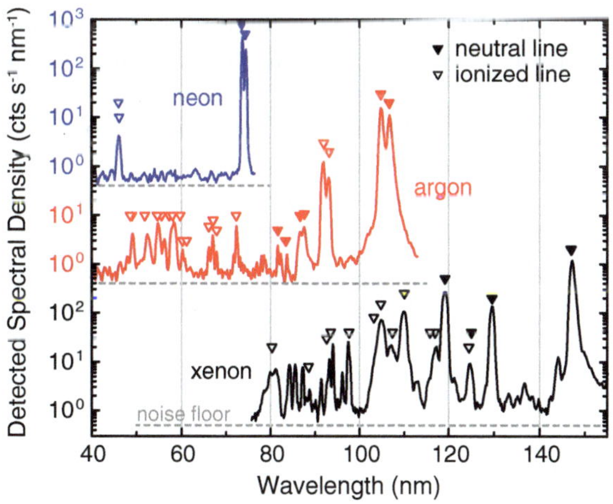

Fig. 10.15 Plasmon-enhanced EUV spectra from noble gases (argon, xenon and neon) using a tapered hollow waveguide (cf. Fig. 10.11b). The argon and neon spectra are up-shifted for better visibility. *Dotted gray lines* indicates the noise floor at 0.5 count/s/nm. The wavelength positions of the expected fluorescence lines from neutral and singly ionized atoms are indicated by the *triangles*. Graph adapted from [18]

nal levels are about one order of magnitude larger than those in the bow-tie measurements. This is mainly caused by the approximately 10 times larger field-enhanced generation volume in the waveguide. Furthermore, even clear fluorescent signals from neutral and singly ionized neon atoms are detectable, which is an indicator for very high local field strengths, considering the ionization threshold of 21.5 eV (corresponding to the energy of 13 infrared photons) of neutral neon.

The above results show that strong-field EUV light generation in plasmonic nanostructures is generally possible under the examined conditions. However, the spectra also clearly identify the origin of the EUV emission as incoherent fluorescence stemming from multiphoton excitation and strong-field ionization of the gas atoms. Previously reported high-harmonic emission in such implementations [13, 14, 16] could not be observed, although the experiments were reproduced multiple times using different nanostructure geometries, gas pressures and incident laser intensities. Accurate detection alignment with respect to an efficient collection of coherent signals was proven by surface-enhanced low-order harmonic generation [17, 18, 78] in the bare nanostructures (without gas). Special care was also taken to rule out insufficient local intensities as a possible reason for the absence of HHG signals. A novel means to gauge local fields by employing intensity-dependent relative fluorescence line strengths in the measured spectra was used to determine the intensity enhancements within the nanostructures [17]. Therefore, local intensities of up to 60 TW/cm^2, generally sufficient for HHG, were confirmed. Furthermore, the observation of significant fluorescence signals, in particular from singly ionized gas atoms, over several hours of operation evidence that the structures remain intact upon incident laser intensities up to 0.15 TW/cm^2 [17].

A physical reason for the lack of HHG in these experiments can be found by considering the nanometric generation volume in the used nanostructures. Specifically, in implementations based on localized surface plasmons, the very small excited gas volume prohibits an efficient coherent build up of the harmonic radiation. In other words, the quadratic scaling of the HHG output with the number of emitting dipoles—or more generally with the pressure-length product—is diminishing the likelihood to observe any significant coherent contributions in the generated EUV emission.

Taking this attribute of the coherent HHG process into account, estimates have shown that the high-harmonic signal levels in these plasmon-enhanced implementations are expected to be several orders of magnitude below that of the simultaneously excited incoherent fluorescence [17, 79]. Thus, it will be very difficult to discriminate potential coherent signals from the dominant fluorescence without employing special detection procedures.

The findings in this Section constitute a clear setback for plasmon-assisted HHG, but they also illustrate that other highly nonlinear effects, such as strong-field gas ionization and fluorescence, or low-order harmonic generation, are present. As a further example for such an effect, we have found a new kind of optical bistability [80–82] in the waveguide experiments. In particular, an intensity-dependent hysteresis in the fluorescence signal strength indicates the formation of a nano-plasma in the waveguides [18]. All of these effects warrant deeper investigation and could help to determine the ultimate frontiers of highly nonlinear strong-field nano-optics.

However, the severe limitations for HHG in such implementation calls for alternative approaches to re-open this important research field with respect to coherent EUV signal generation.

References

1. P. Hommelhoff et al., Field emission tip as a nanometer source of free electron femtosecond pulses. Phys. Rev. Lett. **96**, 077401 (2006)
2. M. Krüger et al., Attosecond control of electrons emitted from a nanoscale metal tip. Nature **475**, 78–81 (2011)
3. P. Dombi et al., Ultrafast strong-field photoemission from plasmonic nanoparticles. Nano Lett. **13**, 674–678 (2013)
4. G. Herink et al., Field emission at terahertz frequencies: AC-tunneling and ultrafast carrier dynamics. New Journal of Physics **16**, 123005 (2014)
5. S.M. Teichmann et al., Strong-field plasmonic photoemission in the mid-IR at < 1 GW/cm2 intensity. Sci. Reports **5** (2015)
6. C. Ropers et al., Localized multiphoton emission of femtosecond electron pulses from metal nanotips. Phys. Rev. Lett. **98**(4), 043907 (2007)
7. M. Schenk et al., Strong-field above-threshold photoemission from sharp metal tips. Phys. Rev. Lett. **105**(25), 257601 (2010)
8. R. Bormann et al., Tip-enhanced strong-field photoemission. Phys. Rev. Lett. **105**(14), 147601 (2010)
9. S. Yalunin et al., Strong-field photoemission from surfaces: theoretical approaches. Phys. Rev. B. **84**, 195426 (2011)
10. S. Zherebtsov et al., Controlled near-field enhanced electron acceleration from dielectric nanospheres with intense few-cycle laser fields. Nat. Phys. **7**(8), 656–662 (2011)
11. G. Herink et al., Field-driven photoemission from nanostructures quenches the quiver motion. Nature **483**(7388), 190–193 (2012)
12. L. Wimmer et al., Terahertz control of nanotip photoemission. Nat. Phys. **10**(6), 432–436 (2014)
13. S. Kim et al., High-harmonic generation by resonant plasmon field enhancement. Nature **453**(7196), 757–760 (2008)
14. I.-Y. Park et al., Plasmonic generation of ultrashort extreme-ultraviolet light pulses. Nat. Photonics **5**(11), 677–681 (2011)
15. M. Sivis et al., Nanostructure-enhanced atomic line emission. Nature **485**(7397), E1–E2 (2012)
16. I.-Y. Park et al., Generation of EUV radiation by plasmonic field enhancement using nanostructured bowties and funnel-waveguides. Ann. Phys. **525**(1–2), 87–96 (2013)
17. M. Sivis et al., Extreme-ultraviolet light generation in plasmonic nanostructures. Nat. Phys. **9**(5), 304–309 (2013)
18. M. Sivis, C. Ropers, Generation and bistability of a waveguide nanoplasma observed by enhanced extreme-ultraviolet fluorescence. Phys. Rev. Lett. **111**(8), 085001 (2013)
19. N. Pfullmann et al., Bow-tie nano-antenna assisted generation of extreme ultraviolet radiation. New J. Phys. **15**(9), 093027 (2013)
20. S.V. Yalunin et al., Field localization and rescattering in tip-enhanced photoemission. Ann. Phys. **525**(1–2), L12–L18 (2013)
21. A. Einstein, Über einen die Erzeugung und Verwandlung des Lichtes betreffenden heuristischen Gesichtspunkt. Ann. Phys. **322**(6), 132–148 (1905)
22. L.V. Keldysh, Ionization in the field of a strong electromagnetic wave. Sov. Phys. JETP **20**, 1307–1314 (1965)
23. B. Barwick et al., Laser-induced ultrafast electron emission from a field emission tip. New J. Phys. **9**(5), 142 (2007)

24. R.H. Fowler, The analysis of photoelectric sensitivity curves for clean metals at various temperatures. Phys. Rev. **38**, 45–56 (1931)
25. L.A. DuBridge, Theory of the energy distribution of photoelectrons. Phys. Rev. **43**, 727–741 (1933)
26. L.A. DuBridge, R.C. Hergenrother, The effect of temperature on the energy distribution of photoelectrons. i. normal energies. Phys. Rev. **44**, 861–865 (1933)
27. P. Agostini et al., Free-free transitions following six-photon ionization of xenon atoms. Phys. Rev. Lett. **42**, 1127–1130 (1979)
28. F. Banfi et al., Experimental evidence of above-threshold photoemission in solids. Phys. Rev. Lett. **94**, 037601 (2005)
29. R. Kopold et al., Channel-closing effects in high-order above-threshold ionization and high-order harmonic generation. J. Phys. B: Atom. Mol. Opt. Phys. **35**(2), 217 (2002)
30. W. Schottky, Über kalte und warme Elektronenentladungen. Z. Angew. Phys. **14**(1), 63–106 (1923)
31. R. Gomer. *Field Emission and Field Ionization* (Harvard University Press, Cambridge, 1961)
32. R.H. Fowler, L. Nordheim, Electron emission in intense electric fields. R. Soc. Lond. Proc. Ser. A **119**, 173–181 (1928)
33. G. Farkas et al., Influence of optical field emission on the nonlinear photoelectric effect induced by ultrashort laser pulses. Phys. Lett. A **39**(3), 231–232 (1972)
34. P.B. Corkum, Plasma perspective on strong field multiphoton ionization. Phys. Rev. Lett. **71**(13), 1994 (1993)
35. J.L. Krause et al., High-order harmonic generation from atoms and ions in the high intensity regime. Phys. Rev. Lett. **68**, 3535–3538 (1992)
36. G.G. Paulus et al., Rescattering effects in above-threshold ionization: a classical model. J. Phys. B: Atom. Mol. Opt. Phys. **27**(21), L703 (1994)
37. G.G. Paulus et al., Plateau in above threshold ionization spectra. Phys. Rev. Lett. **72**, 2851–2854 (1994)
38. F. Kelkensberg et al., Attosecond streaking in a nano-plasmonic field. New J. Phys. **14**(9), 093034 (2012)
39. J. Itatani et al., Attosecond streak camera. Phys. Rev. Lett. **88**, 173903 (2002)
40. P. Agostini et al., Direct evidence of ponderomotive effects via laser pulse duration in above-threshold ionization. Phys. Rev. A **36**, 4111–4114 (1987)
41. E. Goulielmakis et al., Direct measurement of light waves. Science **305**(5688), 1267–1269 (2004)
42. E. Goulielmakis et al., Attosecond control and measurement: lightwave electronics. Science **317**(5839), 769–775 (2007)
43. M.I. Stockman et al., Attosecond nanoplasmonic-field microscope. Nat. Photon **1**(9), 539–544 (2007)
44. E. Skopalová et al., Numerical simulation of attosecond nanoplasmonic streaking. New J. Phys. **13**(8), 083003 (2011)
45. F. Süßmann, M.F. Kling, Attosecond nanoplasmonic streaking of localized fields near metal nanospheres. Phys. Rev. B **84**, 121406 (2011)
46. A.G. Borisov et al., Attostreaking with metallic nano-objects. New J. Phys. **14**(2), 023036 (2012)
47. D.J. Cook, R.M. Hochstrasser, Intense terahertz pulses by four-wave rectification in air. Opt. Lett. **25**(16), 1210–1212 (2000)
48. H.G. Roskos et al., Broadband thz emission from gas plasmas induced by femtosecond optical pulses: from fundamentals to applications. Laser Photonics Rev. **1**(4), 349–368 (2007)
49. K. Wang et al., Antenna effects in terahertz apertureless near-field optical microscopy. Appl. Phys. Lett. **85**(14), 2715–2717 (2004)
50. R. Kersting et al., Terahertz microscopy with submicrometre resolution. J. Opt. A: Pure Appl. Opt. **7**(2), S184 (2005)
51. J.A. Schuller et al., Plasmonics for extreme light concentration and manipulation. Nat. Mater. **9**(3), 193–204 (2010)

52. M. Kauranen, A.V. Zayats, Nonlinear plasmonics. Nat. Photonics **6**(11), 737–748 (2012)
53. D.K. Gramotnev, S.I. Bozhevolnyi, Nanofocusing of electromagnetic radiation. Nat. Photonics **8**(1), 13–22 (2013)
54. A. McPherson et al., Studies of multiphoton production of vacuum-ultraviolet radiation in the rare gases. J. Opt. Soc. Am. B **4**(4), 595 (1987)
55. X. Li et al., Multiple-harmonic generation in rare gases at high laser intensity. Phys. Rev. A **39**(11), 5751–5761 (1989)
56. A. Husakou et al., Theory of plasmon-enhanced high-order harmonic generation in the vicinity of metal nanostructures in noble gases. Phys. Rev. A **83**(4), 043839 (2011)
57. M.F. Ciappina et al., Enhancement of high harmonic generation by confining electron motion in plasmonic nanostructures. Opt. Express **20**(24), 26261 (2012)
58. M.F. Ciappina et al., High order harmonic generation in noble gases using plasmonic field enhancement. Ann. Phys. **525**(1–2), 97–106 (2013)
59. I. Yavuz et al., Generation of a broadband xuv continuum in high-order-harmonic generation by spatially inhomogeneous fields. Phys. Rev. A **85**(1), 013416 (2012)
60. Y.-Y. Yang et al., High-harmonic and single attosecond pulse generation using plasmonic field enhancement in ordered arrays of gold nanoparticles with chirped laser pulses. Opt. Express **21**(2), 2195 (2013)
61. X. Cao et al., Generation of isolated sub-10-attosecond pulses in spatially inhomogenous two-color fields. Opt. Express **22**(21), 26153 (2014)
62. P.B. Corkum, F. Krausz, Attosecond science. Nat. Phys. **3**(6), 381–387 (2007)
63. P.H. Bucksbaum, The future of attosecond spectroscopy. Science (New York, N.Y.) **317**(5839), 9–766 (2007)
64. R.L. Sandberg et al., High numerical aperture tabletop soft x-ray diffraction microscopy with 70-nm resolution. Proc. Natl. Acad. Sci. U.S.A. **105**(1), 7–24 (2008)
65. T. Brabec, F. Krausz, Intense few-cycle laser fields: Frontiers of nonlinear optics. Rev. Mod. Phys. **72**, 545–591 (2000)
66. P.M. Paul et al., Observation of a train of attosecond pulses from high harmonic generation. Science (New York, N.Y.) **292**(5522), 92–1689 (2001)
67. M. Hentschel et al., Attosecond metrology. Nature **414**(6863), 509–513 (2001)
68. J. Krause et al., High-order harmonic generation from atoms and ions in the high intensity regime. Phys. Rev. Lett. **68**(24), 3535–3538 (1992)
69. E.A. Gibson et al., Coherent soft x-ray generation in the water window with quasi-phase matching. Science (New York, N.Y.) **302**(5642), 8–95 (2003)
70. T. Popmintchev et al., Phase matching of high harmonic generation in the soft and hard X-ray regions of the spectrum. Proc. Natl. Acad. Sci. U.S.A. **106**(26), 21 10516 (2009)
71. C. Gohle et al., A frequency comb in the extreme ultraviolet. Nature **436**(7048), 7–234 (2005)
72. A. Cingöz et al., Direct frequency comb spectroscopy in the extreme ultraviolet. Nature **482**(7383), 68–71 (2012)
73. P. Laporte et al., Pulsed-laser-generated rare-gas plasma as a light source in the vacuum ultraviolet. Opt. Lett. **12**(12), 9–987 (1987)
74. J.K. Crane et al., Coherent and incoherent XUV emission in helium and neon, laser-driven plasmas. IEEE Trans. Plasma Sci. **21**(1), 82–89 (1993)
75. R. Coffee, G. Gibson, Vuv fluorescence from selective high-order multiphoton excitation of N_2. Phys. Rev. A **69**(5), 053407 (2004)
76. C. Kern et al., Comparison of femtosecond laser-induced damage on unstructured vs. nano-structured Au-targets. Appl. Phys. A **104**(1), 15–21 (2011)
77. J.E. Sansonetti, W.C. Martin, Handbook of basic atomic spectroscopic data. J. Phys. Chem. Ref. Data **34**(4), 1559 (2005)
78. Y.R. Shen, Surface properties probed by second-harmonic and sum-frequency generation. Nature **337**(6207), 519–525 (1989)
79. M.B. Raschke, High-harmonic generation with plasmonics: feasible or unphysical? Ann. Phys. **525**(3), A40–A42 (2013)
80. H.M. Gibbs, *Optical Bistability: Controlling Light with Light* (Academic, New York, 1985)

81. G. Wurtz et al., Optical bistability in nonlinear surface-plasmon polaritonic crystals. Phys. Rev. Lett. **97**(5), 057402 (2006)
82. Y. Shen, G.P. Wang, Optical bistability in metal gap waveguide nanocavities. Opt. express **16**(12), 6–8421 (2008)

Chapter 11
Attosecond XUV Pulses and Surface Plasmon Polaritons: Two Case Studies

Mattia Lupetti and Armin Scrinzi

Abstract Surface plasmons are collective oscillations of the surface electrons in resonance with an external driving field. In this chapter we investigate their combination with attosecond pulses in two distinct scenarios. First, a high harmonic source using the enhanced field of a surface plasmon is analyzed and characterized in detail. As typical for such sources, one finds high repetition rates and good coherence properties, however, at low photon yields. Conversely, an experimental technique for the spatial and temporal imaging of plasmonic excitations on nanostructured surfaces is presented. Here attosecond pulses allow to image plasmon buildup and localization.

11.1 Introduction

Attosecond pulses are ultrashort radiation bursts produced via high harmonic generation (HHG) during a highly nonlinear excitation of atomic or molecular systems by a near infrared (NIR) laser pulse. Conversely, the attosecond pulses so generated can be also used to probe the ultrafast electron dynamics in the transient regime of the excitation process, with the unprecedented resolution of the attosecond time scale (1 as = 10^{-18} s). In this chapter we show that both the generation of attosecond pulses (AS pulses) and probing of ultrafast processes by means of AS pulses, can be extended to cases in which the respective driving and streaking fields are produced by surface plasmons excited on nanostructured solid surfaces at NIR wavelengths.

Plasmonic excitations are surface optical modes generated by a collective surface electron oscillation in resonance with an external driving source. From a microscopic point of view, they are a coherent superposition of all the possible electron transitions with the same momentum transfer induced by the pump laser. This microscopic

M. Lupetti (✉) · A. Scrinzi
Ludwig Maximilians Universität, Theresienstraße 37, 80333 Munich, Germany
e-mail: mattia.lupetti@physik.uni-muenchen.de

A. Scrinzi
e-mail: armin.scrinzi@physik.uni-muenchen.de

M. Kitzler and S. Gräfe (eds.), *Ultrafast Dynamics Driven by Intense Light Pulses*, Springer Series on Atomic, Optical, and Plasma Physics 86, DOI 10.1007/978-3-319-20173-3_11

oscillation gives rise to typical macroscopic features, most notably to an electromagnetic field confined well below the diffraction limit and at the same time strongly enhanced in proximity of the supporting surface. For these reasons, plasmons are attractive for a number of technological application, ranging from plasmonic circuitry for plasmonic-based computer chips [34], to Surface Enhanced Raman Scattering (SERS) for biological molecule sensing [44].

Here we illustrate the generation of AS pulses from a plasmonic field, and the probing of plasmonic fields by AS pulses. First, we show how to generate AS pulses from the plasmonic field enhancement of a NIR laser launched in a tapered hollow waveguide [28], secondly we introduce a scheme for imaging surface plasmon polaritons in their transient generation stage by mean of AS pulses [27].

The idea of plasmonic generation of AS pulses was first reported in [36]. The experimental concept is to couple the NIR laser pulse into a tapered nanoplasmonic waveguide in such a way that the excited plasmon reaches its maximum amplitude at the propagation turning point, where the field intensity is amplified up to three orders of magnitude. Therefore, with a commercial 10^{11} W/cm^2 laser one can generate harmonics well into the extreme ultraviolet (XUV) regime, while maintaining the original high repetition rate of the commercial source. Thanks to the small enhancement volume, the harmonics naturally add up coherently to generate single AS pulses.

The other work we discuss is an experimental proposal where AS pulses are used to photoionize surface electrons which act as probe of surface collective dynamics. This concept constitutes an extension of the attosecond streak camera [20] to "Attosecond Photoscopy" [27], which allows space- and time-resolved imaging of the excitation of a surface mode, enabling access to its transient buildup dynamics.

11.2 Surface Plasmon Polaritons

Surface Plasmons Polaritons (SPP) are surface electron density oscillations coupled with an external optical field. The electromagnetic field associated with the SPP is described by a wave propagating along the metal/dielectric interface, with an amplitude which decays exponentially with the distance from the surface. Such evanescent waves are always present when dealing with materials discontinuities (see Weyl's expansion [53]). SPPs are a subset of the evanescent waves in the sense that they have amplitudes whose magnitude *exceeds* that of the excitation source. This amplification phenomenon occurs whenever the driving source satisfies a matching condition, a relation involving frequency and angle of incidence [29]. From a microscopical point of view, a plasmon is pole of the polarization propagator $\Pi(k, \omega)$, corresponding to an eigenstate composed of the coherent superposition of all the possible electron-hole transitions with the same momentum transfer k. This microscopic property allows to identify plasmons not only in solid state system, but also in small clusters and even in molecules [23]. SPPs are often viewed as resonances, because at macroscopic level the fulfillment of the matching condition is typically associated with a peak (or dip) in the transmission (or reflection) spectrum of the supporting material. In correspondence of the excitation of an SPP, a portion of the electromagnetic energy is confined in the vicinity of the interaction interface.

This property of amplifying and spatially confining the electromagnetic field in proximity of the surface is what makes SPPs so attractive for many physical, biological and technological applications (see for example [4]). Some physical properties of these modes can be inferred from the Fresnel reflection and transmission coefficients. Assume a TM polarized plane wave of wavelength k impinging on a flat mirror at incidence angle θ. The Fresnel reflection and transmission coefficients are given by [5]:

$$r(\alpha) = \frac{\mathrm{n}^2\beta - \gamma}{\mathrm{n}^2\beta + \gamma} \qquad t = \frac{2\mathrm{n}^2\beta}{\mathrm{n}^2\beta + \gamma} \tag{11.1}$$

where $\alpha = k_\parallel = k_0 \sin\theta$ and $\beta = k_\perp = \sqrt{k_0^2 - \alpha^2}$ are the wavevector components in the vacuum side, while $\gamma = \sqrt{\mathrm{n}^2 k_0^2 - \alpha^2}$ is the component inside the material perpendicular to the interface. If we assume for simplicity that $\mathrm{n}^2 = \epsilon_r$ is real, $\mathrm{n}^2 < 0$ and γ becomes purely imaginary. Thus, analytically continuing the Fresnel coefficients to the complex plane, poles appear in the denominator of (11.1), which correspond to a resonance, i.e. a surface plasmon.

The condition $\mathrm{n}^2\beta + \gamma = 0$ is fulfilled by $\alpha_{\mathrm{spp}} = k_0\mathrm{n}/\sqrt{1+\mathrm{n}^2}$ from which we get the dispersion relation of the mode:

$$k_\parallel = \frac{\omega}{c}\sqrt{\frac{\epsilon(\omega)}{1+\epsilon(\omega)}} \equiv k_{\mathrm{spp}}(\omega). \tag{11.2}$$

Equation (11.2) also provides the *matching condition* which needs to be satisfied by the source for the excitation of an SPP. In fact, the right hand side is the plasmonic wavevector supported by the material at the frequency ω of the source. This has to be equal to component of the source wavevector parallel to the surface.

For gold at a wavelength of 800 nm the dielectric constant is $\epsilon_{\mathrm{Au}} = -24.06+1.51i$. Neglecting the imaginary part, one obtains $\alpha_{\mathrm{spp}}/k_0 > 1$, with the relative β_{spp} and γ_{spp} purely imaginary. These coefficients describe a wave propagating on the surface and *evanescent* in the orthogonal direction. Because of the evanescent behavior, a field gradient perpendicular to the surface exists, such that $\nabla_\perp E_\perp \neq 0$, which means that the field lines close around the surface. However, since $k_{\mathrm{spp}} \geq k_0$, such a resonance can never be excited by a plane wave. This is true for every material displaying plasmonic properties, as illustrated in the dispersion relation graph in Fig. 11.1. The dielectric function is assumed to be that of a perfect electron gas $\epsilon(\omega) = 1 - \omega_p^2/\omega^2$. The coupling never occurs as the dispersion curve always lies out of the light cone. Thus, the fundamental problem is to overcome the situation of a flat surface where $k_{\mathrm{spp}} > k_\parallel$ for each value of ω.

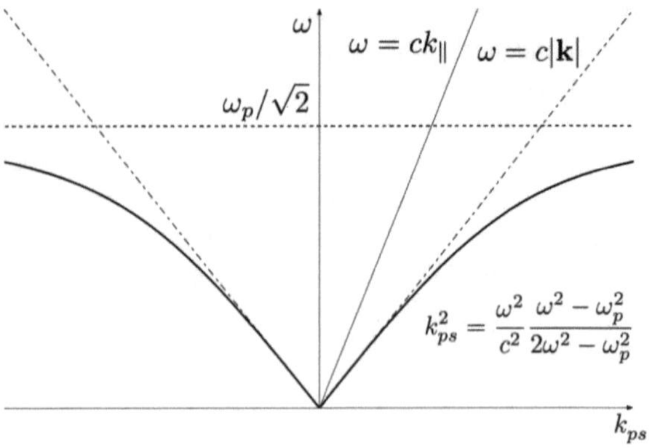

Fig. 11.1 Dispersion relation of SPP on a flat surface of a metal described by the free electron model, for which $\epsilon(\omega) = 1 - \omega_p^2/\omega^2$. The SPP lies outside the light cone, so no matching with a plane wave can occur

11.2.1 Excitation of SPPs

To excite an SPP the strategy is to augment the parallel k-vector of the exciting source, to achieve matching with the surface eigenmode wave vector. One way of "coupling" a source to an SPP is to overlay a dielectric material on the metallic surface. Often used are the Otto [33] and Kretschmann [24] configurations, which differ in the dielectric position in contact or closely above the surface. In each case the SPP is excited at the metal-vacuum surface, see Fig. 11.2. The additional k-vector needed to satisfy the matching condition is provided by passage through the dielectric, which enhances the parallel component by $k_\parallel \rightarrow \mathrm{n_{diel}} k_\parallel$, assuming that $\mathrm{n_{diel}} \sin\theta > 1$.

Another way is the grating coupling: the source is shone on a diffraction grating with lattice constant d. The additional k-vector here is provided by the periodicity of the grating. In fact, the discrete translational invariance along the grating grooves "folds" the SPP dispersion relation back into the first Brillouin zone (FBZ), making the matching possible. The matching condition in this setup is $k_{\mathrm{spp}} = k_\parallel + nq$, where n is the diffraction order and q is the wave number associated to the grating periodicity. Figure 11.3 shows the effect of the discrete periodicity on the SPP dispersion relation. Apart from the formation of Brillouin zones, it is interesting to notice the appearance of plasmonic band gaps, originating from the resolution of the degeneracies of the modes $(n, -n)$ at the center of the FBZ, and $(n, n \pm 1)$ at the borders.

11.2.2 Standard SPP Imaging Techniques

For the experimental characterization of the electromagnetic field of the surface plasmon, the most ubiquitous technique is the Scanning Near-Field Optical Microscopy (SNOM), see [18] for a review. The technique consists in placing a microscopic tip (probe) in the vicinity of the surface field via a positive feedback technique, like Atomic Force Microscopy (AFM). The probe tip, a tapered optical fiber, couples the

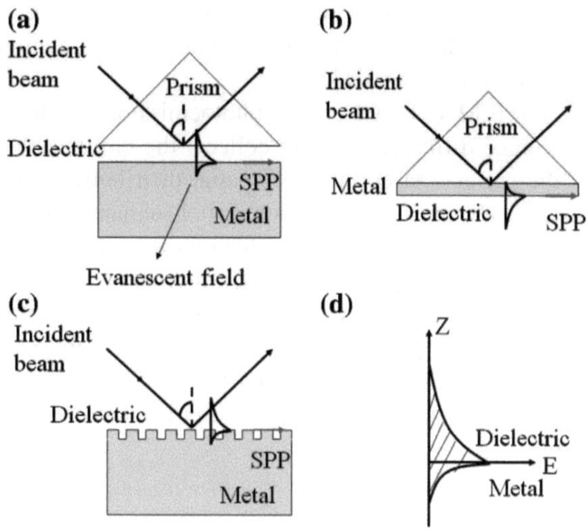

Fig. 11.2 Schematics field coupling into SPP modes: (**a**) Otto configuration, (**b**) Kretschmann configuration, (**c**) grating coupling. In (**d**) is shown the field decay over the perpendicular to the surface. Figure reproduced from [54]

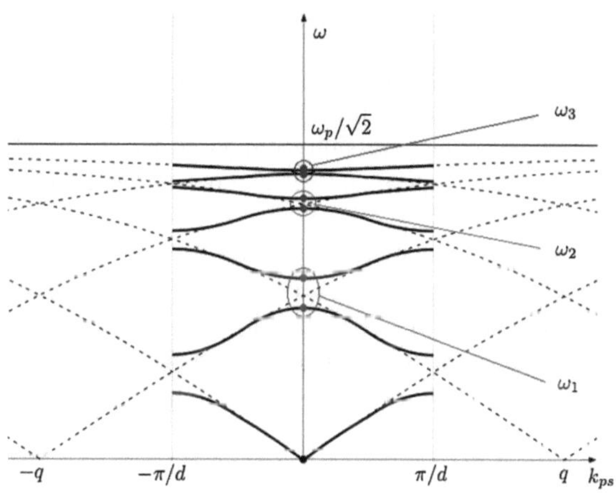

Fig. 11.3 Plasmonic band structure produced in the grating coupling mechanism. The non-zero amplitude of the grating grooves is responsible for the gap formation. Note that the gap opening mechanism is the same as in the band structure theory: the grating groove amplitude plays the same role of the solid periodic potential

tail of the evanescent plasmonic field with a propagating waveguide mode, allowing the measurement of a photon current.

Another way of imaging plasmons is by fluorescence emitters as markers of the plasmonic fields, as in [12]. Since the fluorescence signal is directly proportional to the plasmon intensity at the place of the emitter, by covering the propagation range of the plasmon one can retrieve information on its localization. Alternatively, one can directly tailor the radiation losses of the plasmon into a dielectric substrate, by exploiting the prism coupling mechanism [13]. This "leakage radiation" allows to measure the spatial intensity profile of a plasmon, or, in different arrangements, to directly observe the plasmon dispersion relation.

It has been recently demonstrated [8, 25] that a non-invasive plasmon diagnostic tool is provided by employing the photo-electron emission microscope (PEEM) for the plasmonic near field.

A PEEM is based on the photoemission of electrons from the metal surface. A voltage difference system collects the emitted electrons, which form an image of the surface based on their spatial distribution. Since photoelectron emission is increased because of the plasmonic field enhancement, it is possible to map the spatial distribution of the plasmonic field intensity without perturbing the system. A proposal for an attosecond pump-probe measurement combined with a PEEM was reported in [49], with the aim of creating a nanometer-scale attosecond-time resolution imaging technique. The proposed technique is very complex and its demonstration is, to the author knowledge, not yet been performed. In Sect. 11.4 we present a recent proposal (see [27]) in which the temporal characterization of SPPs is performed using AS pulses.

11.3 A Plasmon Enhanced Attosecond Extreme Ultraviolet Source

Attosecond (AS) pulse sources are employed in the observation of electron dynamics at the time scales of atomic valence electrons. Such sources are mostly based on high harmonic emission from gases and are employed, for example, in attosecond streaking experiments of photoelectrons ionized in atoms [14, 41], and solid surfaces [6].

AS pulse generation is an extremely non-linear frequency up-conversion process of the fundamental NIR driver pulse, which imprints its time-structure onto the harmonic radiation. The core physical features of the process are captured by the classical three-step re-collision model [10, 26]. It consists in separating the electron dynamics in three phases: ionization, quiver motion and recollision. The first step occurs by tunnel-ionization, where the tunneling barrier appears as a result of the bending of the nuclear potential by the strong NIR laser field. Once ionized, the electron is pulled away from the nucleus and then pushed back to collide with the nucleus. If the collision results in the recombination of the electron with the ion, high harmonic radiation is emitted up to a maximal photon energy of $\hbar\omega_c = I_p + 3.2\,U_p$, where I_p is the ionization potential of the gas and $U_p = \frac{e^2}{4m_e}\frac{E_0^2}{\omega^2}$ is the driver ponderomotive potential of a laser of amplitude E_0 and frequency ω.

The non-linearity of the process is such that the highest part of the harmonic spectrum is produced only around the maxima of the NIR pulse, such that bursts of high frequency radiation as short as 67 as can be produced [55]. For photon-energies $\gtrsim$40 eV with Ti:sapphire NIR pulses at 800 nm wavelength, intensities $\gtrsim 10^{14}$ W/cm^2 are needed.

The radiation bursts are time-delayed relative to the maxima of the driver field by $\sim$0.2 optical cycles, i.e. emission occurs near the *nodes* of the driver field. The time-locking of the AS pulse to the NIR pulse allows to perform pump-probe experiments by controlling the relative time delay with a precision of 10 as. With NIR driver pulse of duration ~20 fs, a short sequence of AS pulses is generated, separated from each other by half the driver optical period T_L. While generating trains of AS pulses is relatively easy, using them for time resolution of processes longer than half the optical period (~1.3 fs) is difficult and requires deconvolution of the overlapping signals from neighboring pulses.

For attosecond pump-probe experiments delays beyond the laser half-period, one needs to use isolated single AS pulses. They are typically generated by using optical pulses which are so short, that only one field maximum effectively contributes to the generation of the desired photon energies [2]. Alternative techniques like polarization gating [47], where the laser polarization is manipulated in order to suppress high harmonic generation for all but one field peak, are rather complex to use in spectroscopic experiments. In any case, while few femtosecond lasers at 80 MHz repetition rate are commercially available [40], the amplification required to reach the intensity of 10^{14} W/cm^2 reduces the repetition rate to few kHz.

To overcome this difficulty, it was proposed in [36] to use directly the few cycle pulses delivered by commercial oscillators and exploit plasmonic enhancement to avoid the amplification chain and the consequent reduction of the repetition rate. The "amplification" scheme consists in the excitation of Surface Plasmon Polaritons (SPP) by focusing a laser pulse into a tapered waveguide. As the plasmonic field propagates towards the narrow end of the waveguide, its wavelength decreases, while the field intensity increases. With this enhancement mechanism [36] reported the detection of extreme ultraviolet (XUV) radiation. The observation has been debated in successive works, see Sect. 11.3.7. In a following theoretical study [7], attosecond time structure was found in the response of an isolated atom to the plasmon field.

Here we present an analysis of the macroscopic XUV harmonic propagation to determine whether the high harmonics emitted from the gas atoms add up to form a usable beam with time structure of an isolated AS pulse. We find indeed that the time structure of the generated XUV radiation is that of an isolated AS pulse. Furthermore, being generated by a single plasmonic mode, the pulse shows a very clean spherical wavefront and a good collimation degree, besides being stable under variation of the driver pulse duration and incidence angle.

On the other hand, the emission through the narrow end of the tapered waveguide is severely diffracted, while a well collimated attosecond beam propagates in the reverse direction. Moreover, the beam intensity of the generated AS pulses is several order of magnitude smaller than what found in the original experiment in [36], in agreement with what predicted theoretically in [30].

We compare the "plasmon enhanced attosecond XUV source" (PEAX) to standard AS pulses produced from gas harmonic sources. Despite the significantly lower yield of the PEAX, the extremely clean spatial profile and the high repetition rate make it an attractive alternative source to standard gas harmonics.

We perform our analysis by numerically solving the three-dimensional Maxwell equations for the plasmonic response of the tapered waveguide to the driver NIR pulse, the atomic time-dependent Schrödinger equation for the single atom high harmonic response and then again the Maxwell equations for the propagation of the high harmonics in the waveguide.

11.3.1 Spatial Structure of the Plasmonic Field

For the solution of the boundary value problem on a metallic cylindrical surface we refer to the derivation in [50], Chap. IX. The dispersion relation of the cylinder cavity modes is obtained by setting to zero the determinant of the matrix of the electromagnetic field amplitudes at the cylinder boundary:

$$\left[\frac{1}{u}\frac{I'_m(u)}{I_m(u)} - \frac{1}{v}\frac{K'_m(v)}{K_m(v)}\right]\left[\frac{\epsilon_d}{u}\frac{I'_m(u)}{I_m(u)} - \frac{\epsilon_m}{v}\frac{K'_m(v)}{K_m(v)}\right] = -m^2 n_{\text{eff}}^2\left(\frac{1}{u^2} - \frac{1}{v^2}\right)^2 \tag{11.3}$$

Here $u = \kappa_1 R = Rk_0\sqrt{n_{\text{eff}}^2 - \epsilon_d}$, $v = \kappa_2 R = Rk_0\sqrt{n_{\text{eff}}^2 - \epsilon_m}$, with $n_{\text{eff}} = h/k_0$.

For a mode of order m it is not possible to isolate a transverse electric or magnetic mode, except in the case $m = 0$, where we can speak of TE and TM mode. In this case the dispersion relation simplifies to:

$$\frac{\epsilon_d}{\kappa_d}\frac{I'_0(\kappa_d R)}{I_0(\kappa_d R)} = \frac{\epsilon_m}{\kappa_m}\frac{K'_0(\kappa_m R)}{K_0(\kappa_m R)} \tag{11.4}$$

which has an explicit solution [48]:

$$n_{\text{eff}} = \frac{1}{k_0 R}\left(-\frac{2\epsilon_d/\epsilon_m}{\log(-4\epsilon_d/\epsilon_m)/2 - \gamma}\right)^{1/2}, \tag{11.5}$$

where $\gamma = 0.57721$ is the Euler-Mascheroni constant. It is interesting to notice that this mode exists for any radius R of the cylinder and its propagation wavelength is proportional to the cylinder radius. It is called the *nanowire mode*, since it is the only mode that does not experience any cut-off for $R \to 0$.

For modes with $m \neq 0$ the right hand side in (11.3) is not zero, and couples TE and TM terms. Thus, higher order modes have a cutoff radius after which propagation is forbidden, and the cylindrical SPP is reflected back along the waveguide axis. Since HHG requires a linearly polarized laser, the cylindrical symmetry of the system is broken, and the first available mode is the $m = 1$, which cannot be focused to arbitrarily small radius. This turns out to be favorable, because one can in principle exploit the finite cutoff radius of the mode to separate the harmonic radiation propagating through the waveguide's smaller end from the NIR plasmonic field.

11.3.2 Geometry of the Tapered Nanoplasmonic Waveguide

One finds out that a tapered waveguide with elliptic cross section at fixed axis ratio provides a higher field enhancement at the cone tip. The geometry is shown in

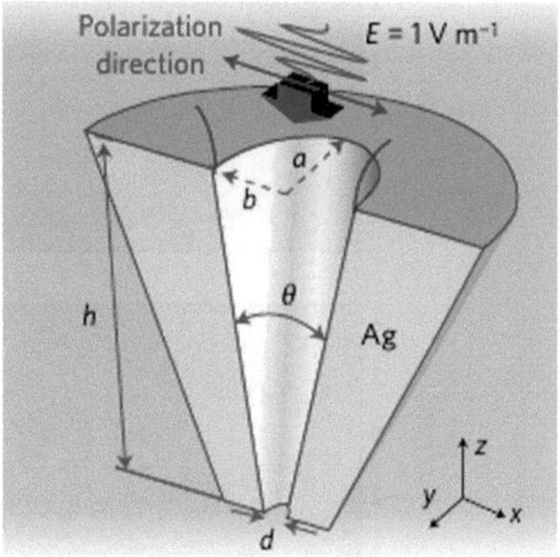

Fig. 11.4 Geometry of the tapered nanoplasmonic waveguide. Figure reproduced from [36]

Fig. 11.4. The ratio between the minor axis b and major axis a of the ellipse is $b/a = 0.25$. Similar to [36], we chose for our simulations a 9 μm long silver cone and opening angles of 14 and 3.5° along the major (x) and minor (y) axis of the ellipse, respectively. These parameters uniquely define the cone geometry. We assume a 5 fs FWHM Gaussian driver pulse at wavelength $\lambda_0 = 800$ nm, beam waist $w_0 = 2.5$ μm, and focused intensity $I_0 \approx 4 \times 10^{11}$ W/cm^2. The focus is at the large opening of the tapered waveguide (see Fig. 11.5).

One finds that coupling is rather robust with variations of the focus position by ± 1 μm causing intensity changes of less than 5 % in peak plasmon intensity. An enhancement factor of $\sim$500 in peak intensity is found for an eccentricity value of $\epsilon = 0.25$. For $\epsilon = 0.5$, as used in [36], the enhancement reduced by about a factor 3. A similar dependence on ellipticity was reported in [7], where pulse durations between 4 and 10 fs were investigated.

Since the peak field is reached at the surface, any surface roughness could intro duce modifications of the exact maxima. This however does not invalidate our analysis, since harmonics produced in proximity of the waveguide surface will be absorbed in the metal, while the dominant contribution coming from the central region is less subject to sub-wavelength structures of the waveguide.

Further optimization of the cone geometry allows for even weaker driver pulses, keeping in mind that ultimately the field inside the cone is limited by thermoelectric damage. Reference [36] claims that silver can sustain much higher fields than expected, probably because of the few femtosecond duration of the employed pulse [38]. In our work we neglect the incoherent radiation emitted by photo-ionized silver atoms.

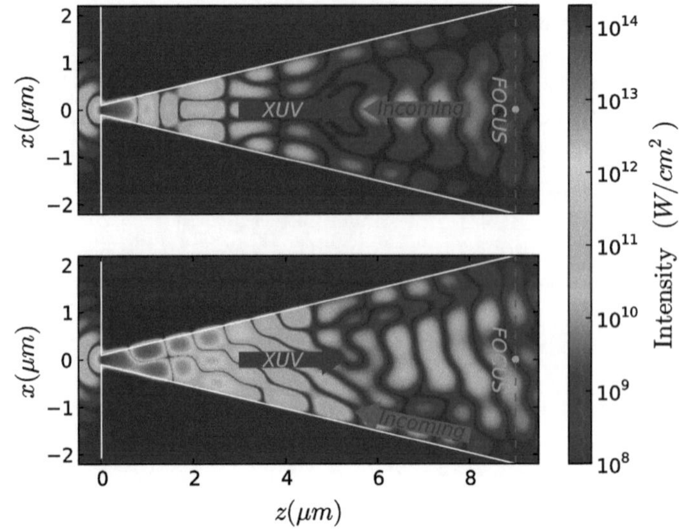

Fig. 11.5 Plasmonic field distribution in the xz-plane at peak plasmon field of 2×10^{14} W/cm^2. The calculation is performed with the Finite-Difference Time-Domain (FDTD) method. The *upper panel* shows the normal incidence case, the lower the oblique case. Polarization is in y-direction perpendicular to the plane and independent of the incidence angle, which in the *lower panel* is of 7° in the zx-plane parallel to the inner cone surface. A remarkable property of PEAX is that the XUV harmonic beam is emitted in the direction of the cone axis for any incidence angle. This is due to the symmetry of the $m = 1$ mode which is excited, irrespective of the angle of incidence. Figure from [28]

11.3.3 Wave-Guiding of XUV Pulses by the Tapered Waveguide

The plasmonic field generated by the driver NIR pulse is computed with MEEP [32], an open source C++ library implementing the FDTD method [52]. In MEEP, the dielectric function must be of the form

$$\epsilon(\omega) = \epsilon_\infty - \frac{\omega_p^2}{\omega(\omega - i\gamma_0)} + \sum_n \frac{f_n \omega_n^2}{\omega_n^2 - \omega^2 + i\omega\gamma_n} \tag{11.6}$$

for algorithmic reasons. In the visible region, the parameters which best fit the dielectric constant of silver are taken from [39]. As for the XUV range, which is relevant for the HHG process, the Drude-Lorentzian modeling reproduces the frequency dependency of the dielectric function poorly. We gave preference to an accurate fit of the imaginary part of the dielectric response against data taken from [17], since we found that the real part of the dielectric response has little influence on the propagation of the harmonic radiation inside the waveguide. By changing the real part of the fitted dielectric function by a factor of 4 (cf. Fig. 11.6), we observed a change in the signal

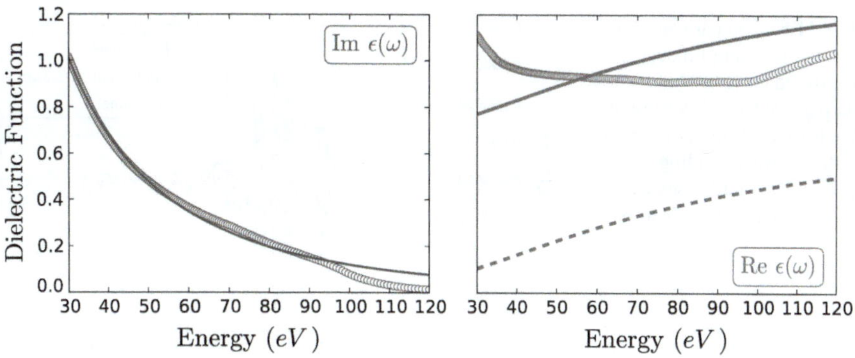

Fig. 11.6 Dielectric functions used for the simulation of high harmonic propagation. Imaginary part of the experimental values (*left*, *blue circles*) are well approximated by a Lorentzian shape (*blue line*). The fit (*right*, *red line*) of the experimental real values (*circles*) is comparatively poor. The deliberately bad fit (*dashed red line*) was used for checking robustness of the simulation. Experimental data from [17]

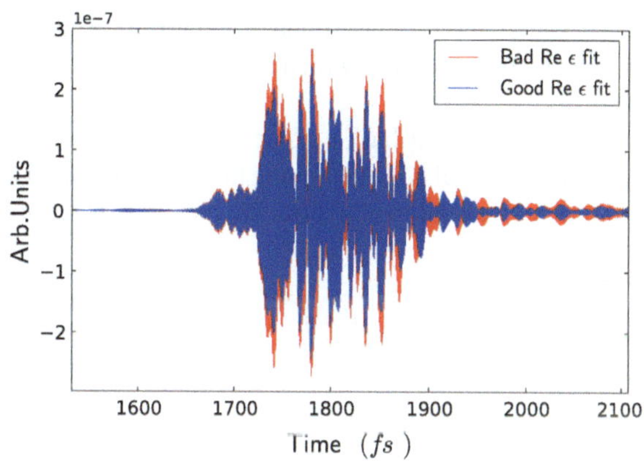

Fig. 11.7 Comparison of the electric field amplitude measured far from the waveguide, in the backwards direction, simulated using different fit of the real part of the dielectric function for the XUV range. In both cases we employed the same Zr filter

intensity by less than 15 %, while the time-structure remained unaffected, as it is shown in Fig. 11.7.

The high harmonics were let propagate in the same FDTD code used for the plasmonic response, but with a sampling spacing of 2.5 nm, well below the characteristic wavelength of 800 nm of the driver plasmon in the waveguide and below the relevant harmonic wavelength of ~27 nm. The dipole responses of the gas atoms were calculated by solving the time-dependent Schrödinger equation by the irECS method [43] using a single-electron model with the ionization potential of Argon. In these calculations we do not include fluorescence, which would contribute incoherently to XUV radiation. The harmonic generation process is strongly dependent on the driver pulse intensity: from 2×10^{14} to 1×10^{14} W/cm^2 the harmonic yield near the cutoff photon energy of ~60 eV drops by ~3 orders of magnitude. For this reason the region of calculation of the atomic dipole responses was limited to the volume

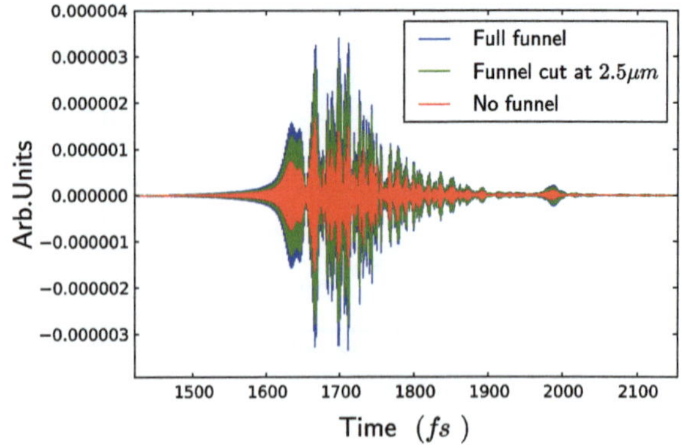

Fig. 11.8 Comparison of the electric field amplitudes measured far from the waveguide larger aperture in the backwards direction, for different wave-guiding conditions. In all cases we employed the same Zr filter

where the total field intensity exceeds 10^{14}W/cm^2, corresponding to an active volume $V_a = \Delta x \times \Delta y \times \Delta z \approx 240 \times 60 \times 500\,\text{nm}^3$.

The harmonic emission occurs also for coupling of the driver over a range of incidence angles. For $\theta = 7^\circ$ around the polarization (y) axis, the plasmonic enhancement is reduced by a factor 4, which can be compensated by an increase of the input filed intensity.

The far field distribution of the harmonic radiation was computed by means of the Kirchhoff integral using as source the plasmonic field at the surface $z = 2.5\,\mu$m from the smaller waveguide aperture. We recall the Kirchhoff formula for monochromatic waves:

$$E(\xi, k) = \int_S dS \left[E(S) \frac{\partial}{\partial n} \left(\frac{e^{i\mathbf{k}\cdot\mathbf{s}}}{|\mathbf{s}|} \right) - \frac{e^{i\mathbf{k}\cdot\mathbf{s}}}{|\mathbf{s}|} \frac{\partial E}{\partial n} \right], \tag{11.7}$$

where $S(x, y)$ is the 2D source surface cut from the FDTD simulation, ξ is the observation point and $\mathbf{s} = \xi - \mathbf{x}$, with $x \in S$. We verified the validity of this procedure by comparing the guiding effect of a wedge-shaped waveguide on the harmonic propagation, where the translational symmetry reduces the problem to 2 dimensions. We found that beyond the distance of 2.5 μm further guiding of the XUV radiation is very small. On the other hand, a comparison with the "free" propagation, where the waveguide has been artificially removed, shows that the XUV intensity is enhanced by a factor 4, as shown in Fig. 11.8. We employed a Zirconium-like filter in order to remove only the fundamental and lower harmonics from the generated harmonic spectrum. Figure 11.9 shows the angular distribution of harmonic emission out of the wide side of the cone for a range of harmonic frequencies at oblique driver incidence.

The incident intensity was adjusted to obtain a maximum plasmonic field near the waveguide tip of 2×10^{14} W/cm^2. As can be clearly observed in Fig. 11.9, the spatial divergence decreases with increasing photon energy, were we define the divergence as the full-width half maximum (FWHM) in the XUV intensity. The normal incidence case with equally adjusted peak intensity gives similar results to oblique incidence.

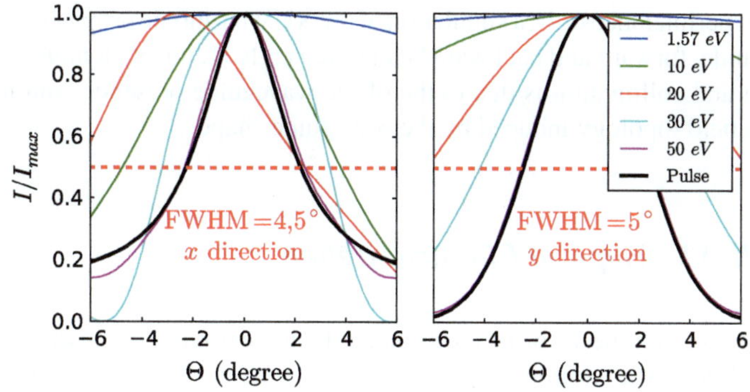

Fig. 11.9 Angular distribution of harmonic radiation for different photon energies reported in the inset legend. At oblique incidence with driver polarization along y, the x-distribution of the harmonics is asymmetric (*left panel*), while it remains symmetric in the y-direction (*right panel*). The harmonics' spatial profiles are taken in the far field, at 1 mm distance from the waveguide large opening in the backwards plasmon propagation direction. The *black line* is the sum of all the harmonics above 45 eV, and therefore gives a measure of the AS pulse collimation angle. Taking as reference the FWHM of the XUV pulse intensity, the beam divergence is about 5° in both x- and y-directions. Thus, the projection of the AS pulse on the plane orthogonal to the waveguide axis is a *circle* centered on the axis, despite the elliptic cross-section of the waveguide and the oblique incidence of the driver NIR pulse. Figure reproduced from [28]

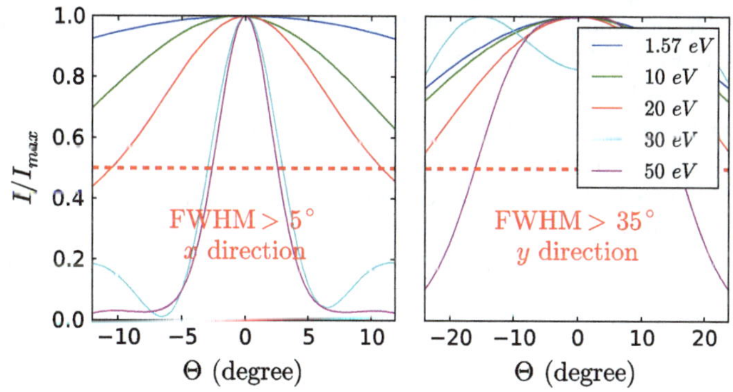

Fig. 11.10 Beam divergence of the XUV pulse propagating through the funnel tip

The reason is to be found in the single mode nature of the plasmonic field, whose topology in its maximal value region is insensitive to the angle of incidence.

Analysis of the attosecond beam propagation through the waveguide smaller aperture shows that in this case the funnel causes diffraction rather than a collimation. The last point could be tested in the full three dimensional case by using the electric field at the waveguide smaller aperture as source surface for the Kirchhoff diffraction integral. The harmonics' spatial profile is shown in Fig. 11.10.

Although the XUV power is comparable in forward and backward emission, the beam divergence is 35° in y-direction, such that the resulting harmonic beam is hard to focus for experimental use. In a simulation of the XUV harmonic propagation without the waveguide, forward and backward beams are nearly identical, demonstrating that diffraction and collimation is due to the of the waveguide presence, and not to the plasmonic field topology induced by the waveguide shape.

11.3.4 PEAX Temporal Characterization

Since the active volume V_a of HHG is smaller than the driving laser wavelength, the phase matching of the harmonics is automatically guaranteed. Geometrically induced phase shifts are automatically included in the simulation. With negligible dephasing between driver and harmonics across V_a, harmonic intensities grow quadratically with the gas density. Atomic dispersion is also expected to remain small, and, if needed, may be controlled by choosing a target gas suitable for a given harmonic wavelength.

In Table 11.1 the parameters of the XUV-AS pulses are obtained at gas pressure of 0.3 bar (density $7.8 \cdot 10^{18}$ cm^{-3}), a value typically used in standard HHG. To check the influence of the atomic species on the spectrum of the HH radiation produced by the plasmonic field, we considered Argon, Neon and Xenon.

Details of the spectral structures of the different atomic species are reported in Fig. 11.11. Despite the peculiarity of the exciting inhomogeneous plasmonic field, the spectra in Fig. 11.11 closely resemble the single-atom responses. For the driver intensity of 2×10^{14} W/cm^2, Argon shows a cutoff similar to Neon, but an order of magnitude larger spectral intensity, while Neon has a higher cutoff and lower spectral intensity. These characteristics reduce the achievable high harmonic yield for both Neon and Xenon, a feature which affects in the same way the PEAX and the standard gas harmonic source.

The PEAX harmonics above 45 eV form an isolated AS pulse. Because of the rapid decay of the spectral intensity with harmonic energy, the AS pulse central

Table 11.1 Harmonic beam characteristics for oblique incidence PEAX and a standard harmonic source using a Gaussian beam (see text for parameters)

		PEAX		Gauss
Gas	Ne	Ar	Xe	Ar
ω_γ (eV)	53	45	45	45
Δt (as)	250	300	250	300
Rep. Rate		80 MHz		3 kHz
γ/pulse	$0.94 \cdot 10^{-3}$	$6.7 \cdot 10^{-3}$	$0.94 \cdot 10^{-3}$	$3 \cdot 10^4$
γ/s	$7.6 \cdot 10^4$	$5.4 \cdot 10^5$	$6.5 \cdot 10^4$	$9 \cdot 10^7$
Divergence		5°		1°
V_a (μm^3)		$\sim 4 \times 10^{-3}$		$\sim$16

Yields and photon flux are integrated over the beam divergence angles

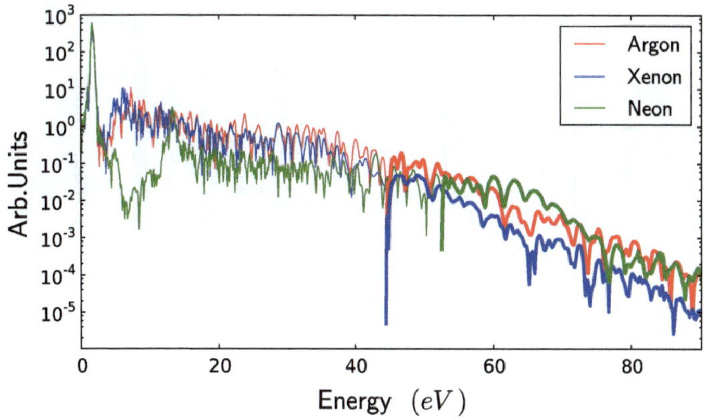

Fig. 11.11 Spectrum of the harmonic radiation produced by different atomic species in plasmonic waveguide. The *thicker lines* show the part of the spectrum which must be kept to produce an isolated AS pulse with maximum photon yield

frequency nearly coincides with the lower cutoff frequency of the harmonics. The pulse contrast, defined as the energy ratio between the main pulse and any satellite pulses, is satisfactory with 85 % of the energy in the main pulse.

11.3.5 PEAX Spatial Properties

The AS pulse emerging from the backwards propagation direction has a perfectly spherical wavefront (cf. Fig. 11.12). The deviation from the spherical shape remains below the central XUV wavelength over the whole front. This can be ascribed to the very small and well-defined generation volume, which is a consequence of the excitation of the single plasmonic mode corresponding to $m = 1$.

Differently from standard gas harmonics, this attosecond source distinguishes itself for the clean wavefront, which allows to focus the pulse without compromising its time structure. Moreover, the AS pulse is emitted on the cone axis, while the reflected driver and lower harmonics are emitted into wider angles with a modulated intensity profile (cf. Fig. 11.9). This is particularly useful because it allows a simple geometric separation of the incident driver pulse from the harmonic pulse. The remaining on-axis reflected driver pulse and low harmonics can be filtered out by standard multilayer mirrors. Furthermore, a shorter driver pulse of 4 fs brings no extra advantage: the AS pulse duration and divergence remain the same, while the pulse intensity is ~20 % less. Similarly, the pulse parameters are not improved by exciting the cylinder plasmon at normal incidence, because the higher field enhancement is compensated by a reduction of the active volume V_a, as it can be observed in Fig. 11.5.

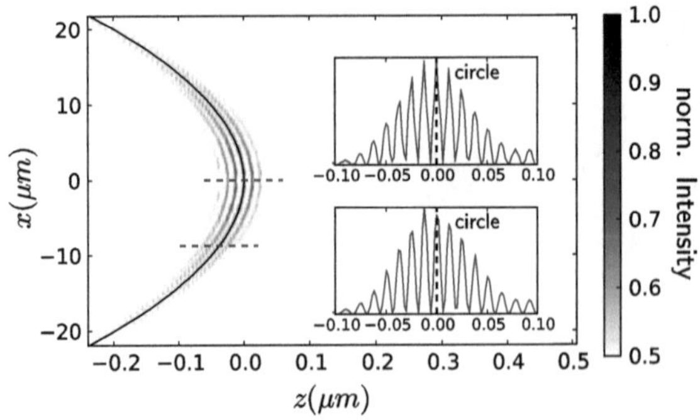

Fig. 11.12 Spatial shape of the XUV AS pulse wavefront. The *color map* shows the part exceeding half the peak intensity. The *black line* indicates the position of a *circle* having as center the waveguide tip, and radius the distance of observation. The *inset* figures show sections of the wavefront taken for two different radii. The *dashed lines* show position of the cuts and the spherical front, in the color map and in the insets, respectively. Figure reproduced from [28]

11.3.6 Comparison with Traditional Gas Harmonics

In Table 11.1, the rightmost column contains the pulse parameters for a traditional harmonic source, simulated with our numerical techniques for benchmarking the PEAX. We used a Gaussian beam with 4 fs FWHM pulse duration and tight focus of $w_0 = 1\,\mu$m in a gas jet with peak intensity equal to the peak plasmonic intensity of $2 \times 10^{14}\,$W/cm^2. The HHG active volume, given by the prolate ellipsoid with w_0 as minor axis and the Rayleigh length $z_r = 3.9\,\mu$m as major axis, is three orders of magnitude bigger than with PEAX. In our configuration, the PEAX source produces slightly longer pulses with larger beam divergence. For photon energies of ∼45 eV and at the given gas pressure, the photon yield per shot is almost 7 orders of magnitude larger compared to PEAX. This corresponds to a difference of a bit more than 3 orders of magnitude in field strength, which is consistent with the ratio of the active volumes of the two sources.

Since no amplification chain is required for a PEAX source, the repetition rate remains that of the oscillator, near 100 MHz, reducing the difference in the photon fluxes with the standard source to less than 500. With an active volume of sub-wavelength dimension, the gas density can be increased by many orders of magnitude before phase matching and coherence problems arise. Assuming tenfold pressure for the PEAX, photon flux could be boosted by 2 orders of magnitude, basically equaling the standard harmonic source performances. Moreover, a further increase of the XUV pulse power could be achieved by using a rasterized arrangement of many PEAX sources as in [35], which is possible thanks to the high quality of the attosecond pulse spatial profile.

On the other hand, the volumes for coherent gas harmonic generation in the standard configuration, and consequently the power of the AS pulses, can be significantly

Table 11.2 Attosecond beam parameters for state-of-the-art standard harmonic generation [16], compared to the parameters obtained in our numerical benchmark case

	Experimental	Theoretical
Atom	Neon	Argon
I_0 (W/cm^2)	5×10^{14}	2×10^{14}
$\omega_\gamma(eV)$	80	45
Δ t (as)	100	250
Rep- Rate	3 KHz	3 KHz
γ/s	10^{11}	9×10^7
Power P_{XUV}	$\sim$1 μW	0.65 nW
V_a	$\sim 10^{-1}$ mm^3	16 μm^3

larger than what was assumed in our comparisons. It is demonstrated (see [16]) that phase matching can be maintained over about 1 mm propagation length at a beam cross section of few hundred μm^2, giving an active volume, and thus an output power of about 3–5 orders of magnitude larger than our reference values. In Table 11.2 we give the parameters provided by experimental results from [16] and compare them with our simulation.

This last consideration can be generalized to all the proposed high harmonic generation schemes based plasmonic enhancement: since the required intensity is achieved in an active volume of sub-wavelength scale, the amount of obtainable photon fluxes is limited compared to the much larger diameters and phase matching length which can be realized in the standard generation scheme.

11.3.7 Discussion and Experimental Issues

We illustrated in our discussion the advantages of PEAX sources in term of beam collimation, pulse front quality, and high repetition rate, which can be used for spatio-time-resolved surface spectroscopy [49].

From an experimental point of view, PEAX sources should be quite feasible, no pulse amplification chain is needed. Moreover, the intrinsic geometrical separation of higher from lower harmonics can be exploited to simplify the experimental setup. The main difficulty is the rather low yield to be expected from a PEAX source. Also, the high harmonic generation for the bow-tie shaped nano-devices reported in [21], has not so far seen independent verification.

A first severe criticism was formulated in 2012, when in [45] it was shown that the observed XUV emission was not a coherent process, but rather a multi-photon atomic line emission (ALE) process. The main criticism was related to the conversion efficiency of the laser power into harmonics: in [21] it is reported to be of the same order of magnitude as in standard harmonic generation, where however the conversion volume is 8 orders of magnitude bigger. It was argued that the ratio between the conversion efficiency of nano-HHG and gas HHG should scale as:

$$\frac{C_{\text{nano}}}{C_{\text{conv}}} = \frac{R_{\text{nano}}}{R_{\text{conv}}} \left(\frac{V_{\text{nano}}}{V_{\text{conv}}}\right)^2 \left(\frac{F_{\text{nano}}}{F_{\text{conv}}}\right)^2 \sim 10^{-8}. \tag{11.8}$$

where $R_{\text{nano}}/R_{\text{conv}} = 10^5$ is the repetition rate ratio, $V_{\text{nano}}/V_{\text{conv}} = 10^{-8}$ is the volume ratio, and as phase matching factor $F^2_{\text{conv}} = 10^{-3}$ is used, while in the plasmonic case is set to $F^2_{\text{nano}} = 1$.

In the response in [22], it was agreed that bow-tie HHG is not a viable technique, and the focus should be shifted to the funnel technique discussed here, where the enhancement volume is 2 orders of magnitude bigger than in the bow-ties case.

Following a reasoning similar to what leads to (11.8), in [30] it was further shown that by comparing the conversion efficiencies in [21–24, 26–29, 31–36] with those achievable in the cavity-enhancement HHG framework [9], it is possible to extrapolate the absolute power of the high harmonic radiation.

In Table 11.1 we compared the PEAX characterization obtained with Xenon, which is used in [30], and Argon, which is used in our work [28]. Comparing the values of the PEAX characterization with Xenon to what obtained in [30], we notice that the value for Xenon is correct within one order of magnitude. The difference is probably due to the small guiding effect provided by the waveguide in the backward direction. What is puzzling is that our theoretical prediction and the one in [30] are 5 orders of magnitude smaller than what extrapolated from the experimental results reported in [36]. As a final remark, during the review for [28], in [46] it was established that in the bow-ties case the HHG signal is a factor 10^{-3} weaker than the signal from multi-photon ALE.

Since the bow-tie plasmonic device and the nanoplasmonic tapered waveguides investigated in our work are closely related, the perspectives of this last technique are not very encouraging at the moment. However, the higher enhancement volume w.r.t. bow-ties structures and the guiding effect in the backward direction may narrow the gap in terms of power with standard harmonic generation. If we add to this the excellent spatial properties, the weak dependence on the incidence angle in the waveguide, and the absence of any amplification chain, the PEAX sources can compete with the standard ones in term of practicality and affordability.

11.4 Attosecond Photoscopy of Surface Excitations

From using SPPs for AS pulse generation we now proceed to a scheme for probing SPP dynamics by AS pulses. Surface plasmons are widely used in many cross-disciplinary fields for their properties of light confinement in the vicinity of metallic supports. In this context, the nanoplasmonic branch is a promising candidate for the development of plasmonic based all-optical processors, since the field confinement property can combine the high operational speed of photonics (PHz scale) with the miniaturization provided by electronics (nm scale). For a comprehensive review see for instance [34]. In this sense, it is interesting to investigate the temporal transient properties of the surface electron excitation. Although the plasmon lifetime can be deduced from the plasmonic resonance width observable in the reflection or transmission spectrum, the formation process of the resonant oscillation cannot be studied in the framework of frequency analysis.

Here we discuss an experimental proposal to image the transient dynamics of a plasmonic mode, based on the so-called attosecond streak camera [20]. The technique was already successfully applied to solid surfaces in [6] for the measurement of time delays among photoelectrons ionized from different bands of a tungsten surface. The attosecond streak camera is a two-color pump-probe scheme, where a weak XUV-AS pulse ionizes electrons from the solid, and a collinear, few-cycle (~5 fs FWHM) NIR pulse serves as the probe, which accelerates the XUV photo-electrons after their escape from the solid.

We study the excitation of two counter-propagating SPPs on a grating structure. A time-delay controlled arrangement of NIR and XUV beams is used to excite the SPPs and to emit photoelectrons which move in the plasmonic field. The temporal structure of the plasmic field is reflected in the photoelectron spectrum. In principle it is possible to spatially separate the pump and probe beams, allowing the imaging of plasmonic modes in different surface regions, thus providing spatio-temporal information. To distinguish this arrangement from standard attosecond streaking experiments, we named our setup "Attosecond Photoscopy".

11.4.1 Experimental Setup

Isolated AS pulses are produced from high harmonic radiation emitted by noble gases irradiated with few-cycle carrier envelope phase (CEP) stabilized NIR laser pulses [1, 6, 10, 19]. The generated high harmonic radiation co-propagates with the NIR pulse. The pulses are focused onto the grating structure with a two part mirror composed of a XUV multilayer mirror in the inner part, designed to reflect only the highest parts of the harmonic spectrum, and a broadband NIR mirror in the outer part. This arrangement allows to produce an isolated AS pulse, timed with a precision of $\lesssim$10 as to the NIR pulse.

The experimental setup is illustrated in Fig. 11.13. The NIR and XUV beams propagate along the direction orthogonal to the grating plane, which we refer to as y-direction, with polarizations along the grating grooves, in the x-direction. In the NIR focus two SPPs are excited, counter-propagating along x with polarizations along y. Although at normal incidence only one of the two plasmonic branches, the bright mode [3], is visible with plane waves, the tight focusing of about 5° of angular dispersion allows the coupling with the second branch, the "dark" mode (cf. Fig. 11.19).

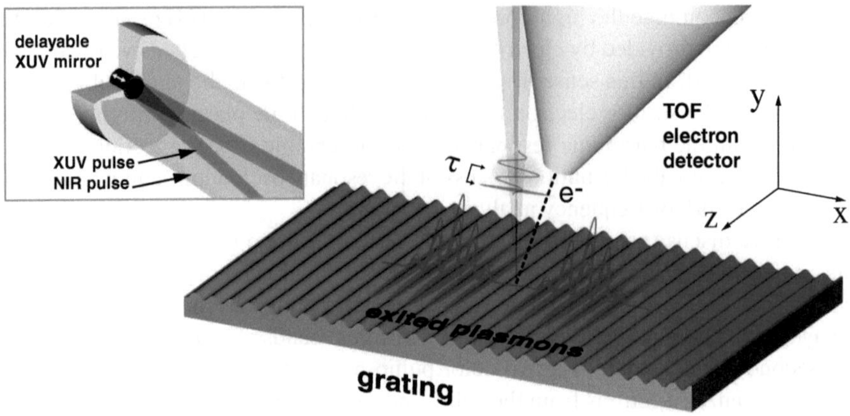

Fig. 11.13 Proposed experimental arrangement for performing attosecond photoscopy of surface plasmon excitations. A cloud of photoelectrons ionized by the XUV AS pulse are accelerated in the SPP fields, which is in turn excited by a short NIR pulse. Attosecond precise control of the time delay between the NIR and XUV pulses enables the determination of SPP transient properties. Figure reproduced from [27]

The photo-electrons emitted by the XUV-AS pulse are then collected by a detector in the y-direction. Following [6], the resulting *spectrogram* of final momenta recorded as a function of the time delay between NIR and XUV pulses, is a convolution of the photoemission process with electron acceleration by the surface fields. Depending on the time delay between the driving NIR pulse and the XUV-AS pulse, the plasmonic field recorded by the emitted photoelectrons differs in amplitude and phase, leading to a modulation of the kinetic energy distribution in the photoscopic spectrogram. We will show in Sect. 11.4.2 that the plasmonic field at the time and position of the photoelectron emission contributes to the final electron energy.

With this technique, it is possible to resolve in the time domain the energy gap between dark and bright plasmonic modes, which appears in the spectrogram as a "transition" from the bright ω_b to the dark ω_d mode frequencies. This feature is measurable in our setup thanks to the attosecond resolution.

In the following we study the properties of the photoscopic spectrogram using a classical analytical model and we compare its predictions with numerical solutions of the electron dynamics in the plasmonic field together with a Monte Carlo simulation of the photoemission process. In the end, we show how the plasmonic field at the surface can be recovered from the spectrogram analysis.

11.4.2 Theory of Attosecond Photoscopy

Typical streaking setups in attosecond metrology use atoms or molecules, which can be considered point-like electron sources on the scale of the laser wavelength.

Hence it is possible to use the dipole approximation: $\mathbf{A}(\mathbf{r}, t) \simeq \mathbf{A}(t)$. The canonical momentum along the pulse polarization is conserved, $\mathbf{P}(t) = \mathbf{P}_i$, which can be rewritten as $\mathbf{p}(t) + \frac{e}{c}\mathbf{A}(t) = \mathbf{p}_i + \frac{e}{c}\mathbf{A}(t_i)$, where $|\mathbf{p}_i| = \sqrt{2m(E_{xuv} - W_f)}$ is the initial momentum of the electron released at time t_i.

As $A(t \to \infty) = 0$, the measured final momentum is

$$\mathbf{p}_f = \mathbf{p}_i + \mathsf{a}(t_i), \tag{11.9}$$

where we absorbed the constants into the new variable $\mathsf{a} = \frac{e}{c}\mathbf{A}$.

The spectral width of the XUV-AS pulse is reflected in a momentum-broadening of the initial electron distribution $n_e = n_e(\mathbf{p}_i, t_i)$. For simplicity we assume Gaussian distributions, both in time and momentum, centered around momentum $\mathbf{p}_0$ and time t_0, respectively. With t_0 we label the time of XUV peak intensity on the target. The time-integrated final momentum distribution is

$$\sigma(\mathbf{p}_f) = \int_{-\infty}^{\infty} dt_i\, n_e(\mathbf{p}_i, t_i) = \int_{-\infty}^{\infty} dt_i\, n_e(\mathbf{p}_f - \mathsf{a}(t_i), t_i), \tag{11.10}$$

where we inserted (11.9) into the initial electron distribution dependence over $\mathbf{p}_i$. The spectrogram at varying time delay τ becomes:

$$\sigma(\mathbf{p}_f, \tau) = \int_{-\infty}^{\infty} dt_i\, n_e(\mathbf{p}_f - \mathsf{a}(t_i), t_i - \tau). \tag{11.11}$$

From this, the NIR pulse can be reconstructed via a Center-Of-Energy (COE) analysis of the average momentum of the streaking spectrogram:

$$\langle \mathbf{p} \rangle(\tau) = \frac{\int_{-\infty}^{\infty} d\mathbf{p}_f\, \mathbf{p}_f\, \sigma(\mathbf{p}_f, \tau)}{\int_{-\infty}^{\infty} d\mathbf{p}_f\, \sigma(\mathbf{p}_f, \tau)}. \tag{11.12}$$

If the detector collects only the photoelectrons emitted along the laser polarization, the vector quantities in (11.12) become scalar. Taking $n_e(p_i, t_i) = N(p_0, dp) \times N(t_0, dt)$ as initial distribution, where $N(a, b)$ is a Gaussian distribution of mean value a and variance b, the integral to solve reads:

$$\begin{aligned}\langle \mathbf{p} \rangle(\tau) &= \int_{-\infty}^{\infty} dt_i\, \mathrm{e}^{-\frac{(t_i - t_0 + \tau)^2}{2\Delta t_{\mathrm{xuv}}^2}} \int_{-\infty}^{\infty} dp\, p\, \mathrm{e}^{-\frac{(p - p_0 - \mathsf{a}_{\parallel}(t_i))^2}{2\sigma_p^2}} \\ &= \int_{-\infty}^{\infty} dt_i\, (p_0 - \mathsf{a}_{\parallel}(t_i))\mathrm{e}^{-\frac{(t_i - t_0 + \tau)^2}{2\Delta t_{\mathrm{xuv}}^2}} \underbrace{\simeq}_{\Delta t_{\mathrm{xuv}} \ll 2\pi/\omega_{\mathrm{L}}} p_0 - \mathsf{a}_{\parallel}(t_0 - \tau),\end{aligned} \tag{11.13}$$

where we defined $p = \mathbf{p}_{f,\parallel}$.

From (11.13) is clear that averaging the streaking spectrogram yields the vector potential of the NIR pulse. The vector potential reconstruction is possible only if Δt_{xuv} is much smaller than the optical period of the NIR, which is guaranteed by the ultrashort duration of the AS pulse.

The streaking technique can be generalized to systems involving inhomogeneous fields. Previous work of attosecond streaking on metal nanoparticles [51] showed that for increasing radius, the photoelectron spectrogram is smeared out by the overlap of photoelectron trajectories ionized by the XUV in different plasmonic field configuration. Therefore, the field retrieved with (11.13) is distorted by the local plasmonic field acting on the photoelectrons. It is thus clear that if we want to apply this method to plasmonic excitations on extended objects, we have to consider that the surface plasmon acting as the streaking field not only is spatially inhomogeneous but also propagates along the surface. Thus, we need to include the position dependence into our initial electron distribution: $n_e(\mathbf{p}_i, t_i) \rightarrow n_e(\mathbf{r}_i, \mathbf{p}_i, t_i)$. From the Liouville theorem, we know that the evolution of the distribution function is governed by

$$\frac{D}{Dt} n_e(\mathbf{r}_i, \mathbf{p}_i, t_i) = 0, \tag{11.14}$$

where D/Dt is the convective derivative, i.e. the derivative along the particle trajectories $D_t = \partial_t + \frac{\mathbf{p}_i}{m_i} \cdot \nabla_i$. Equation (11.14) is formally solved by $n_e(\mathbf{r}_i(t), \mathbf{p}_i(t), t) = n_e(\mathbf{r}_i, \mathbf{p}_i, t_i)$, i.e. the photoelectron distribution function is constant along the photoelectron trajectories. Thus, the problem reduces to solving each single photoelectron trajectory in the plasmonic field. In order to find the trajectory, we need the momentum of the electrons accelerated in the plasmon field, which is given by

$$\mathbf{p}(t) = \mathbf{p}_i - e \int_{-\infty}^{t} \mathbf{E}(\mathbf{r}(t'), t')\, dt'. \tag{11.15}$$

11.4.3 Low-Speed Approximation

We develop an approximation which allows to easily solve (11.15). The carrier photon energy of the XUV-AS pulses considered here is 80 eV, thus the average initial speed of a photoelectron can be estimated to be v_i ~5 nm/fs. For a driving NIR laser pulse of 4 fs duration and 5 μm focal spot, the excited plasmonic field duration can be assumed to not exceed a few tens of femtoseconds. In this time interval, the photoelectrons move in the plasmonic field by less than 100 nm. Since the additional velocity change due to the acceleration in the plasmonic field is small compared to the initial velocity, and the plasmonic evanescent field extension is of the order of the NIR wavelength (800 nm), we can approximate $\mathbf{r}(t') \simeq \mathbf{r}_i$ in (11.15) and write $\mathbf{E}(\mathbf{r}(t'), t') \simeq \mathbf{E}(\mathbf{r}_i, t')$, where $\mathbf{r}_i = \mathbf{r}(t = t_i)$. This is the zero order approximation of the Taylor series:

$$\mathbf{E}(\mathbf{r}(t), t) \simeq \mathbf{E}(\mathbf{r}_i, t) + \left.\frac{\partial \mathbf{E}}{\partial \mathbf{r}(t)}\right|_{\mathbf{r}(t)=\mathbf{r}_i} (\mathbf{r}(t) - \mathbf{r}_i) + \ldots$$

It is equivalent to neglecting the transport effects in the vector potential time derivative:

$$\frac{d}{dt}\mathbf{A}(\mathbf{r}(t), t) = (\mathbf{v} \cdot \nabla_{\mathbf{r}} + \partial_t)\, \mathbf{A}(\mathbf{r}(t), t) \simeq \partial_t \mathbf{A}. \tag{11.16}$$

The integral equation (11.15) now reads:

$$\mathbf{p}(t) = \mathbf{p}_i + \frac{e}{c}\left[\mathbf{A}(\mathbf{r}_i, t) - \mathbf{A}(\mathbf{r}_i, t_i)\right]. \tag{11.17}$$

Assuming $\mathbf{A}(\mathbf{r}_i, t \to \infty) \to 0$, we get a position corrected version of (11.9):

$$\mathbf{p}_f = \mathbf{p}_i + \frac{e}{c}\mathbf{A}(\mathbf{r}_i, t_i). \tag{11.18}$$

Since the photoelectron detector does not provide spatial resolution, the photoscopic spectrogram is the integral over time *and space* covered by the XUV spot on the grating surface:

$$\sigma(\mathbf{p}_f, \tau) = \int_{\mathbb{R}^3} d^3 r_i \int_{-\infty}^{\infty} dt_i\, n_e(\mathbf{r}_i, \mathbf{p}_f - \mathbf{a}(\mathbf{r}_i, t_i), t_i - \tau). \tag{11.19}$$

Since the spatial or temporal integral of a propagating pulse is negligible in this case (exactly zero in free space, valid as long as the polarization is orthogonal to the propagation direction), the average momentum is independent of the time-delay. For extracting time information from the photoscopic spectrogram, we have to analyze the momentum variance

$$S(\tau) = \frac{\int d\mathbf{p}_f\, |\mathbf{p}_f|^2\, \sigma(\mathbf{p}_f, \tau)}{\int d\mathbf{p}_f\, \sigma(\mathbf{p}_f, \tau)} - |\langle \mathbf{p}_f \rangle|^2, \tag{11.20}$$

where $\langle \mathbf{p}_f \rangle$ was defined in (11.12).

11.4.4 Approximation of the Photoelectron Distribution Function

We consider the photoemission process as instantaneous, because the XUV pulse duration is short compared to the NIR period and its interaction with the solid is weak.

Moreover, we discard any possible effect of electron transport in the solid and take into account only the photoelectrons coming from the first few layers of material, an assumption consistent with what was reported in [31]. Furthermore, the photoelectron spatial distribution along the grating surface is a replica of the XUV pulse intensity profile. Thanks to these considerations we can factorize n_e as:

$$n_e(\mathbf{r}_i, \mathbf{p}_i, t_i - \tau) \simeq g_\text{x}(x_i) n_e(\mathbf{p}_i)\delta(y_i - y_s)\delta(t_i - \tau - t_0), \tag{11.21}$$

where y_s is the grating surface average position and g_x is the shape function of width w_x of the XUV focal spot.

The last point needing discussion concerns the nature of the photoemission angular dependence. We take into consideration only the two extremal cases where the initial momenta are either all orthogonal to the grating plane, or are all emitted in random directions. The conjecture behind this reasoning is that the true situation falls between the two cases of unidirectional or isotropic photoemission, and has to be determined in a measurement without NIR field. We will show later that the unidirectional case can be "filtered out" from the isotropic one by choosing a proper measurement geometry, and that for either distribution, the reconstructed times closely reproduce the actual dynamics.

11.4.4.1 Unidirectional Distribution of the Photoelectrons

A unidirectional initial distribution can be expressed as $n_e(\mathbf{p}_i) = n_e(p_i\, \hat{\mathbf{n}}_s)$, where $p_i = |\mathbf{p}_i|$ and $\hat{\mathbf{n}}_s$ is the direction perpendicular to the grating plane. Substitution into (11.19) gives

$$\sigma(p_f, \tau) = \int_{-\infty}^{\infty} dx_i\, g_\text{x}(x_i) n_e \left(p_f - \hat{\mathbf{n}}_s \cdot \mathbf{a}(x_i, t_0 - \tau)\right),$$

where $\hat{\mathbf{n}}_s$ indicates the surface normal. In proximity of the grating, the plasmonic field is mostly perpendicular to the surface, hence the quantity $\hat{\mathbf{n}}_s \cdot \mathbf{a} = \mathsf{a}_y$ is approximately equal to the entire potential a_spp. Computation of the variance in (11.20) for a Gaussian distribution of the initial electron momenta yields

$$S(\tau) = \Delta p^2 + \int_{-\infty}^{\infty} dx_i\, g_\text{x}(x_i) \mathsf{a}^2_\text{spp}(x_i, t_0 - \tau). \tag{11.22}$$

11.4.4.2 Isotropic Distribution of the Photoelectrons

For the case of isotropic XUV photo-electron emission, the initial distribution can be written as: $n_e(\mathbf{p}_i) = \frac{1}{\pi} n_e(p_i) = \frac{1}{\pi} n_e(|\mathbf{p}_f - \mathbf{a}|)$, where we used $p_i = |\mathbf{p}_i|$. We further assume that $|\mathbf{a}| \ll |\mathbf{p}_f|$ to approximate $|\mathbf{p}_f - \mathbf{a}| \simeq p_f - \mathbf{a} \cdot \hat{\theta}$, where θ is the angle between the final momentum and the surface normal. The spectrogram can be then cast into

$$\sigma(p_f, \tau) = \frac{1}{\pi} \int_{-\infty}^{\infty} dx_i\, g_x(x_i)\, n_e(p_f - \mathbf{a} \cdot \hat{\theta}). \qquad (11.23)$$

A lengthy but straightforward calculation for the angular integrations leads to the expression

$$S(\tau) = \Delta p^2 + \frac{1}{\pi} \int_{-\infty}^{\infty} dx_i\, g_x(x_i) |\mathbf{a}(x_i, \tau)|^2. \qquad (11.24)$$

In any case, measuring the variance of the photoscopic spectrogram by use of (11.22) or (11.24) provides direct access to the space-averaged vector potential $\mathbf{a}^2$ at the surface in the direction of the photoelectron detector. The vector potential $|\mathbf{a}|^2 = \mathsf{a}_x^2 + \mathsf{a}_{\text{spp}}^2$ also includes a_x, the NIR field incident on the grating surface.

11.4.5 Numerical Simulation of the Photoscopic Spectrogram

As in Sect. 11.3, simulations of the plasmonic field were performed using MEEP [32]. The properties of the grating material were included by modeling the optical constants with Drude and Lorentz response functions as given in (11.6), with parameters taken from [39]. We assumed a NIR pulse of temporal Gaussian profile, with 4 fs FWHM duration at a central wavelength of 800 nm. The grating parameters were optimized for maximal absorption of the NIR pulse. Beam waists of NIR and XUV were 5 and 10 μm, respectively.

The XUV photoemission process is modeled by a Monte Carlo process, where the random ejection time, position and momentum are assigned according to the appropriate unidirectional or isotropic probability distribution described in Sect. 11.4.4. The electrons final momenta are calculated solving the Lorentz equations in the plasmonic field, for the corresponding initial conditions generated in the Monte Carlo simulation.

The key point is to understand whether the result of the low-speed approximation contained in (11.24), is accurate or not. Since we are able to directly access the simulated plasmonic field, we can check if the variance of the numerically simulated photoscopic spectrogram is in good agreement with (11.24) computed using directly the FDTD field. The spectrogram variance obtained by Monte Carlo simulation is compared in Fig. 11.14 with the space integral of the squared vector potential along the y-direction from the FDTD simulation. We assumed isotropic initial momentum distribution and a time-of-flight (TOF) detector of acceptance angle 5° centered around the direction perpendicular to the grating.

It is worth noting that the variance calculated from the Monte Carlo simulation directly images the integral of the surface plasmonic field squared, without further assumptions or approximation from the theory.

The results are analogous for the unidirectional emission, meaning that the agreement is robust with respect to the angular distribution of the photoelectrons. There-

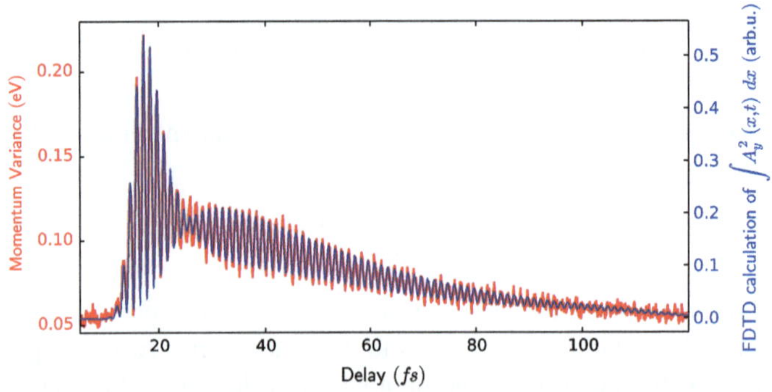

Fig. 11.14 The figure shows a comparison of the momentum variance calculated from the spectrogram in the filtered isotropic emission case of photoelectron streaked by the FDTD field, and the integral of the superficial potential squared $\int |\mathsf{a}_y|^2$ calculated directly from the FDTD field dataset. The offset of the momentum variance is due to the XUV pulse bandwidth. The perfect overlap of the curves proves the validity of our theoretical model and in particular of low-speed approximation, cf. Sect. 11.4.3. Figure reproduced from [27]

fore, knowledge of the precise photoelectron spectrum of the grating is not crucial for the proposed experiment.

In Fig. 11.15 the calculated photoscopic spectrograms for a unidirectional and isotropic initial electron distribution are compared. In the isotropic case the contributions of both the parallel and perpendicular components of the total field are captured. To identify their origins, we simulated the recording process by using a time-of-flight (TOF) spectrometer placed respectively at normal and grazing incidence incidence w.r.t. the grating surface. The resulting "filtered" spectrograms are shown in Fig. 11.16. With an angular resolved TOF measurement, it is possible to isolate the parallel component, dominated by the NIR pulse reflected at the surface, from the perpendicular one, containing the excited plasmonic field. Note that the unidirectional case, showing only the plasmonic contribution, can be obtained by appropriate measurement also in the case of isotropic emission.

For these reasons, imaging of the fields in the isotropic case provides an in situ diagnosis of both the plasmonic field and the driving NIR pulse at the surface, allowing the possibility to investigate the distortion of the NIR pulse undergoing reflection on the grating surface.

11.4.6 Analytic Model for the SPP Field on a Grating

We define a model for the excited plasmonic field in order to extract the buildup- and life-times, as well as the bright and dark mode contributions to the photoscopic spectrogram. The field envelope is assumed to be a Gaussian: $\mathsf{a}_{\mathrm{spp}} = \exp[i\varphi]\exp[-\varphi^2/2\omega_{\mathrm{spp}}^2 T^2]$, with $\varphi = k_{\mathrm{spp}}x - \omega_{\mathrm{spp}}t$. There are two

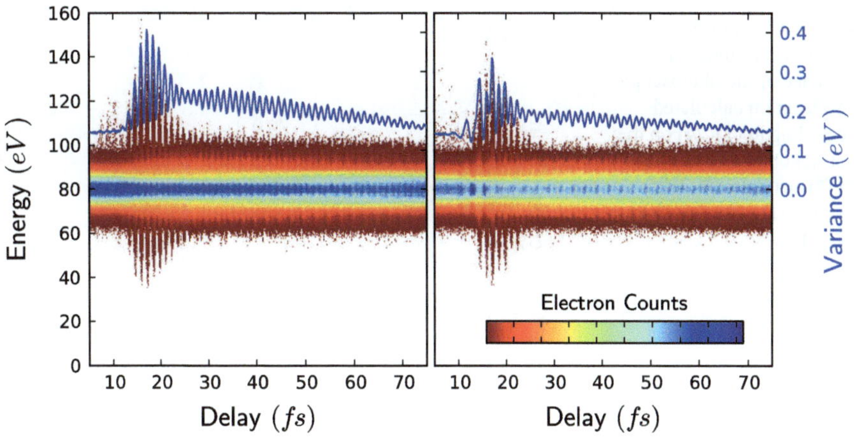

Fig. 11.15 Photoscopic spectrograms obtained with a unidirectional (*left*) and isotropic (*right*) photoelectron distribution, with *solid lines* representing the relative variances

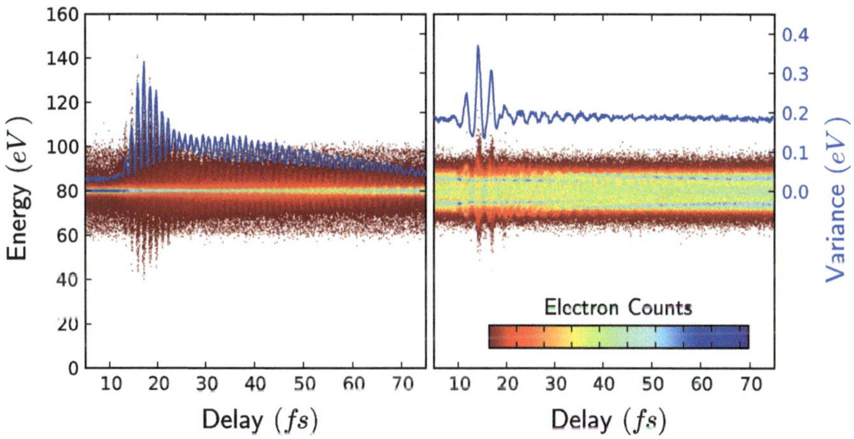

Fig. 11.16 Photoscopic spectrograms at perpendicular (*left*) and grazing (*right*) electron emission. The filtering is performed by simply setting a TOF detector at 90° and $\tilde{0}$° with respect to the grating plane. The *solid lines* are the relative momentum variances, with which one can retrieve plasmonic and NIR field, respectively. Figure reproduced from [27]

counter-propagating SPP wave-packets, each containing a bright ω_b and a dark ω_d frequency. In our configuration both the bright and the dark mode can be excited at frequencies ω_b and ω_d, respectively. Here "bright" and "dark" refer to the coupling properties of the modes: with plane waves only the bright mode would be excited, and thus visible, but thanks to the k-vector dispersion of tightly focused beam also the dark mode contributes to the spectrogram (cf. Sect. 11.4.7). As shown in Fig. 11.17, the two frequencies are well separated. Each counter-propagating plasmonic wavepacket contains both. In addition, we add a term describing the ringing

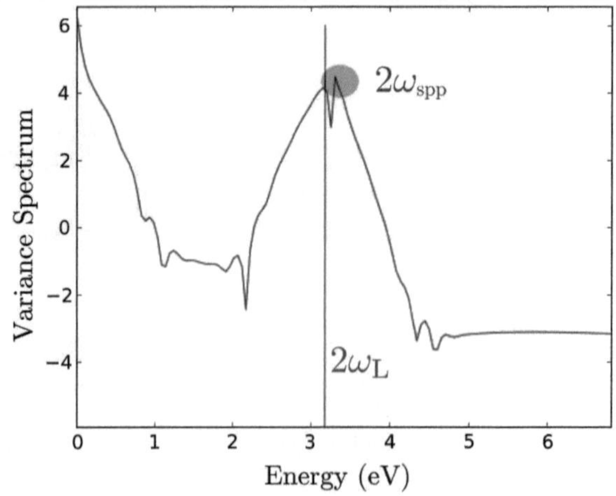

Fig. 11.17 Fourier Transform of momentum variance of the photoscopic spectrogram calculated according to (11.24) on the dataset obtained from the FDTD calculation, with parameters given in Sect. 11.4.5

of a localized mode excited in the focus of the NIR pulse (see [15]). This term can be the dominant one for very deep gratings, but in our configuration is smaller than the propagating components. The contributions of each mode $m = b, d$ to the plasmonic wavepacket are

$$P_m^{(\pm)} = e^{i\varphi_{\pm m}} e^{-\frac{\varphi_{\pm m}^2}{2\omega_m^2 T_m^2}}, \tag{11.25}$$

with the phase of the propagating plasmon

$$\varphi_{\pm m} = \pm k_m x - \omega(t - t_m) \tag{11.26}$$

and

$$P_m^{(0)} = \cos(\omega_m (t - t_m))\, e^{-x^2/2w_{\rm nir}^2}, \tag{11.27}$$

for the localized excitation.

Buildup and decay are assumed to obey a simple rate equation where a Gaussian-shaped buildup of width σ is depleted by decay at a constant rate τ:

$$\dot{f}(t) = e^{-\frac{t^2}{2\sigma^2}} - \frac{1}{2\tau} f(t). \tag{11.28}$$

The resulting time-distribution is

$$\begin{aligned} f(t, \sigma, \tau) &= \int_0^t e^{-\frac{t'^2}{2\sigma^2}} e^{-\frac{t-t'}{2\tau}}\, dt' \\ &= e^{\frac{\sigma^2 - 4\tau t}{8\tau^2}} \left[1 - \mathrm{erf}\left(\frac{\sigma^2 - 2\tau t}{2\sqrt{2}\tau\sigma}\right)\right]. \end{aligned} \tag{11.29}$$

With these assumptions, the complete field is parametrized by

$$\mathsf{a}_y(x,t) = \sum_{m=b,d} f(t-t_m,\sigma_m,\tau_m) \times \left\{\mathsf{a}_m \left[P_m^{(+)} - P_m^{(-)}\right] + \mathsf{c}_m P_m^{(0)}\right\}. \quad (11.30)$$

In practice, we find that the bright mode decays so fast that its propagation can be neglected in the spectrogram variance and we set $\mathsf{a}_b \equiv 0$.

The buildup time of each mode is defined as $\xi_m = \sigma_m\sqrt{\ln(2)}$, the half-width half-maximum of the Gaussian function in (11.29), which allows for a direct comparison with the NIR pulse FWHM duration.

We find that the dark mode plasmon duration T has a measurable effect only during generation, when counter-propagating SPPs have not separated yet and form a standing wave. Since this process is superposed by the bright mode, it cannot be reliably retrieved from the fit. On the other hand, T is only weakly correlated with the dynamical parameters ξ_m, τ_m and ω_m. We can set a conservative lower bound of T to the diameter of the NIR spot size, and upper bound to that size plus the plasmon propagation during excitation. The dynamical parameters for variations of T are shown in Table 11.3 over the range of [10, 20] fs (FWHM).

The amplitudes of the respective plasmon modes are the remaining fitting parameters, which are not reported here, since the relevant free parameters in this study are the excitation buildup times ξ_b, ξ_d, the plasmon decay times τ_b, τ_d and the plasmon frequencies ω_b, ω_d for the bright and dark modes, respectively. The result of the fitting procedure for $T = 15$ fs (FWHM) is shown in Fig. 11.18.

Table 11.3 Buildup-, life-time, and frequency of the bright and dark modes as obtained by fitting (11.24) with the parameterization (11.30), for a range of plasmon durations T_m

T	6	7	8	9	10	11	12	Var (%)
T_{FWHM}	9.99	11.66	13.32	14.99	16.65	18.32	19.98	
ξ_b	1.933	1.964	1.987	2.004	2.016	2.025	2.031	5
τ_b	3.285	3.137	3.031	2.964	2.924	2.903	2.896	13
ξ_d	5.941	5.649	5.430	5.286	5.202	5.160	5.149	15
τ_d	34.18	34.41	34.57	34.63	34.62	34.57	34.47	<1
ω_b	1.613	1.615	1.616	1.617	1.617	1.616	1.615	<1
ω_d	1.645	1.645	1.645	1.645	1.645	1.645	1.645	<1

Times in fs, frequencies in eV, $T_{\mathrm{FWHM}} = 2\sqrt{\ln 2}T$

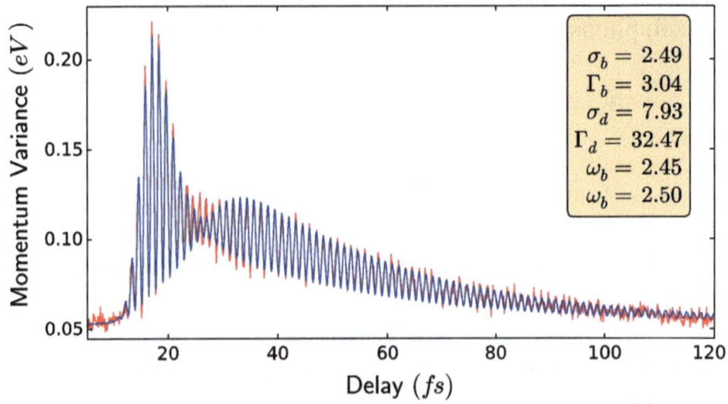

Fig. 11.18 Best fit optimization of the model parameters against the variance of the photoscopic spectrogram in the filtered isotropic case

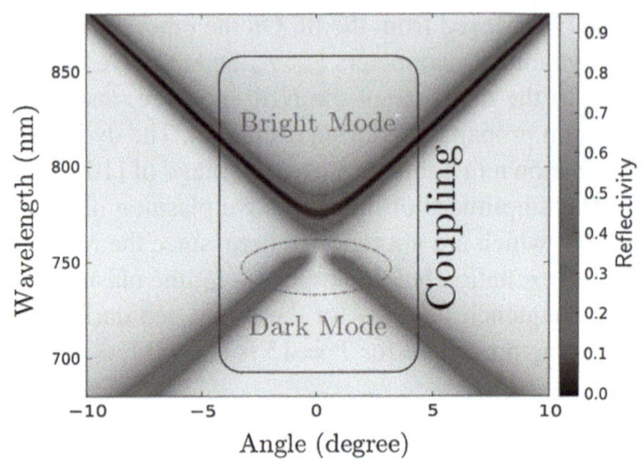

Fig. 11.19 SPP excitation mechanism on a gold grating. When the excitation source is an ultrashort Gaussian beam, the wave-vector and frequency broad bands allow coupling with the dark mode even at normal incidence. The beam divergence for a FWHM spot size of 5 μm at 800 nm is about 5°

11.4.7 Origin of Plasmon Dark and Bright Modes

Concerning the bright and dark mode frequencies, we find the values of $\hbar\omega_d = 1.62\,\text{eV}$ and $\hbar\omega_b = 1.65\,\text{eV}$, which are consistent with the plasmonic band gap of 14 nm given in [42].

It is interesting to comment the origin of the bright and dark frequencies. In Fig. 11.19 the plasmonic band structure of the grating is displayed. A plane wave at normal incidence would be able to couple only with the upper branch of the structure. However, the NIR source pulse is a tightly focused Gaussian pulse having a non-zero Δk spectral width, besides the $\Delta\omega$ bandwidth. Thus, the dark mode that would be otherwise impossible to excite, becomes visible due to k-components non-orthogonal to the grating plane. The coupling area is shown as a black rounded box in Fig. 11.19.

The result of the large beam divergence is that the excitation spectrum is not simply a cut of Fig. 11.19 at $\theta = 0$, but a convolution of the reflectivity with the angular spectrum of the Gaussian pulse. Thus, also the dark mode becomes visible.

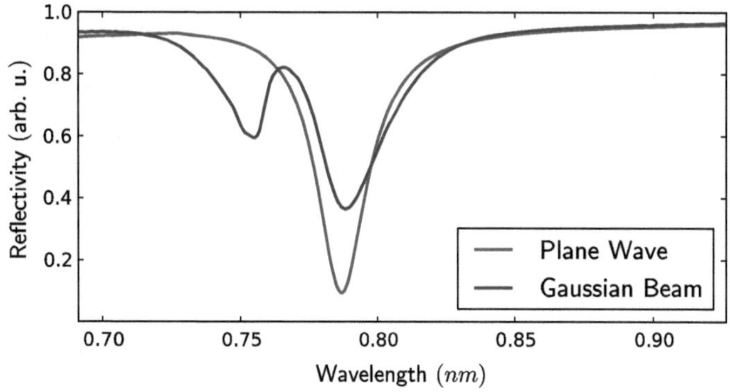

Fig. 11.20 Comparison between the reflectivity spectra of the grating structure illuminated respectively with a plane wave and Gaussian beam

In Fig. 11.20 is reported a comparison between the reflectivity spectrum of the grating calculated for a normal incident plane wave and Gaussian beam of FWHM spot size of 4 μm. The pulse duration is 4 fs FWHM for both cases. The extra peak appearing in Fig. 11.20 is consistent with the dark mode frequency obtained from the best fit optimization.

11.4.8 Results of the Plasmon Imaging

In Table 11.4 we report the results for the buildup- and life-times, obtained with $T =$ 15 fs (FWHM). The plasmon pulse extension T has little influence on the variance, because of the spatial integration. Variations in the range of $T = 10$ and 20 fs have only a small effect on buildup and decay times. With a variation of about 0.7 fs, the effect is largest on the bright and dark mode decay times, due to their overlapping. For any given value of T in this interval, the buildup and decay extracted from the FDTD surface field and from the spectrogram variance are in good agreement.

The plasmonic field enhancement can be evaluated from the comparison of the NIR versus the plasmonic field in the two spectrograms shown in Fig. 11.16. In the present case, it is ∼1. From the spectrogram at grazing direction, we get a NIR pulse duration of $\Delta t_{\mathrm{fwhm}} = 4.5$ fs, in good agreement with the 4.6 fs from the FDTD code. This measurement constitutes an in situ diagnosis of the distortions of the NIR pulse while being reflected from the grating.

Summarizing, this method allows to image the plasmonic field of SPPs, using only already existing equipment typical of attosecond metrology. The technique can be easily extended to any kind of surface plasmonic excitation, by suitably reformulating the model for the plasmonic field. Recent development of the experimental apparatuses indicates the possibility to spatially separate the XUV and NIR pulse while keeping the attosecond precise synchronization. In this scenario, our technique

Table 11.4 Best fit of the carrier frequency ω_m, buildup time ξ_m and lifetime τ_m to the numerically simulated data

	Filtered isotropic	Unidirectional	FDTD
ξ_b	2.07	2.06	2.01
τ_b	3.0	3.1	2.96
ξ_d	6.6	6.2	5.3
τ_d	32.5	33.3	34.6
ω_b	1.61	1.62	1.62
ω_d	1.65	1.65	1.65

The cases of isotropic emission with perpendicular detection ("filtered"), unidirectional emission, as well as values extracted directly from the FDTD calculation are shown. (Times in fs. Frequencies in eV), table taken from [27]

could provide time *and space* resolved imaging of virtually any surface phenomenon: by exciting a surface mode with the NIR pulse in some region, one can trace the dynamics of the excitation along complex nanostructured components by simply pointing the attosecond XUV pulse on the region of interest. The basic parameters important for the chosen configuration can then be determined by controlling the relative pulse delay. This allows the extraction of parameters such as buildup- and life-times. Within the same experiment, also an in situ diagnostics of the driver NIR pulse can be performed.

11.5 Conclusions

We have demonstrated how to extend the attosecond physics and metrology to applications involving surface plasmons. It is conceptually possible to generate isolated AS pulses from high harmonic generation in gas mediated by plasmonic enhancement of a NIR infrared laser pulse in a tapered hollow nanoplasmonic waveguide [28]. AS pulses can also be used to probe the dynamics of surface plasmon excitations in the time domain, allowing a retrieval of the buildup time of the excitation, as well as other optical properties obtainable with traditional imaging methodologies [27]. In the latter case, it is shown how "attosecond photoscopy" is capable of tailoring in situ the dynamics of plasmonic excitations on complex nanostructured surfaces, a feature which is very attractive in view of the currently expanding field of plasmon-based optoelectronics.

Speaking more generally, plasmonic phenomena are one obvious step for extending attosecond metrology from atoms and molecules to solid systems. In this framework the study of a metal and its rather simple electronic properties, which can be fully described within the linear response theory and the random phase approximation (RPA) [37], might be considered an overture in the imaging of the transient dynamics of nonlinear systems or strongly correlated materials.

References

1. L.F. DiMauro, The physics of attosecond light pulses. Reports Prog. Phys. **67**(6), 813–855 (2004)
2. A. Baltuška, Th. Udem, M. Uiberacker, M. Hentschel, E. Goulielmakis, Ch. Gohle, R. Holzwarth, V. Yakovlev, A. Scrinzi, T.W. Hänsch, F. Krausz, Attosecond control of electronic processes by intense light fields. Nature **421**, 611 (2003)
3. W. Barnes, Physical origin of photonic energy gaps in the propagation of surface plasmons on gratings. Phys. Rev. B **54**(9), 6227–6244 (1996)
4. W.L. Barnes, A. Dereux, T.W. Ebbesen, Surface plasmon subwavelength optics. Nature **424**(6950), 824–830 (2003)
5. M. Born, E. Wolf, *Principles of Optics: Electromagnetic Theory of Propagation, Interference and Diffraction of Light*, 7th edn. (Cambridge University Press, Cambridge, 1999)
6. A.L. Cavalieri, N. Müller, V.S.Th. Uphues, A. Baltuska Yakovlev, B. Horvath, B. Schmidt, L. Blümel, R. Holzwarth, S. Hendel, M. Drescher, U. Kleineberg, P.M. Echenique, R. Kienberger, F. Krausz, U. Heinzmann, Attosecond spectroscopy in condensed matter. Nature **449**(7165), 1029–1032 (2007)
7. J. Choi, S. Kim, I.-Y. Park, D.-H. Lee, S. Han, S.-W. Kim, Generation of isolated attosecond pulses using a plasmonic funnel-waveguide. New J. Phys. **14**(10), 103038 (2012)
8. M. Cinchetti, A. Gloskovskii, S. Nepjiko, G. Schönhense, H. Rochholz, et al. Photoemission Electron Microscopy as a tool for the investigation of optical near fields. Phys. Rev. Lett.**95**(4), 047601 (2005)
9. A. Cingoz, D.C. Yost, T.K. Allison, A. Ruehl, M.E. Fermann, I. Hartl, J. Ye, Direct frequency comb spectroscopy in the extreme ultraviolet. Nature **482**(7383), 68–71 (2012)
10. P. Corkum, Plasma perspective on strong field multiphoton ionization. Phys. Rev. Lett. **71**(13), 1994–1997 (1993)
11. R.A. Depine, S. Ledesma, Direct visualization of surface-plasmon bandgaps in the diffuse background of metallic gratings. Optics Lett. **29**(19), 2216 (2004)
12. H. Ditlbacher, J.R. Krenn, G. Schider, A. Leitner, F.R. Aussenegg, Two-dimensional optics with surface plasmon polaritons. Appl. Phys. Lett. **81**(10), 1762 (2002)
13. H. Ditlbacher, J.R. Krenn, A. Hohenau, A. Leitner, F.R. Aussenegg, Efficiency of local light-plasmon coupling. Appl. Phys. Lett. **83**(18), 3665 (2003)
14. M. Drescher, M. Hentschel, R. Kienberger, M. Uiberacker, V. Yakovlev, A. Scrinzi, Th Westerwalbesloh, U. Kleineberg, U. Heinzmann, F. Krausz, Time-resolved atomic inner-shell spectroscopy. Nature **419**, 803 (2002)
15. F.J. Garcia-Vidal, J. Sanchez-Dehesa, A. Dechelette, E. Bustarret, T. Lopez Rios, T. Fournier, B. Pannetier, Localized surface plasmons in lamellar metallic gratings. J. Lightwave Technol. **17**(11), 2191–2195 (1999)
16. E. Goulielmakis, M. Schultze, M. Hofstetter, V.S. Yakovlev, J. Gagnon, M. Uiberacker, A.L. Aquila, E.M. Gullikson, D.T. Attwood, R. Kienberger, F. Krausz, U. Kleineberg, Single-cycle nonlinear optics. Science **320**(5883), 1614–1617 (2008)
17. B.L. Henke, E.M. Gullikson, J.C. Davis, X-Ray interactions: photoabsorption, scattering, transmission, and reflection at E = 50–30,000 eV, Z = 1–92. Atom. Data Nucl. Data Tables **54**(2), 181–342 (1993)
18. H. Heinzelmann, D.W. Pohl, Scanning near-field optical microscopy. Appl. Phys. A Solids Surfaces **59**(2), 89–101 (1994)
19. M. Hentschel, R. Kienberger, C. Spielmann, G.A. Reider, N. Milosevic, T. Brabec, P. Corkum, U. Heinzmann, M. Drescher, F. Krausz, Attosecond metrology. Nature **414**(6863), 13–509 (2001)
20. R. Kienberger, E. Goulielmakis, M. Uiberacker, A. Baltuska, V. Yakovlev, F. Bammer, A. Scrinzi, U. Th Westerwalbesloh, U.Heinzmann Kleineberg, M. Drescher, F. Krausz, Atomic transient recorder. Nature **427**(6977), 21–817 (2004)
21. S. Kim, J. Jin, Y.-J. Kim, I.-Y. Park, Y. Kim, S.-W. Kim, High-harmonic generation by resonant plasmon field enhancement. Nature **453**(7196), 757–760 (2008)

22. S. Kim, J. Jin, Y.-J. Kim, I.-Y. Park, Y. Kim, S.-W. Kim., Kim et al. reply. Nature, **485**(7397):E1–E3 (2012)
23. C.M. Krauter, J. Schirmer, C.R. Jacob, M.Pernpointner, A. Dreuw. Plasmons in molecules: microscopic characterization based on orbital transitions and momentum conservation. J. Chem. Phys. **141**(10), 104101 (2014)
24. E. Kretschmann, Die Bestimmung optischer Konstanten von Metallen durch Anregung von Oberflächenplasmaschwingungen. Zeitschrift für Physik **241**(4), 313–324 (1971)
25. A. Kubo, K. Onda, H. Petek, Z. Sun, Y. S. Jung, et al. Femtosecond imaging of surface plasmon dynamics in a nanostructured silver film. Nano Lett. **5**(6), 1123–1127 (2005)
26. K.C. Kulander, K.J. Schafer, J.L. Krause, Dynamics of short-pulse excitation, ionization and harmonic conversion, ed. by B. Piraux. *Proceedings of the Workshop, Super Intense Laser Atom Physics (SILAP) III* (Plenum, New York, 1993)
27. M. Lupetti, J. Hengster, T. Uphues, A. Scrinzi, Attosecond photoscopy of plasmonic excitations. Phys. Rev. Lett. **113**(11), 113903 (2014)
28. M. Lupetti, M.F. Kling, A. Scrinzi, Plasmon-enhanced-attosecond-extreme ultraviolet source. Phys. Rev. Lett. **110**(22), 223903 (2013)
29. S.A. Maier, *Plasmonics: Fundamentals and Applications* (Springer, New York, 2007)
30. B. Markus, High-harmonic generation with plasmonics: feasible or unphysical? Annalen der Physik Raschke **525**(3), A40–A42 (2013)
31. S. Neppl, R. Ernstorfer, E.M. Bothschafter, A.L. Cavalieri, D. Menzel, J.V. Barth, F. Krausz, R. Kienberger, P. Feulner, Attosecond time-resolved photoemission from core and valence states of magnesium. Phys. Rev. Lett. **109**(8), 087401 (2012)
32. A.F. Oskooi, D. Roundy, M. Ibanescu, P. Bermel, J.D. Joannopoulos, S.G. Johnson, Meep: a flexible free-software package for electromagnetic simulations by the FDTD method. Comput. Phys. Commun. **181**(3), 687–702 (2010)
33. A. Otto, Excitation of nonradiative surface plasma waves in silver by the method of frustrated total reflection. Zeitschrift für Physik A Hadrons Nuclei **216**(4), 398–410 (1968)
34. E. Ozbay, Plasmonics: merging photonics and electronics at nanoscale dimensions. Science **311**(5758), 189–193 (2006)
35. I.-Y. Park, J. Choi, D.-H. Lee, S. Han, S. Kim, S.-W. Kim, Generation of EUV radiation by plasmonic field enhancement using nano-structured bowties and funnel-waveguides. Ann. Phys. **525**(1), 87 (2013)
36. I.-Y. Park, S. Kim, J. Choi, D.-H. Lee, Y.-J. Kim, M.F. Kling, M.I. Stockman, S.-. Kim, Plasmonic generation of ultrashort extreme-ultraviolet light pulses. Nat. Photon. **5**(11), 677–681 (2011)
37. J.M. Pitarke, V.M. Silkin, E.V. Chulkov, P.M. Echenique, Theory of surface plasmons and surface-plasmon polaritons. Rep. Progr. Phys. **70**(1), 1–87 (2007)
38. A. Plech, V. Kotaidis, M. Lorenc, J. Boneberg, Femtosecond laser near-field ablation from gold nanoparticles. Nat. Phys. **2**(1), 44–47 (2005)
39. A.D. Rakić, A.B. Djurišić, J.M. Elazar, M.L. Majewski, Optical properties of metallic films for vertical-cavity optoelectronic devices. Appl. Opt. **37**(22), 5271–5283 (1998)
40. S. Rausch, T. Binhammer, A. Harth, J. Kim, R. Ell, F.X. Kaertner, U. Morgner, Controlled waveforms on the single-cycle scale from a femtosecond oscillator. Optics Express **16**(13), 9739 (2008)
41. T. Remetter, P. Johnsson, J. Mauritsson, K. Varju, Y. Ni, F. Lepine, E. Gustafsson, M. Kling, J. Khan, R. Lopez-Martens, K.J. Schafer, M.J.J. Vrakking, A. L'Huillier, Attosecond electron wave packet interferometry. Nat. Phys. **2**(5), 323 (2006)
42. C. Ropers, T. Elsaesser, G. Cerullo, M. Zavelani-Rossi, C. Lienau, Ultrafast optical excitations of metallic nanostructures: from light confinement to a novel electron source. New J. Phys. **9**(10), 397 (2007)
43. A. Scrinzi, Infinite-range exterior complex scaling as a perfect absorber in time-dependent problems. Phys. Rev. A **81**(5), 1–10 (2010)
44. B. Sharma, R.R. Frontiera, A.-I. Henry, E. Ringe, R.P. Van Duyne, SERS: materials, applications, and the future. Mater. Today **15**(1–2), 16–25 (2012)

45. M. Sivis, M. Duwe, B. Abel, C. Ropers, Nanostructure-enhanced atomic line emission. Nature **485**(7397), E1–E3 (2012)
46. M. Sivis, M. Duwe, B. Abel, C. Ropers, Extreme-ultraviolet light generation in plasmonic nanostructures. Nat. Phys. **9**(5), 304–309 (2013)
47. I.J. Sola, E. Mevel, L. Elouga, E. Constant, V. Strelkov, L. Poletto, P. Villoresi, E. Benedetti, J.P. Caumes, S. Stagira, C. Vozzi, G. Sansone, M. Nisoli, Controlling attosecond electron dynamics by phase-stabilized polarization gating. Nat. Phys. **2**(5), 319 (2006)
48. M. Stockman, Nanofocusing of optical energy in tapered plasmonic waveguides. Phys. Rev. Lett. **93**(13), 137404 (2004)
49. M.I. Stockman, M.F. Kling, U. Kleineberg, F. Krausz, Attosecond nanoplasmonic-field microscope. Nat. Photon. **1**(9), 539 (2007)
50. J.A. Stratton, Electromagnetic theory (1941)
51. F. Süßmann, M.F. Kling, Attosecond nanoplasmonic streaking of localized fields near metal nanospheres. Phys. Rev. B, **84**(12) (2011)
52. A. Taflove, *Computational Electrodynamics: The Finite—difference Time—domain Method* (Artech House, Norwood, 1995)
53. H. Weyl, Ausbreitung elektromagnetischer Wellen über einem ebenen Leiter. Annalen der Physik **365**(21), 481–500 (1919)
54. D. Zhang, L. Men, Q. Chen, Microfabrication and applications of opto-microfluidic sensors. Sensors (Basel, Switzerland) **11**(5):5360–82 (2011)
55. K. Zhao, Q. Zhang, M. Chini, W. Yi, X. Wang, Z. Chang, Tailoring a 67 attosecond pulse through advantageous phase-mismatch. Opt. Lett. **37**(18), 3891–3893 (2012)

Chapter 12
Ultrafast Control of Strong-Field Electron Dynamics in Solids

Vladislav S. Yakovlev, Stanislav Yu. Kruchinin, Tim Paasch-Colberg, Mark I. Stockman and Ferenc Krausz

Abstract We review theoretical foundations and some recent progress related to the quest of controlling the motion of charge carriers with intense laser pulses and optical waveforms. The tools and techniques of attosecond science enable detailed investigations of a relatively unexplored regime of nondestructive strong-field effects. Such extremely nonlinear effects may be utilized to steer electron motion with precisely controlled optical fields and switch electric currents at a rate that is far beyond the capabilities of conventional electronics.

12.1 Introduction

It has long been realized that intense few-cycle laser pulses provide unique conditions for exploring extremely nonlinear phenomena in solids [1, 2], the key idea being that a sample can withstand a stronger electric field if the duration of the interaction is shortened. Ultimately, a single-cycle laser pulse provides the best conditions for studying nonperturbative strong-field effects, especially those where the properties of a sample change within a fraction of a laser cycle. The recent rapid development of the tools and techniques of attosecond science [3] not only creates new opportunities for detailed investigations of ultrafast electron dynamics in solids, but it also opens exciting opportunities for controlling electron motion in solids with unprecedented speed and accuracy. Conventional nonlinear phenomena that accompany the interaction of intense laser pulses with solids have already found a vast number of

V.S. Yakovlev (✉) · S.Y. Kruchinin · T. Paasch-Colberg · F. Krausz
Max-Planck-Institut für Quantenoptik, Hans-Kopfermann-Straße 1,
85748 Garching, Germany
e-mail: vladislav.yakovlev@mpq.mpg.de

V.S. Yakovlev · F. Krausz
Ludwig-Maximilians-Universität, Am Coulombwall 1, 85748 Garching, Germany

V.S. Yakovlev · M.I. Stockman
Center for Nano-Optics (CeNO) and Department of Physics and Astronomy,
Georgia State University (GSU), Atlanta 30303, GA, USA

M. Kitzler and S. Gräfe (eds.), *Ultrafast Dynamics Driven by Intense Light Pulses*, Springer Series on Atomic, Optical, and Plasma Physics 86, DOI 10.1007/978-3-319-20173-3_12

applications in spectroscopy, imaging, laser technology, transmitting and processing information [4]. It can be expected that the less conventional nonperturbative nonlinearities may also find important applications [5, 6]. The main purpose of this chapter is to review theoretical foundations and some recent progress in this field.

Most of the relevant physical phenomena are well known, such as interband tunneling, Franz–Keldysh effect, Bloch oscillations, and Wannier–Stark localization. However, new experiments put these phenomena in a new context, which often leads to nontrivial observations, such as the generation of nonperturbative high-order harmonics in a solid due to Bragg-like scattering at the edges of the Brillouin zone [7, 8], a nearly instantaneous change in extreme-ultraviolet absorptivity and near-infrared reflectivity of a dielectric in the presence of a laser field as strong as several volts per ångström [9], or the induction of electric current in an unbiased dielectric by similarly intense laser pulses [10]. Several decades of research on strong-field phenomena in solids and mesoscopic structures provide a solid ground for developing new theoretical models adapted for new experimental conditions. At the same time, a description of extremely nonlinear processes that unfold during just a few femtoseconds requires approximations that may be different from those established for longer and less intense laser pulses. Both *ab initio* calculations that attempt to capture all the complexity of many-electron dynamics and "toy models" designed to deepen our understanding of basic phenomena are going to play an important role in extending our ability to control the optical and electric properties of solids with controlled light fields.

The key parameters that determine the regime of strong-field light–matter interaction with dielectrics and semiconductors are the Keldysh parameter

$$\gamma_{\mathrm{K}} = \frac{\omega_{\mathrm{L}}\sqrt{m E_g}}{e F_{\mathrm{L}}}, \tag{12.1}$$

the Bloch frequency

$$\omega_{\mathrm{B}} = \frac{e F_{\mathrm{L}} a}{\hbar}, \tag{12.2}$$

and the Rabi frequency

$$\Omega_{\mathrm{R}} = \frac{d_{\mathrm{cv}} F_{\mathrm{L}}}{\hbar}. \tag{12.3}$$

Here, F_{L} is the amplitude of a linearly polarized electric field oscillating at an angular frequency ω_{L}, $e > 0$ is the absolute value of the electron charge, m is the reduced mass of an electron and a hole ($m^{-1} = m_{\mathrm{e}}^{-1} + m_{\mathrm{h}}^{-1}$), a is the lattice period in the polarization direction of the electric field, and d_{cv} is the dipole matrix element responsible for transitions between valence- and conduction-band states. Conditions that have to be fulfilled for perturbation theory to be applicable include $\gamma_{\mathrm{K}} \gg 1$, $\omega_{\mathrm{B}} \ll \omega_{\mathrm{L}}$, and $\Omega_{\mathrm{R}} \ll \omega_{\mathrm{L}}$. Correspondingly, the main physical effects that make the interaction nonperturbative are interband tunneling ($\gamma_{\mathrm{K}} \lesssim 1$), Bloch oscillations ($\omega_{\mathrm{B}} \geq \pi\omega_{\mathrm{L}}$), and carrier–wave Rabi flopping ($\Omega_R \gtrsim \omega_L$), the last effect being particularly important for resonant excitations ($\hbar\omega_L \approx E_g$). For a band gap of several electronvolts and a laser frequency in the near-infrared spectral range ($\lambda_L = 2\pi c/\omega_L \sim 1\ \mu m$), these effects become essential for $F_L \gtrsim 1$ V/Å, which corresponds to a laser intensity of $I_L \gtrsim 10^{13}\ \mathrm{W/cm^2}$ [11].

12.2 Main Theoretical Concepts

For a theoretical description of phenomena that take place on a few-femtosecond time scale, it is common (although not necessarily correct) to neglect dephasing. In this case, the time-dependent Schrödinger equation (TDSE)

$$i\hbar\partial_t\psi(t) = \hat{H}(t)\psi(t) \tag{12.4}$$

fully describes nonrelativistic electron dynamics. In situations where phase and energy relaxation processes are important, one has to use the more general formalisms of density matrices or nonequilibrium Green's functions, which are beyond the scope of this chapter. In the following, we review some theoretical concepts developed for the case of a single charged particle moving in a periodic potential. Furthermore, since the wavelength of visible or infrared light is much larger than the size of a unit cell, and electron velocities are much smaller than the speed of light, we can use the dipole approximation, which neglects the spatial dependence of the laser field while solving the TDSE: $\mathbf{F}_L = \mathbf{F}_L(t)$.

The first step to solve the TDSE is to choose a gauge and a basis. The exact solution does not depend on this choice, but the chosen gauge and basis dictate approximations that one may wish to make, and they influence the physical interpretation of results. In the dipole approximation, the two main options are the velocity and length gauges, which, in the following, will be abbreviated as "VG" and "LG", respectively. These two gauges are related to each other by the following unitary transformation of the respective wave functions ψ_{VG} and ψ_{LG}:

$$\psi_{VG}(t) = \exp\left[-\frac{i}{\hbar}e\mathbf{A}_L(t)\mathbf{r}\right]\psi_{LG}(t), \tag{12.5}$$

where

$$\mathbf{A}_L(t) = -\int^t \mathbf{F}_L(t')\,dt' \tag{12.6}$$

is the vector potential of the laser field.

The Hamilton operators in these two gauges take the following forms:

$$\hat{H}_{\mathrm{VG}} = \frac{\left[\hat{\mathbf{p}} + e\mathbf{A}_{\mathrm{L}}(t)\right]^2}{2m} + U(\mathbf{r}), \tag{12.7}$$

$$\hat{H}_{\mathrm{LG}} = \frac{\hat{\mathbf{p}}^2}{2m} + U(\mathbf{r}) + e\mathbf{F}_{\mathrm{L}}(t)\mathbf{r}. \tag{12.8}$$

A big advantage of the velocity gauge for numerical simulations is that a homogeneous external field does not destroy the spatial periodicity of the Hamiltonian, so that the Bloch theorem applies even in the presence of the field. At the same time, transformation (12.5) can be interpreted as a transition to a moving coordinate system, in which electrons acquire an additional momentum $-e\mathbf{A}_{\mathrm{L}}(t)$. In order to accurately account for such superficial dynamics in a time-independent basis, a sufficient number of basis states and exact transition matrix elements are required. Calculations in the velocity gauge are particularly problematic in the limit of a static field ($\omega_{\mathrm{L}} \to 0$).

These problems are circumvented in the length gauge. The price for this is the fact that the term $e\mathbf{F}_{\mathrm{L}}(t)\mathbf{r}$, which is responsible for the interaction with an external field, destroys the spatial periodicity of the Hamiltonian. Also, when periodic boundary conditions are applied in the length gauge, the interaction potential becomes discontinuous at the boundaries of the unit cell. Nevertheless, once these difficulties are addressed, the length gauge becomes an appropriate choice for numerical simulations [12–16].

When the field of a laser pulse is nonresonant ($\hbar\omega_{\mathrm{L}} \ll E_{\mathrm{g}}$) and strong, it is frequently convenient to use a time-dependent basis of quantum states that adiabatically "adapt" themselves to the external field. In the length gauge, such an adiabatic basis is given by Wannier–Stark states. In the velocity gauge, this role is played by accelerated Bloch states, also known as Houston functions [17]. The following two subsections summarize the most important properties of these states.

12.2.1 Wannier–Stark Resonances

For simplicity, let us consider a one-dimensional problem. For a constant external field parallel to the z-axis, the stationary Schrödinger equation reads

$$\hat{H}_{\mathrm{LG}}\psi \equiv \left(\frac{\hat{p}^2}{2m} + U(z) + eF_{\mathrm{L}}z\right)\psi = E\psi. \tag{12.9}$$

The potential here is periodic: $U(z+a) = U(z)$ with a lattice period a. Leaving the question of the existence of such eigenstates aside, one can ask which properties the solutions of (12.9) possess if they exist. From the periodicity of the potential, we immediately conclude that if $\psi(z)$ is an eigenstate with an energy E, then $\psi(z-a)$

is also an eigenstate with the energy $E + eF_{\mathrm{L}}a$. The additional term $eF_{\mathrm{L}}a$ is the energy required to move an electron against the laser field by one lattice period. This suggests that the eigenstates $\psi(z)$ should be localized functions. They are referred to as Wannier–Stark states.

Wannier found approximate solutions to (12.9) by defining an auxiliary problem

$$\left(\frac{\hat{p}^2}{2m} + U(z) + eF_{\mathrm{L}}\left[z + \mathrm{i}\frac{\partial}{\partial k}\right]\right) b_i(z,k) = E_i(k) b_i(z,k), \tag{12.10}$$

where i is a band index, k is the crystal momentum, and $E_i(k)$ is the energy of the unperturbed Bloch state. Wannier showed [18] that the solutions of this problem, known as Wannier–Bloch states, decouple different bands. Specifically, knowing $b_i(z,k)$ that satisfies the periodic boundary condition in reciprocal space $b_i(z, k + 2\pi/a) = b_i(z,k)$, an approximate solution of the time-dependent Schrödinger equation with the Hamiltonian H_{LG} can be written as

$$\psi_i(z,t) = b_i\left(z, k_0 - \frac{eF_{\mathrm{L}}}{\hbar}t\right) \exp\left[-\frac{\mathrm{i}}{\hbar}\int_{t_0}^{t} E_i\left(k_0 - \frac{eF_{\mathrm{L}}}{\hbar}t'\right) \mathrm{d}t'\right], \tag{12.11}$$

where the approximation consists in restricting the electron motion to a single band, that is, neglecting interband transitions. Obviously, $|\psi_i(z,t)|^2$ is a periodic function of time, the period being equal to the period of Bloch oscillations: $T_{\mathrm{B}} = 2\pi/\omega_{\mathrm{B}} = 2\pi\hbar/(eF_{\mathrm{L}}a)$. The Wannier–Stark states for a bulk crystal $\psi_{i,\ell}^{\mathrm{WS}}(z)$ are defined via the expansion

$$\psi_i(z,t) = \sum_{\ell} \psi_{i,\ell}^{\mathrm{WS}}(z) \exp\left[-\frac{\mathrm{i}}{\hbar} E_{i,\ell}^{\mathrm{WS}} t\right]. \tag{12.12}$$

A state $\psi_{i,\ell}^{\mathrm{WS}}(z)$ is localized at a lattice site ℓ. Explicit expressions for the Wannier–Stark states and their energies are

$$\psi_{i,\ell}^{\mathrm{WS}}(z) = \frac{a}{2\pi}\int_{-\pi/a}^{\pi/a} \mathrm{d}k\, b_i(z,k) \mathrm{e}^{-\mathrm{i}\ell ak}, \tag{12.13}$$

$$E_{i,\ell}^{\mathrm{WS}} = \overline{E}_i + \ell ea F_{\mathrm{L}}, \tag{12.14}$$

where $\overline{E}_i$ is the mean energy of band i:

$$\overline{E}_i = \frac{a}{2\pi}\int_{-\pi/a}^{\pi/a} \mathrm{d}k\, E_i(k).$$

The energies $E_{i,\ell}^{\mathrm{WS}}$ form the so-called "Wannier–Stark ladder"—plotted against F_{L}, they are a set of straight lines, where the slope of each line $\mathrm{d}E_{i,\ell}^{\mathrm{WS}}/\mathrm{d}F_{\mathrm{L}} = \ell ea$ is determined by the lattice index ℓ.

The localization length of a Wannier–Stark state is given by

$$L_i^{\mathrm{WS}} = \frac{\Delta_i}{e|F_{\mathrm{L}}|}, \tag{12.15}$$

where Δ_i is the energy interval covered by band i.

Our introduction to the Wannier–Stark states has so far followed the one given by Wannier [19]. The existence of Wannier–Stark states had been a subject of numerous disputes for three decades until they were experimentally observed in superlattices [20]. The core of these disputes was the question whether the Wannier–Stark states retain their physical significance when interband transitions are accounted for [21–23]. According to modern treatments [24–27], Wannier–Stark states should be viewed as resonances (metastable states) with lifetimes $\tau_i^{\mathrm{WS}} = 1/\Gamma_i$, and (12.14) should be generalized as

$$E_{i,\ell}^{\mathrm{WS}} = \overline{E}_i + \left(\ell - \frac{\gamma_i^{\mathrm{Zak}}}{2\pi}\right) e a F_{\mathrm{L}} - \mathrm{i}\hbar\frac{\Gamma_i}{2}. \tag{12.16}$$

Here, Γ_i is the decay rate due to interband transitions, and

$$\gamma_i^{\mathrm{Zak}} = \oint_{\mathrm{BZ}} \mathrm{d}\mathbf{k} \cdot \boldsymbol{\xi}_{ii}(\mathbf{k}) \tag{12.17}$$

is called Zak's phase [25, 28], where the integral is taken over a smooth closed path across the entire Brillouin zone, and $\boldsymbol{\xi}_{ii}(\mathbf{k})$ is the Berry connection or geometric vector potential [25–27]. An explicit expression for $\boldsymbol{\xi}_{ij}(\mathbf{k})$ is given by (12.26) in the next section. Zak's phase plays an important role in the "modern theory of polarization" [29]; it is equal to either 0 or π for crystals that possess inversion symmetry, and it can assume any value for other crystals. Direct measurements of Zak's phase were performed for cold atoms in optical lattices [30].

Much of the mathematical complexity related to Wannier–Stark states is avoided in finite systems, where the electron motion is restricted. In this case, it is common to refer to the exact length-gauge eigenstates of $\hat{H}_{\mathrm{LG}}$ as Wannier–Stark states (without neglecting interband transitions by evaluating Wannier–Bloch states). These states have properties similar to those of the states introduced by Wannier. One of the most important differences is that the exact eigenstates of a Hamiltonian in one spatial dimension may not be degenerate. While the states defined by (12.14) are strictly linear functions of F_{L}, and, for a certain value of F_{L}, some of these energies $E_{i_1,\ell_1}^{\mathrm{WS}}$ and $E_{i_2,\ell_2}^{\mathrm{WS}}$ may be equal to each other, the corresponding exact energies of a confined quantum system will have avoided crossings (anticrossings). An example of such a Wannier–Stark ladder is shown in Fig. 12.5.

12.2.2 Accelerated Bloch States

In Sect. 12.2.1, we saw that the instantaneous eigenstates of the length-gauge Hamiltonian serve as a convenient basis for developing approximate solutions to the TDSE. In situations where the difficulties related to the length gauge outweigh its advantages, the velocity gauge may be a better choice for either numerical or analytical approximations, and the instantaneous eigenstates of the velocity-gauge Hamiltonian (12.7) may provide a more useful time-dependent basis.

Let $\phi_{i,\mathbf{k}}$ be a Bloch state with a band index i and a crystal momentum $\mathbf{k}$:

$$\left(\frac{\hat{\mathbf{p}}^2}{2m} + U(\mathbf{r})\right)\phi_{i,\mathbf{k}} = E_i(\mathbf{k})\phi_{i,\mathbf{k}}. \tag{12.18}$$

In the coordinate representation, $\phi_{i,\mathbf{k}}(\mathbf{r})$ is a product of a plane wave and a lattice-periodic envelope function:

$$\phi_{i,\mathbf{k}}(\mathbf{r}) = \mathrm{e}^{\mathrm{i}\mathbf{k}\mathbf{r}} u_{i,\mathbf{k}}(\mathbf{r}), \tag{12.19}$$

where $u_{i,\mathbf{k}}(\mathbf{r}+\mathbf{R}) = u_{i,\mathbf{k}}(\mathbf{r})$ for all $\mathbf{R}$ from the Bravais lattice. Let us now consider the instantaneous eigenstates of $\hat{H}_{\mathrm{VG}}$ in the presence of a homogeneous external field:

$$\left(\frac{\left[\hat{\mathbf{p}} + e\mathbf{A}_{\mathrm{L}}(t)\right]^2}{2m} + U(\mathbf{r})\right)\varphi(t) = \widetilde{E}(t)\varphi(t). \tag{12.20}$$

Since the Hamiltonian is periodic in space, the Bloch theorem is applicable. Equation (12.20) has the same form as (12.18), the momentum operator being substituted with $\hat{\mathbf{p}} + e\mathbf{A}_{\mathrm{L}}(t)$. The requirement that the solutions of (12.20) satisfy the Born–von Kármán boundary conditions yields [31]

$$\varphi_{i,\mathbf{k}_0}(\mathbf{r},t) = \exp\left[-\frac{\mathrm{i}}{\hbar}e\mathbf{A}_{\mathrm{L}}(t)\mathbf{r}\right]\phi_{i,\mathbf{k}(t)}(\mathbf{r}), \tag{12.21}$$

$$\widetilde{E}_{i,\mathbf{k}_0}(t) = E_i\big(\mathbf{k}(t)\big), \tag{12.22}$$

where the time-dependent crystal momentum

$$\mathbf{k}(t) = \mathbf{k}_0 + \frac{e}{\hbar}\mathbf{A}_{\mathrm{L}}(t) \tag{12.23}$$

satisfies the acceleration theorem: $\hbar\,\mathrm{d}\mathbf{k}/\mathrm{d}t = -e\mathbf{F}_{\mathrm{L}}(t)$. Here, $\mathbf{k}_0$ is the initial crystal momentum, which the electron possessed prior to the interaction with the laser pulse.

The states $\varphi_{i,\mathbf{k}_0}(\mathbf{r},t)$ are called accelerated Bloch states or Houston functions. We note that there is a close analogy between these functions and Volkov solutions of the TDSE [32], which are frequently used for describing the strong-field ionization of atoms and molecules [33].

To use Houston functions as a basis for solving the TDSE, it is convenient to make the ansatz

$$\psi_{\mathbf{k}_0}(\mathbf{r},t) = \sum_i \alpha_{i,\mathbf{k}_0}(t) \exp\left[-\frac{\mathrm{i}}{\hbar}\int_{t_0}^{t} \mathrm{d}t'\, E_i\big(\mathbf{k}(t')\big)\right] \varphi_{i,\mathbf{k}_0}(\mathbf{r},t), \tag{12.24}$$

which leads to the following system of differential equations [31, 34]:

$$\frac{\mathrm{d}}{\mathrm{d}t}\alpha_{i,\mathbf{k}_0}(t) = \frac{e}{\mathrm{i}\hbar}\sum_j \alpha_{j,\mathbf{k}_0}(t)\mathbf{F}_\mathrm{L}(t)\cdot \boldsymbol{\xi}_{ij}\big(\mathbf{k}(t)\big) \exp\left[\frac{\mathrm{i}}{\hbar}\int_{t_0}^{t} \mathrm{d}t' \Delta E_{ij}\big(\mathbf{k}(t')\big)\right]. \tag{12.25}$$

Here, $\Delta E_{ij}(\mathbf{k}) = E_i(\mathbf{k}) - E_j(\mathbf{k})$, and the matrix elements

$$\boldsymbol{\xi}_{ij}(\mathbf{k}) \equiv \langle i,\mathbf{k}\,|\mathrm{i}\nabla_\mathbf{k}|\, j,\mathbf{k}\rangle_\mathrm{cell} = \frac{\mathrm{i}}{\Omega}\int_\Omega \mathrm{d}^3 r\, u^*_{i,\mathbf{k}}(\mathbf{r})\nabla_\mathbf{k} u_{j,\mathbf{k}}(\mathbf{r}) \tag{12.26}$$

describe the optical transitions between bands, where the integration is performed over the volume Ω of a unit cell.

The relation between $\boldsymbol{\xi}_{ij}(\mathbf{k})$ and the momentum matrix elements

$$\mathbf{p}_{ij}(\mathbf{k}) \equiv \big\langle i,\mathbf{k}\,\big|\hat{\mathbf{p}}\big|\, j,\mathbf{k}\big\rangle_\mathrm{cell} = -\frac{\mathrm{i}\hbar}{\Omega}\int_\Omega \mathrm{d}^3 r\, u^*_{i,\mathbf{k}}(\mathbf{r})\nabla_\mathbf{r} u_{j,\mathbf{k}}(\mathbf{r}) \tag{12.27}$$

is given by

$$\boldsymbol{\xi}_{i\neq j}(\mathbf{k}) = \frac{\mathrm{i}\hbar\mathbf{p}_{ij}(\mathbf{k})}{m_0\Delta E_{ij}(\mathbf{k})}, \tag{12.28}$$

where m_0 is the electron rest mass. This expression can only be applied as long as $\Delta E_{ij}(\mathbf{k})$ in the denominator is not equal to zero. The case of degenerate bands presents additional mathematical challenges [27, 35, 36]; in particular, the transition matrix elements $\boldsymbol{\xi}_{ij}(\mathbf{k})$ are singular at degeneracies [36].

For reference, we also give the relation between $\boldsymbol{\xi}_{ij}(\mathbf{k})$ and the matrix elements of the position operator between the Bloch functions [15, 37, 38]:

$$\big\langle i,\mathbf{k}'\,\big|\hat{\mathbf{r}}\big|\, j,\mathbf{k}\big\rangle_\infty = \int_{\mathbb{R}^3} \mathrm{d}^3 r\, \phi^*_{i,\mathbf{k}'}(\mathbf{r})\mathbf{r}\phi_{j,\mathbf{k}}(\mathbf{r}) = \big[\mathrm{i}\delta_{ij}\nabla_\mathbf{k} + \boldsymbol{\xi}_{ij}(\mathbf{k})\big]\delta(\mathbf{k}-\mathbf{k}'). \tag{12.29}$$

In the case $\alpha_{i,\mathbf{k}_0}(t_0) = \delta_{ij}$, where an electron is in band j before the external field is turned on, a simple approximate solution to (12.25) is [31]

$$\alpha_{i\neq j,\mathbf{k}_0}(t) \approx \frac{e}{\mathrm{i}\hbar}\int_{t_0}^{t} \mathrm{d}t'\, \mathbf{F}_\mathrm{L}(t')\cdot\boldsymbol{\xi}_{ij}\big(\mathbf{k}(t')\big)\exp\left[\frac{\mathrm{i}}{\hbar}\int_{t_0}^{t'} \mathrm{d}t'' \Delta E_{ij}\big(\mathbf{k}(t'')\big)\right], \tag{12.30}$$

provided that the excitation probabilities are small ($|\alpha_{i,\mathbf{k}_0}|^2 \ll 1$ for $i \neq j$).

This equation is a good starting point for numerous analytical approximations. In the case of a constant external field ($\mathbf{F}_\mathrm{L} = \mathrm{const}$), it is convenient to rewrite the right-hand side as an integral over the crystal momentum $\mathbf{k}$. Alternatively, the integral over time can be approximately evaluated using the saddle-point method, which is especially useful in the case of a monochromatic external field. Such approximations can be used to obtain analytical expressions for the rate of strong-field-induced transitions between valence and conduction bands of a dielectric or a semiconductor. These transitions belong to the most important strong-field effects in solids, and they are discussed in the next subsection.

12.2.3 Nonresonant Interband Transitions

In 1928, Zener [39] used semiclassical arguments to show that a constant external field F makes valence-band electrons of a dielectric tunnel to the conduction band at a rate (per unit volume)

$$\Gamma_\mathrm{Zener} = \frac{e|F|a}{2\pi\hbar}\exp\left[-\frac{\pi}{2}\frac{m^{1/2}E_\mathrm{g}^{3/2}}{e\hbar|F|}\right].$$

The prefactor of the exponential function was found to be rather sensitive to a chosen method of approximation, but all such methods yield the same argument of the exponential function. For example, Kane derived [40] the tunneling rate to be equal to

$$\Gamma_\mathrm{Kane} = \frac{e^2F^2m^{1/2}}{18\pi\hbar^2E_\mathrm{g}^{1/2}}\exp\left[-\frac{\pi}{2}\frac{m^{1/2}E_\mathrm{g}^{3/2}}{e\hbar|F|}\right],$$

while Keldysh obtained [41]

$$\Gamma_\mathrm{Keldysh} = \frac{2|eF|^{5/2}m^{1/4}}{9\pi^2\hbar^{3/2}E_\mathrm{g}^{5/4}}\exp\left[-\frac{\pi}{2}\frac{m^{1/2}E_\mathrm{g}^{3/2}}{e\hbar|F|}\right]. \quad (12.31)$$

It is common to state that such expressions for the tunneling rate have an "exponential accuracy" [42].

If the external field is not constant but oscillating at a constant frequency, interband transitions may also occur as a result of absorbing a number of photons sufficient to overcome the band gap. It must be emphasized that there is no sharp distinction between nonresonant interband tunneling and multiphoton transitions. These are two asymptotic cases of interband excitations, which are distinguished by the Keldysh parameter (12.1): $\gamma_\mathrm{K} \gg 1$ for multiphoton excitations and $\gamma_\mathrm{K} \ll 1$ for tunneling. Equation (12.31) is valid in the latter case, where the laser field is strong and its frequency is small. In the intermediate regime ($\gamma_\mathrm{K} \sim 1$), it is impossible to unambiguously distinguish between contributions from multiphoton absorption and tunneling. This is reminiscent of the situation in atomic physics where the electron motion under the potential barrier (tunneling) is known to be important even in the multiphoton regime [43]. Keldysh also derived a more general expression for the transition rate averaged over a laser cycle, which is applicable in all the three regimes [41]:

$$\Gamma_{\mathrm{GKF}} = \frac{2\omega_{\mathrm{L}}}{9\pi}\left[\frac{m\omega_{\mathrm{L}}}{\hbar\beta}\right]^{3/2} Q(\gamma_{\mathrm{K}}, \widetilde{N}) \exp\left[-\pi\lfloor\widetilde{N}+1\rfloor\frac{K(\beta)-E(\beta)}{E(\alpha)}\right], \quad (12.32)$$

$$Q(\gamma_{\mathrm{K}}, \widetilde{N}) = \left[\frac{\pi}{2K(\alpha)}\right]^{1/2} \sum_{n=0}^{\infty} \exp\left[-\frac{\pi n[K(\beta)-E(\beta)]}{E(\alpha)}\right] \times \Phi\left\{\left[\frac{\pi^2(2\lfloor\widetilde{N}+1\rfloor - 2\widetilde{N}+N)}{2K(\alpha)E(\alpha)}\right]^{1/2}\right\},$$

$$\alpha = (1+\gamma_{\mathrm{K}}^2)^{-1/2}, \quad \beta = \gamma_{\mathrm{K}}\alpha, \quad \widetilde{N} = \frac{\widetilde{E}_{\mathrm{g}}}{\hbar\omega_{\mathrm{L}}}, \quad \widetilde{E}_{\mathrm{g}} = \frac{2E(\alpha)}{\pi\beta}E_{\mathrm{g}}, \quad N = \frac{E_{\mathrm{g}}}{\hbar\omega_{\mathrm{L}}}.$$

Here, $\widetilde{E}_{\mathrm{g}}$ is an effective ionization potential, the functions $K(z)$ and $E(z)$ are the complete elliptic integrals of first and second kind, $\Phi(z)$ is the Dawson function, and $\lfloor x \rfloor$ denotes the integer part of x. We refer to (12.32) as the general Keldysh formula (GKF).

Are these formulas, which were obtained decades ago for a monochromatic external field, still useful in the case of few-cycle laser pulses? To address this question, we present, in Fig. 12.1, the outcomes of a simulation where a 4-fs 800-nm laser pulse interacts with a one-dimensional model medium that has properties resembling those of α-quartz [44]: a band gap of $E_{\mathrm{g}} = 9$ eV, a lattice period of 5 Å, and a reduced mass of $m = 0.38m_0$. To compare the numerically evaluated excitation probabilities (solid curve) with the GKF predictions, we evaluated the excitation rate $\Gamma_{\mathrm{GKF}}(t)$ using the real-valued pulse envelope in the place of the electric field F_{L}, which enters (12.32) via γ_{K}. From $\Gamma_{\mathrm{GKF}}(t)$, we estimated the excitation probability as $p_{\mathrm{GKF}} \propto 1 - \exp[-\int \Gamma_{\mathrm{GKF}}(t)\,\mathrm{d}t]$ (thick dashed curve). The overall agreement is surprisingly good, given the fact that the laser pulse is shorter than two optical periods. Both the numerical and GKF results exhibit an oscillatory behavior, which appears due to closing and opening of multiphoton channels. It is analogous to the channel closing phenomenon in atomic physics [45–47]. The thin dashed curve in Fig. 12.1 represents the excitation probability evaluated with (12.31). Since $\Gamma_{\mathrm{Keldysh}}$ is the tunneling rate for a constant field, we used $F_{\mathrm{L}}(t)$ (rather than the pulse envelope) to evaluate the excitation probability by the laser pulse. This procedure is known as the quasistatic approximation [48]. In this example, the tunneling formula (12.31) is inaccurate for $F_{\mathrm{L}} \lesssim 1.2$ V/Å, which corresponds to $\gamma_{\mathrm{K}} \gtrsim 0.9$.

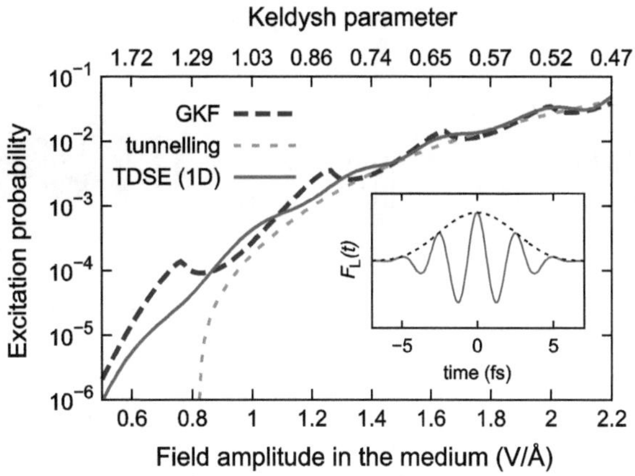

Fig. 12.1 The probability that a 4-fs laser pulse excites a valence-band electron of SiO_2 to one of the conduction bands. The laser pulse, shown in the *inset*, has a central wavelength of $\lambda_L =$ 800 nm. The *dashed curves* show probabilities evaluated by integrating analytical excitation rates: $1 - \exp[-\int \Gamma(t)\,dt]$. For the *curve* marked as 'GKF', the general Keldysh formula (12.32) was used as a cycle-averaged excitation rate; evaluating $\Gamma_{GKF}(t)$ we substituted F_L with the envelope of the laser pulse (*dashed line* in the *inset*). For the *curve* marked as 'tunneling', we used the quasistatic approximation in the tunneling limit $\Gamma(t) = \Gamma_{Keldysh}(t)$ by substituting F_L with the electric field $F_L(t)$ in (12.31). The *solid curve* is a numerical result obtained by solving the TDSE in a 1D model [44]. The transition rates (12.31) and (12.32) were multiplied with constant factors to match the TDSE result for $\gamma_K \lesssim 1$

The pioneering work by Keldysh was followed by numerous investigations. A few that we would like to point out here are analytical results obtained using the adiabatic approach, where parity selection rules were accounted for [34, 49], derivation of expressions for arbitrary N-photon transition probabilities [50], and development of the Keldysh-like theory for cosine-shaped bands [51].

12.3 Strong-Field-Driven Electron Dynamics in Crystals

12.3.1 A Numerical Example

A strong electric field drives interband transitions, it accelerates charge carriers, and it can also cause transient changes in the optical properties of a medium without necessarily exciting electrons to conduction bands. Combined with controlled optical fields, these three classes of physical phenomena enable the manipulation of the electric and optical properties of a medium over time intervals much shorter than a period of optical oscillations. The main examples of such controlled fields are laser pulses with the stabilized carrier–envelope phase (CEP) and optical waveforms [52].

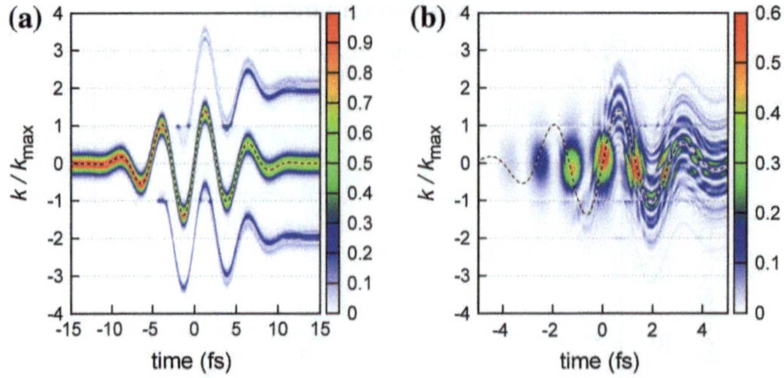

Fig. 12.2 The time-dependent population distribution of conduction-band electrons in the presence of a few-cycle laser pulse. These plots were obtained by solving (12.25) for a 1D model of SiO_2 with the same lattice potential as that used in [44]. The normalized conduction-band population ($\max n(t, k) = 1$) is plotted against time and crystal momentum k in the extended-zone scheme, where the range of crystal momenta covered by the nth lowest conduction band is $n - 1 \leq |k|/k_{\max} \leq n$ with $k_{\max} = \pi/a$. The *dashed curves* show $A_L/k_{\max}$. **a** An electron wave packet is initially placed in the lowest conduction band $\left(\alpha_{n_0,k}(t_0) = \exp[-10(k/k_{\max})^2]\right)$; an 8-fs laser pulse with a central wavelength of $\lambda_L = 1.6$ μm and a peak field of $F_L = 0.7$ V/Å is sufficiently strong to accelerate the electrons out of the first Brillouin zone. **b** All electrons are initially in the valence bands; a 4-fs pulse with $\lambda_L = 800$ nm and $F_L = 1.4$ V/Å excites them to the conduction bands, where they are accelerated by the laser field

We begin this section by illustrating such effects in a simulation where the TDSE was solved in one spatial dimension for a periodic potential. In Fig. 12.2, we show $|\alpha_{i,k(t)}(t)|^2$ obtained by solving the Houston-basis equations (12.25) and representing the conduction-band populations in the extended-zone scheme, where the range of crystal momenta covered by the n-th conduction band ($n \geq 1$) is $n - 1 \leq |k|/k_{\max} \leq n$ with $k_{\max} = \pi/a$. According to the acceleration theorem, the ballistic motion of an electron wave packet is described by $\hbar \mathrm{d}\langle k\rangle/\mathrm{d}t = -eF_L(t)$, where $\langle k\rangle$ is the mean crystal momentum. A continuous change of $\langle k\rangle$ at the borders $|k|/k_{\max} = n$ corresponds to transitions between different conduction bands. Such interband transitions are closely related to Landau–Zener transitions, as they occur at crystal momenta where the energy gap between two adjacent bands is particularly small. If a charge carrier remains within its current band as it crosses a Brillouin zone edge, its crystal momentum changes abruptly. This can be interpreted as a Bragg-like reflection of an electron wave off the crystal lattice. Figure 12.2 shows that both Bragg reflections and interband transitions play an important role when the electron motion is driven by a near-infrared field.

In Fig. 12.2a, the electron wave packet is initially placed in the lowest conduction band of the 1D model of a solid, and even though the laser pulse with $\lambda_L = 1.6$ μm is strong enough to accelerate electrons out of the first Brillouin zone, its amplitude is insufficient to induce transitions between conduction and valence bands. The figure illustrates the importance of transitions between different conduction bands, and shows that they predominantly occur at $k = \pi n/a$, $n \in \mathbb{Z}$.

For the simulations presented in Fig. 12.2b, all electrons were initially placed in the valence bands. The laser pulse had the same peak value of the vector potential as in Fig. 12.2a, but it had a shorter wavelength $\lambda_L = 800$ nm, so that the electric field of the pulse was twice as strong, and it was strong enough to excite electrons from the uppermost valence band. One can see that the population of conduction bands does not constantly increase with time, as one would expect from rate models discussed in Sect. 12.2.3. Instead, there is a transient increase of the population at the extrema of the electric field (at the zero crossings of the vector potential). Such excitations are called "virtual" [53] because they only exist as long as the external field is present. If a laser pulse is so weak that the final excitation probability can be neglected, such virtual excitations represent the distortion of bound states of the crystal—the distorted valence-band states have nonzero projections onto the field-free conduction-band Bloch functions. In the strong-field regime, no clear distinction between the "virtual" and "real" excitation can be made. Nevertheless, in both weak- and strong-field regimes, the transient increase of conduction-band population has an effect on experimentally observable quantities [53, 54].

Currently, there is no measurement technique that would reveal all the details of strong-field-driven electron dynamics like those illustrated by Fig. 12.2. However, time-resolved measurement techniques do provide indirect access to this information. Some evidence of subcycle dynamics in the strong-field excitation of electron in SiO_2 was presented in [55] by measuring the polarization rotation of an elliptically polarized pulse transmitted through a thin glass plate, as well as in [56] using a noncollinear pump–probe measurement scheme. Bragg-like scattering of electrons was found to contribute to the generation of nonperturbative high-order harmonics in solid samples [7, 8]. The reversible field-induced change of absorption in the extreme ultraviolet spectral range was observed by probing the effect of an intense near-infrared field on a thin silica plate using an attosecond pulse of extreme ultraviolet radiation as a probe [9]. A subcycle turn-on of electric current in a dielectric and its manipulation with CEP-stabilized pulses was demonstrated by measuring the residual polarization induced by the laser light [10]. This last effect is most closely related to the topic of this chapter, so it is discussed in more detail in the following section.

12.3.2 Ultrafast Injection and Control of Current in Dielectrics

According to Fig. 12.2b, an optical field that is strong enough to excite valence-band electrons of a dielectric to its conduction bands is also strong enough to significantly accelerate the created charge carriers, thus driving electric current. One of the most important findings in [10] was that such electric current can be turned on within a fraction of a half-cycle of a short intense laser pulse. This was demonstrated by irradiating a SiO_2 sample placed between two gold electrodes with CEP-stabilized laser pulses (see Fig. 12.3) and measuring the current induced by the pulses in an external circuit. It was found that a short laser pulse with the electric field directed

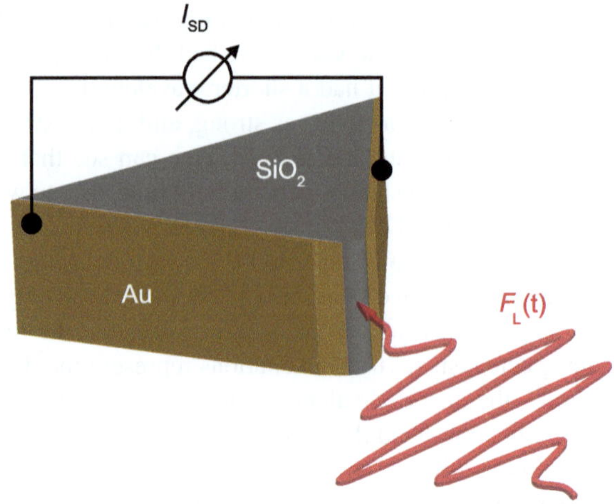

Fig. 12.3 A schematic representation of the experimental arrangement used in [10]. An intense few-cycle laser pulse induces electric current in the dielectric (SiO_2) placed between two gold electrodes. The measured signal is proportional to the net charge displaced by the laser pulse, and it is controlled by the carrier-envelope phase of the laser pulse

perpendicularly to the electrodes was able to leave the sample in a polarized state, implying that a certain electric charge was displaced by the pulse. By varying the CEP of the pulse, it was possible to control the amount of the displaced charge. This fact alone already suggests that the observed effect should be controlled by the electric field (rather than the envelope) of the laser pulse, but the most convincing evidence for the subcycle turn-on of electric current was provided by pump–probe measurements, where the sample was irradiated by a pair of pulses: an intense "injection pulse" polarized parallel to the electrodes and a relatively weak "drive" pulse polarized perpendicularly to them. By observing how the displaced charge depends on the delay between the two pulses, it was possible to conclude that the injection pulse makes the SiO_2 sample conductive within a time interval $\lesssim 1$ fs. These measurements were well reproduced by simulations: the one-dimensional tight-binding simulations in the original paper [10], a model that used a one-dimensional pseudopotential [44], and recent *ab initio* three-dimensional simulations [57]. Nevertheless, these observations permit several interpretations, which we present in the rest of this section.

12.3.2.1 Semiclassical Interpretation

The most intuitive interpretation is based on the observation that the strong-field-driven motion of an electron wave packet in the conduction bands of a dielectric largely obeys the acceleration theorem (see Fig. 12.2a)—apart from interband transitions and occasional Bragg scattering, which splits and reunites electron wave packets, each wave packet moves as a classical particle with an effective mass that depends on the mean crystal momentum. This suggests that the residual polarization induced by the laser pulse may be interpreted in semiclassical terms. A rigorous approach to this interpretation consists in writing the current density averaged over a unit cell

$$\mathbf{J}(t) = \int_{\mathrm{BZ}} \frac{\mathrm{d}^3 k_0}{(2\pi)^3} \mathbf{j}_{\mathbf{k}_0}(t) \tag{12.33}$$

in the basis of Houston functions, where the contribution from an electron with an initial crystal momentum $\mathbf{k}_0$ is given by

$$\mathbf{j}_{\mathbf{k}_0}(t) = -e \sum_i \left|\alpha_{i,\mathbf{k}_0}(t)\right|^2 \mathbf{v}_i\big(\mathbf{k}(t)\big) - \frac{2e}{m_0} \sum_{i,j<i} \mathrm{Re}\left\{\alpha^*_{i,\mathbf{k}_0}(t)\alpha_{j,\mathbf{k}_0}(t)\mathbf{p}_{ij}\big(\mathbf{k}(t)\big) \exp\left[\frac{\mathrm{i}}{\hbar}\int_{t_0}^{t} \mathrm{d}t' \Delta E_{ij}\big(\mathbf{k}(t')\big)\right]\right\}. \tag{12.34}$$

Here, $\mathbf{k}(t)$ is defined by (12.23), m_0 is the electron rest mass, $\mathbf{p}_{ij}(\mathbf{k})$ are the momentum matrix elements (12.27), and $\mathbf{v}_i(\mathbf{k}) = \nabla_{\mathbf{k}} E_i(\mathbf{k})/\hbar = \mathbf{p}_{ii}(\mathbf{k})/m_0$ is the group velocity in band i. The first sum on the right-hand side of (12.34) is responsible for the current due to the ballistic motion of charge carriers. In the semiclassical interpretation, the contribution from this term to the residual polarization (displaced charge density)

$$\mathbf{P}(t_{\max}) = \int_{-\infty}^{t_{\max}} \mathbf{J}(t)\,\mathrm{d}t$$

is assumed to be much larger than that from the second sum, which describes effects related to interband coherences.

In the tunneling regime ($\gamma_{\mathrm{K}} \lesssim 1$), charge carriers are predominantly created at the extrema of the electric field, each of which launches an electron wave packet. For a wave packet launched at a time t_0, it is convenient to introduce a *semiclassical displacement*:

$$\mathbf{s}(t_0) = \int_{t_0}^{t_{\max}} \mathbf{v}\big(\mathbf{k}(t)\big)\,\mathrm{d}t, \tag{12.35}$$

where the group velocity $\mathbf{v}(\mathbf{k})$ should correspond to the most probable quantum path of the wave packet in reciprocal space. As long as $\mathbf{k}(t)$ is not limited to the first Brillouin zone, this approach is most useful if the probabilities of Bragg scattering at the edges of the Brillouin zone are either negligibly small or close to 100 %. The contribution from each wave packet to the final polarization $\mathbf{P}(t_{\max})$ is the product of the charge carried by the wave packet and its semiclassical displacement $\mathbf{s}(t_0)$.

This kind of semiclassical analysis explains some outcomes of numerical simulations [58]. In particular, it explains the observation that, for moderate laser intensities, the residual polarization scales as $P \propto F_{\mathrm{L}}^{2N+1}$ [44], where $N = E_{\mathrm{g}}/(\hbar\omega_{\mathrm{L}})$ is the ratio of the band gap to the photon energy. As long as the Keldysh parameter is sufficiently large to view interband excitations as the result of absorbing N photons, the excitation probability scales as $p \propto I_{\mathrm{L}}^N \propto F_{\mathrm{L}}^{2N}$, I_{L} being the peak laser intensity, while the semiclassical displacement is proportional to F_{L}. Thus, the product of the charge and the displacement is proportional to F_{L}^{2N+1}.

12.3.2.2 Interference of Multiphoton Pathways

For moderate laser intensities, where perturbation theory is expected to yield at least qualitatively correct predictions, it is also possible to interpret the optically controlled electric current in terms of interference between different multiphoton excitation pathways. This interpretation is a generalization of ideas developed in the field of coherent control, where irradiating a semiconductor by two monochromatic laser beams with frequencies ω_1 and $\omega_2 = 2\omega_1$ was found to induce an electric current sensitive to the relative phase between the two beams [59]. This phase sensitivity is due to the interference between single- and two-photon absorption processes. The density $n(\mathbf{k})$ of electrons excited to the conduction band at a crystal momentum $\mathbf{k}$ is determined by the phase parameter $\Delta\varphi = 2\varphi_{\omega_1} - \varphi_{\omega_2}$. Furthermore, $n(\mathbf{k})$ is, in general, an asymmetric function of $\mathbf{k}$ because the transition amplitudes for the one- and two-photon channels have different symmetries with respect to the transformation $\mathbf{k} \to -\mathbf{k}$ [59]. Injecting currents through interfering photoexcitation pathways was investigated in theory and experiments for semiconductors [59–66] and molecular wires [67].

In the case where charge carriers are excited by a few-cycle pulse with a central frequency that is much smaller than the band gap, the interfering photoexcitation pathways are multiphoton excitation channels. It is convenient to analyze them in the velocity gauge using the basis of Bloch states, where an external homogeneous electric field only induces transitions between states with the same the crystal momentum. Due to the large bandwidth of ultrashort laser pulses, it is possible to make the same transition by absorbing different numbers of photons, as it is schematically shown in Fig. 12.4.

Within this picture, the scaling law $Q \propto F_{\mathrm{L}}^{2N+1}$ may be interpreted as a result of interference between quantum pathways that involve absorbing N and $N+1$ photons. The corresponding probability amplitudes are proportional to F_{L}^{N} and F_{L}^{N+1}, respectively. When the two pathways interfere, the excitation probability, which is the squared modulus of the sum of the probability amplitudes, is described by an expression that contains the product of the two amplitudes. This product, which is proportional to F_{L}^{2N+1}, determines the induced electric current. Even though these arguments by no means form a rigorous proof, and these considerations are only applicable in the multiphoton regime, the interpretation in terms of interfering multiphoton channels was shown to be a plausible one, explaining not only the scaling with intensity, but also the fact that a CEP-sensitive displaced charge can only be observed if the laser pulse is sufficiently broadband [44].

12.3.2.3 Adiabatic Metallization

Another interpretation of the optical-field-induced current in dielectrics accompanied the first publication of these experimental results [10]. This mechanism relies on the effect of "adiabatic metallization", predicted theoretically for dielectric nanofilms [68, 69]. It was found that a strong electric field can significantly and

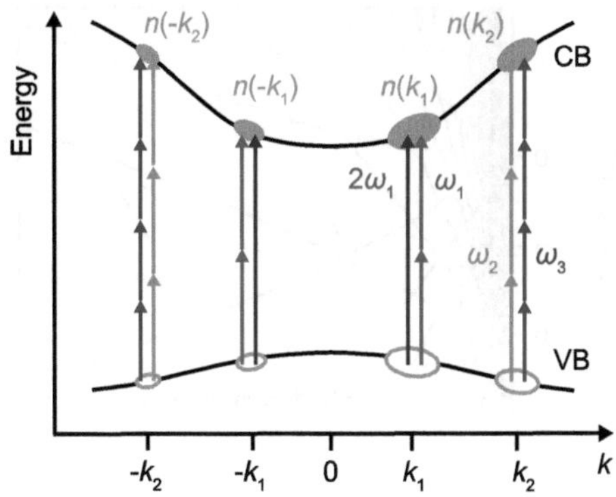

Fig. 12.4 Current injection via interference of multiphoton excitation channels. Charge carriers can be injected from a valence band (*VB*) into a conduction band (*CB*) via different multiphoton excitation pathways if the spectral bandwidth of the applied laser light is sufficiently large. For example, the excitation at crystal momenta k_2 and $-k_2$ may be the outcome of absorbing either three high-energy or four low-energy photons from the same ultrashort laser pulse. Quantum interference between the odd- and even-numbered contributions in each pathway determines the population in the final state. An asymmetry in the conduction-band population $\left(n(k) \neq n(-k)\right)$ results in a net current density is formed inside the material

reversibly change the optical and electric properties of a sufficiently thin dielectric nanofilm. During the interaction with the field, the real part of the dielectric constant may even become negative, which is a property attributed to metals. These effects are best understood in the basis of instantaneous eigenstates of the length-gauge Hamiltonian, which are the Wannier–Stark states discussed at the end of Sect. 12.2.1. Even though these results were obtained for nanofilms, similar effects may be expected in bulk solids [70], provided that the localization length of Wannier–Stark states does not exceed a few lattice sites, which is indeed the case for field strengths $F_{\mathrm{L}} \gtrsim 1$ V/Å.

Using the Wannier–Stark states as a time-dependent basis, the interaction with an intense pulse can be analyzed in terms of adiabatic and diabatic transitions at avoided crossings. This analysis [10, 71] shows that the strong field may cause a reversible quantum transition from an insulating state to a state with an increased conductivity. This transition occurs when the energy gaps at avoided crossings between Wannier–Stark states become sufficiently large for electrons to adiabatically pass such an anticrossing (see Fig. 12.5). The anticrossing gap takes its largest value when the relevant Wannier–Stark states are localized at neighboring lattice sites, which happens when the electric field in the medium is as strong as $F_{\mathrm{crit}} = E_{\mathrm{g}}/(ea)$. This condition is equivalent to $\hbar\omega_{\mathrm{B}} = E_{\mathrm{g}}$. For SiO_2, $F_{\mathrm{crit}} \approx 3$ V/Å, which is above the damage threshold, but even the penultimate anticrossing, which occurs at $F_{\mathrm{crit}} \approx 1.5$ V/Å, was found to have a sufficiently large energy gap. This interpretation of optical-field-controlled current as a result of adiabatic metallization was also supported by the observation of the optical-field-induced transient reflectivity in the strong-field regime [9].

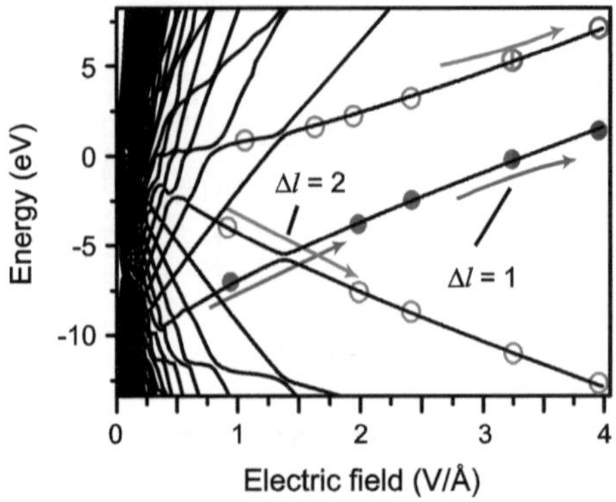

Fig. 12.5 Eigenstates of the Wannier-Stark Hamiltonian (12.8) for a 50-nm nanofilm of SiO_2 in dependence of the electric field. The labels $\Delta l = 1$ and $\Delta l = 2$ indicate the anticrossings that correspond to Wannier-Stark states being localized one and two lattice sites apart, respectively. *Closed circles* represent occupied states, while *open circles* represent unoccupied states. The figure is adapted from [71]

12.4 Summary and Outlook

The main message of this chapter is that the control over the electric field of light pulses can be transformed into the control over strong-field-driven electron dynamics, which may open attractive opportunities for both basic research and applications in ultrafast signal processing. One example of such an application is a solid-state CEP detector [72], and more advanced applications may emerge in the future. Attosecond science provides powerful tools and techniques for studying strong-field dynamics in solids, which may lead to important insights into fundamentally important phenomena. So far, only the first few steps have been done in this direction, and many questions remain open. Further research is necessary to verify conjectures made in the interpretation of recent measurements, and it is yet to be understood how electron–electron interaction, scattering, and dephasing phenomena affect our ability to launch and manipulate electron wave packets. One of the most important goals is to identify those extremely nonlinear effect that are largely reversible on a few-femtosecond time scale, as only such effect may serve as a basis for applications in signal processing. On the experimental side, there is a large potential for exploiting various laser sources, intense optical waveforms, and (nano) structures designed for strong-field measurements.

Acknowledgments We acknowledge useful discussions with and help from A. Schiffrin and V. Apalkov. We thank the Munich Centre for Advanced Photonics for support. The work of M.I.S. and V.S.Y. was supported primarily by the Chemical Sciences, Biosciences and Geosciences Division (grant no. DEFG02-01ER15213) and additionally by the Materials Sciences and Engineering Division (grant no. DE-FG02-11ER46789) of the Office of the Basic Energy Sciences, Office of Science, US Department of Energy and by a MURI grant from the US Air Force Office of Scientific Research.

References

1. M. Lenzner, J. Krüger, S. Sartania, Z. Cheng, C. Spielmann, G. Mourou, W. Kautek, F. Krausz, Phys. Rev. Lett. **80**, 4076 (1998)
2. S.S. Mao, F. Qur, S. Guizard, X. Mao, R.E. Russo, G. Petite, P. Martin, Appl. Phys. A **79**(7), 1695 (2004)
3. F. Krausz, M. Ivanov, Rev. Mod. Phys. **81**, 163 (2009)
4. E. Garmire, Opt. Express **21**(25), 30532 (2013)
5. F. Krausz, M.I. Stockman, Nat. Photon. **8**(3), 205 (2014)
6. S. Ghimire, G. Ndabashimiye, A.D. DiChiara, E. Sistrunk, M.I. Stockman, P. Agostini, L.F. DiMauro, D.A. Reis, J. Phys. B **47**(20), 204030 (2014)
7. S. Ghimire, A.D. DiChiara, E. Sistrunk, P. Agostini, L.F. DiMauro, D.A. Reis, Nat. Phys. **7**(2), 138 (2011)
8. O. Schubert, M. Hohenleutner, F. Langer, B. Urbanek, C. Lange, U. Huttner, D. Golde, T. Meier, M. Kira, S.W. Koch, R. Huber, Nat. Photon. **8**(2), 119 (2014)
9. M. Schultze, E.M. Bothschafter, A. Sommer, S. Holzner, W. Schweinberger, M. Fiess, M. Hofstetter, R. Kienberger, V. Apalkov, V.S. Yakovlev, M.I. Stockman, F. Krausz, Nature **493**(7430), 75 (2013)
10. A. Schiffrin, T. Paasch-Colberg, N. Karpowicz, V. Apalkov, D. Gerster, S. Mühlbrandt, M. Korbman, J. Reichert, M. Schultze, S. Holzner, J.V. Barth, R. Kienberger, R. Ernstorfer, V.S. Yakovlev, M.I. Stockman, F. Krausz, Nature **493**, 70 (2013)
11. M. Wegener, *Extreme Nonlinear Optics* (Springer, Berlin, 2005)
12. C. Aversa, J.E. Sipe, Phys. Rev. B **52**, 14636 (1995)
13. R. Resta, Phys. Rev. Lett. **80**, 1800 (1998)
14. I. Souza, J. Íñiguez, D. Vanderbilt, Phys. Rev. B **69**, 085106 (2004)
15. K.S. Virk, J.E. Sipe, Phys. Rev. B **76**, 035213 (2007)
16. M. Springborg, B. Kirtman, Phys. Rev. B **77**, 045102 (2008)
17. W.V. Houston, Phys. Rev. **57**, 184 (1940)
18. G.H. Wannier, Phys. Rev. **117**, 432 (1960)
19. G.H. Wannier, Rev. Mod. Phys. **34**, 645 (1962)
20. P. Voisin, J. Bleuse, C. Bouche, S. Gaillard, C. Alibert, A. Regreny, Phys. Rev. Lett. **61**, 1639 (1988)
21. J. Zak, Phys. Rev. Lett. **20**, 1477 (1968)
22. G.H. Wannier, Phys. Rev. **181**, 1364 (1969)
23. J.E. Avron, J. Zak, A. Grossmann, L. Gunther, J. Math. Phys. **18**(5), 918 (1977)
24. M. Glück, A.R. Kolovsky, H.J. Korsch, Physics Reports **366**(3), 103 (2002)
25. R. Resta, J. Phys.: Condensed Matter. **12**(9), R107 (2000). http://stacks.iop.org/0953-8984/12/i=9/a=201
26. G. Sundaram, Q. Niu, Phys. Rev. B **59**, 14915 (1999)
27. D. Xiao, M.C. Chang, Q. Niu, Rev. Mod. Phys. **82**, 1959 (2010)
28. J. Zak, Phys. Rev. Lett. **62**, 2747 (1989)
29. R.D. King-Smith, D. Vanderbilt, Phys. Rev. B **47**, 1651 (1993)
30. M. Atala, M. Aidelsburger, J. Barreiro, D. Abanin, T. Kitagawa, Nat. Phys. **9**(12), 795 (2013)

31. J.B. Krieger, G.J. Iafrate, Phys. Rev. B **33**, 5494 (1986)
32. D.M. Volkov, Z. Physik **94**, 250 (1935)
33. H.D. Jones, H.R. Reiss, Phys. Rev. B **16**, 2466 (1977)
34. Yu.A. Bychkov, A.M. Dykhne, Sov. Phys. JETP **58**, 1734 (1970)
35. D. Culcer, Y. Yao, Q. Niu, Phys. Rev. B **72**, 085110 (2005)
36. B.A. Foreman, J. Phys.: Condensed Matter **12**(34), R435 (2000). http://stacks.iop.org/0953-8984/12/i=34/a=201
37. E.I. Blount, in *Solid State Physics: Advances in Research and Applications*, ed. by F. Seitz, D. Turnbull (Academic Press, New York, 1962), pp. 305–373
38. B. Gu, N.H. Kwong, R. Binder, Phys. Rev. B **87**, 125301 (2013)
39. C. Zener, Proceedings of the Royal Society of London, Series A **145**(855), 523 (1934). http://www.jstor.org/stable/2935519
40. E.O. Kane, J. Phys. Chem. Solids **12**(2), 181 (1960)
41. L.V. Keldysh, Sov. Phys. JETP **20**(5), 1307 (1965)
42. V.S. Popov, Physics-Uspekhi **47**(9), 855 (2004). http://stacks.iop.org/1063-7869/47/i=9/a=R01
43. M. Ivanov, M. Spanner, O. Smirnova, J. Mod. Opt. **52**(2–3), 165 (2005)
44. S.Yu. Kruchinin, M. Korbman, V.S. Yakovlev, Phys. Rev. B **87**, 115201 (2013)
45. J.G. Story, D.I. Duncan, T.F. Gallagher, Phys. Rev. A **49**, 3875 (1994)
46. G.G. Paulus, F. Grasbon, H. Walther, R. Kopold, W. Becker, Phys. Rev. A **64**, 021401 (2001)
47. R. Kopold, W. Becker, M. Kleber, G.G. Paulus, J. Phys. B **35**(2), 217 (2002). http://stacks.iop.org/0953-4075/35/i=2/a=302
48. M. Ivanov, O. Smirnova, Chem. Phys. **414**, 3 (2013)
49. H. Minasian, S. Avetisian, Phys. Rev. B **34**, 963 (1986)
50. V.A. Kovarskii, E.Yu. Perlin, Phys. Stat. Sol. B **45**, 47 (1971)
51. V.E. Gruzdev, Phys. Rev. B **75**, 205106 (2007)
52. A. Wirth, M.T. Hassan, I. Grguras, J. Gagnon, A. Moulet, T.T. Luu, S. Pabst, R. Santra, Z.A. Alahmed, A.M. Azzeer, V.S. Yakovlev, V. Pervak, F. Krausz, E. Goulielmakis, Science **334**(6053), 195 (2011)
53. E. Yablonovitch, J.P. Heritage, D.E. Aspnes, Y. Yafet, Phys. Rev. Lett. **63**, 976 (1989)
54. A.V. Kuznetsov, C.J. Stanton, Phys. Rev. B **48**, 10828 (1993)
55. M. Gertsvolf, M. Spanner, D.M. Rayner, P.B. Corkum, J. Phys. B: At. Mol. Opt. Phys. **43**(13), 131002 (2010)
56. A.V. Mitrofanov, A.J. Verhoef, E.E. Serebryannikov, J. Lumeau, L. Glebov, A.M. Zheltikov, A. Baltuška, Phys. Rev. Lett. **106**, 147401 (2011)
57. G. Wachter, C. Lemell, J. Burgdörfer, S.A. Sato, X.M. Tong, K. Yabana, Phys. Rev. Lett. **113**, 087401 (2014)
58. P. Földi, M.G. Benedict, V.S. Yakovlev, New J. Phys. **15**(6), 063019 (2013)
59. T.M. Fortier, P.A. Roos, D.J. Jones, S.T. Cundiff, R.D.R. Bhat, J.E. Sipe, Phys. Rev. Lett. **92**, 147403 (2004)
60. G. Kurizki, M. Shapiro, P. Brumer, Phys. Rev. B **39**, 3435 (1989)
61. R. Atanasov, A. Haché, J.L.P. Hughes, H.M. van Driel, J.E. Sipe, Phys. Rev. Lett. **76**, 1703 (1996)
62. A. Haché, Y. Kostoulas, R. Atanasov, J.L.P. Hughes, J.E. Sipe, H.M. van Driel, Phys. Rev. Lett. **78**, 306 (1997)
63. J.M. Fraser, A.I. Shkrebtii, J.E. Sipe, H.M. van Driel, Physica B **272**(1–4), 353 (1999)
64. H. Zhao, E.J. Loren, A.L. Smirl, H.M. van Driel, J. Appl. Phys. **103**(5), 053510 (2008)
65. L. Costa, M. Betz, M. Spasenovic, A.D. Bristow, H.M. van Driel, Nature Physics **3**(9), 632 (2007)
66. J. Rioux, J.E. Sipe, Physica E **45**(0), 1 (2012)
67. I. Franco, M. Shapiro, P. Brumer, Phys. Rev. Lett. **99**, 126802 (2007)
68. M. Durach, A. Rusina, M.F. Kling, M.I. Stockman, Phys. Rev. Lett. **105**, 086803 (2010)
69. M. Durach, A. Rusina, M.F. Kling, M.I. Stockman, Phys. Rev. Lett. **107**, 086602 (2011)
70. V. Apalkov, M.I. Stockman, Phys. Rev. B **86**, 165118 (2012)

71. A. Schiffrin, T. Paasch-Colberg, N. Karpowicz, V. Apalkov, D. Gerster, S. Muhlbrandt, M. Korbman, J. Reichert, M. Schultze, S. Holzner, J.V. Barth, R. Kienberger, R. Ernstorfer, V.S. Yakovlev, M.I. Stockman, F. Krausz, Nature **507**(7492), 386 (2014)
72. T. Paasch-Colberg, A. Schiffrin, N. Karpowicz, S. Kruchinin, O. Sağlam, S. Keiber, O. Razskazovskaya, S. Mühlbrandt, A. Alnaser, M. Kübel, V. Apalkov, D. Gerster, J. Reichert, T. Wittmann, J.V. Barth, M.I. Stockman, R. Ernstorfer, V.S. Yakovlev, R. Kienberger, F. Krausz, Nat. Photon. **8**(3), 214 (2014)

Part IV
Atoms and Molecules Driven and Probed by Intense X-Ray Pulses

Chapter 13
Atomic and Molecular Systems Under Intense X-Ray Radiation

Maria Krikunova, Nicusor Timneanu and Jakob Andreasson

Abstract The review covers recent progress in the development of X-ray pulse metrology essential for experiments at Free Electron Lasers. The scientific section is focused on time-resolved studies of ionization dynamics of atoms, molecules and (bio-)nanoparticles. We discuss the role of ionization dynamics for high resolution imaging of bio- and bio-like nanoparticles and illustrate the potential for multi-directional imaging of unique non-reproducible samples.

13.1 Introduction

The new generation of linear-accelerator based Free Electron Lasers (FELs) operating from soft (XUV) to hard X-ray photon energies has triggered a wide range of experiments to study the dynamics of electronic and nuclear rearrangement in small quantum systems like atoms, molecules, and nano-particles [1]. Especially, the unique combination of coherent intense radiation with a short pulse duration opens a new parameter regime where non-linear interactions between X-rays and matter can be studied. One of the most exciting applications of X-ray lasers is Coherent

M. Krikunova (✉)
Institut Für Optik und Atomare Physik, Technische Universität Berlin,
Strasse des 17. Juni 135, ER 1-1, 10623 Berlin, Germany
e-mail: maria.krikunova@physik.tu-berlin.de

N. Timneanu · J. Andreasson
Molecular Biophysics Department of Cell and Molecular Biology (ICM),
Uppsala University, Husargatan 3, Box 596, 75124 Uppsala, Sweden
e-mail: nicusor@xray.bmc.uu.se

N. Timneanu
Department of Physics and Astronomy, Uppsala University,
Box 516, 75120 Uppsala, Sweden

J. Andreasson
ELI-Beamlines, Institute of Physics, Academy of Sciences of the Czech Republic,
Na Slovance 2, 18221 Prague, Czech Republic
e-mail: jakoba@xray.bmc.uu.se

M. Kitzler and S. Gräfe (eds.), *Ultrafast Dynamics Driven by Intense Light Pulses*, Springer Series on Atomic, Optical, and Plasma Physics 86, DOI 10.1007/978-3-319-20173-3_13

Diffractive Imaging (CDI) of non-periodic nano-scale objects, such as large biomolecules within their functional environment. In the so called *diffraction before destruction* approach, high irradiation intensity is needed to create high resolution diffraction pattern within a single shot. At the same time, the FEL pulse has to be short enough to avoid significant nuclear displacement during the exposure [2]. Otherwise radiation damage that triggers electronic rearrangements and nuclear displacements in the nanometer sized (bio-)particle will unavoidably limit the maximum resolution of the image. Therefore, a comprehensive understanding of the physical processes that govern the response of matter under extreme irradiation conditions is essential to optimize the conditions for the CDI application.

The dynamics of matter exposed to X-ray radiation is very complex and proceeds on several different time-scales [3]. The initial step in the interaction of X-ray light with the sample under study depends strongly on sample composition, namely on the absorption cross-sections of individual atoms. Absorption of an X-ray photon will preferably create an inner-shell vacancy, which is strongly localized. The de-excitation processes is accompanied by the emission of secondary Auger electrons or X-ray fluorescence photons. In extended systems like molecules and clusters, the relaxation dynamics will involve electron rearrangement within the entire atomic or molecular system. This is followed by atomic movement due to accumulated positive charge and increasing temperature leading to a complete fragmentation of the sample.

In the high intensity X-ray energy regime the diversity of available excitation and de-excitation channels is substantially increased. Among thoroughly studied processes are multi-photon absorption, dynamics of electron trapping, and the formation of a nano-plasma within the nanometer sized sample. These initial dynamics trigger secondary processes like multi-electron collisions with energy exchange and electron impact ionization followed by electron-ion recombination (see e.g. [1] and references therein). Moreover, an unusual collective plasma-like state has been reported recently [4] rising the question about the role of atomic resonances in the excitation processes.

According to theoretical calculations [5, 6] and recent experiments [7] the ionization dynamics of a single-particle in the focus of an FEL beam is strongly influenced by the FEL pulse parameters. However, most studies on the ionization dynamics in clusters and nano-particles reported so far average experimental data over both the pulse intensity profile and particle size distribution [8–13]. Notably, ion yield spectra obtained from a single Xe cluster at highest power density reveal substantial difference with respect to the focal volume integrated ion spectra [7]. These studies demonstrate very impressively that some excitation channels might be hidden by the averaging of experimental data. Thus, tracing the evolution of highly-excited non-equilibrium states of matter directly in the time domain is challenging because it requires experiments with high temporal resolution and spectral selectivity. Moreover, experiments have to be performed at very well defined and reproducible experimental parameters.

One aspect that affects the reproducibility of FEL pulse parameters from shot-to-shot is the self-amplified spontaneous emission (SASE) mode of most currently operated FEL sources [1]. The statistical nature of the FEL radiation causes substantial shot-to-shot fluctuations of the pulse energy and spectrum, as well as the spatial and temporal profile. Additionally, the implementation of two-color pump-probe techniques like optical-pump/X-ray-probe or X-ray-pump/optical-probe require a perfect synchronization of FEL pulses with respect to the pulses from an external laser system.

Recently developed techniques capable of generating close to transform-limited, temporally and spatially coherent XUV or X-ray pulse are using a seed pulse for the light amplification process. Several injection seeding schemes have been implemented. Fermi in Trieste uses seeding based on high gain harmonic generation (HGHG) starting from the external UV laser pulse [14] whereas FLASH in Hamburg implements seeding from the 21st harmonic of a near-infrared (NIR) laser pulse generated in a gas medium [15]. Moreover, self-seeding using an X-ray pulse from the first half of the undulator have been realized at the LCLS in Stanford [16].

For SASE-FELs the pulse-to-pulse diagnostic techniques represent alternatives to complex seeding schemes. In the so called data-tagging concept the properties of individual FEL pulses are measured independently. Subsequently, data from simultaneously performed experiment can be sorted according to essential pulse parameters. Several pulse-to-pulse diagnostic tools have been developed recently [17] and are under continuous further improvement [18]. The feasibility of data-tagging has been proven in pump-probe experiments using an FEL and an optical laser combination [19]. The use of an XUV/laser cross-correlator [20] for shot-to-shot delay measurement allows for a complete compensation of the arrival time jitter between FEL and laser pulses and, thus, significantly improves time-resolution [19].

The availability of free electron lasers provides a unique opportunity to study the interaction of intense X-ray light with matter. These fundamental studies require the development of new experimental techniques and are intimately linked to the development of theoretical approaches able to provide a complete view of very complex and intertwined dynamics of matter under intense irradiation conditions. The current chapter is organized as follows: in the methodological section we review the recent progress in the development of X-ray pulse metrology for the measurement of the temporal profile of individual FEL-pulses and their relative timing with respect to optical laser pulses. These techniques are essential to fullfill the requirement of well defined and reproducible experimental parameters. The scientific section is focused on the time-resolved studies of ionization dynamics of small quantum systems as well as the applicability of the *diffraction before destruction* approach for high resolution CDI of bio- and bio-like nano-particles. In the outlook section we discuss the perspectives of time-resolved single-shot single-particle experiments and additional challenges regarding the processing of large data in a multi-parameter experimental space [21]. We underline a particular importance of a simultaneous implementation of X-ray light scattering and complementary diagnostic techniques like electron-, ion-, and fluorescence spectroscopy. At the end of our chapter we review further activities on the atomic, molecular, and optical (AMO) physics at light sources, which are currently under development.

The most comprehensive review to date on achievements of AMO physics at FEL sources in Europe, Japan, and USA can be found in the special issue of J. Phys. B [1]. This work covers the details about the performance of the machines as well as the development of new experimental and theoretical methods.

13.2 Temporal Diagnostics of Individual FEL Pulses

A commonly used pump-probe scheme allows the evolution of excited states to be followed by measuring the system response to two subsequent light pulses. Depending on the process under study, pump and probe pulses with the same or different photon energy can be used, i.e. one- or two-color pump-probe configuration. For one-color experiments using XUV pulses several technical concepts for split-and-delay units are brought into operation at FLASH. One possible configuration to create a pair of pulses is based on a back-reflecting focusing multilayer mirror which is split into two parts [22]. Another approach relies on grazing incidence optics in a Mach-Zehnder geometry which covers almost the whole spectral range at FLASH with reasonable transmission [23, 24]. Time delays up to 18 ps between the pump and probe pulses are realized this way. The implementation of two-color pump-probe schemes at SASE-FELs like FLASH is technically more challenging because it requires simultaneous measurement of the relative timing between XUV and optical pulses, which is subject to shot-to-shot fluctuations. It should be mentioned, that at seeded FELs like FERMI jitter-free two-color pump-probe experiments are possible by using optical pulses from the seed laser [25]. Moreover, a novel twin-pulse seed scheme has been realized recently, which allows two color jitter-free pump-probe experiments in the extreme ultraviolet region [26].

Generally, the temporal profile of the transient pump-probe signal is described by the convolution integral over the response function from the sample and the instrument response function. The sample response is defined by the lifetime of the intermediate excited states involved. The instrument response determines the experimental time resolution and can, in the simplest case, be approximated by a Gaussian with a width σ

$$\sigma = \sqrt{\sigma_{pump}^2 + \sigma_{probe}^2 + \sigma_{res}^2} \tag{13.1}$$

where σ_{pump}, σ_{probe}, and σ_{res} are Gaussian widths of the pump and probe pulses and, in the case of two-color pump-probe experiment, the residual timing uncertainty caused by jitter. Additionally, since important processes will be dominated by the peak intensity rather than the average intensity, probing the nonlinear response of matter to an intense FEL pulse may critically depend on the distribution of energy within the pulse envelope. Thus, the resolution of ultra-fast studies will be limited by the shot-to-shot variations of the temporal pulse characteristics. In the following we

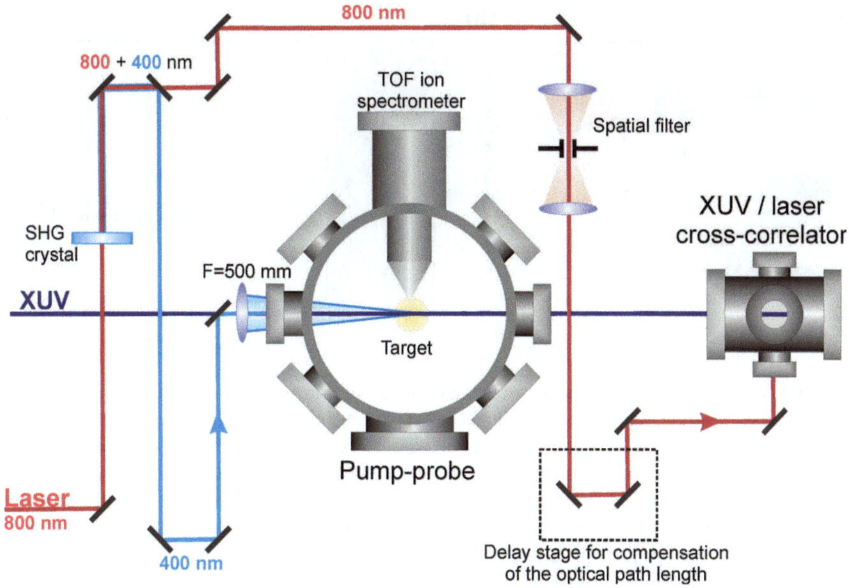

Fig. 13.1 Scheme of jitter-compensated pump-probe experiment (*top view*). The cross-correlator setup is installed directly behind the pump-probe experiment. Thus, both experiments share the same XUV beam. A delay stage before the cross-correlator compensates the optical pathway difference between both experiments. Adopted from [17]

review the experimental techniques for the temporal diagnostics of individual FEL pulses with proven single-shot capability.

13.2.1 Solid Surface Cross-Correlation Technique

A scheme of the experimental geometry for a jitter-compensated pump-probe experiment is shown in Fig. 13.1. This setup has been realized at FLASH for studies of ionization dynamics of atoms [19] and molecules [27, 28] using XUV and near-infrared (NIR) laser pulses with the highest time resolution presently possible. The XUV/NIR laser cross-correlator is installed directly behind the main interaction chamber, thus both experiments share the same XUV beam. An optical delay line available at FLASH scans the arrival time of the NIR pulse with respect to the XUV pulse within a time-window of several nanoseconds. Another delay stage introduced before the cross-correlator setup compensates the optical pathway difference between both experiments.

The measurement of the relative timing between NIR and XUV pulses is based on the time-to-space mapping principle schematically shown in Fig. 13.2a. The pulse-fronts of the XUV and NIR beams are superimposed on the solid surface under a

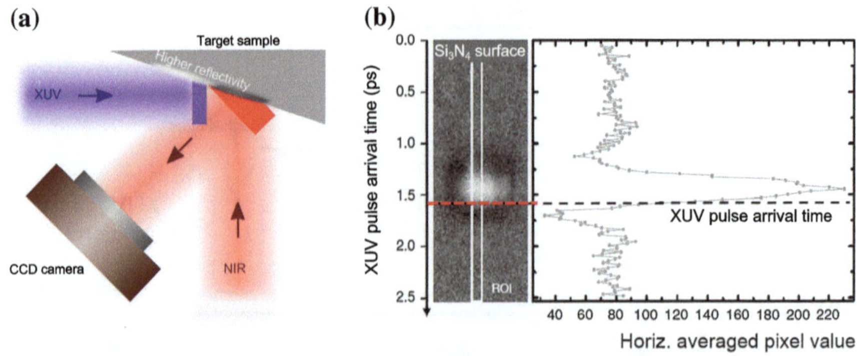

Fig. 13.2 **a** An unfocused NIR pulse is reflected from a solid surface onto a CCD camera. The XUV pulse hits the substrate at different times in different places and changes the reflectivity for the NIR pulse. The area where the XUV pulse hits the Si_3N_4 surface before the NIR pulse appears brighter. **b** Single-shot cross-correlation image (*left*) and the reflectivity change (*right*) plotted within the region of interest (ROI). One pixel of the CCD-image corresponds to 17 fs. The inflection point of the reflectivity change curve is used as a marker for the relative timing. Adopted from [17, 20]

certain angle. The XUV/NIR time delay dependent observable is the transient reflectivity change of the surface since the spatial coordinates at the surface correspond to different arrival-times of the tilted wave fronts.

Figure 13.2b, left panel, shows a typical image of a single-shot cross-correlation signal obtained at a Si_3N_4 surface. The regions where the XUV pulse hits the surface prior to the NIR pulse appear brighter. This effect is understood as a creation of a high electron density within the conduction band of the solid through the absorption of XUV photons [29]. The transient reflectivity change integrated within the region-of-interest (ROI) is shown in Fig. 13.2b, right panel. The inflection point of the rising edge of the signal is used as an arrival-time marker for the XUV pulse (dashed line). Simple geometrical considerations define the relationship between the spatial coordinate of the inflection point determined in pixels on the CCD chip and the arrival time in fs [20]. For the case shown in Fig. 13.2 one pixel of the CCD-image corresponds to 17 fs.

The feasibility of this data tagging concept has been demonstrated in studies of the ionization dynamics of Xe atoms [19] and iodine molecules [27, 28]. Figure 13.3 shows ion-charge-state spectra measured in the experimental geometry of Fig. 13.1, i.e. with the data collected with the optical delay stage in a fixed position while the XUV arrival-time is simultaneously measured by the XUV/NIR cross-correlator. It should be mentioned that the cross-correlator provides the information on relative timing between both XUV and NIR pulses. The time-delay zero defined as the maximum overlap between XUV and NIR pulses has to be set from the data analysis.

In the example shown in Fig. 13.3 the time-delay zero is defined by the inflection point of the temporal profile corresponding to the Xe^{2+*} intermediate state, presumably in $5s^{-1}5p^{-2}6p$ and $5p^{-3}nl$ configuration [19]. This state is long lived compared to the timescale of the experiment and is assumed to be quasi-instantaneously

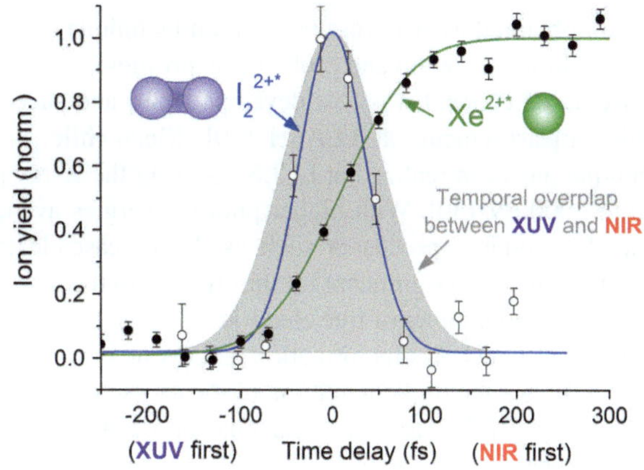

Fig. 13.3 Transient profiles (background subtracted and normalized) of the intermediate Xe^{2+*} atomic (*closed circles*) and $I_2{}^{2+*}$ molecular ionic (*opened circles*) states and the corresponding fits to the data. The data sets were measured within separate experiments. The ion charge-state-spectra were sorted according to the XUV arrival-time and averaged subsequently within 30 fs time bins. Each data point represents an average over about 200 single measurements. *Gray filled curve* represents the temporal overlap between XUV and NIR pulses approximated by a Gaussian profile with 125 fs FWHM. Adopted from [19, 28]

populated by the XUV pulse of 92 eV photon energy upon the ionization of the 4d shell of Xe followed by the ultra-fast (within several fs) Auger decay transition. The optical laser pulse can ionize the Xe^{2+*} intermediate state, inducing an additional time-dependent increase in the Xe^{3+} ion yield. Thus, the fit of the transient Xe^{3+} ion yield (green curve) with the smoothed step function allows the extraction of the convolution integral between the XUV and NIR pulse profiles (gray filled curve) as well as the inflection point which sets the time-zero in the experiment as shown in Fig. 13.3.

Another valuable example shown in the same Fig. 13.3 is the transient profile of the $I_2{}^{2+*}$ molecular ionic state (opened circles). Note that both Xe^{2+*} and $I_2{}^{2+*}$ temporal profiles were measured in separate but similar experiments. This state is populated by the NIR pulse and probed by the XUV pulse [28]. The corresponding temporal profile is fitted with a Gaussian function of (75 ± 15) fs FWHM (blue curve) and is considerably shorter than the (125 ± 21) fs FWHM NIR/XUV temporal overlap (gray filled curve). The narrow temporal width of the $I_2{}^{2+*}$ intermediate state is explained by the temporal confinement of this molecular cationic state within the NIR pulse envelope. This means that the intensity threshold of about 5×10^{13} W/cm^2 is needed to produce doubly charged molecular ion $I_2{}^{2+*}$ by the field ionization. Assuming a Gaussian intensity distribution for the NIR pulse envelope, the temporal profile of the $I_2{}^{2+*}$ intermediate state then reveals the time range where the electrical field strength is sufficient for over-the-barrier ionization as explained in [28]. Thus, the nonlinearity of the NIR induced strong-field ionization demonstrates that the temporal resolution obtained by the cross-correlation technique [20] could be even better than the convolution of pump and probe pulse profiles.

The reflectivity based concept was first developed [29] and successfully implemented for timing measurements at FLASH [20]. Meanwhile, a similar cross-correlation technique has been realized at LCLS covering the X-ray photon energy range from 500 to 2000 eV [30]. With higher photon energies available at LCLS, the contribution of secondary processes such as Auger cascade transitions and impact ionization become more significant for the creation of the dense electron-hole plasma [31]. The increased number of free electrons that can be created by one incident photon allows an efficient plasma formation far beyond the damage threshold of the sample. This facilitates the implementation of the cross-correlation concepts on thin Si_3N_4 membranes [32]. In particular, transmission geometries based on detection of transient changes either through spectral [33] or transmission [34] encoding are promising for further improvements of the temporal resolution in pump-probe measurements on absorbing samples like solids or liquids.

The signal based on transient changes of the reflectivity or transmission represents the convolution of the FEL and probe laser pulse profiles. A recently developed theoretical model for the transmission curve [35] extends the implementation of this timing concept to work also as a diagnostic of the FEL pulse duration. The pulse duration measurement can be implemented for well characterized optical laser pulses as long as the rise-time of the electron density due to secondary processes is fast compared to the FEL pulse duration [31]. Another powerful temporal diagnostic of individual FEL pulses is the light field streaking [36] which is discussed below.

13.2.1.1 Single-Shot FEL Pulse Characterization with a THz-Field-Driven Streak-Camera

Light-field-streaking represents a new time-resolved technique developed for metrology of attosecond pulses [37]. In this approach, electrons generated from photoionization by the XUV pulse acquire an additional momentum in the presence of an intense linearly polarized light field. The electron kinetic energy distribution (a streaking spectrogram) is then obtained as a function of the relative timing between the ionizing pulse and the phase of the streaking field. In this so called streak-camera scheme, the temporal profile of the XUV pulse is mapped onto the photoelectron kinetic energy distribution if the XUV pulse is short compared to the oscillation period of the streaking field. The temporal characteristics of the pulse can then be retrieved using special reconstruction algorithms [38].

In attosecond metrology a NIR field is used for streaking. The implementation of the streak-camera principle to femtosecond FEL pulses requires streaking light fields with a longer oscillation period i.e. frequencies in the terahertz (THz) range. A review covering the recent development of sources able to produce THz-pulses with very strong electromagnetic fields can be found in [39]. The feasibility of THz streaking was first demonstrated at FLASH utilizing the ionization of the 4p valence electrons in krypton by 92 eV FEL pulses [36]. In this experiment, the THz streaking field (frequency of 3.3 THz corresponding to a 92 μm wavelength) is produced from the same electron bunch that gives the FEL pulse using a dedicated undulator. This guarantees a high level of intrinsic synchronization between the FEL and THz pulses. Alternatively, the bending magnet used to dump the electron bunches was also shown to produce THz pulses close to single-cycle [39].

Analysis of the individual FEL pulses reveals a pronounced variation in their temporal properties as expected due to SASE processes. The approach introduced in [36] uses an approximation with a Gaussian envelope and enables the detection of linear changes of the instantaneous frequency within the FEL pulse of the order of several meV/fs. This seems to be a characteristic feature of the FEL pulses due to the linear acceleration and bunch compression scheme. Moreover, classification procedures show that a certain fraction of pulses (about 10 % in [36]) carry a complex substructure attributed to two or more well separated sub-pulses. More details on data analysis can be found in the tutorial [40]. Therefore, simultaneous operation of a streak-camera for pulse structure determination with experiments that are especially sensitive to multiphoton absorption processes like [7], will help sorting out active excitation channels according to different temporal pulse properties.

A back-focusing geometry with a multilayer mirror has to be used in an experiment like [36] to compensate for the optical path difference of about 4 m between the THz and the FEL branches. For some experiments this geometry might not be beneficial because it complicates the use of a downstream scattering detector for CDI. An alternative way is to produce the THz-radiation directly at the experimental end station by using e.g. the frequency conversion within a nonlinear crystal pumped with a NIR-laser [39]. This approach avoids the need to delay the FEL beam with respect to the THz beam. The THz-light field streaking in combination with a laser-driven source has successfully been implemented in laboratory HHG sources [41] as well as with FEL pulses from FLASH [42]. In the latter case the approach has been mainly used for measurements of the relative timing jitter between FEL and NIR-laser pulses as an alternative to reflectivity based schemes discussed in previous section.

13.3 Ultrafast Ionization Dynamics of Small Quantum Systems

Time-domain observation of relaxation processes followed by subsequent structural rearrangements requires a pump pulse to transform the system to a non-equilibrium state and a delayed probe pulse for detecting the transition state of the evolving system. The classical transition-state spectroscopy utilizes this pump-probe scheme to follow the dynamics of chemical reactions like molecular bond formation and bond breaking on femtosecond time and sub-Ångström length scales [43]. High temporal resolution is obtained through the short pulse duration. The sub-Ångström spatial resolution is guaranteed by an exact selection of the wavelength of the pulses from the optical laser as well as by their narrow spectral band width. Both criteria are required to selectively excite the system into a very well defined transient state and to probe this transient state at a certain internuclear separation [43].

The high photon energies of (hard or soft) X-ray FEL pulses enable them to penetrate deeply into the atomic electron shells, thereby delivering detailed information about the electronic structure. However, in contrast to the visible range, where the system can be excited to the transient state of a specific configuration, the preferable X-ray excitation of inner-shell core-levels produces a multitude of highly unstable states giving rise to different relaxation pathways. Recently, the Auger decay following ionization and excitation of the 3d-subshell in Kr by 94 eV photons has been studied using an attosecond metrology approach [44]. The results of this study show that a single photon absorption event creates a $3d^{-1}$ vacancy that decays rapidly through single or cascaded Auger transitions leading to the production of ionic charge states up to 3+. Extensive theoretical calculations including a large number of possible electron configurations for intermediate and end states were needed to verify the experimental results, in particular to explain the unexpectedly long life time of several Auger photolines (up to 70 fs) in the low part of the photoelectron spectrum [44]. The results of this study clearly demonstrate that a strong theoretical support is required to understand the complex and intertwined multi-electron dynamics already in atomic samples.

In systems composed of many atoms, like molecules and clusters, the complexity of possible decay channels is further increased. Indeed, in such systems due to the accumulated positive charge, the electron dynamics will be coupled to the nuclear dynamics. In molecules and small clusters the disintegration will proceed mainly through Coulomb explosion, while larger clusters disintegrate also through hydrodynamic expansion mechanisms (see e.g. for recent review [3]). Therefore, an implementation of X-ray pulses in pump-probe schemes requires a development of spectroscopic techniques able to separate distinct pathways out of a multitude of possible reaction channels. In the following subsection we provide some representative examples of time-domain measurements where the creation of excited valence states is used to probe the multi-electron relaxation dynamics.

13.3.1 XUV Pump—NIR Probe Experiments of Multi-electron Relaxation Dynamics

This approach is based mainly on the different interaction mechanisms of the pump and probe pulses on atomic (a), molecular (b), and cluster (c) targets as illustrated in Fig. 13.4. An XUV pulse acting as a pump produces inner-core-hole states localized on individual atoms. The created core-hole vacancy will undergo rapid de-excitation through several channels. Some of the relaxation channels can end up with a system in an excited electronic configuration. The temporally delayed NIR pulse probes the creation of these weakly bound states through further ionization by one or few

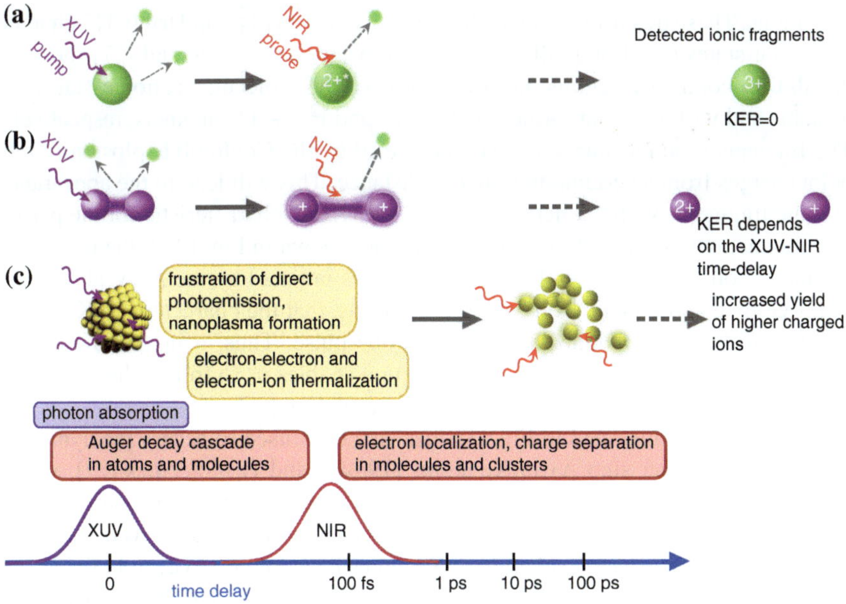

Fig. 13.4 Illustration of XUV pump—NIR probe experiment to reveal ionization dynamics in atoms (**a**), molecules (**b**), and clusters (**c**). Electron dynamics is triggered by the XUV pulse through the absorption of single photons by individual atoms. Subsequently, a temporally delayed optical laser pulse of moderate intensity ionizes weakly bound electrons and, thereby, probes the creation of the excited states. The approach allows access to the dynamics of Auger decay cascades in atoms (**a**) and molecules (**b**), as well as the disintegration dynamics of molecules (**b**) and clusters, (**c**) on femtosecond and hundreds of femtosecond to picosecond time scales, respectively

photons. Therefore, looking for an increased yield of highly charged ions as a function of the delay between XUV (pump) and optical (probe) pulses provides insight into the relaxation dynamics of the system.

As a first application of this technique, a relaxation process in xenon atoms [19, 45] and iodine molecules [27] in the vicinity of 4d$\rightarrow$ εf giant resonance is considered. In the case of Xe the absorption of one ~90 eV photon will preferably create a vacancy in the $4d^{-1}$ inner-shell. This state will decay through a single A_1 or a cascade of two A_1+A_2 Auger decay transitions resulting in Xe^{2+} and Xe^{3+} charge states. Some of the populated Xe^{2+} states have an excited electron configuration and can be ionized further by NIR pulses of moderate intensity through multi-photon absorption. Figure 13.4a illustrates this approach. The first step of Auger decay transition can be traced by the NIR pulse through the ionization of Xe^{2+*} intermediates and a detection of the transient yield of Xe^{3+} ions. From the shape of the Xe^{3+} transient profile, measured with sub-fs resolution at a HHG source, the life time of the $4d^{-1}$ core-hole has been extracted $\tau_{A1} = 6.0 \pm 0.7$ fs [45].

In molecular iodine I_2 the $4d^{-1}$ vacancy created by the XUV radiation in one of the iodine atoms will also decay through subsequent A_1 or A_1+A_2 Auger decay transitions. These transitions lead to the creation of doubly I_2^{2+} and triply I_2^{3+} charged molecular states (see Fig. 13.4b for the single Auger decay case and [27] for more details). In contrast to atoms, however, most of the molecular cationic states are unstable and will fragment further via $I^+ + I^+$ and $I^{2+} + I^+$ channels, respectively. The fragmentation mechanism is well described by the Coulomb explosion of two point charges from a certain inter-nuclear distance. This will lead to the appearance of ionic fragments with a kinetic energy release (KER) characteristic for the particular fragmentation channel. Note that in the case shown in Fig. 13.4b the molecular fragmentation via the $I_2^{2*} \rightarrow I^+ - I^{+*}$ channel is initiated by the XUV absorption followed by a subsequent (A_1) decay. This means, that the characteristic KER will depend on the time of ionization by the probe pulse. Thus, using the KER as a measure for the current inter-nuclear separation enables us to follow the molecular explosion in the time domain with Ångström resolution [27].

The approach illustrated in Fig. 13.4 has also been used to access the dynamics of the second Auger step A_2 in Xe atoms [19, 45] and I_2 molecules [27], which proceeds on a time scale of several tens of fs. Moreover, in experiments performed at FLASH, a signature of the creation of double core-hole states $4d^{-2}$ has been found, demonstrating a nonlinear excitation of Xe atoms in an intense FEL-light field [19].

Ionization of large systems like (bio-)molecules and clusters by intense XUV pulses leads to electron trapping and the build-up of a nanoplasma inside the extended Coulomb potential. The system disintegrates due to the hydrodynamic pressure of hot electrons as well as Coulomb repulsion of uncompensated positive charge. Extensive theoretical [5, 6] and steady-state experimental [8, 10, 11] studies provide evidence for a crucial role of electron-ion recombination in the disintegration processes. Experiments on Xe-core/Ar-shell clusters show that in spite of a much higher atomic absorption cross-section of Xe with respect to Ar at 13.5 nm excitation wavelength, a very small number of Xe ions is detected [10]. On the other hand, fluorescence emission studies performed on the same system at similar irradiation conditions show that charge states up to at least 11+ are formed inside the Xe core [11]. Experiments on methane (CH_4) clusters, which represent an important system to model the response of the most common constituents of bio-molecules (i.e. carbon and hydrogen atoms) to the intense soft x-ray radiation, reveal a dominant presence of recombination products like protonated methane (CH_5^+) and higher adducts ($C_nH_m^+$, $n \geq 2$) [8]. Moreover, systematically higher maximum kinetic energies observed for deuterons compared to protons show an isotope effect related to inertial confinement. This indicates that heavier structurally important ions maintain their positions in the sample while protons escape early, carrying excess energy and heat away from the sample.

Recently a number of pump-probe studies have been undertaken to address the dynamics of electron-ion recombination in clusters [12, 13] and large samples with low-Z composition [46]. In experiments on Xe clusters, FEL pulses with 92 eV photon energy are split in to pump and probe paths with an intensity ratio of 20 % to 80 %, respectively [12]. The ion charge state spectra show an increase for the average charge of the ionic fragments with increasing delay, providing experimental access to the nanoplasma relaxation dynamics on the subpicosecond time scale. The results are interpreted as a lowering of the cluster Coulomb potential during the expansion, leading to a more efficient direct electron emission. Recent quasi-classical molecular dynamics simulations motivated by this study also show that the reduced efficiency of the electron-ion recombination contributes to the observed temporal evolution of the average charge state [47].

XUV pump/XUV probe experiment at 6.8 nm wavelength and an intensity approaching 5×10^{16} W/cm^2 on individual sucrose particles directly demonstrates that the sample fragmentation is heavily dominated by recombination effects [46]. In this study diffraction images were collected in coincidence with the ion spectra. This combination of pump-probe ion spectroscopy and simultaneous CDI provides control of important experimental parameters like irradiation intensity and sample size distribution on single shot basis. The analysis focuses on the relative amount of detected high and low charge states of carbon. Results show that sucrose particles in the size range between 0.5 and 1 micron in diameter stay relatively intact for the first 500 fs following the exposure to the pump pulse, and that significant sample expansion happens between 0.5 and 2 ps after the initial exposure.

The one color XUV/XUV pump-probe scheme discussed above uses an intense probe pulse relative to the pump pulse [12, 46]. Suppressed electron-ion recombination due to the expansion of the system is manifested by the shift of detected ion distributions towards higher charge states. Another more direct possibility to access the electron-ion recombination dynamics is the reionization of atoms in excited states by a weak probe field as shown in Fig. 13.4c. In recent experiments on Xe, Ar, and Xe-core/Ar-shell clusters ionized by the intense XUV pulses from an HHG source, spectroscopic signatures for atoms in high-lying Rydberg states have been found [13, 48]. The reionization of these loosely bound electrons by a weak optical laser field leads to the production of very slow electrons providing a state-specific probe for the electron-ion recombination dynamics [13, 48].

13.4 The Role of Ionization Dynamics for High Resolution Imaging of Bio- and Bio-like Nanoparticles

An understanding of the interaction of high-intensity X-ray pulses with matter is of fundamental importance for the structure-determination of bio-molecules where ultra-fast radiation induced processes can be assumed to become increasingly important with increasing intensity. At the extreme photon intensity required for single particle CDI these processes inevitably lead to either a Coulomb explosion or to a hydrodynamic expansion of the sample. The actual outcome depends on pulse length, wavelength, and pulse intensity as well as the size, composition, and density of the sample. As a result, for high-resolution CDI to be possible, the sample structure must remain sufficiently intact during the laser pulse. Simulations suggest that CDI can be successfully used for single large molecules of biological origin if radiation damage from the high X-ray dose is outrun by ultra-short exposures. In such cases interpretable diffraction patterns should be obtainable from molecules, nano-crystals, virus particles and even living cells [2, 49]. Traditionally in imaging applications (classical microscopy) the wavelength used must be of the order of the requested resolution making X-rays suitable for high-resolution imaging. However, at intensities required for high resolution imaging X-rays cause significant radiation damage. To reach nanometer resolution on a cell a dose in excess of 100,000,000 Gray (1 Gy=1 J/kg) is required. Considering that 25,000 Gy is lethal to all known organisms, 60–800 Gy kills most cells and 20 Gy causes certain death in humans, no cell can survive this type of imaging [50].

To improve on the general possibility of imaging bio-particles to high resolution, highly invasive procedures are undertaken such as labeling, freezing or even slicing the sample. As a consequence, all high-resolution information we have on cells comes from dead cells. Femtosecond X-ray pulses are predicted to be able to outrun key damage processes and deliver molecular-level snapshots of cells which are alive at the time of image formation [2, 49]. Furthermore, it has been shown that cells in aerosol droplets can survive the transport into the interaction zone of a CDI experiment [50]. This means that the *diffraction before destruction* approach provides the first opportunity to obtain high resolution data from complex bio-particles, such as small living cells and the recently discovered giant viruses [51, 52].

Following the initial simulations [2] the *diffraction before destruction* approach has been verified in subsequent experiments to high resolution for nano-crystals [53–56] and to low resolution for single particles [50, 57, 58]. Serial nano-crystallography has rapidly developed into a comparably mature technique. Initial experiments were performed with the commissioning of the LCLS, the first X-ray FEL in 2009 [53] and recently the first "molecular movie" of dynamics in Photosystem II was reported [59]. Furthermore it has been shown that damage in the nanocrystals develops on atomic scale within a few 100 fs and that this damage effectively causes a self-gating of the diffraction so that the main contribution to the observed Bragg peaks comes from the unperturbed regions of the sample [54]. However, the self-gating effect is not as easily applicable to the analysis of diffraction patterns from non-crystalline sample, since the incoherent addition to the diffraction pattern from the damaged part of the sample is of the same order as the diffraction signal from the unperturbed sample. As a consequence complementary techniques to gain information about the onset of sample damage are being developed. Figure 13.5 shows an experimental set-up used in recent femtosecond CDI studies [21]. Secondary diagnostic techniques like ion, electron, and plasma emission spectrometers can be integrated into the interaction chamber and operated in coincidence with the scattering detector. Such simultaneous experiments using complementary techniques opens up the possibility of getting detailed information about the electronic transitions happening in the sample during and shortly after the exposure to the imaging pulse.

The very first results with sample fragmentation data recorded in coincidence with a CDI experiment (combined ion spectra and CDI images) were reported in 2008 [60]. Since then methods are under continuous refinement [21, 46, 61]. The most resent results from aerosolized nanoparticles show that at low intensities (so

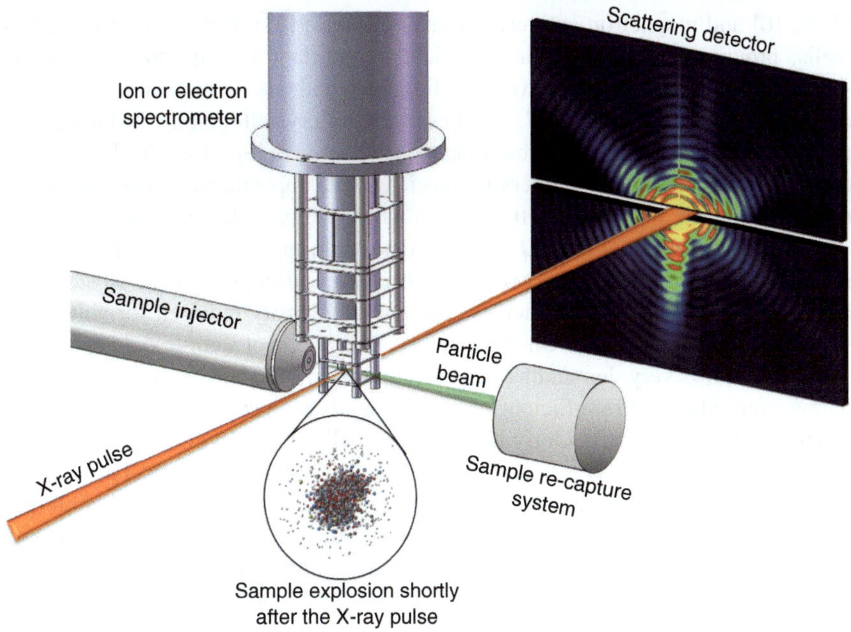

Fig. 13.5 Setup for a femtosecond CDI experiment with a secondary diagnostic. An Aerodynamic lens aerosol injector is used to deliver the (bio-) nano-particles into the vacuum environment of the experiment. In the interaction region the particle beam is overlapped with the X-ray beam. A fast detector is read out after each FEL shot. The strength of the scattering signal depends on X-ray power density, detector efficiency, particle size, and its location within the intensity profile of the FEL beam. The distribution of electrons or ionic fragments is measured by the spectrometer, which is operated in coincidence with the scattering detector. Adopted from [21]

called weak hits) the signal from the background gas outweighs the signal from the target particle [21, 46]. This is despite the fact that the sample is expected to be entirely ionized (at least one ionization per atom) under the given interaction parameters [54]. This observation shows that the ions formed in the interaction largely recombine before they can be extracted by the ion time-of-flight spectrometer and that an understanding of the recombination process is critical to the interpretation of the ion spectra. With increasing hit strength new dynamics allow highly charged and/or highly energetic ions to escape the recombination processes present in the expanding plasma [7, 46]. This observation of highly accelerated fragments from biological sample are likely related to the observation of unexpectedly high charge states at the highest intensities in cluster experiments discussed above [7].

Another important point which motivates the simultaneous operation of secondary diagnostic techniques in CDI experiments is connected to the characterization of the sample delivery methods. For efficient sample delivery an array of methods optimized for different purposes has been developed. In cluster research commonly used methods are supersonic expansion from the gas phase (rare gas and water clusters)

[7, 8, 10] and magnetron sputtering (metal clusters) [62]. For single particle CDI (cells, proteins, and viruses) it is critically important to minimize background scattering from any sample delivery medium and methods for aerosol sample delivery have been developed [50, 57, 58]. In serial nano-crystallography the background problem is less profound and samples can be delivered in liquid jets [63]. At present cluster and aerosol sample delivery is associated with opposite problems with respect to sample density in the interaction region where the particle beam has to be overlapped with the femtosecond FEL pulse (and possible pump pulses). In cluster experiments the main problem is to avoid multiple hits, and in general the cluster beam has to be heavily skimmed to achieve the particle density required for single-shot single-particle experiments. On the contrary, using aerosol sample delivery in bio-particle CDI it is very demanding to generate the particle density in the interaction region required to reach a sufficient hit-rate. Presently available techniques for aerosol generation make this particularly difficult for small samples with a diameter below approximately 100 nm. This situation indicates that it may be a fruitful approach to combine cluster and aerosol sample delivery techniques to reliably reach a hit rates in the range between 0.1 and 40 % (depending on the focal spot size) for wide range of sample types and sizes. Secondary diagnostics can be efficiently used to characterize the injection process since the data obtained by e.g. an ion detector can be read out and analyzed much more rapidly that the data from an imaging detector. This means that secondary diagnostics have a great potential for on-line use in the task to optimize sample delivery settings at high repetition-rate sources like the European XFEL [21].

The methods for sample delivery (the random way in which the particles are hit [7, 21]) contribute to the fact that parameter space for post-measurement data sorting will become extremely large, in particular for particle beams with a distribution in particle size and structure. Moreover, it will increase further in pump-probe experiments since the position and orientation of the particle within both the pump and probe volume will influence the measurement. In the following section we review recently developed experimental concepts regarding the challenges associated with processing of large data-sets sampling a multi-parameter space.

13.5 Automated and Unsupervised Identification and Classification of Single-Shot Single-Particle CDI Data

The ultimate goal to produce high-resolution 3D movies of proteins performing actual biological functions will require enormous amounts of data to be stored and analyzed. Thus the perspective of using single-shot single-particle experiments for time-resolved studies of molecular dynamics (also in pump-probe experiments) is intimately linked to the development of concepts for storage and processing of the extremely large amount of accumulated experimental data.

Current X-ray FEL facilities operate at repetition rates of 10–120 Hz and, in the future, the European XFEL is expected to produce up to 27,000 pulses per second. Using a very rough estimate of 10 Mb of data generated for each saved shot, this means that the maximum data rate could approach 1 Pb per hour with an imaging detector capable of reading out data for each shot (although no such detector is presently available). This kind of data rate is clearly out of scope of what can be realistically treated in a foreseeable future. Nevertheless, since most shots will not produce optimal hits (as discussed above), the high repetition rate is required to produce the necessary amount of data but must be complemented by sophisticated data reduction schemes to be implemented in several layers, both on- and off-line. This requires a very rapid but also reliable analysis in order to ensure that the right fraction of the data (the best hits) is saved.

To reduce the huge data stream automated data analysis methods are necessary. The development of real-time frame rejection strategies based on ion time-of-flight spectroscopy with subsequent data sorting according to several selection criteria is introduced in [21]. It is suggested that the read-out (or buffering) of dark frames of the X-ray scattering detector can be limited by the use of a secondary diagnostic tool like an ion or electron time-of-flight spectrometer which can be operated at much higher read-out rates than the imaging detector. In principle this kind of on-line hit-finding can be performed also in the 200 ns interval between pulses at the European XFEL but further research is needed to optimize this function [21].

As discussed above the particles hit by the FEL-beam will be transformed into a highly excited non-equilibrium plasma state [2] with characteristics depending on the nature of the hit. Therefore, it is possible that very selective data rejection and sorting based on fast spectroscopy methods can be done already on-line as a support for subsequent off-line analysis of the selected diffraction patterns. Once an on-line selection of the most likely good hits is done a more detailed analysis of the saved diffraction patterns can be undertaken. For the structure analysis of samples with inhomogeneous size distribution a selection criteria based on particle size as well as the possibility of sorting out multiple particle hits are crucial. In [21] a calculation of the autocorrelation function, which represents the inverse Fourier transform of the detected scattering image, is proposed as a selection method for sorting out single particle hits.

Another important point for analysis of scattering images is the strength of the diffraction signal. At high scattering angles this is generally weak (single photons per pixel) and is superimposed on a low but continually fluctuating background signal, originating from stray light from the beam line and electronic noise from the detector. This necessitates the development of advanced methods for efficient, automated and unsupervised sorting of diffraction data based on rapid image analysis. A recently applied powerful approach which is able to classify CDI images from the same particle at different orientations is based on spectral clustering [64]. Further development of this approach demonstrates that diffraction data from single particle hits on Carboxysomes can be "purified in silico" through automated analysis [58].

13.6 Future Perspectives of AMO Science at Novel Light Sources

The availability of free electron lasers provides a unique opportunity for studies of interactions of intense X-ray light with matter on the fs time and nm size scales. The parameters of the machine, pulse diagnostics techniques as well as synchronization schemes to external lasers are under continuous further development. To fully exploit unique characteristics of the FEL radiation for studies of ionization dynamics of atoms, molecules and nano-particles, end-stations like CAMP [65] or LDM [66] have been developed for user experiments. These instruments are designed to accommodate different types of target preparation and delivery systems like atomic and molecular beams, pure and doped droplets and clusters, nano-crystals, and bio-particles. The electrons, ions and photons produced in the interaction between the FEL pulse and the target are ideally analyzed by various spectrometers and imaging detectors operated in coincidence mode. In particular, coincident detection of characteristic charged fragments together with the diffraction patterns helps overcome the problem of averaging over irradiation intensity and size distribution and allows studies of spatially restricted nanoplasma under well-defined conditions [7]. The very first reports of single-shot single-particle experiments combined with pump-probe techniques have appeared [46]. Further development of this research area will provide new insight into the evolution of the new excited states of matter created in CDI experiments and help to develop control strategies to minimize the effects of radiation damage.

High repetition rate X-ray FEL sources are ideal for the development of CDI experiments aiming at 3D imaging of reproducible samples. Combining online and offline data reduction and classification this can be done based on a large number of diffraction patterns obtained from many identical copies of a reproducible sample exposed in different orientations. Complementary to the present availability of X-ray FELs, driven by km-long conventional electron accelerators, compact laser-driven alternatives are under development. Single shot CDI has been performed using a high power HHG source [67] and recent simulations indicate that both laser driven FELs [68, 69] (based on wake field acceleration) and short pulse soft X-ray plasma lasers [70] with sufficient pulse parameters are feasible. A variety of these state-of-the-art laser driven sources are presently under development within the framework of a new generation of high power laser facilities being built to operate as users facilities (ELI Beamlines and ELI ALPS) [71]. At this point none of these sources are capable of generating pulse parameters sufficient for high resolution CDI of weakly scattering bio-samples. However, if the suitable pulse parameters can be obtained the relative compactness of these sources would allow 3D information to be obtained also from non-reproducible samples exposed simultaneously by multiple sources from multiple directions. Simulations [49] indicate that using radiation in the water window, 3D imaging with sub-nm resolution from non-reproducible samples, like living cells (or the recently discovered giant viruses [51, 52]), may be possible using this type of multi-beam CDI. This would allow e.g. individual protein molecules and the DNA structure to be resolved inside of living cells thus taking us a large step in the direction of understanding the fundamental processes necessary to support independent life.

References

1. Ph. Bucksbaum, Th. Möller, K. Ueda (eds.), Frontiers of free-electron laser science. J. Phys. B **46**(16), 164027 (2013) (Special issue)
2. R. Neutze, R. Wouts, D. van der Spoel, E. Weckert, J. Hajdu, Nature **406**, 752–757 (2000)
3. J. Feldhaus, M. Krikunova, M. Meyer, Th Möller, R. Moshammer, A. Rudenko, Th. Tschentscher, J. Ullrich, J. Phys. B **46**, 164002 (2013)
4. Y. Ovcharenko et al., Phys. Rev. Lett. **112**, 073401 (2014)
5. M. Arbeiter, T. Fennel, New J. Phys. **13**, 053022 (2011)
6. U. Saalmann, J. Phys. B **43**, 0194012 (2010)
7. T. Gorkhover et al., Phys. Rev. Lett. **108**, 245005 (2012)
8. B. Iwan et al., Phys. Rev. A **86**, 033201 (2012)
9. H. Thomas et al., Phys. Rev. Lett. **108**, 133401 (2012)
10. M. Hoener et al., J. Phys. B **41**, 181001 (2008)
11. L. Schroedter et al., Phys. Rev. Lett. **112**, 183401 (2014)
12. M. Krikunova et al., J. Phys. B **45**, 105101 (2012)
13. B. Schütte, F. Campi, M. Arbeiter, Th. Fennel, M.J.J. Vrakking, A. Rouzée, Phys. Rev. Lett. **112**, 253401 (2014)
14. E. Allaria et al., Nat. Photonics **6**, 699–704 (2012)
15. S. Ackermann et al., Phys. Rev. Lett. **11**, 114801 (2013)
16. J. Amann et al., Nat. Photonics **6**, 693–698 (2012)
17. M. Drescher, U. Frühling, M. Krikunova, Th Maltezopoulos, M. Wieland, J. Phys. B **43**, 194010 (2010)
18. N. Hartmann et al., Nat. Photonics **8**, 706–709 (2014)
19. M. Krikunova, Th. Maltezopoulos, A. Azima, M. Schlie, U. Frühling, H. Redlin, R. Kalms, S. Cunovic, N. Kabachnik, M. Wieland, M. Drescher, New J. Phys. **11**, 123019 (2009)
20. Th. Maltezopoulos, S. Cunovic, M. Wieland, M. Beye, A. Azima, H. Redlin, M. Krikunova, R. Kalms, U. Frühling, F. Budzyn, W. Wurth, A. Föhlisch, M. Drescher, New J. Phys. **10**, 033026 (2008)
21. J. Andreasson, A.V. Martin, M. Liang, N. Timneanu, A. Aquila, F. Wang, B. Iwan, M. Svenda, T. Ekeberg, M. Hantke, J. Bielecki, D. Rolles, A. Rudenko, L. Foucar, R. Hartmann, B. Erk, B. Rudek, H.N. Chapman, J. Hajdu, A. Barty, Opt. Express **22**, 2497–2510 (2014)
22. Y.H. Jiang, T. Pfeifer, A. Rudenko, O. Herrwerth, L. Foucar, M. Kurka, K.U. Kühnel, M. Lezius, M.F. Kling, X. Liu, K. Ueda, S. Düsterer, R. Treusch, C.D. Schröter, R. Moshammer, J. Ullrich, Phys. Rev. Lett. **82**, 041403(R) (2010)
23. F. Sorgenfrei, W.F. Schlotter, T. Beeck, M. Nagasono, S. Gieschen, H. Meyer, A. Föhlisch, M. Beye, W. Wurth, Rev. Sci. Instr. **81**, 043107 (2010)
24. M. Wöstmann, R. Mitzner, T. Noll, S. Roling, B Siemer1, F. Siewert, S. Eppenhoff, F. Wahlert, H. Zacharias. J. Phys. B **46**, 164005 (2013)
25. T. Mazza et al., Nat. Commun. **5**, 3648 (2014)
26. E. Allaria et al., Nat. Commun. **4**, 2476 (2013)
27. M. Krikunova, Th. Maltezopoulos, Ph. Wessels, M. Schlie, A. Azima, M. Wieland, M. Drescher, J. Chem. Phys. **134**, 024313 (2011)
28. M. Krikunova, Th. Maltezopoulos, Ph. Wessels, M. Schlie, A. Azima, T. Gaumnitz, T. Gebert, M. Wieland, M. Drescher, Phys. Rev. A **86**, 043430 (2012)
29. C. Gahl, A. Azima, M. Beye, M. Deppe, K. Döbrich, U. Hasslinger, F. Hennies, A. Melnikov, M. Nagasono, A. Pietzsch, M. Wolf, W. Wurth, A. Föhlisch, Nat. Photonics **2**, 165–169 (2008)

30. O. Krupin, M. Trigo, W.F. Schlotter, M. Beye, F. Sorgenfrei, J.J. Turner, D.A. Reis, N. Gerken, S. Lee, W.S. Lee, G. Hays, Y. Acremann, B. Abbey, R. Coffee, M. Messerschmidt, S.P. Hau-Riege, G. Lapertot, J. Lüning, P. Heimann, R. Soufli, M. Fernández-Perea, M. Rowen, M. Holmes, S.L. Molodtsov, A. Föhlisch, W. Wurth, Opt. Express **20**, 11396 (2012)
31. B. Ziaja, A. London, J. Hajdu, Appl. Phys. Lett. **97**, 064905 (2005)
32. S. Schorb, T. Gorkhover, J.P. Cryan, J.M. Glownia, M.R. Bionta, R.N. Coffee, B. Erk, R. Boll, C. Schmidt, D. Rolles, A. Rudenko, A. Rouzee, M. Swiggers, S. Carron, J.-C. Castagna, J.D. Bozek, M. Messerschmidt, W.F. Schlotter, C. Bostedt, Appl. Phys. Lett. **100**, 121107 (2012)
33. M.R. Bionta, H.T. Lemke, J.P. Cryan, J.M. Glownia, C. Bostedt, M. Cammarata, J.-C. Castagna, Y. Ding, D.M. Fritz, A.R. Fry, J. Krzywinski, M. Messerschmidt, S. Schorb, M.L. Swiggers, R.N. Coffee, Opt. Express **19**, 21855 (2011)
34. M. Beye, O. Krupin, G. Hays, A.H. Reid, D. Rupp, S. de Jong, S. Lee, W.-S. Lee, Y.-D. Chuang, R. Coffee, J.P. Cryan, J.M. Glownia, A. Föhlisch, M.R. Holmes, A.R. Fry, W.E. White, C. Bostedt, A.O. Scherz, H.A. Durr, W.F. Schlotter, Appl. Phys. Lett. **100**, 121108 (2012)
35. R. Riedel, A. Al-Shemmary, M. Gensch, T. Golz, M. Harmand, N. Medvedev, M.J. Prandolini, K. Sokolowski-Tinten, S. Toleikis, U. Wegner, B. Ziaja, N. Stojanovic, F. Tavella, Nat. Commun. **4**, 1731 (2013)
36. U. Frühling, M. Wieland, M. Gensch, Th. Gebert, B. Schuette, M. Krikunova, R. Kalms, F. Budzyn, O. Grimm, Oliver, J. Rossbach, E. Ploenjes, M. Drescher, Nat. Photonics **3**, 523 (2009)
37. R. Kienberger, E. Goulielmakis, M. Uiberacker, A. Baltuska, V. Yakovlev, F. Bammer, A. Scrinzi, Th. Westerwalbesloh, U. Kleineberg, U. Heinzmann, M. Drescher, F. Krausz, Nature **427**, 817–821 (2004)
38. F. Quere, Y. Mairesse, J. Itatani, J. Mod. Opt. **52**, 339–360 (2005)
39. N. Stojanovic, M. Drescher, J. Phys. B **46**, 192001 (2013)
40. U. Frühling, J. Phys. B **44**, 243001 (2011)
41. B. Schütte, S. Bauch, U. Frühling, M. Wieland, M. Gensch, E. Plönjes, T. Gaumnitz, A. Azima, M. Bonitz, M. Drescher, Phys. Rev. Lett. **108**, 253003 (2012)
42. I. Grguras et al., Nat. Photonics **6**, 852–857 (2012)
43. A.H. Zewail, Science **242**, 1645–1653 (1988)
44. A.J. Verhoef, A.V. Mitrofanov, X.T. Nguyen, M. Krikunova, S. Fritzsche, N.M. Kabachnik, M. Drescher, A. Baltuska, New J. Phys. **13**, 113003 (2011)
45. M. Uiberacker et al., Nature **446**, 627–632 (2007)
46. A. Rath et al., Opt. Express **22**, 28914–28925 (2014)
47. M. Arbeiter, C. Pelz, Th. Fennel, Phys. Rev. A **89**, 043428 (2014)
48. B. Schütte, T. Oelze, M. Krikunova, M. Arbeiter, Th. Fennel, M.J.J. Vrakking, A. Rouzée, J. Phys. **17**, 033043 (2015)
49. M. Bergh, G. Huldt, N. Timneanu, F.R.N.C. Maia, J. Hajdu, Q. Rev. Biophys. **41**, 181–204 (2008)
50. G. van der Schot et al., Nat. Commun. **6**, 5704 (2015)
51. B. La Scola et al., Science **299**, 2033 (2003)
52. N. Philippe et al., Science **341**, 281–286 (2013)
53. H.N. Chapman et al., Nature **470**, 73–77 (2011)
54. A. Barty et al., Nat. Photonics **6**, 35–40 (2012)
55. S. Boutet et al., Science **337**, 362 (2012)
56. L. Redecke et al., Science **339**, 227 (2013)
57. M.M. Seibert et al., Nature **470**, 78–81 (2011)
58. M. Hantke et al., Nat. Photonics **8**, 943–949 (2014)
59. C. Kupitz et al., Nature **513**, 261–265 (2014)
60. M.J. Bogan et al., Nano Lett. **8**, 310–316 (2008)
61. D. Loh et al., Nature **486**, 513–517 (2012)
62. J.T. Lau, J. Rittmann, V. Zamudio-Bayer, M. Vogel, K. Hirsch, Ph. Klar, F. Lofink, T. Möller, B. von Issendorff. Phys. Rev. Lett. **101**, 153401 (2008)
63. D.P. DePonte et al., J. Phys. D **41**, 195505 (2008)
64. C.H. Yoon et al., Opt. Express **19**, 16542 (2008)

65. L. Strüder et al., Nucl. Instrum. Methods Phys. Res. A **614**, 483–496 (2010)
66. V. Lyamayev et al., J. Phys. B **46**, 164007 (2013)
67. A. Ravasio et al., Phys. Rev. Lett. **103**, 028104 (2009)
68. M. Fuchs et al., Nature Phys. **5**, 826 (2009)
69. A.R. Maier et al., Phys. Rev. X **2**, 031019 (2012)
70. E. Oliva et al., Nat. Photonics **6**, 764–767 (2012)
71. G.A. Mourou, G. Korn, W. Sandner, J.L. (eds.), *Collier Extreme Light Infrastructure ELI Whitebook—Science and Tehcnology with Ultra-Intense Lasers* (CNRS, Paris, 2011)

Chapter 14
Probing Molecular Photoexcited Dynamics by Soft X-Rays

Markus Gühr

Abstract This chapter presents a short introduction into probing photoexcited dynamics of isolated molecules by ultrashort x-ray pulses. It describes the basic concepts of electronic and nuclear dynamics in a molecule after optical excitation and shows basic concepts like element and site specific probing associated with x-rays. It discusses ultrafast x-ray sources for molecular spectroscopy and presents some of the very first gas phase molecular dynamics results that this young field has produced.

14.1 Introduction

We are currently in the second decade of ultrafast x-rays. Since 2000, ultrafast x-ray sources have been operating at synchrotrons [1]. With the first light from the free electron laser in Hamburg/Germany (FLASH) [2], the Linac coherent light source in SLAC National Accelerator lab [3], and SACLA in Japan [4] the brilliance of ultrafast extreme ultraviolet and x-ray sources has been increased by several orders of magnitude [5]. This chapter outlines the new possibilities using the brilliant and short soft x-ray pulses for probing the photodynamics of isolated molecules in the gas phase. While many interesting applications are found in solution phase, the photoexcited dynamics of isolated molecules offer a great chance for fundamental understanding [6, 7] since the highest accuracy simulations can be performed for these systems.

The key question in the molecule-light interaction is about the conversion of light energy into other forms of energy. The light does not provide any information on how its energy needs to get converted; different molecules however show very different conversion pathways after excitation. While some transform a large fraction of the

M. Gühr (✉)
PULSE, SLAC National Accelerator Laboratory and Stanford University,
2575 Sand Hill Road, Menlo Park, CA 94025, USA
e-mail: mguehr@stanford.edu

M. Kitzler and S. Gräfe (eds.), *Ultrafast Dynamics Driven by Intense Light Pulses*, Springer Series on Atomic, Optical, and Plasma Physics 86, DOI 10.1007/978-3-319-20173-3_14

light energy into making different bonds resulting in a molecular geometry change, others distribute the photon energy over many vibrational modes leading to heating without bond change.

A few examples show the diversity of the molecular energy conversion after interaction with light. Many molecules undergo a major structural change localized around few bonds under light illumination [8]. Prominent examples are azobenzene [9, 10], stilbene [11, 12], rhodopsin, which is important for vision [13, 14] and in bacterial light harvesting [15, 16], and green fluorescent protein (GFP), used as fluorescent marker in genetics [17]. A prominent class of molecules that does not show any major structural change after photoexcitation is the nucleobases. Here, light energy is not directed towards the large amplitude motion of one particular molecular coordinate, but many molecular modes are "heated". It is argued, that the light-to-heat-conversion in nucleobases contributes to the photoprotection of our genetic code, by avoiding ultraviolet induced dimerization of neighboring bases in the DNA strand [18–20].

How does this efficient and selective usage of light energy emerge? The optical light is absorbed by valence electrons which are delocalized and mediate chemical binding. Upon light absorption, the valence electrons change their shape and a concerted motion of electrons and nuclei follows. In this complex dynamics on the photoexcited state, some channels for energy conversion are preferred over others and the fastest (accessible) process dominates the energy conversion [21–25]. Dictated by the Einstein rate coefficients, radiative processes on photoexcited states typically happen on the nanosecond time scale for ultraviolet to visible light. Much faster and thus more efficient reaction channels are provided by radiationless processes resulting from the coupling of nuclear and electronic motion in the photoexcited state. Typical timescales for these processes lie in the ultrafast domain of pico- to femtoseconds, requiring ultrafast tools for their observation.

The rest of this chapter will give and account on the use of ultrafast x-ray probes to for photoexcited dynamics. The topic is approached for isolated molecules, having the above mentioned advantage of very high level simulation capabilities. Moreover, gas phase spectroscopy allows a wide variety of probe methods. Besides time and energy resolved ion or electron spectroscopic methods, also coincidence methods for charged particles (ions and electrons) can be used giving very detailed information on the molecular state at the time of the probe interaction [26]. While demonstrating the main statements in this chapter on a general level, more detail will be presented for the case of nucleobases. These molecules have been studied by ultrafast optical methods for a long time and a rich experimental and theoretical literature is evidence of the interesting processes occurring in their photoexcited states. The nucleobases have also been studied thoroughly with many x-ray methods and in addition a time resolved study on the photoexcited dynamics has been performed recently with ultrafast soft x-rays at the LCLS [27].

The following section will describe some basics of potential energy surfaces (PESs) and nonradiative processes that cannot be described using the Born-Oppenheimer approximation (BOA) [28]. It is believed, that most optically excited polyatomic molecules posses geometries at which nonradiative processes become determining in the energy conversion process [25, 29].

The third section of this chapter will introduce the concept of element and site selective probing and chemical shifts observed in x-ray spectroscopy [30–33], which can be used to address particular sites within the molecule. In the fourth section, crucial parameters of ultrafast x-ray sources will be presented and their usability for different x-ray spectroscopic methods discussed. The final section discusses the first examples of ultrafast x-ray probe experiments on photoexcited molecular states with an emphasis on the ultrafast x-ray Auger probing of nucleobase dynamics.

Ultrafast x-rays provide an important extension in the ultrafast toolbox due to their element and site selective probe character. The core wave functions involved in the x-ray probe transitions are extremely tightly bound which results in a high the local sensitivity within a molecule [34]. While the purpose of this chapter is to show how useful these new ultrafast x-ray probes can be and what promise they show for the future, it is mandatory to say that the new methods stand on the shoulders of giants. Ultrafast optical probe studies of light induced dynamics have been extremely successful and it is impossible in this context to give an overview of the most important works contributing to our understanding of photoexcited dynamics. I would understand the new ultrafast x-ray probe methods in this context: they are a valuable extension and maybe a crucial tool for many processes. Their real value however will lie in the complementarity to the optical ultrafast methods.

14.2 Molecular Processes

The interaction of optical light with molecules and the following molecular dynamics is the topic of many excellent textbooks [35–38]. In the following, the light induced dynamics is described, first using simple orbital arguments followed by a discussion of adiabatic potential energy surfaces. We then turn to vibrational relaxation and dynamics that cannot be described in the framework of the Born-Oppenheimer approximation.

In the beginning of any optically excited molecular dynamics is the interaction of valence electrons with light. The valence electrons are delocalized, that means they keep the molecule together by screening the repulsive forces between the positively charged nuclei, which are only in part screened by the core electrons. The absorption process results in a valence electron structure which is different from the "equilibrated" electronic ground state. This in turn will change the forces between the nuclei, resulting in nuclear dynamics. For the simple case of diatomic molecules, the light induced changes in the nuclear structure can be read from the change in equilibrium internuclear distance from ground to excited state. For Br_2, the ground state $^1\Sigma_g$ has a $R_e = 2.28$ Å. The optical excitation promotes one electron from a π_u orbital with no node orthogonal to molecular axis to a σ_u orbital with nodes [39, 40]. The reduced electron density in between the nuclei results in an expanded excited state ($^3\Pi_u$) with $R_e = 2.66$ Å

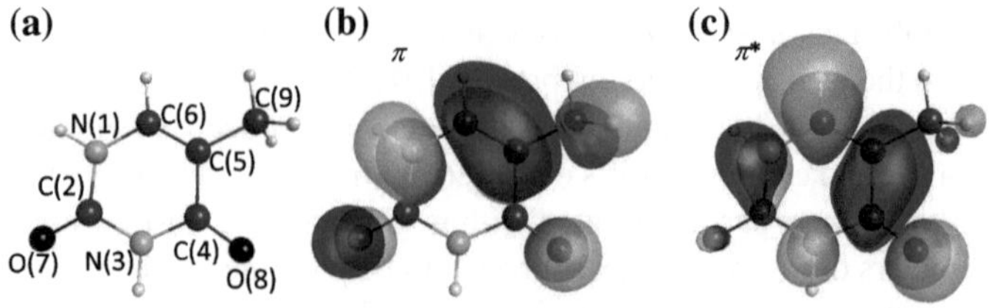

Fig. 14.1 Nuclear structure (**a**), second highest occupied molecular orbital (HOMO-1) π and lowest unoccupied molecular orbital (LUMO) π* in (**b**) and (**c**) respectively. Ultraviolet radiation induces the so called ππ* transition, in which one electron is promoted from the π into the π* orbital. Due to the nodal structure of the orbitals, one expects an elongation of the C(5)–C(6) and C(4)–O(8) bond as well as a contraction of the C(4)–C(5) bond after photoexcitation

For the case of thymine, the excitation scheme is displayed in Fig. 14.1. The light accomplishes a so called π–π* excitation, promoting and electron from a π orbital as shown in Fig. 14.1b into the π* orbital in Fig. 14.1c. The π orbital shows high electron density in between atoms C(5)–C(6), and a node in the C(4) atom. The newly populated π* orbital shows a node in between C(5)–C(6) as well as in between C(4)–O(8) leading to reduced electron density triggering bond elongation. On the C(4)–C(5) bond, the π* orbital increases the electron density leading to a bond contraction in the excited state. These coordinates are said to be Frank-Condon active, since they are directly coupled to the electronic change upon light excitation. Calculation on a higher level confirm this simple picture [41, 42].

The light induced nuclear motion is described in terms of potential energy surfaces (PES) which are intimately connected to the Born-Oppenheimer approximation simplifying the molecular Schrödinger equation. First, the eigenvalues of the electronic part of the Hamiltonian are calculated for fixed nuclei. This can be done because the light electrons can adapt instantaneously to any change in the slow motion of the heavy nuclei. The electronic spectrum is repetitively calculated for many nuclear geometries; connecting them leads to adiabatic PES. Based on these potentials, the complete Hamiltonian is solved. During this step, the Born-Oppenheimer approximation [28] neglects coupling terms among different electronic states based on the fact that they scale with the ratio of electronic over nuclear mass.

In order to simulate the nuclear dynamics following optical excitation, a nuclear wavepacket is constructed by calculating the dipole matrix element of the ground state with the excited PES. Once more, the electronic and nuclear wavefunction are separated giving rise to an electronic transition dipole and a Franck-Condon matrix element between the vibrational eigenstates of the electronic ground state and those of the electronically excited state. For harmonic oscillator PES, the nuclear wavepacket undergoes repetitive motion, with periodic recurrence in position and shape. An harmonic oscillator in one nuclear dimension leads to wavepacket dephasing followed by revivals of the wavepacket or of multiple copies at different positions [43–45]. For molecules with more than two nuclei an additional phenomenon called intermolecular vibrational redistribution (IVR) occurs. The initially excited Franck-Condon active vibrational modes are anharmonically coupled to other vibrational modes. An excellent introduction to IVR has been presented by Nesbitt and Field [46].

The Born-Oppenheimer approximation is often not sufficient to describe photoexcited molecular dynamics. As two PES come close, the separation of electron and nuclear motion breaks down, which can be understood using an intuitive approach by Herzberg [47]. The key argument allowing for separation of electronic and nuclear degrees of freedom is that light and therefore fast electrons are adapting instantaneously to any changing geometry of the heavy and slow nuclei. In the quantum world, the timescale of motion for any quantum object is given by the inverse of the energy spacing ($1/\Delta E$) of the eigenstates. Imagine just two electronic states being involved. Their ΔE changes as a function nuclear geometry. In the most extreme case, the nuclei can drive the quantum system to geometries where ΔE becomes very small or even zero. The electrons move slowly at these nuclear geometries and thus the electron-nuclear separation is not justified any more. A more formal derivation for example in [22, 28, 48] shows that the so called non-BOA coupling elements, which are based on the nuclear momentum operator, couple different adiabatic electronic states. The non-BOA couplings scale by the inverse energy spacing of PESs, which means that close lying adiabatic states are strongly coupled and wavepacket population is nonradiatively transferred from one adiabatic state to another. The nonradiative non-BOA transitions are fast, which means they are very efficient in driving excited state populations into particular channels.

A special class of topological constructs in which two adiabatic surfaces are close and even degenerate are the so called conical intersections (CIs). The PESs describe the shape of a cone around a point of degeneracy between the adiabatic states, leading to the nomenclature. The CIs were first proposed by Teller in 1937 [49] and afterwards mostly ignored as a kind of exotic topology. In the recent two decades however, conical intersections are intensely studied for their role in photoexcited dynamics [25, 29, 48, 50]. The dimensionality of a conical intersection is explained by Atchity et al. [51]: In an M dimensional nuclear geometry space, two particular geometry vectors g and h [50] span the CI and the point of degeneracy is a M-2 dimensional subspace, with certain exceptions based on symmetry arguments. Thus, in most molecules a "seam of conical intersections" [52] is lined up the nuclear geometry space.

Conical intersections are relevant for the photoexcited dynamics of thymine, the main example pursued in this contribution. Figure 14.2 presents a sketch of the ultraviolet excitation that initially populates the $\pi\pi^*$ state. Population is funneled into the lower lying $n\pi^*$ and ground state through conical intersections as the nuclear wave packet propagates. Directly after the photoexcitation, the planar relaxation coordinates described in the context of in Fig. 14.1 drive the wavepacket out of the Franck-Condon range. A ring folding coordinate gets activated which promotes the molecular wavepacket close to the conical intersections. Several intersections are predicted among the different states. The lowest energy intersections predicted by Hudock et al. [41] and Szymcak et al. [42] are sketched in Fig. 14.2. The g and h vectors for the intersection with $n\pi^*$ from [42] are shown in Fig. 14.2b; they illustrate the complexity of these coordinates. While g has a dominant contribution from

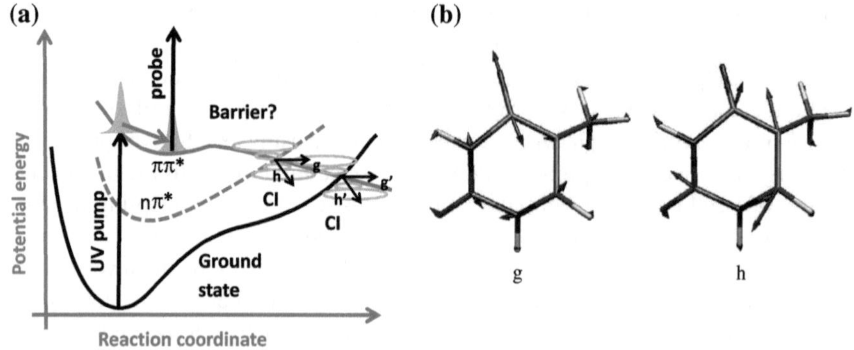

Fig. 14.2 **a** Reaction path from the Franck-Condon region to conical intersections with the ground and nπ* states. A barrier along the relaxation coordinate is object of discussion in the literature. **b** g and h vectors sketched by motion direction of certain atoms within the thymine molecule (reprinted with permission from [42]. Copyright 2009 American Chemical Society). These two vectors span the space for the conical intersection π π*-nπ*

the Franck-Condon active C–O stretch, the h vector possesses a strong out of plane component. The path from the Franck-Condon (FC) region to the conical intersections sketched here proceeds via an elongation of the C(5)–C(6) and C(4)–O(8) bond. A barrier in between the FC region and the CIs is indicated in Fig. 14.2. This barrier is currently matter of a debate and it is crucial for the molecular dynamics. Hudock et al. [41] and Szymcak et al. [42] predict a barrier that slows down electronic relaxation on the π π* state by trapping population in a minimum. According to these simulations, the molecular population in the π π* state needs several picoseconds to arrive at the CIs sketched in Fig. 14.2. Other groups use similar methods but slightly different basis sets and arrive at fundamentally different conclusions. Merchan et al. [53], Perun et al. [54] as well as Asturiol et al. [55] argue that the relaxation out of the π π* state occurs without any trapping of population behind a barrier, and thus electronic relaxation is expected to happen on a ~100 fs timescale.

14.2.1 Experimental Work on Molecular Dynamics Outside the BOA Framework

Almost all ultrafast photoexcited dynamics measurements on molecules are done with light in the visible and ultraviolet domain. For the pump pulse, it is necessary to use this wavelength range if one wants to draw connections to molecular excitation by sunlight. Since the pump-pulse is optical, it is relatively practical to also use optical light for probing (sometimes the same color) which however is not necessary. In fact, the probe pulse is observing the phenomenon and its spectral range can be chosen according to whichever method is best suited.

In the gas phase most ultrafast experiments on neutral photoexcited molecules are performed by measuring ion yields or electron kinetic energy spectra as a function of pump-probe time delay, with additional parameters such as angular distribution. Multi-photon probing combined with ion counting have been among the first ultrafast probes of photoexcited dynamics [56, 57] and the strong field tunneling picture allows certain conclusions about the shape of the ionized orbital [58, 59]. In the following we concentrate on time resolved photoelectron spectroscopy, for which an excellent introduction has been provided by Stolow and Neumark [60, 61]. Energy and time resolved photoelectron spectroscopy is often able to distinguish ultrafast vibrational and electronic dynamics, as shown by Stolow and coworkers [62, 63]. The analysis of those spectra is based on the notion that ionization of a photoexcited neutral state occurs most likely upon removal of one only electron. If several cationic states are energetically accessible by the probe pulse, photoelectrons will reflect predominantly a transition to the ionic state requiring only removal of one electron.

In order to separate electronic from nuclear relaxation it is however necessary that: (1) the change in photoelectron energy due to competing vibrational relaxation does not infer with the change in photoelectron energy from the nonadiabatic transition and (2) the probe pulse needs to have a photon energy high enough to ionize the molecule despite vibrational and electronic relaxation [64]. As the molecular dynamics distributes energy into several nuclear degrees of freedom and at the same time also other electronic states, an optical photon is not energetic enough for many molecules to provide ionization. The general decay on the photoelectron signal is then not sufficient any more to distinguish vibrational and electronic relaxation. Barbatti and Ullrich recently elaborated on these aspects of optical time resolved photoelectron spectroscopy for the case of the nucleobase adenine [65].

For the example of the nucleobase thymine, past experiments are not interpreted in a coherent way, mostly due to the ambiguity between electronic and vibrational relaxation. Several experimental studies in liquids and isolated molecules confirm that the excited state dynamics occurs in the sub-picosecond and picosecond regime [19]. The isolated studies are best suited to be compared to the single molecule ab-initio theory discussed above. Time resolved ion yield experiments in the gas phase show a very short sub 100 fs decay constant followed by additional 5–7 ps decay [66–68]. The short decay has been interpreted by some authors as ultrafast electronic relaxation from the $\pi\,\pi^*$ state to the electronic ground state [67]. However, the fast initial dynamics can also be attributed to ultrafast nuclear relaxation as done in photoemission experiments [69]. The 500 fs time constant found there was interpreted to display nuclear dynamics followed by population trapping in the $\pi\,\pi^*$ minimum [41]. As described above, the simulations are divided into two different classes: one set is predicting the ultrafast electronic relaxation into the electronic ground state on a few 100 fs [53–55] not slowed down by a reaction barrier. The other class is predicting a reaction barrier, slowing down nonadiabatic electronic relaxation favoring the interpretation of fast decays as nuclear relaxation [41, 42]. Combining the experimental and theoretical studies in thymine, an 'ideal' experiment needs to be able to distinguish nuclear relaxation on the $\pi\,\pi^*$ state from electronic relaxation. In the latter case it should also be able to distinguish nonadiabatic transitions to the nπ* state from those to the electronic ground state. Given the essential theoretical disagreement about the reaction barrier and the experimental dispute about analysis and interpretation of transients, it will be advantageous to measure the photoexcited dynamics with a drastically different method. We will describe such an attempt in the last section of this contribution. We choose ultrafast x-rays as probes of the molecular dynamics and we will therefore introduce basic concepts and advantages of molecular x-ray spectroscopy below.

14.3 Probing Molecular Electronic Structure by Soft X-Rays

Probing in the photoexcited states by ionization in the optical and vacuum ultraviolet range uses exclusively the mostly delocalized valence electrons. In the extreme ultraviolet (EUV, 30–280 eV) and soft and hard x-ray spectral domain however, light can couple to localized *core* electrons of distinct elements inside the molecule. Figure 14.3 shows the binding energies of important elements in the EUV and SXR region. As can be immediately seen, these energies are very distinct, which allows to individually address those elements by interaction with light. In the extreme ultraviolet, the M edges of the 3d transition metals provide element sensitivity. In the soft x-ray range, the K edges of oxygen, nitrogen and carbon as well as the L edges

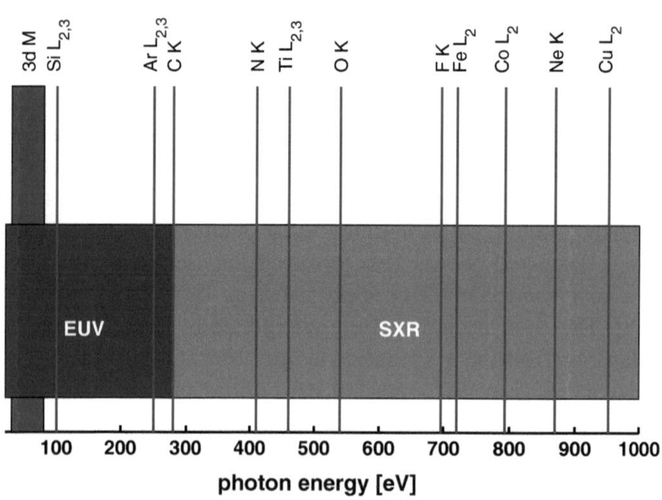

Fig. 14.3 Elemental edges in the extreme ultraviolet and soft x-ray spectral regions. The M edges of the 3d transition metals are due to 3p transitions, the K edges due to 1s absorption and the L edges due to 2s and 2p absorption.

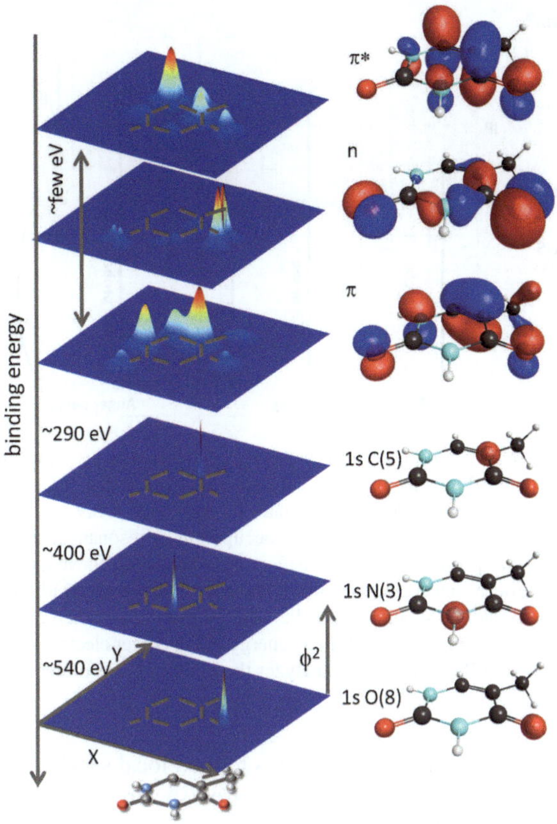

Fig. 14.4 Orbitals of the nucleobase thymine calculated using GAMESS and the 6–311 basis set. The *right hand column* shows isosurfaces of selected orbitals. On the upper part, three selected valence orbitals are shown. The π* and π orbitals are mostly delocalized over the molecule whereas the lone pair n orbital consists of p like lobes at the two oxygen atoms. In the lower part of the picture, three representative 1s orbitals of carbon nitrogen and oxygen are shown with their binding energies. The core hole wavefunctions are extremely well localized at the respective atoms. The *left hand column* shows 2d representations of the orbitals. It is obtained by cutting through it in a plane parallel to the ring plane and taking the absolute

of 3d transition metals provide this element contrast due to the element specific energy of core levels. An overview of different core level binding energies is given in [30, 31, 70, 71].

The electronic wavefunctions of core electrons are extremely well localized in contrast to the valence wave functions that are in general delocalized over large parts of the molecule. Figure 14.4 shows two-dimensional cuts of some C, N, and O 1s orbitals of the thymine molecule together with some typical valence orbitals. The enormous difference in localization properties shows that any transition (or quantum mechanical matrix element) involving core electrons is necessarily very confined in

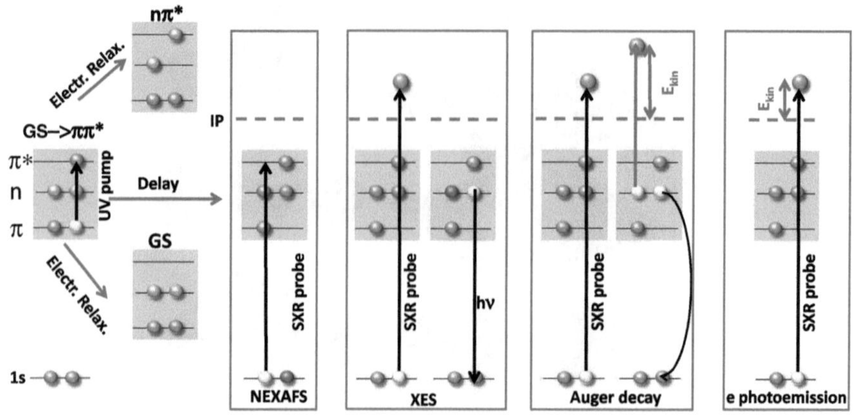

Fig. 14.5 Different schemes to probe transient electronic structure by x-ray interactions, exemplified on the important orbitals of thymine. The ultraviolet excitation pulse transfers population from the electronic ground state to the π π* state. In near edge x-ray absorption fine structure (NEXAFS) spectroscopy, one probes the valence occupation by resonant absorption from the core levels. In x-ray emission spectroscopy, the core electron is promoted above the ionization limit and the emitted photons from valence electrons filling the core hole are spectrally resolved. In the soft x-ray region, photoemission is small; instead the Auger emission sketched in the third field is the prominent process. In this technique, the kinetic energy of the Auger electrons is analyzed. In x-ray photoelectron emission (XPS) one directly analyzes the kinetic energy of the emitted core electron

space. In the x-ray community this aspect is termed local sensitivity. Moreover, we will show below in the section on photoelectron spectroscopy that the core binding energies of an element depend on the nearest environment of binding partners. Although chemical shift is only fraction of an eV up to a few eV as compared to many 10 and 100 eV element shift, it is visible in spectra and can be used in time resolved studies. Generally the term site selectivity is reserved for this phenomenon. In the following we will discuss the element and site selectivity of x-ray absorption, x-ray photoemission, Auger electron emission and x-ray photoelectron spectroscopy. An excellent overview of different techniques for gas phase targets is given in [72, 73]. A graphical overview of the different techniques is shown in Fig. 14.5.

14.3.1 X-Ray Absorption

The progress in x-ray absorption spectroscopy until 1992 is documented in J. Stöhr's monograph [74]. The interpretation of near edge x-ray absorption fine structure (NEXAFS) spectra in the soft x-ray domain was systematically elaborated from small molecules [72, 73], to small molecules on surfaces [74, 75] and complex systems in solution [76, 77]. The interaction of light with a quantum system during absorption is described by the matrix element $\langle \phi_i | \mu | \phi_f \rangle$, where $\phi_{i,f}$ are the initial and final states of the system and μ is the dipole operator. The initial state usually has completely filled shells up to the valence levels; the final state is described as a core hole and an additional electron in a valence or continuum state accompanied by some restructuring of valence electron due to electron correlation. The one electron approximation used from now on neglects the coupling among the electrons, which simplifies the description. Within the one electron approximation, the initial state of the absorption process can be represented by a core electron orbital. The final state in absorption spectroscopy is an *unoccupied* or *partially* occupied valence or a continuum orbital. Within the linear combination of atomic orbital (LCAO) approximation, the transition strength between core electron and unoccupied molecular valence states is then proportional to the atomic orbital coefficient at the core hole site of the respective molecular orbital [34, 74, 78]. Thus, one probes a highly local electronic property of valence electronic states. This picture can be helpful thinking about the applications of NEXAFS spectroscopy for photoexcited molecular dynamics. As pointed out in the previous section, the electronic character of photoexcited states undergoes transient changes. In the case of thymine, the $\pi\,\pi^*$ decays into an $n\pi^*$ state upon electronic relaxation. Thereby the occupation changes from $(\pi^1 n^2\,\pi^{*1})$ to $(\pi^2 n^1\,\pi^{*1})$. Referring to Fig. 14.4, the transition between one of the oxygen 1 s electrons and the highly localized n orbital is must be very strong. The absorption 1s-n is not possible in the ground and $\pi\,\pi^*$ states since the n orbital is completely filled; it becomes allowed with the nonadiabatic transition to the $n\pi^*$ state, offering a great chance for the direct detection of the non-BOA transition. It is worth mentioning that any molecule containing lone pair orbitals n among the highest occupied or lowest unoccupied orbitals provides highly localized valence orbitals. The processes involving the lone pair orbitals do not necessarily need to be restricted to the photorelaxation of a single molecule like in the nucleobase. For instance, the [2+2] photocycloaddition of carbonyl compounds, which is a bond formation, proceeds via an $n\pi^*$ photoexcited state.

We demonstrate the content of NEXAFS spectra for thymine, discussing data from Plekan et al. [79] in Fig. 14.6. One identifies the sharp transitions to unoccupied molecular orbitals, which are of π^* character. At the oxygen absorption edge, the two different oxygen sites O(7) and O(8) lead to a splitting of the lowest energy feature. The corresponding transitions involve two different π^* orbitals, $\pi_6{}^*$ corresponds to the orbital shown in Fig. 14.1c. As mentioned in the introduction to this section, we can clearly distinguish the O(7) and O(8) sites with $\sim$1 eV splitting. In contrast, the two lowest energy features in the nitrogen K absorption spectrum of thymine are not easily separated in terms of the two nitrogen atoms N(1) and N(3). The lines from 401 to 403 eV photon energy are mostly of mixed nature. At the carbon K edge however, the two lowest lines are due to a specific carbon atoms, the higher lying lines are of mixed character [79]. Towards higher energies, transitions between core electrons and molecular Rydberg states are excited and finally above the ionization limit an unstructured and very broad feature sets in. This feature is called multiple scattering or σ^* resonance. It originates from the multiple electron scattering in the molecular potential made up by Coulomb forces and centrifugal barriers for states with high angular momentum [80–84].

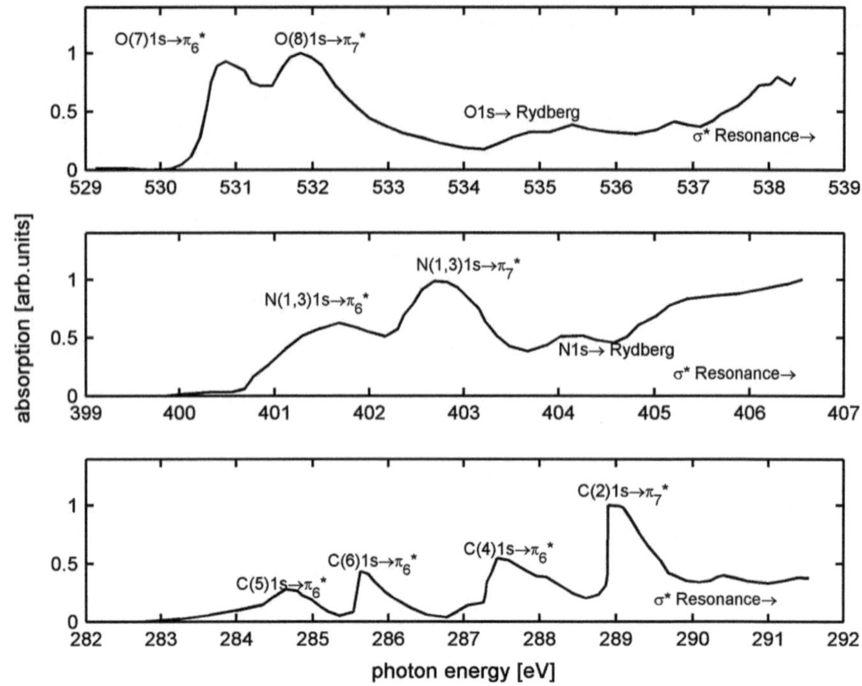

Fig. 14.6 Near edge x-ray absorption fine structure (NEXAFS) spectra of thymine at the oxygen (*top*), nitrogen (*middle*) and carbon (*bottom*) edges (reproduced by digitizing data from [79]). The labeled peaks correspond to transitions from ground state molecules to a state with a 1s core electron missing and an electron created in an initially unoccupied valence orbital. The orbital π_6* corresponds to the orbital shown in Fig. 14.1c, π_7* is not shown here but can be found in [79]. At the oxygen and carbon edges one identifies peaks that correspond to core hole creation at particular sites, for instance O(7) or O(8)

14.3.2 X-Ray Emission

X-ray (photo)emission spectroscopy (XES) can be described by the same matrix element as absorption; however the initial state is an *occupied* valence orbital the final state is a core orbital. Thus, XES possesses the same element and site selectivity as NEXAFS, with the big difference that XES monitors occupied orbitals, which makes the two methods ideally complementary. In the soft x-ray region, the decay of core ionized states via XES is in fierce competition with radiationless Auger decay. The time needed for an Auger decay into 1s states is typically in the domain of a few femtoseconds [85]. The photoemission time is governed by the Einstein rate coefficient for spontaneous decay, which is much slower than the Auger decay in the SXR range. Thus the relative yield of photoemission compared to Auger decay is low [85]. With increasing XES photon energy, the Einstein rate coefficient for spontaneous emission increases, leading to a much higher fluorescence yield in the hard x-ray domain [86]. The low SXR yield, together with low apparatus efficiency due to limited angular acceptance of grazing incidence gratings [87], make XES applications for dilute systems very challenging. For small molecules, this problem can be easily be compensated using high pressures, as demonstrated for N_2 [88], O_2 [89], CO [90], and CO_2 [91]. Here, XES spectra are to first order attributed to certain molecular orbitals. For more complex molecules like transition metal complexes, multiplet effects need to be taken into account [92]. For nucleobases, the only XES spectra documented were obtained from solid samples at the nitrogen edge [93].

To use the soft x-ray emission technique for time resolved measurements on diluted isolated molecules, sources will have to provide much more average flux. Boosting single pulse energy will only lead to non-useful sequential molecular ionization. Instead, the repetition rate will have to be increased, which is exactly matched in the strategy for future FEL x-ray sources.

14.3.3 Auger Electron Emission and Fragmentation

The matrix element for Auger emission is much more complex compared to the dipole matrix element. The Auger decay of a core hole is described by a two electron operator $1/r_{12}$, whose eigenvalue reflects the inverse distance of two electrons. The Auger matrix element can be simplified to $\langle \phi_k \, \phi_l \, |1/r_{12}| \, \phi_{core} \, \phi_{cont} \rangle$, where $\phi_k \, \phi_l$ are the two valence electron orbitals that fill the core hole and emit the Auger electron, ϕ_{core} is the initially ionized core orbital and ϕ_{cont} is the continuum orbital that is filled by the Auger electron [94, 95]. In summing over the different valence channels kl, the spin multiplicity needs to be taken into account properly [94]. If we neglect the electron interaction, the kinetic energy E_{kin} of an Auger electron is given as $E_{core} - E_k - E_l$, where $E_{core,k,l}$ are the binding energies of the core and two valence electrons. It is important to note here, that the photon energy does not influence the Auger kinetic energy, as long as complete core hole ionization into the continuum occurs. This makes this technique very attractive for FEL sources based on self amplified spontaneous emission (SASE). For those FELs, lasing starts from noise, which results in a large spectral shot-to-shot fluctuation [96].

The x-ray typical element selectivity and local sensitivity still holds, even for the more complex Auger decay. Since the core hole energy differs strongly depending on the element, the Auger decay energies from different elements within a molecule are separable (see Moddeman et al. [97] for element selected Auger spectra CO, NO, H_2O, and CO_2). The local sensitivity comes from the fact that the very narrow core orbital ϕ_{core} is included in the spatial integral of the matrix element. Early Auger spectroscopy on small di-and triatomic molecules [94, 98, 99] demonstrated that the Auger spectrum amplitude can be constructed reasonably enough by just tracking the atomic orbital coefficients at the core hole site of the LCAO constructed molecular orbitals $\phi_k \, \phi_l$. The molecular Auger spectra of 2p elements like carbon, nitrogen and oxygen separate into three different energy ranges. Figure 14.7 shows the oxygen

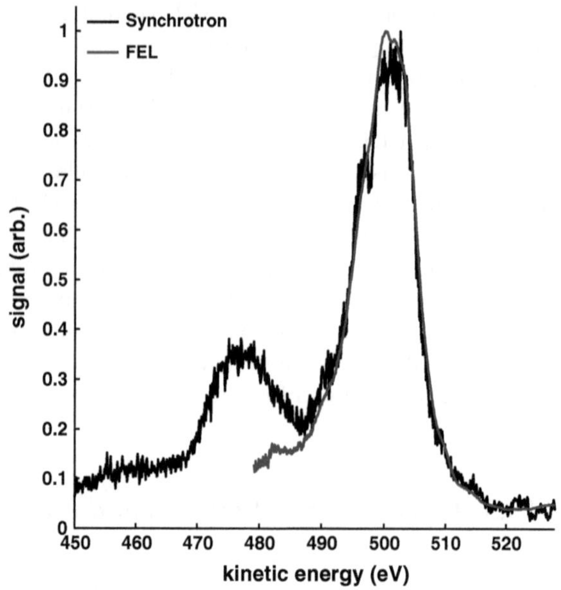

Fig. 14.7 Auger spectra of thymine in the electronic ground state. The *black* spectrum was measured at a synchrotron (advanced light source beamline 8), the *grey* spectrum at the free electron laser (Linac coherent light source)

Auger spectrum of thymine. The highest kinetic energies around 500 eV result from Auger decay involving mostly two 2p valence orbitals, since those have the lowest binding energy. The group of Auger decay around 475 eV is due to valance orbitals of mixed character, one 2s and another 2p. The third and lowest kinetic energy group hardly visible from 450 to 460 eV is due to decays involving two 2s orbitals.

All of the Auger lines are lifetime broadened and identifying one particular channel or a particular decay site among equal elements is often impossible. Auger decay spectra measured and simulated by Storchi et al. on pyrimidine [100] (which is essentially the ring of without the methyl group and the oxygen atoms) exemplify this point. Figures 14.2 and 14.3 in [100] show Auger spectra due to nitrogen 1s core hole decay and carbon 1s core hole decay respectively. The kinetic energies of the different core hole Auger decays are very well separated by about 100 eV. The spectra consist of many lines, calculated here on the algebraic diagrammatic construction (ADC(2)) level [101, 102] and the isolation of one particular nitrogen or carbon core hole decay from the experimental spectrum is practically impossible.

The involvement of valence orbitals ϕ_k ϕ_l in the Auger decay makes Auger spectroscopy interesting for time resolved spectroscopy on photoexcited molecular states. Electronic relaxation changes the valence orbital occupation, SXR core hole ionization followed by Auger decay offers an observation opportunity for this process. Together with the element selectivity and local sensitivity, Auger decays at different core hole sites predominantly probe different orbitals. For the example of thymine in Fig. 14.4 it is obvious that the Auger decay of oxygen 1s vacancies predominantly probes the oxygen lone pair type orbitals. The C and N core hole decays do not involve these lone pair but more likely to the π orbitals, which are delocalized over the ring. An important consideration for time resolved experiments is the fact that the Auger decay takes some time itself. For most practical experiments however, the few femtosecond time constant is short compared to the molecular dynamics.

The Auger decay into dicationic states with two valence electrons missing leads mostly to molecular fragmentation. This can be understood in an orbital picture in which initially binding orbitals become depleted by the Auger decay. Resonant excitation of small molecules into unoccupied molecular orbitals can lead to fragmentation at a particular site [103]. For thymine, fragmentation was induced by C ionization at different sites [104]. The larger molecules like thymine have a high density of dicationic states, all of which are occupied according to the Auger decay matrix element. This manifold of states mix strongly along the fragmentation path and it turns out that most fragment yields do *not* depend on the core hole site [104, 105]. Nevertheless, some thymine fragments show an interesting behavior on the core hole site. For instance, the creation of CO+ fragments is correlated to the core hole on C(2) and C(4), which are the C atoms connected to the two oxygen atoms in the molecule. For C(5) and C(6) core hole excitation, a strong site specific decay into $HNCOH^+$ and $C_4H_4ON^+$ was observed. Itälä et al. associate this decay to the participation of the π orbital (shown in Fig. 14.1b) in the Auger decay as the π orbital has a high amplitude at the C(5) and C(6) positions. The final states with π population missing after the Auger decay are very low in energy. Thus, they can only lead to fragmentation into the most stable fragments, which happen to be the $HNCOH^+$ and $C_4H_4ON^+$ fragments.

14.3.4 X-Ray Photoelectron Spectroscopy

X-ray photoelectron spectroscopy is one of the most advanced methods of all x-ray probes in the gas phase as well as in liquid or solid state [106]. Like in absorption spectroscopy described above, the dipole matrix element induces a transition, here between the core electron and a *continuum* state. The continuum electron possesses kinetic energy, leaving behind a molecule with a core hole. For different elements, core binding energies are very distinct. For one and the same element occurring on different sites within the molecule the photoelectron spectrum shows site splitting on the sub eV to eV scale as well as vibrational features on the meV scale. The binding energy depends on the environment of the core hole site. The valence electrons located close to the core site screen the nuclear potential leading to a lower binding energy. Reduced valence density results in less screening and higher binding energies. The resulting sensitivity on the chemical environment has been called chemical shift and gave rise to the ESCA (electron spectroscopy for chemical analysis) technique invented by Siegbahn [32]. A compilation of typical site specific molecular spectra can be found in the compilation by Jolly, Bomben, and Eyermann [107].

The local chemical environment depends on the exact nuclear geometry and the electronic structure that goes with it. Reduced concepts like electronegativity, partial atomic charge and polarizability are used to express the chemical sensitivity of core hole binding energies [108]. Site effects in the carbon photoelectron spectrum of thymine [79, 104, 109] again exemplify the electronegativity arguments. The electronegativity of atoms is increasing from carbon over nitrogen to oxygen. The C(2) atom, surrounded by two N and one O atoms, shows the lowest kinetic energy of all [79, 109] (note: [109] uses the standard nomenclature as in Fig. 14.1a whereas [79] uses a different one). Atom C(4) shows a slightly lower electronegative environment with one N and one O atom, thus its kinetic energy is higher compared to C(2) but lower compared to C(6) with one nitrogen as closest binding partner, and C(5) and C(9) being bound only to C or H atoms. The overall difference between C(2) and C(5), C(9) kinetic energies (binding energies) is about 5 eV. The two nitrogen and oxygen atoms show negligible chemical shifts that are not resolved in the x-ray photoelectron spectra. The same concept also applies to the interpretation of the site shift observable in the C1s to π^* resonances in the NEXAFS spectrum in Fig. 14.6.

The concept of electronegativity is only a first approximation to local electron density due to binding. Detailed calculations become necessary for quantitative statements and comparisons between different nucleobases [109]. Since the electronic structure is changing as a result of optical light absorption (for example in form of the $\pi\,\pi^*$ excitation sketched in Fig. 14.1), the photoelectron spectrum will change. Moreover, electronic and nuclear relaxation will both change the photoelectron energies and also their amplitudes. Currently however, free electron lasers are only starting to become a useful tool for x-ray photoelectron spectroscopy in the C, N and O 1s region. Typical spectral width and shot-to-shot jitter instabilities spoil the spectral characteristics needed for this type of spectroscopy. New schemes like self seeding are on the horizon and might radically change the prospects for ultrafast soft and hard x-ray photoelectron spectroscopy for photoexcited molecular states [110, 111].

14.4 Sources for Ultrafast X-Ray Spectroscopy

Table 14.1 shows a systematic comparison of different short pulse x-ray sources. It is important to note here, that the usefulness of a particular source depends on the scientific question to be addressed, the experimental method to be implemented, the availability of beamtime, and the complexity of its operation. In general, there is no best source for all experimental schemes and circumstances. Parameters not included in the table like brilliance, longitudinal and transversal coherence as well as bandwidth might be crucial in the selection of the light source. We have chosen to include high harmonic generation and plasma emission in this scheme, although their optimal spectral range is not in the soft x-ray domain, but rather in the extreme ultraviolet for the case of harmonics and the hard x-ray (HXR) for the laser driven plasma sources.

Table 14.1 Parameters for different sources in the extreme ultraviolet and x-ray range

	HHG	Laser driven plasma	Synchrotron	FEL
Photon energy	Typically 20–100 eV—recently up to 1.5 keV at reduced flux [112]	K_α lines and bremsstrahlung of metals in hard x-ray region	Throughout EUV, SXR, HXR range	Throughout EUV, SXR, HXR at different facilities
Pulse duration	~0.1 fs to some 10 fs	>100 fs	100 fs for slicing, 1 ps for special mode, 100 ps normal	Few femtosecond to few 100 fs
Flux (photons/s)	$\sim 10^{12}$ @ 1 kHz	10^{10} @ 1 kHz into full solid angle	10^4–10^6 in slicing at 1 kHz, 10^9 for picoseconds pulses @ MHz [113], $>10^{13}$ for 100 ps @ MHz	About 10^{12} per pulse at repetition rates determined by the facility

Strong field laser based harmonic sources started in the late 80's to early 90's [114–117]. Their temporal structure, interpreted as a train of attosecond pulses resulting from electron ionization and recollision [118–122], made them suitable for attosecond physics in the extreme ultraviolet domain from about 20–100 eV [123, 124]. New developments in mid-infrared laser sources have led to higher photon energies reaching up to 1.5 keV [112], however at reduced flux due to the 20 Hz repetition rate of the drive laser.

Laser driven plasma [125–127] sources are based on strong field ionization and plasma generation on a metal target. The laser accelerates electrons from a metal target and accelerates them back into the target, thereby creating incoherent emission of characteristic recombination lines as well as a bremsstrahlung continuum. The pulse duration is given by the travel time of the electrons through the metal target, which can be as short as 100 fs [128]. Those sources typically create 10^{10} photons/sec into the full spatial angle. The usable portion is given by the sample size or the collection efficiency of hard x-ray optics [129].

Slicing beamlines at synchrotron sources were pioneered in the beginning of the new century at the Advanced Light Source in Berkeley [1, 130]. Now, those beamlines are available in the soft and hard x-ray range at a few synchrotrons around the world achieving ~100 fs time accuracy with an integrated flux of 10^4–10^6 photons/sec [131–133]. At slicing beamlines, an optical laser is used for energy modulation of the synchrotron energy bunches. The modulated part consecutively emits femtosecond pulses during the interaction with bending magnets, wigglers or undulators. The optical laser, which has been used for the electron energy modulation, can be used as a perfectly synchronized sample excitation pulse. Without slicing, synchrotrons typically deliver 100 ps time resolution, which can be further shortened into the picoseconds range [113, 134, 135]. For all laser pump-synchrotron probe experiments, the laser oscillator has to be feedback locked to the radio frequency clock of the synchrotron, as demonstrated for example in [136, 137].

Free electron laser (FEL) sources are now the brightest short pulse EUV and x-ray sources. Historically, the free electron laser in Hamburg-FLASH [2] in Germany was the first FEL in the EUV and soft x-ray domain. The Linac coherent light source (LCLS) [3] at SLAC/USA pushed the spectral range of FELs further into the soft and hard x-ray domains in 2009. The SPring-8 Angstrom Compact free electron Laser (SACLA) [4], which turned on at Riken-Harima/Japan in 2011 is currently the latest SXR/HXR FEL. Due to the high flux of 10^{12} photons/pulse at 120 Hz repetition rate for the example of LCLS, the FELs provide an enormous flexibility for different types of spectroscopy. All three sources mentioned above were designed unseeded, resulting in large pulse to pulse fluctuations in of pulse power and pulse spectrum. In the hard x-ray range, a self seeding scheme was recently accomplished at the LCLS [110]. The resulting pulses have 1/40–1/50 compared to the SASE bandwidth and are much closer to the Fourier transform limit [110]. Similar schemes are planned for narrowing and stabilizing the pulse spectra in the soft x-ray range [111]. The first operational seeded FEL is FERMI in Italy [138] and recent progress by direct seeding with EUV light was demonstrated at FLASH [139].

For seeded FELs, timing to optical excitation lasers is straightforward, since the seeding laser infrastructure can also be used for sample excitation. At unseeded FEL sources delay accuracy between optical excitation and x-ray probe pulses can be improved by additional setups which determine the delay between the two pulses on a shot to shot basis. There are two fundamentally different methods to determine the excitation—x-ray delay. The first scheme uses the fact that the optical reflectivity of a solid sample is changed if pumped by a sufficiently strong x-ray pump pulse. The physical effect was demonstrated at FLASH, and authors pointed out that a single shot tool could be built by spatially resolving the reflectivity for a non-collinear geometry of x-ray and optical probe pulse [140]. A single shot monitor based on spatial encoding was later used at LCLS [141, 142]. Higher accuracy monitors were later realized using spectral encoding by a chirped (that means temporally extended by a spectral phase) continuum and provide down to 10 fs timing accuracy [143–145]. A second scheme is based on strong field interactions with x-ray emitted electrons. When a gas phase target is x-ray ionized in the presence of a strong infrared laser field, the laser field modulates the electron kinetic energy if the pulses overlap in time [146–149]. If the few femtosecond period of the optical laser is further increased to several 100 fs by THz pulse generation, a single shot delay monitor can be accomplished since the electron kinetic energy modulation reflects the THz vector potential strength at the time of x-ray photoemission [150]. This scheme has recently been used to even determine the temporal profile of the x-ray pulses [151, 152].

14.5 Ultrafast X-Ray Probing of Photoexcited Molecular Dynamics

We now start with the discussion of ultrafast optical pump—x-ray probe experiments. The field of ultrafast x-ray molecular spectroscopy with gas phase molecules has been really opened by x-ray FELs due to the high flux needed in these experiments. The sources are young thus there are currently very few experiments.

Cryan et al. [153] had performed the first optical- pump- x-ray probe experiment at the LCLS. They used a nonresonant infrared laser pulse to align nitrogen molecules before core-ionizing them with a soft x-ray pulse. In the experiment, alignment provided means to measure angular distributions of electrons in the molecular frame. The intense and short (<5 fs) soft x-ray pulse created two core hole vacancies in the molecule [154–156]. This process can occur on a single atom within the molecule (single site double core hole, ssDCH) or on two different atoms (double site double core hole, dsDCH). Cryan et al. succeeded to identify the signature of the ssDCH in the high kinetic energy part of the Auger spectrum and measured the Auger electron angular dependence by varying the angle between the molecular alignment and the soft x-ray polarization.

Molecular dynamics on a UV excited molecular state was probed using soft x-ray induced fragmentation by Petrović et al. [157]. Compared to the well established technique of infrared strong field Coulomb explosion, the x-ray induced fragmentation bears several advantages. In the strong infrared field with field strength sufficient needed to ionize, other effects like alignment, bond softening, above threshold ionization, and intermediate multi-photon resonances can alter the ionization yield and kinetic energy. All these effects are negligible for soft x-ray core ionization creating two valence holes via Auger decay. The UV pump pulse triggered a ring opening reaction from cyclohexadiene to hexatriene via nonadiabatic excited state dynamics. Increasing time delay between the UV and soft x-ray pulse resulted in an increased H^+ emission. Further steady state spectroscopy on the pure compounds confirmed that the photoproduct hexatriene leads to higher proton production after soft x-ray ionization compared to cyclohexadiene.

For the case of thymine, the lifetime of the $\pi\,\pi^*$ state is a matter of debate. As discussed in the second section of this chapter on molecular dynamics, two different theoretical approaches predict different $\pi\,\pi^*$ state lifetimes based on the involvement or absence of a reaction barrier. The current ultrafast optical experiments all show very similar time constants, however they can be interpreted by both competing models. This is clearly a good motivation to investigate the molecular excited state dynamics with a fundamentally different probe, as the element selective and locally sensitive x-rays provide.

The experiment discussed in the following was performed at the LCLS by the "LCLS nucleobase collaboration" [27, 158] at the so called AMO instrument [159]. Thymine was evaporated at 150 °C and transported into the interaction region using a hollow capillary oven [160]. An ultraviolet pulse excited thymine from the electronic ground state to its π π* state. The time-delayed soft x-ray pulse had a photon energy above the oxygen K edge and photoionized the molecule. The oxygen 1s Auger spectrum was energy resolved using a 2 m long magnetic bottle spectrometer [161, 162], a spectrum was saved for every shot. In addition, the time jitter between UV pump and soft x-ray probe pulse was determined by x-ray induced changes in the optical reflectivity for every shot [141] as described in the preceding section. This allowed for resorting of all single shot spectra with jitter accuracy around 70–100 fs. More technical details of the experiment are described in [27].

Figure 14.8 shows the experimental data of the UV pump-x-ray probe scan over a range from −0.5 to 20 ps delay. It is worth mentioning that the time points in Fig. 14.8a are not equidistant; for small delays we have a time binning of 70 fs, the largest delays have a relative separation of 15 ps. It is however obvious, that most change occurs in the picosecond after the zero delay. The subtraction of unexcited spectra from UV excited spectra reveals the change between the unexcited and excited molecule, therefore we plot difference spectra of UV excited minus unexcited samples in part a. Blue colors indicate a UV induced decrease of Auger electron yield and the specific kinetic energy and delay, red indicates a UV induced increase of the Auger yield. The signal at negative times before zero delay has a large error due to reduced data acquisition times, which results in random color fluctuations. The first systematic signal occurs right after zero delay at positive times. One can clearly identify a signal decrease in kinetic energy region number II (around 500 eV kinetic energy) that lasts for the full measurement interval. In region number I (around 507 eV kinetic energy), we identify a short lived positive feature, which can be analyzed better in the integrated signals plotted as a function of delay in c. The fit to the integral of

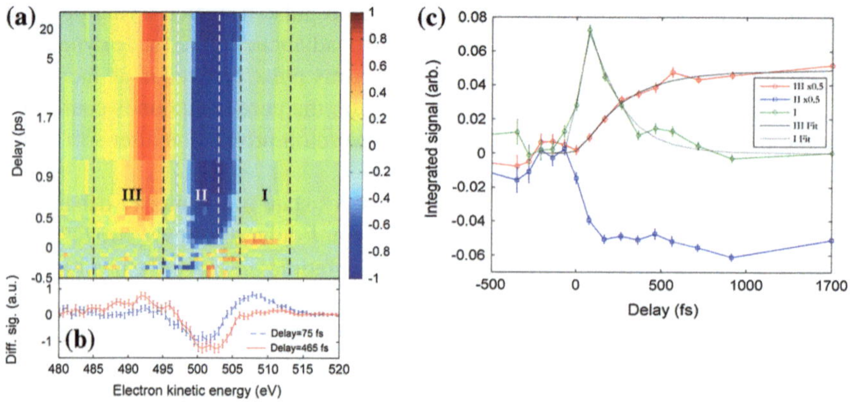

Fig. 14.8 **a** Difference Auger spectra with and without UV excitation versus time delay between UV excitation and SXR probe pulse. The false color plot indicates the sign and amplitude of the difference, *red* indicates a UV induced increase of the Auger yield, *blue* indicates a UV induced Auger yield decrease. Part **b** shows difference spectra at two different delays (75 fs *dashed blue* and 465 fs *solid red*). In **c**, integrations over kinetic energy ranges indicated in (**a**) are scaled by a factor (see legend) and fitted (see main text)

region I shows a 200 fs decay constant. The rise of region I and decay of region II is synchronous. As region I decays, the signal in region III (around 490 eV) rises with a time constant similar to the decay of I. This suggests that some kind of decay in a photoexcited state is accompanied by a rise in a product state.

A cut through the difference spectrum at short times is given in blue in Fig. 14.8b. We clearly identify the decrease in II accompanied by the increase in I. Such a difference spectrum reflects a UV induced shift of the whole ground state Auger band (see Fig. 14.7) towards higher Auger kinetic energies. The immediate appearance with UV excitation indicates the creation of a wave packet in the $\pi\,\pi^*$ excited state.

The reason for the blueshift of the Auger electrons upon $\pi\,\pi^*$ excitation lies in the nuclear dynamics of the molecule, as can be shown by comparison to simulations as done in [27]. However, the signal can as well be interpreted using intuitive arguments which are based on the x-ray typical element and local sensitivity. As explained in Sect. 14.3, the Auger decay of the oxygen 1s core holes involves orbitals with strong oxygen atomic orbital character [94, 98]. This means that the final state after Auger decay is a dicationic state with valence charge missing at one of the two oxygen atoms, indistinguishable for us. The lacking valence electrons at the oxygen are reducing the electron density between the oxygen and the next nearest carbon. Due to this effect, a strong Coulomb repulsion acts between those atoms in the dicationic state. Figure 14.9 shows a sketch of the electronic energies in the neutral $\pi\,\pi^*$ state, the core ionized $1s^{-1}\pi\,\pi^*$ state and a 'band' of dicationic states reached upon Auger decay. As the molecule elongates the C–O coordinate, the dicationic state lowers its energy considerably due to Coulomb repulsion. This results in a increase of the Auger kinetic energy for O1s core hole decay as the C–O coordinate increases.

Simulations predict that one of the two thymine C–O bonds indeed stretches as the molecule is excited to the $\pi\,\pi^*$ state. Upon excitation, the C(4)–O(8) bond, which is close to the methyl group, is Franck-Condon active, as explained in the context of Fig. 14.1. It elongates by about 15 % as the molecule relaxes from the Franck-Condon nuclear geometry to the predicted $\pi\,\pi^*$ state minimum [41, 42]. In contrast, the C(2)–O(7) bond does not change as a consequence of photoexcitation. This fast nuclear relaxation occurs immediately after the UV excitation, and the experimental time resolution is not sufficient to follow the relaxation and resolve the gradual shift. It is interesting to note that the high local sensitivity of the 1s Auger decay allows to really attribute motion to individual bonds. For the current case of thymine, it is not possible to tell from the spectra, which of the two C–O bonds stretches. As explained in Sect. 14.3, sites (here corresponding to O(7) and O(8)) cannot be identified in the broad and complex Auger spectra. For molecules with singly occurring elements however, the x-ray probe combined with Auger decay does provide an opportunity to intuitively explore excited state nuclear dynamics.

About 200 fs after photoexcitation, a decrease in region I is observed together with a rise in region III. The time delay cut at 465 fs in Fig. 14.8b shows a difference spectrum indicating UV induced shift to lower Auger kinetic energies. This is contrary to the nuclear dynamics trend described above. We evaluate the decay of the blueshifted spectrumfrom the $\pi\,\pi^*$ state in the light of past investigations. Simula-

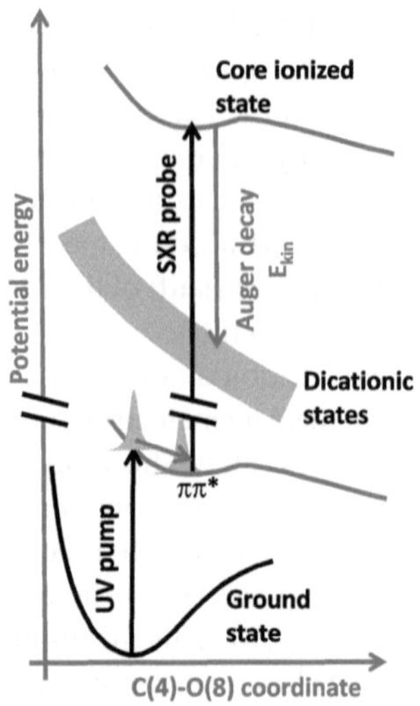

Fig. 14.9 Scheme of potential energy surfaces involved in the excited state Auger probing, plotted over the C(4)–O(8) distance. The UV excitation pulse promotes a molecular wavepacket to the π π* state. The soft x-ray pulse core ionizes the molecule creating a 1s vacancy at oxygen and leaving the valence π π* configuration intact. This core ionized state decays within a few femtoseconds to the dicationic states, which are indicated as a grey band. The kinetic energy of the Auger electron is given by the energy difference between core excited and dicationic state. The dicationic states lower the energy strongly as the C–O bond stretches. This results from the high localization of the double valence hole in the vicinity of the oxygen atom after the Auger decay. The unscreened nuclear repulsion term between O(8) and its bond partner C(4) decreases as the C–O bond stretches, explaining the Auger blueshift from Franck-Condon region to π π* minimum

tions in [41, 42] predict that the molecule is stuck on a minimum in the π π* state for some picoseconds due to a reaction barrier in the path to reach the conical intersection. The C(4)–O(8) bond stays elongated in this minimum. Thus these simulations would predict a blueshifted spectrum for the next couple of picoseconds after excitation. The experimentally observed decay of the blueshifted channel with 200 fs time constant indicates that this minimum can only live for a short time <200 fs. Thus, we assume that the majority of the photoexcited π π* population is not blocked by a barrier and has quick access to the conical intersection region. The redshift can only be explained by considering a transition to a different electronic state. The shift is than induced either by the C–O bond shrinking or a changed electronic structure.

A return of photoexcited population to the vibrationally hot ground state as well as population of the dark nπ* state is possible and was controversially discussed in the theoretical literature [41, 42, 53–55]. Simulations of Auger spectra in [27] show a redshift for the nπ* state and no shift but rather a slight broadening for a return to the hot ground state. While relaxation to the nπ* state can explain the experimental redshift, a statement on ground state relaxation is not possible. Our simulated ground state spectra indicate that the x-ray Auger method has reduced sensitivity on the ground state relaxation channel.

Short decay constants, comparable to the 200 fs trace, have been documented in the optical pump-probe literature [66–68] and were interpreted either as nuclear relaxation on the π π* state or electronic relaxation out of this state. The soft x-ray Auger experiment distinguishes between nuclear and electronic relaxation and we can interpret the 200 fs decrease in region I as an *electronic* reduction of the π π* population. The fast decay of a majority of the π π* state population means barrierless decay of the π π* state. Several simulations point towards the hot ground state as a product of few 100 fs electronic relaxation [53, 54], which was confirmed in liquid phase [163]. Due to reduced Auger sensitivity to this state, we abstain from a statement. The time constant of the lower energy signal indicates a 200–300 fs filling of the nπ* state. In order to show excited state fluorescence as measured in liquid phase according to [164], some molecular population must be trapped behind a π π* reaction barrier. We assume that experimental signal to noise ratio in the x-ray Auger experiment is not yet good enough to deduce such detailed information.

14.6 Outlook

We are only at the beginning of investigating photoexcited molecular dynamics with ultrashort x-ray pulses. The couple of experiments done in the past showed the promise of the field, which is element and site selective probing of valence dynamics. Many more systematic experiments are needed to fully appreciate the strength and weakness of ultrafast x-ray probing and to position it in the context of other ultrafast techniques.

Several exciting new methodological developments are of crucial importance for this new field. Powerful x-ray pulses can now be delivered with stable bandwidth due to self-seeding schemes [110, 111], which enables femtosecond core electron photoemission for molecular dynamics probing. In addition, techniques relying on hitting molecular core-valence resonances will be profiting to a great extend. Having sources with higher average repetition rate like to European XFEL [165] will make emission experiments in dilute samples feasible. In addition, new nonlinear x-ray methods are currently getting developed on the theoretical side for molecular systems [166–169] and at the same time first FEL induced lasing and Raman processes in atoms have shown that these methods can also be translated into an experiment [170, 171]. We currently witness a fast growth in available FEL beamtime as more lasers are coming online in the near future in Germany [165], Switzerland [172] and Korea [173]. At the same time, the existing facilities are becoming more versatile. This should deliver ideal conditions for a broad molecular science program at free electron lasers.

Acknowledgments I would like to acknowledge all my colleagues in the PULSE Institute and at the LCLS for many discussions and fruitful collaborations. I would like to acknowledge especially the whole team of colleagues in the LCLS thymine experiment: B.K. McFarland, J.P. Farrell, S. Miyabe, F. Tarantelli, A. Aguilar, N. Berrah, C. Bostedt, J.D. Bozek, P.H. Bucksbaum, J.C. Castagna, R.N. Coffee, J.P. Cryan, L. Fang, R. Feifel, K.J. Gaffney, J.M. Glownia, T.J. Martinez, M. Mucke, B. Murphy, A. Natan, T. Osipov , V.S. Petrović, S. Schorb, Th. Schultz, L.S. Spector, M. Swiggers, I. Tenney, S. Wang, J.L. White, and W. White. This work was supported by the AMOS program within the Chemical Sciences, Geosciences, and Biosciences Division of the Office of Basic Energy Sciences, Office of Science, U.S. Department of Energy. I acknowledge funding via the Office of Science Early Career Research Program through the Office of Basic Energy Sciences, U.S. Department of Energy.

References

1. R. Schoenlein, S. Chattopadhyay, H. Chong, T. Glover, P. Heimann, C. Shank, A. Zholents, M. Zolotorev, Science **287**, 2237 (2000)
2. W. Ackermann, G. Asova, V. Ayvazyan, A. Azima, N. Baboi, J. Baehr, V. Balandin, B. Beutner, A. Brandt, A. Bolzmann, R. Brinkmann, O.I. Brovko, M. Castellano, P. Castro, L. Catani, E. Chiadroni, S. Choroba, A. Cianchi, J.T. Costello, D. Cubaynes, J. Dardis, W. Decking, H. Delsim-Hashemi, A. Delserieys, G. Di Pirro, M. Dohlus, S. Duesterer, A. Eckhardt, H.T. Edwards, B. Faatz, J. Feldhaus, K. Floettmann, J. Frisch, L. Froehlich, T. Garvey, U. Gensch, C. Gerth, M. Goerler, N. Golubeva, H.-J. Grabosch, M. Grecki, O. Grimm, K. Hacker, U. Hahn, J.H. Han, K. Honkavaara, T. Hott, M. Huening, Y. Ivanisenko, E. Jaeschke, W. Jalmuzna, T. Jezynski, R. Kammering, V. Katalev, K. Kavanagh, E.T. Kennedy, S. Khodyachykh, K. Klose, V. Kocharyan, M. Koerfer, M. Kollewe, W. Koprek, S. Korepanov, D. Kostin, M. Krassilnikov, G. Kube, M. Kuhlmann, C.L.S. Lewis, L. Lilje, T. Limberg, D. Lipka, F. Loehl, H. Luna, M. Luong, M. Martins, M. Meyer, P. Michelato, V. Miltchev, W.D. Moeller, L. Monaco, W.F.O. Mueller, A. Napieralski, O. Napoly, P. Nicolosi, D. Noelle, T. Nunez, A. Oppelt, C. Pagani, R. Paparella, N. Pchalek, J. Pedregosa-Gutierrez, B. Petersen, B. Petrosyan, G. Petrosyan, L. Petrosyan, J. Pflueger, E. Ploenjes, L. Poletto, K. Pozniak, E. Prat, D. Proch, P. Pucyk, P. Radcliffe, H. Redlin, K. Rehlich, M. Richter, M. Roehrs, J. Roensch, R. Romaniuk, M. Ross, J. Rossbach, V. Rybnikov, M. Sachwitz, E.L. Saldin, W. Sandner, H. Schlarb, B. Schmidt, M. Schmitz, P. Schmueser, J.R. Schneider, E.A. Schneidmiller, S. Schnepp, S. Schreiber, M. Seidel, D. Sertore, A.V. Shabunov, C. Simon, S. Simrock, E. Sombrowski, A.A. Sorokin, P. Spanknebel, R. Spesyvtsev, L. Staykov, B. Steffen, F. Stephan, F. Stulle, H. Thom, K. Tiedtke, M. Tischer, S. Toleikis, R. Treusch, D. Trines, I. Tsakov, E. Vogel, T. Weiland, H. Weise, M. Wellhoeffer, M. Wendt, I. Will, A. Winter, K. Wittenburg, W. Wurth, P. Yeates, M.V. Yurkov, I. Zagorodnov, K. Zapfe, Nat. Photonics **1**, 336 (2007)
3. P. Emma, R. Akre, J. Arthur, R. Bionta, C. Bostedt, J. Bozek, A. Brachmann, P. Bucksbaum, R. Coffee, F.-J. Decker, Y. Ding, D. Dowell, S. Edstrom, A. Fisher, J. Frisch, S. Gilevich, J. Hastings, G. Hays, P. Hering, Z. Huang, R. Iverson, H. Loos, M. Messerschmidt, A. Miahnahri, S. Moeller, H.-D. Nuhn, G. Pile, D. Ratner, J. Rzepiela, D. Schultz, T. Smith, P. Stefan, H. Tompkins, J. Turner, J. Welch, W. White, J. Wu, G. Yocky, J. Galayda, Nat. Photonics **4**, 641 (2010)
4. T. Ishikawa, H. Aoyagi, T. Asaka, Y. Asano, N. Azumi, T. Bizen, H. Ego, K. Fukami, T. Fukui, Y. Furukawa, S. Goto, H. Hanaki, T. Hara, T. Hasegawa, T. Hatsui, A. Higashiya, T. Hirono, N. Hosoda, M. Ishii, T. Inagaki, Y. Inubushi, T. Itoga, Y. Joti, M. Kago, T. Kameshima, H. Kimura, Y. Kirihara, A. Kiyomichi, T. Kobayashi, C. Kondo, T. Kudo, H. Maesaka, X.M. Maréchal, T. Masuda, S. Matsubara, T. Matsumoto, T. Matsushita, S. Matsui, M. Nagasono, N. Nariyama, H. Ohashi, T. Ohata, T. Ohshima, S. Ono, Y. Otake, C. Saji, T. Sakurai, T. Sato, K. Sawada, T. Seike, K. Shirasawa, T. Sugimoto, S. Suzuki, S. Takahashi, H. Takebe, K.

Takeshita, K. Tamasaku, H. Tanaka, R. Tanaka, T. Tanaka, T. Togashi, K. Togawa, A. Tokuhisa, H. Tomizawa, K. Tono, S. Wu, M. Yabashi, M. Yamaga, A. Yamashita, K. Yanagida, C. Zhang, T. Shintake, H. Kitamura, N. Kumagai, Nat. Photonics **6**, 540 (2012)
5. J. Ullrich, A. Rudenko, R. Moshammer, Annu. Rev. Phys. Chem. **63**, 635 (2012)
6. F. Plasser, M. Barbatti, A.J.A. Aquino, H. Lischka, Theor. Chem. Acc. **131**, 233902 (2012)
7. J.C. Tully, J. Chem. Phys. **137**, 22A301 (2012)
8. H. Dürr, H. Bouas-Laurent (eds.), *Photochromism: Molecules and Systems*, Rev edn. (Elsevier, Boston, 2003)
9. M. Alemani, M.V. Peters, S. Hecht, K.-H. Rieder, F. Moresco, L. Grill, J. Am. Chem. Soc. **128**, 14446 (2006)
10. M. Comstock, N. Levy, A. Kirakosian, J. Cho, F. Lauterwasser, J. Harvey, D. Strubbe, J. Fréchet, D. Trauner, S. Louie, M. Crommie, Phys. Rev. Lett. **99**, 038301 (2007)
11. R.J. Sension, S.T. Repinec, A.Z. Szarka, R.M. Hochstrasser, J. Chem. Phys. **98**, 6291 (1993)
12. D.H. Waldeck, Chem. Rev. **91**, 415 (1991)
13. R. Schoenlein, L. Peteanu, R. Mathies, C. Shank, Science **254**, 412 (1991)
14. D. Polli, P. Altoe, O. Weingart, K.M. Spillane, C. Manzoni, D. Brida, G. Tomasello, G. Orlandi, P. Kukura, R.A. Mathies, M. Garavelli, G. Cerullo, Nature **467**, 440 (2010)
15. R. Mathies, C. Brito Cruz, W. Pollard, C. Shank, Science **240**, 777 (1988)
16. S. Schenkl, F. van Mourik, G. van der Zwan, S. Haacke, M. Chergui, Science **309**, 917 (2005)
17. G. Groenhof, M. Bouxin-Cademartory, B. Hess, S.P. de Visser, H.J.C. Berendsen, M. Olivucci, A.E. Mark, M.A. Robb, J. Am. Chem. Soc. **126**, 4228 (2004)
18. C.T. Middleton, K. de La Harpe, C. Su, Y.K. Law, C.E. Crespo-Hernandez, B. Kohler, Annu. Rev. Phys. Chem. **60**, 217 (2009)
19. C. Crespo-Hernandez, B. Cohen, P. Hare, B. Kohler, Chem. Rev. **104**, 1977 (2004)
20. W.J. Schreier, T.E. Schrader, F.O. Koller, P. Gilch, C.E. Crespo-Hernandez, V.N. Swaminathan, T. Carell, W. Zinth, B. Kohler, Science **315**, 625 (2007)
21. E. Teller, Isr. J. Chem. **7**, 227 (1969)
22. H. Köppel, W. Domcke, L.S. Cederbaum, Adv. Chem. Phys. **57**, 59 (1984)
23. F. Bernardi, M. Olivucci, M.A. Robb, Chem. Soc. Rev. **25**, 321 (1996)
24. G.A. Worth, L.S. Cederbaum, Annu. Rev. Phys. Chem. **55**, 127 (2004)
25. W. Domcke, D. Yarkony, H. Köppel (eds.), *Conical Intersections Electronic Structure, Dynamics & Spectroscopy* (World Scientific, River Edge, 2004)
26. R. Dörner, V. Mergel, O. Jagutzki, L. Spielberger, J. Ullrich, R. Moshammer, H. Schmidt-Böcking, Phys. Rep.-Rev. Sect. Phys. Lett. **330**, 95 (2000)
27. B.K. McFarland, J.P. Farrell, S. Miyabe, F. Tarantelli, A. Aguilar, N. Berrah, C. Bostedt, J.D. Bozek, P.H. Bucksbaum, J.C. Castagna, R.N. Coffee, J.P. Cryan, L. Fang, R. Feifel, K.J. Gaffney, J.M. Glownia, T.J. Martinez, M. Mucke, B. Murphy, A. Natan, T. Osipov, V.S. Petrović, S. Schorb, T. Schultz, L.S. Spector, M. Swiggers, I. Tenney, S. Wang, J.L. White, W. White, M. Gühr, Nat. Commun. **5**, 4235 (2014)
28. M. Born, R. Oppenheimer, Ann. Phys. **389**, 457 (1927)
29. W. Domcke, D.R. Yarkony, H. Köppel (eds.), *Conical Intersections Theory*, Computation and Experiment (World Scientific Publishing Company, Singapore, 2011)
30. B.L. Henke, E.M. Gullikson, J.C. Davis, At. Data Nucl. Data Tables **54**, 181 (1993)
31. J.J. Yeh, I. Lindau, At. Data Nucl. Data Tables **32**, 1 (1985)
32. K. Siegbahn, *ESCA Applied to Free Molecules* (North-Holland Pub. Co, Amsterdam, 1969)
33. P.S. Bagus, F. Illas, G. Pacchioni, F. Parmigiani, J. Electron Spectrosc. Relat. Phenom. **100**, 215 (1999)
34. R. Manne, J. Chem. Phys. **52**, 5733 (1970)
35. J.M. Coxon, *Organic Photochemistry* (Cambridge University Press, New York, 1974)
36. M. Klessinger, *Excited States and Photochemistry of Organic Molecules* (VCH, New York, 1995)
37. J. Michl, V. Bonačić -Koutecký, Electronic Aspects of Organic Photochemistry (Wiley, New York, 1990)
38. N.J. Turro, Modern Molecular Photochemistry (University Science Books, Mill Valley, 1991)

39. R. Mulliken, Phys. Rev. **46**, 549 (1934)
40. R.S. Mulliken, J. Chem. Phys. **55**, 288 (1971)
41. H.R. Hudock, B.G. Levine, A.L. Thompson, H. Satzger, D. Townsend, N. Gador, S. Ullrich, A. Stolow, T.J. Martínez, J. Phys. Chem. A **111**, 8500 (2007)
42. J.J. Szymczak, M. Barbatti, J.T. Soo, Hoo, J.A. Adkins, T.L. Windus, D. Nachtigallova, H. Lischka, J. Phys. Chem. A **113**, 12686 (2009)
43. D.J. Tannor, *Introduction to Quantum Mechanics: A Time-Dependent Perspective* (University Science Books, Sausalito, 2007)
44. J. Yeazell, T. Uzer (eds.), *The Physics and Chemistry of Wave Packets* (Wiley, New York, 2000)
45. M. Gühr, H. Ibrahim, N. Schwentner, Phys. Chem. Chem. Phys. **6**, 5353 (2004)
46. D.J. Nesbitt, R.W. Field, J. Phys. Chem. **100**, 12735 (1996)
47. G. Herzberg, Molecular Spectra and Molecular Structure: Infrared and Raman Spectra of Polyatomic Molecules (R.E. Krieger Pub. Co., Malabar, 1991)
48. D. Yarkony, Rev. Mod. Phys. **68**, 985 (1996)
49. E. Teller, J. Phys. Chem. **41**, 109 (1937)
50. B.G. Levine, T.J. Martínez, Annu. Rev. Phys. Chem. **58**, 613 (2007)
51. G.J. Atchity, S.S. Xantheas, K. Ruedenberg, J. Chem. Phys. **95**, 1862 (1991)
52. T.J. Martínez, Nature **467**, 412 (2010)
53. M. Merchán, R. González-Luque, T. Climent, L. Serrano-Andrés, E. Rodríguez, M. Reguero, D. Peláez, J. Phys. Chem. B **110**, 26471 (2006)
54. S. Perun, A.L. Sobolewski, W. Domcke, J. Phys. Chem. A **110**, 13238 (2006)
55. D. Asturiol, B. Lasorne, M.A. Robb, L. Blancafort, J. Phys. Chem. A **113**, 10211 (2009)
56. A.H. Zewail (ed.), *Femtochemistry: Ultrafast Dynamics of the Chemical Bond* (World Scientific, River Edge, 1994)
57. T. Baumert, M. Grosser, R. Thalweiser, G. Gerber, Phys. Rev. Lett. **67**, 3753 (1991)
58. M. Spanner, S. Patchkovskii, C. Zhou, S. Matsika, M. Kotur, T.C. Weinacht, Phys. Rev. A **86**, 053406 (2012)
59. B.K. McFarland, J.P. Farrell, P.H. Bucksbaum, M. Gühr, Science **322**, 1232 (2008)
60. A. Stolow, Annu. Rev. Phys. Chem. **54**, 89 (2003)
61. A. Stolow, A.E. Bragg, D.M. Neumark, Chem. Rev. **104**, 1719 (2004)
62. V. Blanchet, M. Zgierski, T. Seideman, A. Stolow, Nature **401**, 52 (1999)
63. V. Blanchet, S. Lochbrunner, M. Schmitt, J. Shaffer, J. Larsen, M. Zgierski, T. Seideman, A. Stolow, Faraday Discuss. **115**, 33 (2000)
64. I.V. Hertel, W. Radloff, Rep. Prog. Phys. **69**, 1897 (2006)
65. M. Barbatti, S. Ullrich, Phys. Chem. Chem. Phys. **13**, 15492 (2011)
66. H. Kang, K.T. Lee, B. Jung, Y.J. Ko, S.K. Kim, J. Am. Chem. Soc. **124**, 12958 (2002)
67. C. Canuel, M. Mons, F. Piuzzi, B. Tardivel, I. Dimicoli, M. Elhanine, J. Chem. Phys. **122**, 074316 (2005)
68. J. González-Vázquez, L. González, E. Samoylova, T. Schultz, Phys. Chem. Chem. Phys. **11**, 3927 (2009)
69. S. Ullrich, T. Schultz, M.Z. Zgierski, A. Stolow, Phys. Chem. Chem. Phys. **6**, 2796 (2004)
70. A.P. Hitchcock, J. Electron Spectrosc. Relat. Phenom. **67**, 1 (1994)
71. E.D. Palik, *Handbook of Optical Constants of Solids* (Academic Press, New York, 1998)
72. K. Ueda, J. Phys. B At. Mol. Opt. Phys. **36**, R1 (2003)
73. U. Becker, D.A. Shirley (eds.), *VUV and Soft X-Ray Photoionization* (Plenum Press, New York, 1996)
74. J. Stöhr, NEXAFS Spectroscopy, 1st edn. corr. print (Springer, New York, 1996)
75. A. Nilsson, L.G.M. Pettersson, in *Chemical Bonding at Surfaces and Interfaces*, ed. by A. Nilsson, L.G.M. Pettersson, J. Norskov (Elsevier Science, Amsterdam, 2011)
76. S. Schreck, G. Gavrila, C. Weniger, P. Wernet, Rev. Sci. Instrum. **82**, 103101 (2011)
77. M. Ukai, A. Yokoya, K. Fujii, Y. Saitoh, Chem. Phys. Lett. **495**, 90 (2010)
78. J. Nordgren, New Directions in Research with Third-Generation Soft X-Ray Synchrotron Radiation Sources (Kluwer Academic, Boston, 1994)

79. O. Plekan, V. Feyer, R. Richter, M. Coreno, M. de Simone, K.C. Prince, A.B. Trofimov, E.V. Gromov, I.L. Zaytseva, J. Schirmer, Chem. Phys. **347**, 360 (2008)
80. J.L. Dehmer, J. Chem. Phys. **56**, 4496 (1972)
81. D. Dill, J. Dehmer, J. Chem. Phys. **61**, 692 (1974)
82. J. Dehmer, D. Dill, Phys. Rev. Lett. **35**, 213 (1975)
83. J.L. Dehmer, D. Dill, J. Chem. Phys. **65**, 5327 (1976)
84. W. Wurth, J. Stöhr, Vacuum **41**, 237 (1990)
85. W. Bambynek, B. Crasemann, R. Fink, H. Freund, H. Mark, C. Swift, R. Price, P. Rao, Rev. Mod. Phys. **44**, 716 (1972)
86. P. Glatzel, U. Bergmann, Coord. Chem. Rev. **249**, 65 (2005)
87. J. Nordgren, G. Bray, S. Cramm, R. Nyholm, J.-E. Rubensson, N. Wassdahl, Rev. Sci. Instrum. **60**, 1690 (1989)
88. P. Glans, P. Skytt, K. Gunnelin, J.-H. Guo, J. Nordgren, J. Electron Spectrosc. Relat. Phenom. **82**, 193 (1996)
89. P. Glans, K. Gunnelin, P. Skytt, J.-H. Guo, N. Wassdahl, J. Nordgren, H. Ågren, F. Gel'mukhanov, T. Warwick, E. Rotenberg, Phys. Rev. Lett. **76**, 2448 (1996)
90. P. Skytt, P. Glans, K. Gunnelin, J. Guo, J. Nordgren, Y. Luo, H. Ågren, Phys. Rev. A **55**, 134 (1997)
91. A. Cesar, F. Gel'mukhanov, Y. Luo, H. Ågren, P. Skytt, P. Glans, J. Guo, K. Gunnelin, J. Nordgren, J. Chem. Phys. **106**, 3439 (1997)
92. I. Josefsson, K. Kunnus, S. Schreck, A. Föhlisch, F. de Groot, P. Wernet, M. Odelius, J. Phys. Chem. Lett. **3**, 3565 (2012)
93. Y. Harada, T. Takeuchi, H. Kino, A. Fukushima, K. Takakura, K. Hieda, A. Nakao, S. Shin, H. Fukuyama, J. Phys. Chem. A **110**, 13227 (2006)
94. H. Siegbahn, L. Asplund, P. Kelfve, Chem. Phys. Lett. **35**, 330 (1975)
95. E.H.S. Burhop, *The Auger Effect and Other Radiationless Transitions* (R. E. Krieger Pub. Co, Huntington, 1980)
96. P. Emma, K. Bane, M. Cornacchia, Z. Huang, H. Schlarb, G. Stupakov, D. Walz, Phys. Rev. Lett. **92**, 074801 (2004)
97. W.E. Moddeman, T.A. Carlson, M.O. Krause, B.P. Pullen, W.E. Bull, G.K. Schweitz, J. Chem. Phys. **55**, 2317 (1971)
98. L. Karlsson, L.O. Werme, T. Bergmark, K. Siegbahn, J. Electron Spectrosc. Relat. Phenom. **3**, 181 (1974)
99. H. Ågren, J. Chem. Phys. **75**, 1267 (1981)
100. L. Storchi, F. Tarantelli, S. Veronesi, P. Bolognesi, E. Fainelli, L. Avaldi, J. Chem. Phys. **129**, 154309 (2008)
101. J. Schirmer, A. Barth, Z. Fuer. Phys. At. Nucl. **317**, 267 (1984)
102. F. Tarantelli, Chem. Phys. **329**, 11 (2006)
103. W. Eberhardt, T. Sham, R. Carr, S. Krummacher, M. Strongin, S. Weng, D. Wesner, Phys. Rev. Lett. **50**, 1038 (1983)
104. E. Itälä, D.T. Ha, K. Kooser, M.A. Huels, E. Rachlew, E. Nõmmiste, U. Joost, E. Kukk, J. Electron Spectrosc. Relat. Phenom. **184**, 119 (2011)
105. E. Itälä, D.T. Ha, K. Kooser, E. Nõmmiste, U. Joost, E. Kukk, Int. J. Mass Spectrom. **306**, 82 (2011)
106. S. Hüfner, *Photoelectron Spectroscopy: Principles and Applications*, 3rd edn. (Springer, New York, 2003)
107. W.L. Jolly, K.D. Bomben, C.J. Eyermann, At. Data Nucl. Data Tables **31**, 433 (1984)
108. L.J. Sæthre, K.J. Børve, T.D. Thomas, J. Electron Spectrosc. Relat. Phenom. **183**, 2 (2011)
109. Y. Takahata, A.K. Okamoto, D.P. Chong, Int. J. Quantum Chem. **106**, 2581 (2006)
110. J. Amann, W. Berg, V. Blank, F.-J. Decker, Y. Ding, P. Emma, Y. Feng, J. Frisch, D. Fritz, J. Hastings, Z. Huang, J. Krzywinski, R. Lindberg, H. Loos, A. Lutman, H.-D. Nuhn, D. Ratner, J. Rzepiela, D. Shu, Y. Shvyd'ko, S. Spampinati, S. Stoupin, S. Terentyev, E. Trakhtenberg, D. Walz, J. Welch, J. Wu, A. Zholents, D. Zhu, Nat. Photonics **6**, 693 (2012)

111. D. Cocco, R. Abela, J.W. Amann, K. Chow, P.J. Emma, Y. Feng, G.L. Gassner, J. Hastings, P. Heimann, Z. Huang, H. Loos, P.A. Montanez, D. Morton, H.-D. Nuhn, D.F. Ratner, L.N. Rodes, U. Flechsig, J.J. Welch, J. Wu, ed. by A. Klisnick, C.S. Menoni, *Proceedings SPIE 8849 X-Ray Lasers Coherent X-Ray Sources: Development and Applications*, vol. X, p. 88490A (2013)
112. T. Popmintchev, M.-C. Chen, D. Popmintchev, P. Arpin, S. Brown, S. Alisauskas, G. Andriukaitis, T. Balciunas, O.D. Mucke, A. Pugzlys, A. Baltuska, B. Shim, S.E. Schrauth, A. Gaeta, C. Hernandez-Garcia, L. Plaja, A. Becker, A. Jaron-Becker, M.M. Murnane, H.C. Kapteyn, Science **336**, 1287 (2012)
113. R. Reininger, D.J. Keavney, M. Borland, L. Young, J. Synchrotron Radiat. **20**, 654 (2013)
114. A. McPherson, G. Gibson, H. Jara, U. Johann, T.S. Luk, I.A. McIntyre, K. Boyer, C.K. Rhodes, J. Opt. Soc. Am. B **4**, 595 (1987)
115. M. Ferray, A. L'Huillier, X.F. Li, L.A. Lompre, G. Mainfray, C. Manus, J. Phys. B At. Mol. Opt. Phys. **21**, L31 (1988)
116. J.J. Macklin, J.D. Kmetec, C.L. Gordon, Phys. Rev. Lett. **70**, 766 (1993)
117. A. L'Huillier, P. Balcou, Phys. Rev. Lett. **70**, 774 (1993)
118. J.L. Krause, K.J. Schafer, K.C. Kulander, Phys. Rev. Lett. **68**, 3535 (1992)
119. K.C. Kulander, K.J. Schafer, J.L. Krause, Laser Phys. **3**, 359 (1993)
120. K.J. Schafer, B. Yang, L.F. DiMauro, K.C. Kulander, Phys. Rev. Lett. **70**, 1599 (1993)
121. P.B. Corkum, Phys. Rev. Lett. **71**, 1994 (1993)
122. M. Lewenstein, P. Balcou, M.Y. Ivanov, A. L'Huillier, P.B. Corkum, Phys. Rev. A **49**, 2117 (1994)
123. P.B. Corkum, F. Krausz, Nat. Phys. **3**, 381 (2007)
124. F. Krausz, M. Ivanov, Rev. Mod. Phys. **81**, 163 (2009)
125. C. Rose-Petruck, R. Jimenez, T. Guo, A. Cavalleri, C.W. Siders, F. Rksi, J.A. Squier, B.C. Walker, K.R. Wilson, C.P.J. Barty, Nature **398**, 310 (1999)
126. K. Sokolowski-Tinten, C. Blome, J. Blums, A. Cavalleri, C. Dietrich, A. Tarasevitch, I. Uschmann, E. Förster, M. Kammler, M. Horn-von-Hoegen, D. von der Linde, Nature **422**, 287 (2003)
127. M. Bargheer, N. Zhavoronkov, Y. Gritsai, J.C. Woo, D.S. Kim, M. Woerner, T. Elsaesser, Science **306**, 1771 (2004)
128. M. Bargheer, N. Zhavoronkov, M. Woerner, T. Elsaesser, Chem. Phys. Chem. **7**, 783 (2006)
129. M. Bargheer, N. Zhavoronkov, R. Bruch, H. Legall, H. Stiel, M. Woerner, T. Elsaesser, Appl. Phys. B **80**, 715 (2005)
130. A. Zholents, M. Zolotorev, Phys. Rev. Lett. **76**, 912 (1996)
131. S. Khan, K. Holldack, T. Kachel, R. Mitzner, T. Quast, Phys. Rev. Lett. **97**, 074801 (2006)
132. A. Streun, A. Al-Adwan, P. Beaud, M. Böge, G. Ingold, S. Johnson, A. Keller, T. Schilcher, V. Schlott, T. Schmidt, EPAC, p. 3427 (2006)
133. P. Prigent, P. Hollander, M. Labat, M.E. Couprie, J.L. Marlats, C. Laulhé, J. Luning, T. Moreno, P. Morin, A. Nadji, F. Polack, S. Ravy, M. Silly, F. Sirotti, J. Phys. Conf. Ser. **425**, 072022 (2013)
134. J. Feikes, K. Holldack, P. Kuske, G. Wustefeld, Proceedings of Ninth European Particle Accelerator Conference EPAC04, Lucerne, CH, pp. 2287–2289 (2004)
135. X. Huang, J. Safranek, J. Corbett, Y. Nosochkov, J. Sebek, A. Terebilo, 2007 IEEE Particle Accelerator Conference, vols. 1–11, IEEE, 345 E 47th St, New York, NY 10017 USA, pp. 3507–3509 (2007)
136. E.M. Dufresne, B. Adams, M. Chollet, R. Harder, Y. Li, H. Wen, S.J. Leake, L. Beitra, X. Huang, I.K. Robinson, Nucl. Instrum. Methods Phys. Res. Sect. Accel. Spectrometers Detect. Assoc. Equip. **649**, 191 (2011)
137. H. Navirian, R. Shayduk, W. Leitenberger, J. Goldshteyn, P. Gaal, M. Bargheer, Rev. Sci. Instrum. **83**, 063303 (2012)
138. E. Allaria, R. Appio, L. Badano, W.A. Barletta, S. Bassanese, S.G. Biedron, A. Borga, E. Busetto, D. Castronovo, P. Cinquegrana, S. Cleva, D. Cocco, M. Cornacchia, P. Craievich, I. Cudin, G. D'Auria, M. Dal Forno, M.B. Danailov, R. De Monte, G. De Ninno, P. Delgiusto,

A. Demidovich, S. Di Mitri, B. Diviacco, A. Fabris, R. Fabris, W. Fawley, M. Ferianis, E. Ferrari, S. Ferry, L. Froehlich, P. Furlan, G. Gaio, F. Gelmetti, L. Giannessi, M. Giannini, R. Gobessi, R. Ivanov, E. Karantzoulis, M. Lonza, A. Lutman, B. Mahieu, M. Milloch, S.V. Milton, M. Musardo, I. Nikolov, S. Noe, F. Parmigiani, G. Penco, M. Petronio, L. Pivetta, M. Predonzani, F. Rossi, L. Rumiz, A. Salom, C. Scafuri, C. Serpico, P. Sigalotti, S. Spampinati, C. Spezzani, M. Svandrlik, C. Svetina, S. Tazzari, M. Trovo, R. Umer, A. Vascotto, M. Veronese, R. Visintini, M. Zaccaria, D. Zangrando, M. Zangrando, Nat. Photonics **6**, 699 (2012)

139. S. Ackermann, A. Azima, S. Bajt, J. Bödewadt, F. Curbis, H. Dachraoui, H. Delsim-Hashemi, M. Drescher, S. Düsterer, B. Faatz, M. Felber, J. Feldhaus, E. Hass, U. Hipp, K. Honkavaara, R. Ischebeck, S. Khan, T. Laarmann, C. Lechner, T. Maltezopoulos, V. Miltchev, M. Mittenzwey, M. Rehders, J. Rönsch-Schulenburg, J. Rossbach, H. Schlarb, S. Schreiber, L. Schroedter, M. Schulz, S. Schulz, R. Tarkeshian, M. Tischer, V. Wacker, M. Wieland, Phys. Rev. Lett. **111**, 114801 (2013)
140. C. Gahl, A. Azima, M. Beye, M. Deppe, K. Doebrich, U. Hasslinger, F. Hennies, A. Melnikov, M. Nagasono, A. Pietzsch, M. Wolf, W. Wurth, A. Foehlisch, Nat. Photonics **2**, 165 (2008)
141. S. Schorb, T. Gorkhover, J.P. Cryan, J.M. Glownia, M.R. Bionta, R.N. Coffee, B. Erk, R. Boll, C. Schmidt, D. Rolles, A. Rudenko, A. Rouzee, M. Swiggers, S. Carron, J.-C. Castagna, J.D. Bozek, M. Messerschmidt, W.F. Schlotter, C. Bostedt, Appl. Phys. Lett. **100**, 121107 (2012)
142. M. Beye, O. Krupin, G. Hays, A.H. Reid, D. Rupp, S. de Jong, S. Lee, W.-S. Lee, Y.-D. Chuang, R. Coffee, J.P. Cryan, J.M. Glownia, A. Foehlisch, M.R. Holmes, A.R. Fry, W.E. White, C. Bostedt, A.O. Scherz, H.A. Durr, W.F. Schlotter, Appl. Phys. Lett. **100**, 121108 (2012)
143. M.R. Bionta, H.T. Lemke, J.P. Cryan, J.M. Glownia, C. Bostedt, M. Cammarata, J.-C. Castagna, Y. Ding, D.M. Fritz, A.R. Fry, J. Krzywinski, M. Messerschmidt, S. Schorb, M.L. Swiggers, R.N. Coffee, Opt. Express **19**, 21855 (2011)
144. M.R. Bionta, D. French, J.P. Cryan, J.M. Glownia, N. Hartmann, D.J. Nicholson, K. Baker, C. Bostedt, M. Cammarrata, M. Chollet, Y. Ding, D.M. Fritz, S.M. Durbin, Y. Feng, M. Harmand, A.R. Fry, D.J. Kane, J. Krzywinski, H.T. Lemke, M. Messerschmidt, D.F. Ratner, S. Schorb, S. Toleikis, D. Zhu, W.E. White, R.N. Coffee, S.P. Moeller, M. Yabashi, S.P. Hau-Riege (Eds.), *Proceedings SPIE*, vol. 8504, p. 85040M (2012)
145. M. Harmand, R. Coffee, M.R. Bionta, M. Chollet, D. French, D. Zhu, D.M. Fritz, H.T. Lemke, N. Medvedev, B. Ziaja, S. Toleikis, M. Cammarata, Nat. Photonics **7**, 215 (2013)
146. M. Meyer, D. Cubaynes, P. O'Keeffe, H. Luna, P. Yeates, E.T. Kennedy, J.T. Costello, P. Orr, R. Taieb, A. Maquet, S. Duesterer, P. Radcliffe, H. Redlin, A. Azima, E. Ploenjes, J. Feldhaus, Phys. Rev. A **74**, 011401 (2006)
147. M. Meyer, D. Cubaynes, J. Dardis, P. Hayden, P. Hough, V. Richardson, E.T. Kennedy, J.T. Costello, S. Duesterer, W.B. Li, P. Radcliffe, H. Redlin, J. Feldhaus, S.I. Strakhova, E.V. Gryzlova, A.N. Grum-Grzhimailo, R. Taieb, A. Maquete, J. Electron. Spectrosc. Relat. Phenom. **181**, 111 (2010)
148. S. Düsterer, P. Radcliffe, C. Bostedt, J. Bozek, A.L. Cavalieri, R. Coffee, J.T. Costello, D. Cubaynes, L.F. DiMauro, Y. Ding, G. Doumy, F. Grüner, W. Helml, W. Schweinberger, R. Kienberger, A.R. Maier, M. Messerschmidt, V. Richardson, C. Roedig, T. Tschentscher, M. Meyer, New J. Phys. **13**, 093024 (2011)
149. M. Meyer, P. Radcliffe, T. Tschentscher, J.T. Costello, A.L. Cavalieri, I. Grguras, A.R. Maier, R. Kienberger, J. Bozek, C. Bostedt, S. Schorb, R. Coffee, M. Messerschmidt, C. Roedig, E. Sistrunk, L.F. Di Mauro, G. Doumy, K. Ueda, S. Wada, S. Düsterer, A.K. Kazansky, N.M. Kabachnik, Phys. Rev. Lett. **108**, 063007 (2012)
150. U. Frühling, M. Wieland, M. Gensch, T. Gebert, B. Schuette, M. Krikunova, R. Kalms, F. Budzyn, O. Grimm, J. Rossbach, E. Ploenjes, M. Drescher, Nat. Photonics **3**, 523 (2009)
151. I. Grguraš, A.R. Maier, C. Behrens, T. Mazza, T.J. Kelly, P. Radcliffe, S. Düsterer, A.K. Kazansky, N.M. Kabachnik, T. Tschentscher, J.T. Costello, M. Meyer, M.C. Hoffmann, H. Schlarb, A.L. Cavalieri, Nat. Photonics **6**, 852 (2012)
152. W. Helml, A.R. Maier, W. Schweinberger, I. Grguraš, P. Radcliffe, G. Doumy, C. Roedig, J. Gagnon, M. Messerschmidt, S. Schorb, C. Bostedt, F. Grüner, L.F. DiMauro, D. Cubaynes,

J.D. Bozek, T. Tschentscher, J.T. Costello, M. Meyer, R. Coffee, S. Düsterer, A.L. Cavalieri, R. Kienberger, Nat. Photonics **8**, 950 (2014)
153. J.P. Cryan, J.M. Glownia, J. Andreasson, A. Belkacem, N. Berrah, C.I. Blaga, C. Bostedt, J. Bozek, C. Buth, L.F. DiMauro, L. Fang, O. Gessner, M. Guehr, J. Hajdu, M.P. Hertlein, M. Hoener, O. Kornilov, J.P. Marangos, A.M. March, B.K. McFarland, H. Merdji, V.S. Petrovic, C. Raman, D. Ray, D. Reis, F. Tarantelli, M. Trigo, J.L. White, W. White, L. Young, P.H. Bucksbaum, R.N. Coffee, Phys. Rev. Lett. **105**, 083004 (2010)
154. L. Young, E.P. Kanter, B. Kraessig, Y. Li, A.M. March, S.T. Pratt, R. Santra, S.H. Southworth, N. Rohringer, L.F. DiMauro, G. Doumy, C.A. Roedig, N. Berrah, L. Fang, M. Hoener, P.H. Bucksbaum, J.P. Cryan, S. Ghimire, J.M. Glownia, D.A. Reis, J.D. Bozek, C. Bostedt, M. Messerschmidt, Nature **466**, 56 (2010)
155. M. Hoener, L. Fang, O. Kornilov, O. Gessner, S.T. Pratt, M. Gühr, E.P. Kanter, C. Blaga, C. Bostedt, J.D. Bozek, P.H. Bucksbaum, C. Buth, M. Chen, R. Coffee, J. Cryan, L.F. DiMauro, M. Glownia, E. Hosler, E. Kukk, S.R. Leone, B. McFarland, M. Messerschmidt, B. Murphy, V. Petrovic, D. Rolles, N. Berrah, Phys. Rev. Lett. **104**, 253002 (2010)
156. L. Fang, M. Hoener, O. Gessner, F. Tarantelli, S.T. Pratt, O. Kornilov, C. Buth, M. Gühr, E.P. Kanter, C. Bostedt, J.D. Bozek, P.H. Bucksbaum, M. Chen, R. Coffee, J. Cryan, M. Glownia, E. Kukk, S.R. Leone, N. Berrah, Phys. Rev. Lett. **105**, 083005 (2010)
157. V.S. Petrovic, M. Siano, J.L. White, N. Berrah, C. Bostedt, J.D. Bozek, D. Broege, M. Chalfin, R.N. Coffee, J. Cryan, L. Fang, J.P. Farrell, L.J. Frasinski, J.M. Glownia, M. Gühr, M. Hoener, D.M.P. Holland, J. Kim, J.P. Marangos, T. Martinez, B.K. McFarland, R.S. Minns, S. Miyabe, S. Schorb, R.J. Sension, L.S. Spector, R. Squibb, H. Tao, J.G. Underwood, P.H. Bucksbaum, Phys. Rev. Lett. **108**, 253006 (2012)
158. B.K. McFarland, J.P. Farrell, N. Berrah, C. Bostedt, J. Bozek, P.H. Bucksbaum, R. Coffee, J. Cryan, L. Fang, R. Feifel, K. Gaffney, J. Glownia, T. Martinez, M. Mucke, B. Murphy, S. Miyabe, A. Natan, T. Osipov, V. Petrovic, S. Schorb, T. Schultz, L. Spector, F. Tarantelli, I. Tenney, S. Wang, W. White, J. White, M. Gühr, EPJ Web Conf. **41**, 07004 (2013)
159. J.D. Bozek, Eur. Phys. J.-Spec. Top. **169**, 129 (2009)
160. B.K. McFarland, N. Berrah, C. Bostedt, J. Bozek, P.H. Bucksbaum, J.C. Castagna, R.N. Coffee, J.P. Cryan, L. Fang, J.P. Farrell, R. Feifel, K.J. Gaffney, J.M. Glownia, T.J. Martinez, S. Miyabe, M. Mucke, B. Murphy, A. Natan, T. Osipov, V.S. Petrovic, S. Schorb, T. Schultz, L.S. Spector, M. Swiggers, F. Tarantelli, I. Tenney, S. Wang, J.L. White, W. White, M. Gühr, J. Phys. Conf. Ser. **488**, 012015 (2014)
161. V. Zhaunerchyk, M. Mucke, P. Salén, P. vd Meulen, M. Kaminska, R.J. Squibb, L.J. Frasinski, M. Siano, J.H.D. Eland, P. Linusson, R.D. Thomas, M. Larsson, L. Foucar, J. Ullrich, K. Motomura, S. Mondal, K. Ueda, T. Osipov, L. Fang, B.F. Murphy, N. Berrah, C. Bostedt, J.D. Bozek, S. Schorb, M. Messerschmidt, J.M. Glownia, J.P. Cryan, R.N. Coffee, O. Takahashi, S. Wada, M.N. Piancastelli, R. Richter, K.C. Prince, R. Feifel, J. Phys. B At. Mol. Opt. Phys. **46**, 164034 (2013)
162. L.J. Frasinski, V. Zhaunerchyk, M. Mucke, R.J. Squibb, M. Siano, J.H.D. Eland, P. Linusson, P. v.d. Meulen, P. Salén, R.D. Thomas, M. Larsson, L. Foucar, J. Ullrich, K. Motomura, S. Mondal, K. Ueda, T. Osipov, L. Fang, B.F. Murphy, N. Berrah, C. Bostedt, J.D. Bozek, S. Schorb, M. Messerschmidt, J.M. Glownia, J.P. Cryan, R.N. Coffee, O. Takahashi, S. Wada, M.N. Piancastelli, R. Richter, K.C. Prince, R. Feifel, Phys. Rev. Lett. **111**, 073002 (2013)
163. P.M. Hare, C.E. Crespo-Hernández, B. Kohler, Proc. Natl. Acad. Sci. **104**, 435 (2007)
164. T. Gustavsson, Á. Bányász, E. Lazzarotto, D. Markovitsi, G. Scalmani, M.J. Frisch, V. Barone, R. Improta, J. Am. Chem. Soc. **128**, 607 (2006)
165. M. Altarelli, Nucl. Instrum. Methods Phys. Res. Sect. B Beam Interact. Mater. At. **269**, 2845 (2011)
166. I. Schweigert, S. Mukamel, Phys. Rev. Lett. **99**, 163001 (2007)
167. S. Mukamel, D. Abramavicius, L. Yang, W. Zhuang, I.V. Schweigert, D.V. Voronine, Acc. Chem. Res. **42**, 553 (2009)
168. J.D. Biggs, Y. Zhang, D. Healion, S. Mukamel, J. Chem. Phys. **136**, 174117 (2012)
169. J.D. Biggs, Y. Zhang, D. Healion, S. Mukamel, J. Chem. Phys. **138**, 144303 (2013)

170. N. Rohringer, D. Ryan, R.A. London, M. Purvis, F. Albert, J. Dunn, J.D. Bozek, C. Bostedt, A. Graf, R. Hill, S.P. Hau-Riege, J.J. Rocca, Nature **481**, 488 (2012)
171. C. Weninger, M. Purvis, D. Ryan, R. London, J. Bozek, C. Bostedt, A. Graf, G. Brown, J. Rocca, N. Rohringer, Phys. Rev. Lett. **111**, 233902 (2013)
172. B.D. Patterson, R. Abela, H.-H. Braun, U. Flechsig, R. Ganter, Y. Kim, E. Kirk, A. Oppelt, M. Pedrozzi, S. Reiche, L. Rivkin, T. Schmidt, B. Schmitt, V.N. Strocov, S. Tsujino, A.F. Wrulich, New J. Phys. **12**, 035012 (2010)
173. E.-S. Kim, M. Yoon, I.E.E.E. Trans, Nucl. Sci. **56**, 3597 (2009)

[illegible]

Index

M. Kitzler and S. Gräfe (eds.), *Ultrafast Dynamics Driven by Intense Light Pulses*, Springer Series on Atomic, Optical, and Plasma Physics 86, DOI 10.1007/978-3-319-20173-3

MIX
Papier aus verantwortungsvollen Quellen
Paper from responsible sources
FSC® C105338

If you have any concerns about our products, you can contact us on
ProductSafety@springernature.com

In case Publisher is established outside the EU, the EU authorized representative is:
Springer Nature Customer Service Center GmbH
Europaplatz 3, 69115 Heidelberg, Germany

Printed by Libri Plureos GmbH
in Hamburg, Germany